# HUMAN HEALTH BENEFITS OF PLANT BIOACTIVE COMPOUNDS

## Potentials and Prospects

*Innovations in Plant Science for Better Health: From Soil to Fork*

# HUMAN HEALTH BENEFITS OF PLANT BIOACTIVE COMPOUNDS

## Potentials and Prospects

*Edited by*

**Megh R. Goyal**

**Hafiz Ansar Rasul Suleria**

APPLE ACADEMIC PRESS

Apple Academic Press Inc.
3333 Mistwell Crescent
Oakville, ON L6L 0A2
Canada

Apple Academic Press Inc.
1265 Goldenrod Circle NE
Palm Bay, Florida 32905
USA


*Exclusive worldwide distribution by CRC Press, a member of Taylor & Francis Group*



International Standard Book Number-13: 978-1-77188-739-7 (Hardcover)
International Standard Book Number-13: 978-0-42945-791-3 (eBook)



**Library and Archives Canada Cataloguing in Publication**

Title: Human health benefits of plant bioactive compounds : potentials and prospects / edited by Megh R. Goyal, Hafiz Ansar Rasul Suleria

Names: Goyal, Megh Raj, editor. | Suleria, Hafiz Ansar Rasul, editor.

Series: Innovations in plant science for better health.

Description: Series statement: Innovations in plant science for better health : from soil to fork | Includes bibliographical references and index.

Identifiers: Canadiana (print) 20190080221 | Canadiana (ebook) 2019008023X | ISBN 9781771887397 (hardcover) | ISBN 9780429457913 (ebook)

Subjects: LCSH: Plant bioactive compounds—Health aspects. | LCSH: Functional foods. | LCSH: Botanical chemistry.

Classification: LCC QK898.B54 H86 2019 | DDC 615.3/21—dc23

**Library of Congress Cataloging-in-Publication Data**

Names: Goyal, Megh Raj, editor. | Suleria, Hafiz Ansar Rasul, editor

Title: Human health benefits of plant bioactive compounds : potentials and prospects / editors: Megh R. Goyal, Hafiz Ansar Rasul Suleria.

Description: Palm Bay, Florida : Apple Academic Press, 2019. | Series: Innovations in plant science for better health | Includes index.

Identifiers: LCCN 2019011704 (print) | LCCN 2019013561 (ebook) | ISBN 9780429457913 () | ISBN 9781771887397 (hardcover : alk. paper)

Subjects: LCSH: Plant bioactive compounds--Health aspects. | Functional foods. | Botanical chemistry.

Classification: LCC QK898.B4 (ebook) | LCC QK898.B4 H86 2019 (print) | DDC 572/.2--dc23

LC record available at https://lccn.loc.gov/2019011704

Apple Academic Press also publishes its books in a variety of electronic formats. Some content that appears in print may not be available in electronic format. For information about Apple Academic Press products, visit our website at **www.appleacademicpress.com** and the CRC Press website at **www.crcpress.com**

# INNOVATIONS IN PLANT SCIENCE FOR BETTER HEALTH: FROM SOIL TO FORK BOOK SERIES

**Series Editor-in-Chief:**

**Hafiz Ansar Rasul Suleria, PhD**
Honorary Fellow at the Diamantina Institute,
Faculty of Medicine, The University of Queensland (UQ), Australia
email: hafiz.suleria@uqconnect.edu.au

The objective of this new book series is to offer academia, engineers, technologists, and users from different disciplines information to gain knowledge on the breadth and depth of this multifaceted field. The volumes will explore the fields of phytochemistry, along with its potential and extraction techniques. The volumes will discuss the therapeutic perspectives of biochemical compounds in plants and animal and marine sources in an interdisciplinary manner because the field requires knowledge of many areas, including agricultural, food, and chemical engineering; manufacturing technology along with applications from diverse fields like chemistry; herbal drug technology; microbiology; animal husbandry; and food science; etc. There is an urgent need to explore and investigate the innovations, current shortcomings, and future challenges in this growing area of research.

We welcome chapters on the following specialty areas (but not limited to):

- Food and function
- Nutritional composition of different foods materials
- Functional and nutraceutical perspectives of foods
- Extraction of bioactive molecules
- Phytomedicinal properties of different plants
- Traditional plants used as functional foods
- Phytopharmacology of plants used in metabolic disorders
- Importance of spices and medicinal and functional foods
- Natural products chemistry
- Food processing waste/byproducts: management and utilization

- Herbals as potential bioavailability enhancer and herbal cosmetics
- Phytopharmaceuticals for the delivery of bioactives
- Alternative and complementary medicines
- Ethnopharmacology and ethnomedicine
- Marine phytochemistry
- Marine microbial chemistry
- Other related includes nuts, seed spices, wild flora, etc.

**About the Book Series Editor-in-Chief**

Dr. Hafiz Suleria is an eminent young researcher in the field of food science and nutrition. Currently, he is an Honorary Fellow at the Diamantina Institute, Faculty of Medicine, The University of Queensland (UQ), Australia. Before joining the UQ, he worked as a lecturer in the Department of Food Sciences, Government College University Faisalabad, Pakistan. He also worked as a Research Associate in a PAK-US Joint Project funded by the Higher Education Commission, Pakistan, and Department of State, USA, with the collaboration of the University of Massachusetts, USA, and the National Institute of Food Science and Technology, University of Agriculture, Faisalabad, Pakistan.

Dr. Suleria's major research focus is on food science and nutrition, particularly in screening of bioactive molecules from different plant, marine, and animal sources, using various cutting-edge techniques, such as isolation, purification, and characterization. He also did research work on functional foods, nutraceuticals, and alternative medicine. He has published more than 60 peer-reviewed scientific papers in different reputed/impacted journals. He is also in collaboration with more than five universities where he is working as a co-supervisor/special member for PhD and postgraduate students and also involved in joint publications, projects, and grants.

**BOOKS IN THE SERIES**

**Bioactive Compounds of Medicinal Plants: Properties and Potential for Human Health**
Editors: Megh R. Goyal, PhD, and Ademola O. Ayeleso, PhD

**Plant- and Marine-Based Phytochemicals for Human Health: Attributes, Potential, and Use**
Editors: Megh R. Goyal, PhD, and Durgesh Nandini Chauhan, MPharm

**Human Health Benefits of Plant Bioactive Compounds: Potentials and Prospects**
Editors: Megh R. Goyal, PhD, and Hafiz Ansar Rasul Suleria, PhD

**Plant Secondary Metabolites for Human Health: Extraction of Bioactive Compounds**
Editors: Megh R. Goyal, PhD, P. P. Joy, and Hafiz Ansar Rasul Suleria, PhD

**Bioactive Compounds from Plant Origins: Extraction, Applications, and Potential Health Benefits**
Editors: Hafiz Ansar Rasul Suleria, PhD, and Colin Barrow, PhD

**Phytochemicals from Medicinal Plants: Scope, Applications, and Potential Health Claims**
Editors: Hafiz Ansar Rasul Suleria, PhD, Megh R. Goyal, PhD, and Masood Sadiq Butt, PhD

**The Therapeutic Properties of Medicinal Plants: Health-Rejuvenating Bioactive Compounds of Native Flora**
Editors: Megh R. Goyal, PhD, PE, Hafiz Ansar Rasul Suleria, PhD, Ademola Olabode Ayeleso, PhD, T. Jesse Joel, and Sujogya Kumar Panda

# ABOUT THE SENIOR EDITOR-IN-CHIEF

**Megh R. Goyal, PhD**

*Retired Professor in Agricultural and Biomedical Engineering, University of Puerto Rico, Mayaguez Campus; Senior Acquisitions Editor, Biomedical Engineering and Agricultural Science, Apple Academic Press, Inc.*

Megh R. Goyal, PhD, PE, is a Retired Professor in Agricultural and Biomedical Engineering from the General Engineering Department in the College of Engineering at the University of Puerto Rico–Mayaguez Campus; and Senior Acquisitions Editor and Senior Technical Editor-in-Chief in Agriculture and Biomedical Engineering for Apple Academic Press, Inc. He has worked as a Soil Conservation Inspector and as a Research Assistant at Haryana Agricultural University and Ohio State University.

During his professional career of 45 years, Dr. Goyal has received many prestigious awards and honors. He was the first agricultural engineer to receive the professional license in Agricultural Engineering in 1986 from the College of Engineers and Surveyors of Puerto Rico. In 2005, he was proclaimed as "Father of Irrigation Engineering in Puerto Rico for the Twentieth Century" by the American Society of Agricultural and Biological Engineers (ASABE), Puerto Rico Section, for his pioneering work on micro irrigation, evapotranspiration, agroclimatology, and soil and water engineering. The Water Technology Centre of Tamil Nadu Agricultural University in Coimbatore, India, recognized Dr. Goyal as one of the experts "who rendered meritorious service for the development of micro irrigation sector in India" by bestowing the Award of Outstanding Contribution in Micro Irrigation. This award was presented to Dr. Goyal during the inaugural session of the National Congress on "New Challenges and Advances in Sustainable Micro Irrigation" on March 1, 2017, held at Tamil Nadu Agricultural University. Dr. Goyal is slated to receive the Netafim Award for Advancements in Microirrigation: 2018 from the American Society of Agricultural Engineers at the ASABE International Meeting in August 2018.

A prolific author and editor, he has written more than 200 journal articles and textbooks and has edited over 62 books. He is the editor of three book series published by Apple Academic Press: Innovations in Agricultural & Biological Engineering, Innovations and Challenges in Micro Irrigation, and Research Advances in Sustainable Micro Irrigation. He is also instrumental in the development of the new book series *Innovations in Plant Science for Better Health: From Soil to Fork.*

Dr. Goyal received his BSc degree in engineering from Punjab Agricultural University, Ludhiana, India; his MSc and PhD degrees from Ohio State University, Columbus; and his Master of Divinity degree from Puerto Rico Evangelical Seminary, Hato Rey, Puerto Rico, USA.

# ABOUT THE CO-EDITOR

**Hafiz Ansar Rasul Suleria, PhD**

Hafiz Anasr Rasul Suleria, PhD, is currently working as the Alfred Deakin Research Fellow at Deakin University, Melbourne, Australia. He is also an Honorary Fellow in the Diamantina Institute, Faculty of Medicine, The University of Queensland, Australia.

Recently he worked as Postdoc Research Fellow in the Department of Food, Nutrition, Dietetic and Health at Kansas State University, USA.

Previously, he has been awarded an International Postgraduate Research Scholarship (IPRS) and Australian Postgraduate Award (APA) for his PhD research at UQ School of Medicine, the Translational Research Institute (TRI) in collaboration with Commonwealth and Scientific and Industrial Research Organization (CSIRO, Australia).

Before joining the UQ, he worked as a Lecturer in the Department of Food Sciences, Government College University Faisalabad, Pakistan. He also worked as a Research Associate in the PAK-US Joint Project funded by the Higher Education Commission, Pakistan, and Department of State, USA, with the collaboration of the University of Massachusetts, USA, and National Institute of Food Science and Technology, University of Agriculture Faisalabad, Pakistan.

He has a significant research focus on food nutrition, particularly in the screening of bioactive molecules—isolation, purification, and characterization using various cutting-edge techniques from different plant, marine, and animal source, and in vitro, in vivo bioactivities, cell culture, and animal modeling. He has also done a reasonable amount of work on functional foods and nutraceutical, food and function, and alternative medicine.

Dr. Suleria has published more than 50 peer-reviewed scientific papers in different reputed/impacted journals. He is also in collaboration with more than ten universities where he is working as a co-supervisor/special member for PhD and postgraduate students and is also involved in joint publications, projects, and grants. He is Editor-in-Chief for the book series on *Innovations in Plant Science for Better Health: From Soil to Fork*, published by AAP. Readers may contact him at: hafiz.suleria@uqconnect.edu.au.

# CONTENTS

*Contributors* ........ *xv*

*Abbreviations* ........ *xix*

*Preface 1 by Megh R. Goyal* ........ *xxvii*

*Preface 2 by Hafiz Ansar Rasul Suleria* ........ *xxix*

**PART I: Functional Foods for Human Health ........ 1**

1. **Functional Foods: Concepts and Their Health Perspectives ........ 3**
   Aamir Shehzad, Asna Zahid, Sana Mahmood, and Hafiz Ansar Rasul Suleria

2. **Plant-Based Functional Foods for Human Nutrition: Current Trends, Future Prospective, and Phytochemical Attributes ........ 21**
   Muhammad Sohaib, Ayesha Zafar, and Azmat Ullah Khan

3. **Natural or Synthetic Antioxidants in Foods ........ 55**
   Tosin Michael Olajide, Md. Ramim Tanver Rahman, Zaixiang Lou, and Jing Qi

**PART II: Pharmacological Aspects of Fruits and Vegetables ........ 67**

4. **Health Benefits of Anthocyanins in Black Carrot *(Daucus carota)* ........ 69**
   Iahtisham-Ul-Haq, Masood Sadiq Butt, Aneela Shamshad, and Hafiz Ansar Rasul Suleria

5. **Olive Oil Phenols: Chemistry, Synthesis, Metabolism, Fate, and Their Allied Health Claims ........ 95**
   Muhammad Imran, Faqir Muhammad Anjum, Shinawar Waseem Ali, Iahtisham-Ul-Haq, and Hafiz Ansar Rasul Suleria

6. **Phytochemicals from Citrus Peel: Perspectives and Allied Health Claims ........ 129**
   Humaira Ashraf, Iahtisham-Ul-Haq, Masood Sadiq Butt, Mir Muhammad Nasir Qayyum, and Hafiz Ansar Rasul Suleria

7. **Phytochemistry of Grapes (*Vitis vinifera* L.): Functional and Nutraceutical Attributes ........ 153**
   Hafiz Ansar Rasul Suleria, Farhan Saeed, Tabussam Tufail, Saira Sultan, and Hafsa Daud

**PART III: Pharmacological Aspects of Natural Products....................181**

**8. Piceatannol: A Review on Natural Sources, Extraction Methods, and Biological Activities ....................183**

Juliane Viganó, Andressa Mara Baseggio, Larissa Akemi Kido, and Julian Martínez

**9. Plant-Based Neurotoxins: Impact on Neural Pathologies....................213**

Seema Patel, Girish Kumar Gupta, Abdur Rauf, Haroon Khan, and Hafiz Ansar Rasul Suleria

**10. Health Benefits of Candy Leaf (*Stevia rebaudiana*): Physiological and Pharmacological Actions ....................229**

Zubaria Ishaq, Haneen Fatima Kirmani, Munawar Abbas, and Farhan Saeed

**PART IV: Pharmacological Aspects of Cereals, Grains, and Tea.......243**

**11. Wheatgrass Juice: A Nutritional Weapon Against Various Maladies....245**

Tabussam Tufail, Farhan Saeed, Muhammad Afzaal, and Hafiz Ansar Rasul Suleria

**12. Flaxseed: A Shield Against Lifestyle-Related Maladies....................255**

Huma Bader-Ul-Ain, Munawar Abbas, Farhan Saeed, and Hafiz Ansar Rasul Suleria

**13. Green Tea for Human Health Benefits: Phytonutrients and Their Therapeutic Potentials ....................281**

Faiza Ashfaq, Masood Sadiq Butt, Ahmad Bilal, Mir Muhammad Nasir Qayyum, and Hafiz Ansar Rasul Suleria

**14. Black Tea Polyphenols: Health-Endorsing Perspectives: A Mechanistic Appraisal....................321**

Ali Imran, Muhammad Umair Arshad, Muhammad Sajid Arshad, and Hafiz Ansar Rasul Suleria

***Index....................355***

# CONTRIBUTORS

**Munawar Abbas, MS**
Research Associate, Institute of Home & Food Sciences, Government College University, Faisalabad, Pakistan. Tel.: +92 332 0677170; E-mail: foodian2007@gmail.com

**Muhammad Afzaal, PhD Scholar**
Lecturer, Institute of Home and Food Sciences, Government College University, Faisalabad, Pakistan. Tel.: +92 3003159543; E-mail: muhammadafzaal@gcuf.edu.pk

**Shinawar Waseem Ali, PhD**
Assistant Professor (Food Technology), Institute of Agricultural Sciences, University of the Punjab, Lahore, Pakistan. Tel.: +923344146456; E-mail: shinawar.iags@pu.edu.pk

**Faqir Muhammad Anjum, PhD**
Vice Chancellor, University of The Gambia, PO Box 3530, Serekunda, The Gambia, West Africa Director General (Retd.), Faculty of Food, Nutrition and Home Sciences, National Institute of Food Science and Technology, University of Agriculture, Faisalabad 38040, Pakistan. Tel.: +2203650001; E-mail: faqir_65@yahoo.com

**Muhammad Sajid Arshad, PhD**
Assistant Professor, Institute of Home & Food Sciences, Government College University, Faisalabad, Pakistan. Tel.: +923338802239; E-mail: sajid_ft@yahoo.com

**Muhammad Umair Arshad, PhD**
Director, Institute of Home & Food Sciences, Government College University, Faisalabad, Pakistan. Tel.: +923137986776; E-mail: umairfood1@gmail.com

**Faiza Ashfaq, PhD**
Faculty of Science and Technology, Government College Women University Faisalabad, Faisalabad, Pakistan. Tel.: +92 3015588259; E-mail: kahloonfazi@yahoo.com

**Humaira Ashraf, PhD**
Assistant Professor, Department of Food Sciences, University of Sargodha Lyallpur Campus, Faisalabad, Pakistan. Tel.: +92 333 5877520; E-mail: humairaagri@gmail.com

**Huma Bader-Ul-Ain, MS**
Research Associate, Institute of Home & Food Sciences, Government College University, Faisalabad, Pakistan. Tel.: +92-3331672734; E-mail: humahums@yahoo.com

**Andressa Mara Baseggio, MSc, PhD Student**
Department of Food and Nutrition, College of Food Engineering, University of Campinas, 80 Monteiro Lobato Street Campinas 13083-862, São Paulo, Brazil. Tel.: +55 19 35214069; E-mail: andressam.baseggio@gmail.com

**Ahmad Bilal, PhD**
Faculty of Science and Technology, Government College Women University Faisalabad, Faisalabad, Pakistan. Tel.: +92 3044346933; E-mail: ahmedbilal1317@yahoo.com

**Masood Sadiq Butt, PhD**
Dean, Faculty of Food, Nutrition and Home Sciences, University of Agriculture, Faisalabad, Pakistan. Tel.: +92-300-6622685; E-mail: drmsbutt@yahoo.com

**Hafsa Daud, MSc (Hons.)**
Department of Food Technology, Pir Mehr Ali Shah Arid Agriculture University Rawalpinid Islamabad, Punjab, Pakistan. Tel.: +61 452615741; E-mail: hafsa.daud87@gmail.com

**Megh R. Goyal, PhD, PE**
Retired Faculty in Agricultural and Biomedical Engineering from the General Engineering Department in the College of Engineering at University of Puerto Rico – Mayaguez Campus
Senior Acquisitions Editor and Senior Technical Editor-in-Chief in Agricultural and Biomedical Engineering for Apple Academic Press Inc.; PO Box 86, Rincon – PR – 006770086, USA.
E-mail: goyalmegh@gmail.com

**Girish Kumar Gupta**
Maharishi Markandeshwar University, (Accreditation by NAAC A Grade), Approved as SIRO by DSIR, Govt. of India, New Delhi Mullana, Ambala-133207, Haryana, India

**Iahtisham-Ul-Haq, PhD**
Department of Diet and Nutritional Sciences, Faculty of Health and Allied Sciences, Imperial College of Business Studies, Lahore, Pakistan
Faculty of Food, Nutrition and Home Sciences, National Institute of Food Science and Technology, University of Agriculture, Faisalabad 38040, Pakistan. Tel.: +92-300-7275255;
E-mail: iahtisham@hotmail.com

**Ali Imran, PhD**
Assistant Professor, Institute of Home and Food Sciences, Government College University, Faisalabad 38040, Pakistan. Tel.: +923336575583; E-mail: dr.aliimran@gcuf.edu.pk

**Muhammad Imran, PhD**
Associate Professor, Faculty of Allied Health Sciences, University Institute of Diet and Nutritional Sciences, The University of Lahore, Pakistan
E-mail: mic_1661@yahoo.com

**Zubaria Ishaq, MS**
Department of Biosciences, COMSATS Institute of Information Technology, Sahiwal, Pakistan. Tel.: +923356271083, E-mail; zubaria.salman.mushtaq@gmail.com

**Azmat Ullah Khan, PhD**
Assistant Professor, Department of Food Science and Human Nutrition, University of Veterinary and Animal Sciences, Lahore 54000, Pakistan. E-mail; azmat.khan@uvas.edu.pk

**Haroon Khan**
Department of Pharmacy, Abdul Wali Khan University, Mardan 23200, Pakistan

**Larissa Akemi Kido, PhD, Post-doctoral**
Department of Structural and Functional Biology, Institute of Biology, University of Campinas, Bertrand Russel Avenue Campinas 13083-865, São Paulo, Brazil. Tel.: +55 19 35216103;
E-mail: larissa.kido@gmail.com

**Haneen Fatima Kirmani, MPhil**
Department of Botany, University of Agriculture, Faisalabad, Pakistan. Tel.: +923356271083,
E-mail: Haneenkirmani26@gmail.com

**Zaixiang Lou, PhD**
State Key Laboratory of Food Science and Technology, School of Food Science and Technology, Jiangnan University, Wuxi, Jiangsu 214122, PR China. Tel.: +86 51085917797;
E-mail: louzaixiang@126.com

**Sana Mahmood, MSc (Hons.)**
Human Nutrition and Dietetics, Faculty of Food, Nutrition and Home Sciences, National Institute of Food Science and Technology, University of Agriculture, Faisalabad 38040, Pakistan.
Tel: +92 (0) 322 7641808; E-mail: sanamehmood4209@yahoo.com

**Julian Martínez, PhD**
Professor, Department of Food Engineering, College of Food Engineering, University of Campinas, 80 Monteiro Lobato Street Campinas 13083-862, São Paulo, Brazil. Tel.: +55 19 35214046;
E-mail: jmartinez94@gmail.com

**Tosin Michael Olajide, MSc**
School of Life Sciences, Shanghai University, 333 Nanchen Road, Shanghai 200444, PR China.
Tel.: +86 18501617340; E-mail: tosmics@gmail.com

**Seema Patel**
Bioinformatics and Medical Informatics Research Center, San Diego State University, 92182, San Diego, CA, USA. E-mail: seemabiotech83@gmail.com

**Mir Muhammad Nasir Qayyum, PhD**
Assistant Professor, Department of Agriculture and Food Technology, Karakoram International University, Gilgit-Baltistan, Pakistan. Tel.: +92 3405694020; E-mail: mir.qayyum@kiu.edu.pk

**Jing Qi, PhD**
College of Food Engineering, Guangxi University of Chinese Medicine, Nanning, Guangxi 530200, PR China. Tel.: +86 18377173047; E-mail: qijinxs@163.com

**Md. Ramim Tanver Rahman, ME**
State Key Laboratory of Microbial Resources, Institute of Microbiology, Chinese Academy of Sciences, Beijing 100101, PR China
University of Chinese Academy of Sciences, Beijing 100049, PR China. Tel.: +86 13051010812;
E-mail: ramimbau@gmail.com

**Abdur Rauf**
Department of Chemistry, University of Swabi, Anbar-23561, K.P.K, Pakistan

**Farhan Saeed, PhD**
Assistant Professor, Institute of Home & Food Sciences, Government College University, Faisalabad (GCUF), Allama Iqbal Road, Faisalabad 38000, Pakistan. Tel.: +92-3338040311;
E-mail: F.saeed@gcuf.edu.pk; far1552@yahoo.com

**Aneela Shamshad, MSc (Hons.)**
Faculty of Food, Nutrition and Home Sciences, National Institute of Food Science and Technology, University of Agriculture, Faisalabad 38040, Pakistan. Tel.: +92-347-6503966;
E-mail: aneela.awaan786@gmail.com

**Aamir Shehzad**
Assistant Professor, Faculty of Food, Nutrition and Home Sciences, National Institute of Food Science and Technology, University of Agriculture, Faisalabad 38040, Pakistan. Tel: +92 (0) 41 9201105;
Fax: +92 (0) 41 9201439; E-mail: draamir@uaf.edu.pk

**Muhammad Sohaib, PhD**
Assistant Professor, Department of Food Science and Human Nutrition, University of Veterinary and Animal Sciences, Lahore, 54000, Pakistan. Tel.: +92-3216843849;
E-mail: muhammd.sohaib@uvas.edu.pk

**Hafiz Ansar Rasul Suleria, PhD**
Department of Food, Nutrition, Dietetics and Health, Kansas State University, Manhattan, KS 66506, USA
Department of Agriculture and Food Technology, Karakoram International University, Gilgit-Baltistan, Pakistan
Postdoc Research Fellow/Honorary Research Fellow (Food and Nutrition), The University of Queensland, School of Medicine, Translational Research Institute, Brisbane, QLD 4072, Australia. Tel.: +61-470439670; E-mail: hafiz.suleria@uqconnect.edu.au

**Saira Sultan, PhD**
National Institute of Food Science and Technology, The University of Agriculture, Faisalabad 38040, Pakistan. Tel.: +61 452 439 670; E-mail: saira.sultan@hotmail.com

**Tabussam Tufail, MSc**
Research Associate, Institute of Home and Food Sciences, Government College University, Faisalabad (GCUF), Allama Iqbal Road, Faisalabad 38000, Pakistan. Tel.: +92-3057678764; E-mail: tabussamtufail@gcuf.edu.pk

**Juliane Viganó, PhD, Post-doctoral**
Department of Food Engineering, College of Food Engineering, University of Campinas, 80 Monteiro Lobato Street Campinas 13083-862, São Paulo, Brazil. Tel.: +55-17-981750208, +55-19-35210080; E-mail: julianevigano@gmail.com

**Ayesha Zafar, MPhil**
Department of Food Science and Human Nutrition, University of Veterinary and Animal Sciences, Lahore 54000, Pakistan. E-mail; aysh.mir95.az@gmail.com

**Asna Zahid MSc (Hons.)**
Human Nutrition and Dietetics, Faculty of Food, Nutrition and Home Sciences, National Institute of Food Science and Technology, University of Agriculture, Faisalabad 38040, Pakistan.
Tel: +92 (0) 323 7755001; E-mail: asna.zahid52@gmail.com

# ABBREVIATIONS

| | |
|---|---|
| AA | arachidonic acid |
| ABTS | 2, 2'-azino-bis (3-ethylbenzothiazoline-6-sulphonic acid) |
| ACAT | acyl CoA: cholesterol *O*-acyltransferase |
| ACC | acetyl-CoA Carboxylase |
| ACE | angiotensin converting enzyme |
| ADI | acceptable daily intake |
| AFE | accelerated fluid extraction |
| AH109A | rat hepatoma cell subline |
| AKT | protein kinase B |
| ALA | α-linolenic acid |
| ALP | alkaline phosphatase |
| ALT | alanine transaminase |
| AMPK | adenosine monophosphate-protein kinase |
| AMPK α | 5'-AMP-activated protein kinase alpha |
| AOM | azoxymethane |
| AP-1 | activator protein 1 |
| ApoB | apolipoprotein B |
| AST | aspartate aminotransferase |
| ATP | adenosine triphosphate |
| AUC | area under curve |
| BFC | bioactive food components |
| bFGF | basic fibroblast growth factor |
| BHA | butylated hydroxyanisole |
| BHT | butylated hydroxytoluene |
| BMI | body mass index |
| BMT | bone marrow transplantation |
| BUN | blood urea nitrogen |
| BW | body weight |
| CaCo-2 | human intestinal epithelial cell line |
| CAT | catalase |
| cAMP | cyclic adenosine monophosphate |
| CCA | cholangiocarcinoma |
| $CCl_4$ | carbon tetrachloride |
| CDK | cyclin-dependent kinases |

| | |
|---|---|
| CE | capillary electrophoresis |
| CHD | coronary heart diseases |
| CIP | CDK-interacting protein |
| CKD | chronic kidney disease |
| CLA | conjugated linoleic acid |
| CM | chylomicron |
| CoA | coenzyme A |
| CoQ10 | coenzyme $Q_{10}$ |
| COX-2 | cyclooxygenase-2 |
| CPK | creatine phosphokinase |
| CPK-MB | creatine phosphokinase-MB |
| CRE | binding cAMP response element-binding |
| CSE | conventional solvent extraction |
| CVD | cardiovascular diseases |
| Cy3XCGG | cyanidin-3-(2-xylose-6-(4-coumaroyl)glucose-galactoside) |
| Cy3XFGG | cyanidin-3-(2-xylose-6-feruloyl-glucose-galactoside) |
| Cy3XG | cyanidin-3-(2-xylose-galactoside) |
| Cy3XGG | cyanidin-3-(2-xylose-6-glucose-galactoside) |
| Cy3XSGG | cyanidin-3-(2- xylose-6-sinapoyl-glucose-galactoside) |
| CYP1A1 | cytochrome p450, subfamily i, polypeptide 1 |
| DADS | diallyl disulfide |
| DAS | diallyl sulfide |
| DFF | defatted flaxseed flour |
| DHA | docosahexaenoic acid |
| DHPE | 1,1-di-(2′,5′-dihydroxy-4′-*tert*butylphenyl)ethane |
| DM | diabetes mellitus |
| DMBA | 7,12-Dimethylbenz[a]anthracene |
| DMSO | dimethyl sulfoxide |
| DPA | docosapentaenoic acid |
| DPP-IV | dipeptidyl peptidase 4 |
| DPPH | 1, 1-diphenyl-2-picrylhydrazyl |
| DSS-CRC | dextran sulfate sodium-induced colorectal cancer |
| DU145 | cell line from prostate metastatic site |
| DW | dried weight |
| DXR | doxorubicin |
| E2 | 17b-estradiol |
| EC | epicatechins |
| ECG | epicatechingallate |
| ED | enterodiol |
| EFAs | essential fatty acids |

| | |
|---|---|
| EGC | epigallocatechin |
| EGCG | epigallocathingallate |
| EIA | enteroinsular axis |
| EL | enterolactone |
| eNOS | endothelial nitric oxide synthase |
| EPA | eicosapentaenoic acid |
| ER | estrogen receptor |
| ERK 1/2 | extracellular signal-regulated kinase ½ |
| F–C | Folin–Ciocalteu method |
| F–D | Folin–Denis method |
| FAO | Food and Agriculture Organization |
| FAS | fatty acid synthase enzyme |
| FGF | fibroblast growth factors |
| FRAP | ferric-ion-reducing antioxidant capacity |
| FSH | follicles-stimulating hormone |
| FTIR | Fourier-transform infrared spectroscopy |
| FW | fresh weight |
| GC | gas chromatography |
| GC-MS | gas chromatography-mass spectrometry |
| GIP | gastric inhibitory polypeptide |
| GIT | gastrointestinal tract |
| GJC | gap junction communication |
| GLP-1 | glucagon-like peptide-1 |
| Glut | glucose transporters |
| Glut2 | glucose transporter 2 |
| GLUT4 | glucose transporter type-4 |
| Glut5 | glucose transporter 5 |
| GPDH | glycerol-3-phosphate dehydrogenase |
| GPSE | Graduate Partners in Science Education |
| GPx | glutathione peroxidase |
| GRAS | generally recognized as safe |
| GSE | grapefruit seed extract |
| GSH | glutathione |
| GSH-Px | glutathione peroxidase |
| GSSH | oxidized glutathione |
| GST | glutathione-S-transferase |
| $H_2O_2$ | hydrogen peroxide |
| HCC | hepatocellular carcinoma |
| HDL | high-density lipoproteins |
| HFD | high-fat diet |

| | |
|---|---|
| HIF | hypoxia inducible factor |
| HMG-CoA | 3-hydroxy-3-methylglutaryl coenzyme A |
| HNE | 4-hydroxynonenal |
| HO-1 | heme oxygenase-1 |
| HPLC | high-pressure liquid chromatography |
| HPLC-MS | high-performance liquid chromatography-mass spectrometry |
| HPLC-NMR | high-performance liquid chromatography-nuclear magnetic resonance |
| HPLC-PDA | high-performance liquid chromatography-photodiode array |
| HT-29 | human colon adenocarcinoma cell line |
| $I_A$ | oxidation induction period for a fat or oil in the presence of an antioxidant |
| ICAM | intracellular adhesion molecule |
| IDF | insoluble dietary fiber |
| IKKβ | inhibitor of nuclear factor kappa-B kinase subunit beta |
| IL | interleukin |
| iNOS | inducible nitric oxide synthase |
| $I_O$ | oxidation induction period of a fat or oil without an antioxidant |
| IRβ | insulin receptor β subunit |
| IUPAC | International Union of Pure and Applied Chemistry |
| JNK | c-Jun NH2-terminal kinase protein |
| Keap1 | Kelch-like ECH-associated protein 1 |
| LA | linoleic acid |
| LAB | labrasol |
| LD | lethal dose |
| LDH | lactate dehydrogenase |
| LDL | low-density lipoprotein |
| LH | luteinizing hormone |
| LPO | lipid peroxides |
| LPS | lipopolysaccharide |
| LTBHQ | lauryl *tert*-butylhydroquinone |
| LTBQ | lauryl *tert*-butylquinone |
| MAE | microwave-assisted extraction |
| MAPKs | mitogen-activated protein kinases |
| MCA | metal chelating activity |
| MCP | monocyte chemotactic proteins |
| MDA | melondialdehyde |
| MGAM | maltase-glucoamylase |

| | |
|---|---|
| MMPs | matrix metalloproteinases |
| MPP | microparticulated protein |
| MNU | methylnitrosourea |
| MS | mass spectrometry |
| mtDNA | mitochondrial deoxyribose nucleic acid |
| mTOR | mammalian target of rapamycin |
| MW | molecular weight |
| NADH | nicotinamide adenine dinuceotide |
| NADPH | nicotinamide adenine dinucleotide phosphate |
| NAFLD | non-alcoholic fatty liver disease |
| NF-κB | nuclear factor-κB |
| NMR | nuclear magnetic resonance |
| NNK | 4-(methylnitrosamino)-1-(3-pyridyl)-1-butanone |
| NO | nitric oxide |
| NQO-1 | NADPH quinone dehydrogenase-1 |
| Nrf2 | nuclear factor (erythroid-derived 2)-like 2 |
| NT | nitrotyrosine |
| O/W | oil-in-water emulsions |
| OHdG | 8-hydroxy-2'-deoxyguanosine |
| ORAC | oxygen radical absorbance capacity |
| P12K | phosphatidylinositol-4,5-bisphosphate 3-kinase |
| p21/WAF1 | cyclin-dependent kinase inhibitor |
| p27KIP1 | cyclin-dependent kinase inhibitor |
| PAPS | 30-phosphoadenoseine-50-phosphosulfate |
| PAs | phenolic acids |
| PANC | Power Association of Northern California |
| PC | phosphatidylcholine |
| PCOS | polycystic ovarian syndrome |
| PCs | protein carbonyls |
| Pf | protection factor |
| PFE | pressurized fluid extraction |
| PG | propyl gallate |
| PG | prostaglandin |
| PI3K | phosphatidylinositol-3-kinase |
| PKC | protein kinase C |
| PLE | pressurized liquid extraction |
| PLs | phospholipids |
| PMFs | polymethoxylated flavones |
| POD | peroxidase |
| PPARγ | peroxisome proliferator-activated receptor gamma |

| | |
|---|---|
| PPARGC1α | peroxisome proliferator-activated receptor gamma coactivator 1-alpha |
| PPO | polyphenol peroxidase |
| PTEN | phosphatase and tensin homolog |
| PUFAs | polyunsaturated fatty acids |
| PVN | paraventricular nucleus |
| QCT | 3-methyl-2-quinoxalin benzenevinylketo-1,4-dioxide |
| QSDR | quantitative structure function relationships |
| RA | rheumatoid arthritis |
| RAS | renin–angiotensin system |
| RBCs | red blood cells |
| RDA | reference daily intake |
| ROS | reactive oxygen species |
| RTD | ready-to-drink |
| SAPK | Stress-activated Protein Kinase |
| SC-$CO_2$ | supercritical carbon dioxide |
| SDF | soluble dietary fiber |
| SDG | secoisolariciresinol diglucoside |
| SDS-PAGE | sodium dodecyl sulfate PAGE |
| SECO | secoisolariciresinol |
| SFE | supercritical fluid extraction |
| SGLT1 | sodium-dependent glucose co-transporters |
| SGOT | serum glutamic oxaloacetic transaminase |
| SGPT | serum glutamic pyruvic transaminase |
| SHR | spontaneously hypertensive rats |
| SI | sucrose-isomaltase |
| SKH | silicon and potassium with humic acid |
| SMC | smooth muscle cell |
| SOD | superoxide dismutase |
| STAT | signal transducer and activators of transcription |
| STZ | streptozotocin |
| SULT | sulfotransferases enzymes |
| SWE | subcritical water extraction |
| TBARS | thiobarbituric reactive substances |
| TBHPC | 3-(*tert*-butyl)-5-methylbenzene-1,2-diol |
| TBHQ | tertiary butyl hydroquinine |
| TBRAS | thiobarbituric acid reactive substances |
| TC | total cholesterol |
| TEAC | trolox equivalent antioxidant capacity |
| TFs | theaflavins |

| | |
|---|---|
| TF2A | theaflavin-3-gallate |
| TF2B | theaflavin-3'gallate |
| TF3 | theaflavin-3, 3'-digallate |
| TG | triglycerides |
| TGAs | triacylglycerols |
| THQ | toluhydroquinone |
| THP-1 | human monocytic cell line |
| TIMP-2 | tissue inhibitor of metalloproteinases 2 |
| TNF | tumor necrosis factor |
| TNFR | tumor necrosis factor receptor |
| TPA | tissue plasminogen activator |
| TRAP | total radical-trapping antioxidant parameter |
| TRBs | thearubigins |
| UAE | ultrasound-assisted extraction |
| UC | ulcerative colitis |
| UDPGT | UDP-glucuronosyltransferase |
| UGT | uridine 5'-diphospho-glucuronosyltransferase |
| uPA | urokinase type plasminogen activator |
| UV-C | ultraviolet C |
| VCAM | vascular cell adhesion molecule |
| VEGF | vascular endothelial growth factor |
| VLDL | very low-density lipoprotein |
| VSMC | vascular smooth muscle cell |

# PREFACE 1

*"25 grams of soy protein a day,*
*As part of a diet low in saturated fat and cholesterol,*
*May reduce the risk of heart disease." [USFDA 21CFR101.82]*
—Ramabhau Patil, PhD

*To be healthy, it is our moral responsibility,*
*Towards Almighty God, ourselves, and our family;*
*Eating fruits and vegetables makes us healthy,*
*Believe and have a faith;*
*Reduction of food waste can reduce the world hunger and*
*can make our planet eco-friendly.*
—Megh R. Goyal

*If wealth is lost, nothing is lost,*
*If health is lost, something is lost, and*
*If character is lost, everything is lost.*
—P. P. Joy, Kerala
<https://www.researchgate.net/profile/Pp_Joy>

We introduce this book volume published under book series *Innovations in Plant Science for Better Health: From Soil to Fork.* This book mainly covers the current scenario of the research and case studies and looks at the importance of phytochemicals from marine and plant sources therapeutics. The volume is broken into four parts: Part I: Functional Foods for Human Health; Part II: Pharmacological Aspects of Fruits and Vegetables; Part III: Pharmacological Aspects of Natural Products; and Part IV: Pharmacological Aspects of Cereals, Grains, and Tea.

This book volume sheds light on the potential of plants for human health for different technological aspects, and it contributes to the ocean of knowledge on food science and technology. We hope that this compendium will be useful for students and researchers as well as for those working in the food, nutraceuticals, and herbal industries.

The contributions by the cooperating authors to this book volume have been most valuable in the compilation. Their names are mentioned in each chapter and in the list of contributors. We appreciate you all for having patience with our editorial skills. This book would not have been written without the valuable cooperation of these investigators, many of whom are renowned scientists who have worked in the field of food engineering and food science throughout their professional careers.

I am glad to introduce Dr. Hafiz Ansar Rasul Suleria, who is a Postdoc Research Fellow/Honorary Research Fellow (Food and Nutrition) at The University of Queensland, School of Medicine, Translational Research Institute, in Australia. With several awards and recognitions, Dr. Suleria brings his expertise and innovative ideas on pharmaceutical sciences in this book.

We will like to thank editorial staff, Sandy Jones Sickels, Vice President, and Ashish Kumar, Publisher and President at Apple Academic Press, Inc., for making every effort to publish the book when the diminishing water resources are a major issue worldwide. Special thanks are due to the AAP Production Staff as well.

We request that readers offer their constructive suggestions that may help to improve the next edition.

We express our admiration to our families and colleagues for understanding and collaboration during the preparation of this book volume. As an educator, there is a piece of advice to one and all in the world: *"Permit that our almighty God, our Creator, provider of all and excellent Teacher, feed our life with Healthy Food Products and His Grace—and Get married to your profession."*

**—Megh R. Goyal, PhD, PE**
Senior Editor-in-Chief

# PREFACE 2

In the recent era, along with technological advancement and food-based strategies, changes in dietary patterns and nutritional awareness are becoming a foremost topic of concern. Globally, the consumption of therapeutic foods, including functional foods, is rising as they are a vital component of dietary interventions for health promotion and disease management. It has been observed that there is tremendous surge in the rate of survival by exploring the secrets of functional food components over the past 200 years. The concept of "proper diet providing health benefits and disease prevention" is emerging as more than basic nutrition.

Presently, due to increased burden of diseases, people are more inclined toward consumption of foods that provide additional health benefits along with fulfillment of nutritional requirements. Functional foods encompass physiologically active components that may or may not have been modified to enhance their bioactivity and provide health benefits in addition to basic nutrition. These foods may help to decrease the risk of diseases, prevent disease development, and enhance an individual's health. Owing to increased demand, the functional food market is growing and expanding at a strange level.

Plant-based functional foods are known to contain compounds (also referred to as phytochemicals) in the leaves, stems, flowers, and fruits that can help to promote human health. Therefore, plant products are drawing the attention of researchers and policymakers because of their demonstrated beneficial effects against diseases with high global burdens such as diabetes, hypertension, cancer, and neurodegenerative diseases. The side effects associated with conventional medicine have awakened the interest of researchers to explore these plants as alternative or complementary medicine. Various plants like onion, garlic, ginger, citrus, flaxseed, broccoli, and other cruciferous vegetables and green and black tea are utilized to develop functional foods that are effectively used to improve human nutrition.

Cereal products, especially oat, soya, and flaxseed, are widely used as a source of cholesterol-lowering dietary fiber β-glucan, whose consumption decreases cholesterol level, reducing the risk of coronary heart disease. Flaxseed being a rich source of linolenic acid and as its consumption is associated with fiber-associated compounds and lignin, it can reduce cholesterol along with platelet aggregation. These plants were discovered by traditional healers to have activities against certain diseases mostly by chance or by testimonies from other users. Hence, there is the need for substantial scientific evidence in terms of efficacy, dosage, and safety in order for traditional herbs to have a place in modern medicine.

This book presents scientific reports on therapeutic values of different plants against diseases. It aims to further encourage the need for the development of plant-based drugs through innovative and ground breaking research studies and, thus, will help to promote the health and economic well-being of people around the world. The understanding of the therapeutic values of these plants will also help to improve their sustainability, as people and governments will be encouraged to preserve and conserve the plants for future generations. The book covers the phytochemistry and health-promoting potentials of the plants against different ailments such as diabetes, hypertension, and microbial infections. Some of the mechanisms by which these plants exert their beneficial effects are also reported.

I thank Dr. Megh R. Goyal for his leadership qualities for inviting me to join his team. He is a world-renowned scientist and engineer with expertise in agricultural and biological engineering. Truly, he is a giver and a model for budding scientists. I am on the board to learn.

**—Hafiz Ansar Rasul Suleria**
Coeditor

# PART I
# Functional Foods for Human Health

CHAPTER 1

# FUNCTIONAL FOODS: CONCEPTS AND THEIR HEALTH PERSPECTIVES

AAMIR SHEHZAD, ASNA ZAHID, SANA MAHMOOD, and HAFIZ ANSAR RASUL SULERIA

## ABSTRACT

Recent technological developments and public awareness regarding health and diseases have brought a tremendous shift in dietary patterns. Presently, due to increased burden of diseases people are more aware of foods that provide additional health benefits along with fulfillment of nutritional requirements. Functional foods encompass physiologically active components, which may or may not have been modified to enhance their bioactivity and provide health benefits. These foods may help to decrease the risk of diseases, prevent from disease development, and enhance individual's health. Owing to increased demand, functional food market is growing and expanding at a strange level. Functional foods can be categorized as: (1) *first class* foods inherently consist of bioactive substance to impart extra benefits to the user. This category includes natural products such as fruits, vegetables, grains, fish, soy products, nuts, and fruit and vegetable juices rich in antioxidants. (2) *The second class* functional foods include the processed foods in which an active health-promoting constituent is injected into the food to give it additional health benefits such as omega-3 rich eggs, fortified margarines, fortified grains, probiotic-enriched yogurts, infant formulas, etc. Moreover, bioactive components in functional foods that impart health benefits usually include essential fatty acids, carotenoids, dietary fiber, polyphenolic compounds, flavonoids, flavonols, flavanols, plant sterol and stanols, prebiotics and probiotics, soy products and proteins, isothiocyanates, minerals and vitamins, and other antioxidants. Phenomenal evidence from in vitro and in vivo trials has revealed that functional foods accompanying functional ingredients can reduce risk of chronic diseases such as cardiovascular diseases and cancer. These diseases preventing properties are due to anti-inflammatory, antithrombotic, antiatherosclerotic, antivasodilatory, antioxidant, antihypertensive, and anticarcinogenic effects of functional foods such as: onion, garlic, soy products, eggs, yogurt, nuts, etc.

## 1.1 INTRODUCTION

In the recent era, along with technological advancement and food-based strategies changes in dietary patterns and nutritional awareness are becoming foremost topic of concern. It has been observed that there is tremendous surge in rate of survival by exploring the secrets of functional food components over the past 200 years. The concept of "proper diet providing health benefits and disease prevention" is emerging than merely focusing on basic nutrition. To address the global burden of diet-related diseases, not only public health nutrition interventions such as nutrition education or food labeling are mandatory but it also requires multidisciplinary and comprehensive approach. It has been estimated that globally 842 million people are undernourished. And concurrently the number of obese and overweight individuals has increased to 2.1 billion in 2013. There is a growing opinion that dietary intervention should focus not only on the safety and quality of food but also on equal access and distribution of food.[4]

Dietary recommendations must include methods which determine the relationship between food security and consumer health. Improvements in health, reduced rates of monotonous diet, well-organized use of facilities such as food in addition to food security, and changes in climate have been observed in many parts of the world.[2] Obesity and other chronic diseases worldwide have resulted in a series of challenging and complex nutrition anomalies. Such challenges include inadequate access to economic, nutritious, and safe diets. Diet modeling is a tool to provide essential nutrients needed by the body along with equal distribution of wide variety of foods from each food groups. It is an approach to explore traditional diet and nutrition-related questions in both developing and industrialized countries where it has already been used to describe persistent and long-term dietary strategy. Moreover, functional foods have ability to modulate normal health perspectives from a comprehensive food security strategy and can also be helpful to recognize reasonable food systems.[1]

Functional foods are foods or part of foods containing physiologically active components, which may or may not have been modified or manipulated to enhance their bioactivity and provide health benefits in addition to basic nutrition. These foods may help to decrease the risk of disease, prevent disease development, or enhance health. Functional foods can be taken in various therapeutic forms or as whole foods. Some functional foods, which are now be identified and linked to positive health outcome, may be consumed as traditional foods with bioactive components. While some functional foods may be fortified or enriched foods are created specially to lower the risk of disease progression and development in different population groups.[24]

The consumer can make selection from a huge range of foods, which contain functionally active ingredients either naturally (e.g., soy beans, cranberries) or through fortification process (e.g., foods fortified with folate and iron). Health benefits of functional foods can be achieved by enhancing the intake of foods, by adding new substances to an individual's diet or through the foods which are the part of an individual's diet already.[3] As new bioactive food substances are being recognized, the chances for development of functional foods will be wider that inherently supplies a bioactive disease ameliorating substance. Level of bioactive substances in foods maybe enhanced to boost up their concentration in specific food (e.g., eggs with elevated levels of omega-3 fatty acids). On the other hand, foods that do not possess an active ingredient naturally can be fortified to provide consumers with a wide range of food sources with a component with specific health benefits (e.g., calcium-fortified orange juice) as mentioned in Table 1.1.

**TABLE 1.1** Current Market Trends Regarding Functional Foods.

| Product type | Market design |
|---|---|
| Altered food products | Present components are replaced with beneficial ones |
| Enhanced food commodities | To alter nutrient component, raw material is changed |
| Enriched food products | Addition of new nutritional components to food products |
| Fortified food products | Food products with increased nutrient content as compared to existing products |
| Nonaltered food products | Naturally existing functional foods in nature |

This chapter focuses on concepts and health perspectives of functional foods.

## 1.2 MARKET TRENDS AND CONSUMER ACCEPTANCE

Functional food acceptance implies "acceptance of any food based upon its taste and physical appearance." It is generally a process of scoring on a scale as high as food is acceptable for good taste and minimum three if the food tastes a bit worse in comparison to its conventional counterpart. During the late 20th century, when consumer's awareness regarding maintaining and achieving optimal health increased, then consumer interest in functional foods also increased. Nowadays, health-conscious people give more importance to health benefits of food rather than merely content of nutritionally active ingredient.[39]

At global level, the growth rate of functional foods market is estimated to be steady and will probably achieve an increase rate of nearly 8% in few years. Furthermore, it has been well demonstrated that until the end of 2021, the potential health-promoting and disease-preventing benefits of functional foods and beverages will increase demand of functional products in the global market. Among well-known ingredients of functional foods are: vitamins, herbs, amino acids, antioxidants, and bacteria and minerals which have physiological health benefits.[28] Major benefits include enhanced mental performance, improved immunity, healthy heart rate, electrolyte and water balance with postexercise benefits, good gastrointestinal health, and better large intestinal and gut microflora. Moreover, it also plays major role in decreasing risk of noncommunicable diseases.[14]

With latest technology and advances in research, consumers are aware of the potential therapeutic and medicinal health benefits linked with specific foods. They have increased intake of substances including fiber, calcium, and soy into their diets.[17] For development of disease specific dietary guidelines or nutritional requirement of different population subgroups perceptive, the role of nutrients at the molecular level is mandatory. With the research area being at its initial stage, role of other bioactive food components requires similar research for their identification. The universal functional foods market size was $129.39 billion in 2015.[8] It is growing and expanding at a strange level significantly among people in relation to their proper diet and health. It is expected to support the overall industry over the next 8 years. Now the key role of food is not just to satisfy consumer's hunger but also to alleviate nutrition and other health-related chronic ailments. In short, functional foods containing biologically active ingredient is a factor, which is manipulated to influence the worldwide industry needs in a positive manner.[4]

## 1.3 FUNCTIONAL FOODS: CATEGORIES AND ORIGIN

In recent era, nutritionists and food technologists are more concerned regarding food security and safety, malnutrition, and diet–health linkages. Basically, there are two main subdivisions of functional foods based upon the presence of active ingredient. The first group contains the functional foods which inherently consist of bioactive substance to impart extra benefits to the end user. The second class of functional foods consists of processed foods in which an active health promoting constituent is injected to the food to impart additional potential health benefits.

### 1.3.1 FOODS CONTAINING FUNCTIONAL INGREDIENTS

Foods of plant origin contain secondary metabolites that provide mechanical protection to plants but are also a source of functional ingredients to provide health benefits to us. Moreover, animal products also contain ingredients that provide a wide range of health benefits. Some products containing functional ingredients are discussed in this section.

#### 1.3.1.1 TOMATO

Tomato is an example of a functional food which naturally consists of the bioactive element lycopene. Lycopene is related with amelioration of prostate cancer proliferation. There is comprehensive research for lycopene's role in reducing the prostate cancer, but it may also help to diminish the risk of many other cancers and cardiac diseases. Its food sources are crushed tomatoes, fresh diced tomatoes, canned tomatoes, tomato paste, low-salt tomato soup, and salsa.[5]

#### 1.3.1.2 OATS

Having functional ingredient phytochemicals and β-glucan, they are customarily called saponins. They have potential ability to reduce serum triglycerides, total blood cholesterol, low-density lipoprotein (LDL) cholesterol, enhance high-density lipoproteins (HDL) cholesterol, and may assist in lowering hypertension. Food sources are whole oat products, oatmeal, whole oats, whole oat bread, etc.[16]

#### 1.3.1.3 FATTY FISH

Natural source of functional ingredients is omega-3 fatty acids such as eicosapentaenoic acid (EPA) and docosahexaenoic acid (DHA). They possess beneficial role in reducing the risk of coronary heart disease and other degenerative health maladies. Foods sources are marine water fish such as salmon, tuna, striped bass, halibut, sardines and trout.[6]

#### *1.3.1.4 SOY*

Soy is the good source of phytochemicals such as: isoflavones, genistein, and soy protein. They may help to attenuate total lipids level and LDL cholesterol. Food sources are tofu, tempeh, miso, and soy nuts.[36]

#### *1.3.1.5 NUTS*

Nuts consist of monounsaturated fatty acids as a bioactive component including healthy fats and vitamin E as antioxidant. Nuts aid in decreasing risk of coronary heart disease, and prevent from cataracts and severity of carcinogenic and mutagenic health disorders. Food origin is walnuts, almonds, pecans, pistachios, peanuts, cashews, hazelnuts, chestnuts, and Brazil nuts.[30]

#### *1.3.1.6 RED WINE/GRAPE JUICE*

Grape juice has functional component resveratrol, which shows heart health promoting effects. Food sources are 100% grape juice or grape juice mixtures (i.e., grape–apple 100% juice mixtures) and any variety of red wine.[22]

#### *1.3.1.7 GREEN LEAFY VEGETABLES*

They contain phytochemicals such as carotenoids, sulforaphanes, apigenin, lutein, and zeaxanthin.[31] Carotenoids can retard the entry of carcinogenic species into cells thus preventing metastasis and tumor cell proliferation; while apigenin and sulforaphanes stimulate the healthy immune system and promote heart protection; lutein is active in reduction of degenerative cataracts especially blindness in the elderly; and zeaxanthin boosts up local immunity. The food sources of these components are collard greens, spinach, broccoli, kale, broccoli sprouts, and other leafy greens.[34] Table 1.2 presents selected examples.

**TABLE 1.2** Functional Ingredients in Plant Food Products.

| Food products | Bioactive ingredients | Physiological action |
|---|---|---|
| Carrot, squash, pumpkin, citrus fruits | β-carotene | Free radical scavenging action and anticarcinogenic |
| Cruciferous vegetables | Isothiocyanates | Anti-inflammatory and vasodilation of arteries |
| Fish oil | EPA and DHA | Healthy brain development and modulating intelligence quotient |
| Grapes, red wine | Resveratrol | Antimutagenic, anti-inflammatory |
| Oat bran | β-glucan | Lowers blood cholesterol, maintains blood glucose level, and enhances insulin response |
| Olive oil | Oleuropein | Reduces the lipid per oxidation and oxidative stress |
| Onion, garlic | Allyl sulfur compounds | Antihypertensive effect |
| Red grapes, citrus fruits, yellow squash, onions | Quercitin | Modulates apoptotic process and antiproliferative action on abnormal cells |
| Soybean and legumes | Isoflavones | Prevent breast cancer, improve cardiac and bone health |
| Tea, berries | Catechins | Promote weight loss, anticarcinogenic potential, lower LDL, and increase HDL |
| Turmeric | Curcumin | Anti-inflammatory, natural healing of wounds, and improves immune health |
| Yogurt and other dairy products | Probiotics | Prevent colorectal cancer, irritable bowel syndrome, piles, and improve the health of large intestine |

DHA, docosahexaenoic acid; EPA, eicosapentaenoic acid.

**Source:** Modified from Ref. [22].

### *1.3.2 FOODS WITH INCREASED FUNCTIONAL COMPONENTS*

Omega-3-enriched eggs contain omega-3 fatty acids. Eggs are not enriched directly with omega-3 fatty acids but these are added in the diet of hens. The feed with huge amounts of flax seed (which is naturally high in omega-3) is fed to egg-laying hens.[9] Commonly, flaxseed, fish oil, or sea algae are main sources of omega-3 fatty acids for hens. Therefore, such eggs consist of large amounts of omega-3 fatty acids and reduced level of saturated fats. They help to lesser cholesterol, triglycerides, and decrease the danger for coronary heart disease. Their origin is whole eggs in the pack, labeled as "DHA/omega-3-enriched." Based upon the brand, normally up to 350 mg omega-3 per egg is present in omega-3 eggs in comparison with 60 mg in a standard egg.[10]

### 1.3.3 FOODS ENRICHED WITH FUNCTIONAL INGREDIENTS

This category usually includes processed foods. Examples of such foods are fortified orange juice at a desirable dosage of vitamin D, added fiber in breads/cereals, and wide variety of other food products.[17]

Food fortification is a cost-effective, flexible, and generally acceptable approach to improve the intake of nutrients in the vulnerable segments, and contributes to provide healthier diet in many developing countries. Processed foods are often fortified with certain micronutrients accessible to large number of population thus playing pivotal role in prevention of deficiency disorder.[11]

#### 1.3.3.1 FORTIFIED MARGARINES

These consist of bioactive health ingredient called plant sterol and stanol esters. They help to attenuate LDL and total cholesterol in those patients with increased serum level of cholesterol and prevent the risk of cardiovascular diseases. They are present in foods including fortified margarines. By replacing the normal margarine or butter and serving with fortified margarines can help to achieve normal health status.[12]

#### 1.3.3.2 PROBIOTICS

Probiotics possess healthy gut microflora to improve human gastrointestinal health and overall these can stimulate the immune system. Generally, the healthy gut microflora belongs to the genera *Lactobacillus* and *Bifidobacterium* (good bacteria). They also help to hinder the growth of abnormal cells and induce apoptosis.[15] The foods with healthy microflora are yogurts, supplemented yogurt with probiotics, fermented vegetables, and fermented soy products.[13]

#### 1.3.3.3 GRAINS FORTIFIED WITH FOLIC ACID

These help lower the risk of heart diseases and neural tube defects, and may also reduce the birth and spinal cord defects in newborn.[3] Micronutrients fortification is one of the most economical means of improving health and survival of mother and children. Most of cereal grains and pulses are fortified with iron and folic acid to reduce incidence of iron deficiency anemia and thus increasing the survival rate.[12] Various intervention strategies are applicable but food fortification with appropriate fortificant can be employed to control this menace. If folic acid level is decreased, then it is also main risk factor for cardiac diseases as low level of folate leads to high level of homocysteine.[32]

#### *1.3.3.4 INFANT FORMULA WITH IRON FORTIFICATION*

Infant formula fortified with iron reduces risk of iron deficiency, enhances intelligence score, and boosts up memory power. Various methods have been used to decrease the under-five child mortality rate. At home, micronutrient powders in complementary foods can be successfully used for iron fortification. This strategy helps to reduce the risk of developing iron deficiency anemia in infants and young children by ensuring the adequate level of iron without changing traditional dietary patterns.[33]

#### *1.3.3.5 MILK WITH ADDED VITAMIN D*

It helps to enhance calcium absorption and reduce the risk of rickets in children and osteoporosis in adults.[40] Some of the related examples are mentioned in Table 1.3.

**TABLE 1.3** Foods with Added Functional Ingredients.

| Added ingredient functional food | Why it is important? |
|---|---|
| Breads and cereals with added fiber | Contain soluble and insoluble fiber which helps in alleviation of constipation |
| Margarine fortified with plant sterols | Source of plant sterols considered helpful in reduction of cholesterol |
| Orange juice with added vitamin D | It is rich in vitamin D which reduces risk of bone diseases such as rickets in children and osteoporosis in adults |
| Yogurt with probiotics | Probiotics yogurt normally maintains proper gastrointestinal health |

**Source:** Adapted from Ref. [4].

## 1.4 FUNCTIONAL FOODS AND NUTRACEUTICALS: HEALTH BENEFITS AND DISEASE PREVENTION

Functional foods and nutraceuticals can attenuate different physiological threats by improving the human health. Functional foods help to fight against some major maladies, for example, cardiovascular problems, cholesterol, diabetes, obesity, osteoporosis, and cancer.[4] Different types of products that are associated with nutraceuticals and functional foods are dietary supplements, medicinal foods, and pharmaceuticals.[26] Inclusively, nutraceutical and functional foods have entirely transformed the food industry. Recently, researches have led to a new era of natural medicines showing a lot of positive effects on human health. Lately, there is an increased demand of functional foods; especially plant-based functional food products, owing to their potential therapeutic effects on human health.[24]

Functional foods provide potential benefits owing to presence of functional ingredients such as: fatty acids, carotenoids, dietary fiber, phenolic acids, flavonoids, plant sterol and stanols, prebiotics and probiotics, soy products and proteins, isothiocyanates, minerals, and vitamins.[10] Prodigious evidence from in vitro and in vivo trials has revealed that foods accompanying functional ingredients can reduce risk of chronic diseases. Some foods with disease-prevention benefits are discussed in this section.

### *1.4.1 OATS*

Oats (*Avena Sativa* L.) possess unique characteristics among whole grain cereals. Even short-term consumption of oats has been linked to reduction in total cholesterol and LDL cholesterol. Oats contain comparatively higher levels of soluble fiber compared to other whole grain cereals. Active soluble fiber present in oats is β-glucan. Soluble fiber present in oats hampers the reabsorption of bile acid in gut and reduces level of cholesterol in the body.[2] As compared to other cereals, oats contain one-third more proteins and four times more fat (good fat) but possess less starch content. Oats contain several phytochemical substances, which possess antioxidant activities due to their radical scavenging property.[21]

#### *1.4.1.1 ANTI-INFLAMMATORY PROPERTIES OF OATS*

Bioactive and functional ingredients present in oats also interact with inflammatory pathways, and with molecular and cell signaling pathways that govern inflammation.[41] This exclusive property of oat cereal has shown to inhibit endothelial and vascular cell expression of cellular adhesion molecules such as intracellular adhesion molecule (ICAM), vascular cell adhesion molecule (VCAM), E-selectin, and others. Suppression of these adhesion molecules results in decreased production of inflammatory chemokines and cytokines such as MCP-1, IL-6, IL-8, and other inflammatory mediators.[20]

#### *1.4.1.2 EFFECTS OF OATS ON VASODILATION*

Research studies have indicated that oats may increase production and endothelial expression of nitric oxide synthase. Owing to vasodilatory properties of oats, consumption of oat or oatmeal cereals in breakfast has been attributed to reduction of blood pressure.[29]

#### *1.4.1.3 ANTIPROLIFERATIVE PROPERTIES OF OATS*

Oats are considered to possess antiproliferative effect on vascular smooth muscle cells (VSMC). In Japan, a synthetic drug was produced to function as an antihistamine which shares a structural similarity with oats.[21,35] Cellular proliferation is considered as a major contributing factor in development of atherosclerosis. Oats, both structurally and mechanically, play a role as antiproliferative substance and have a soothing effect on VSMC.[21]

### *1.4.2 ONION AND GARLIC*

Onion (*Allium cepa*) and garlic (*Allium sativum*) are two widely used foods in culinary preparations; and have been used widely for medicinal purpose for more than 4000 years. Garlic has been used as an antiseptic for wounds and ulcers and for prevention of gangrene; moreover, onion and garlic have also been used to alleviate fever, headache, dysentery, and cholera. Garlic is still being employed as a folk medicine for treatment of several maladies.[7] According to research studies, onion and garlic contain organosulfur compounds that contribute to medicinal and biological benefits.[31] Compounds such as allicin and other lipid-soluble sulfur-containing compounds such as diallylsulpide and diallyldisulphide are also present in onion and garlic. These compounds are responsible for odor, flavor, and functional properties.[18]

#### *1.4.2.1 ANTIATHEROGENIC PROPERTIES OF ONION AND GARLIC*

Several studies and meta-analysis showed that garlic and onion play an imperative role in reducing cholesterol level. Allicin and derivatives of allicin present in onion and garlic have been known to possess hypolipidemic and hypocholesterolemic effects.[16] Moreover, some nonsulfur compounds such as saponins found in garlic have been known to reduce serum cholesterol. Moreover, these compounds inhibit biosynthesis of cholesterol in liver. Further studies showed that saponins inhibit absorption of cholesterol from lumen without any change in HDL.[20]

#### *1.4.2.2 ANTIHYPERTENSIVE PROPERTIES OF ONION AND GARLIC*

Functional properties of onion and garlic also include lowering of blood pressure. Studies suggested that ingredients present in onion and garlic exert hypotensive potential by affecting contraction of smooth muscles thus providing relaxing effect. Furthermore, garlic acts as a vasodilator due to presence of nitric oxide and hydrogen sulfide.[7]

#### *1.4.2.3 ANTIHYPERGLYCEMIC PROPERTIES OF ONION AND GARLIC*

Generally, use of fruits and vegetables along with antidiabetic drugs increases therapeutic potential of herbs in controlling diabetes. Bioactive functional ingredients present in onion and garlic such as methyl and S-allyl cysteine sulfoxide hold antidiabetic potential. Both stimulate production of insulin by pancreas and interfere with glucose absorption.[32]

#### *1.4.2.4 ANTITHROMBOTIC EFFECTS OF ONION AND GARLIC*

Concentration of organosulfur compounds present in onion and garlic determine the antiplatelet potential of onion and garlic. Garlic has been considered to possess 13 times more antithrombotic potential as compared to onion.[27]

### *1.4.3 SOY AND SOY PRODUCTS*

Isoflavones are one of the classes of polyphenol and called phytoestrogens due to their ability to counteract estrogen receptors in cells. Soy isoflavones are powerful plant substances with same chemical nature as the female hormone estrogen and are of greatest importance.[23] Over the last decades, these have been employed in management of cardiovascular disease, osteoporosis, and cancers owing to their potential preventive and curative roles. Foods such as soybeans and legumes contain isoflavones. Foods made from soybeans eaten in Western countries include soy milk, soy cheese, soy yogurt, and soy meat substitutes.[12]

#### *1.4.3.1 ANTICARCINOGENIC EFFECTS OF SOY AND SOY PRODUCTS*

The use of soy and soy products has been linked to lower risk of breast cancer. Reduction in density of breast is seen by isoflavones supplementation in postmenopausal women. Increased breast density is considered as a marker for augmented breast cancer risk. Moreover, many studies have shown that addition of soy in diet from puberty decreases the risk of estrogen-dependent tumors in postmenopausal women[12] by binding to estrogen receptor sites and decreasing the action of endogenous estrogen. Furthermore, antioxidative, antiproliferative, and nonsteroidal actions of isoflavones have also been observed. When phytohormones are consumed as whole foods, they provide protection to uterine and breast tissues from estrogen overstimulation.[13]

#### *1.4.3.2 CARDIOVASCULAR BENEFITS OF SOY AND SOY PRODUCTS*

Daily consumption of soy isoflavones is attributed to reduced risk of cardiovascular diseases and diabetes due to reduction of cholesterol in body and improving vascular functions. Soy contains isoflavones that resemble estradiol in structure. Soy products have been known to decrease postprandial estrogen receptors. Soybean and its products are also associated with reduction in obesity. Obesity is considered as a major contributor of cardiovascular disease.[19]

#### *1.4.3.3 HYPOGLYCEMIC EFFECTS OF SOY AND SOY PRODUCTS*

Soy isoflavones are known to reduce diabetes type-2 risk. Soy isoflavones decrease insulin resistance and oxidation thus reducing hyperinsulinemia. This process increases availability of glucose to cells of the body.[37]

### 1.4.4 PROBIOTICS

The probiotics are one of the most important functional ingredients that are incorporated in designer foods.[42] Probiotics include large number of microorganisms such as bacteria and yeast, which are considered as food supplements and confer potential health benefits to human beings.[1] Generally, probiotics are taken orally and are available in different forms such as tablets, capsules, and powders in food products. The addition of probiotics to different food products is a natural approach to increase their functional aspects.[35]

#### 1.4.4.1 ANTI-INFLAMMATORY AND ANTICARCINOGENIC EFFECTS OF PROBIOTICS

The ability of probiotics to stimulate the gut microbiota, maintenance of gut integrity, and the physiochemical conditions has been shown in many studies.[15] Several studies recommend that promoting the health of large intestinal colon by probiotics is associated with lowering intestinal damage and injuries that are responsible for uncontrolled cell growth and inflammation.[25] Many probiotics microflora are produced through the fermentation of lactic acid, polyphenols with antioxidant potential, and unsaturated fatty acids such as conjugated linoleic acid, a group of linoleic acid isomers, which have anti-inflammatory and anticancer properties.[19]

#### 1.4.4.2 ANTIHYPERTENSIVE POTENTIAL OF PROBIOTICS

The epidemiological studies indicate that gut microbiota improvement by probiotic supplementation might positively help in reducing blood pressure in hypertensive conditions.[8,13] Another mechanism involved in the antihypertensive effect of probiotics is the production of bioactive peptides having inhibitory effect on angiotensin converting enzyme during the fermentation process.[38]

#### 1.4.4.3 HYPOLIPIDEMIC POTENTIAL OF PROBIOTICS

The addition of probiotics to diet can improve vascular oxidative stress.[27] Probiotics can also curtail cholesterol level indirectly by conjugating bile salt with hydrolase and inhibiting enterohepatic circulation of bile salt. Probiotics bind cholesterol to their cell walls and assimilate it during growth.[28]

## 1.5 SUMMARY

Functional foods are chemically defined bioactive constituents of known therapeutic activities, which help to fight against some major maladies, for example, cardiovascular problems, cholesterol, diabetes, obesity, osteoporosis, and cancer. Examples of functional foods include: tomatoes, apples, potatoes, oats, melon, nuts, soy products, carrots, grapes, tea, coffee, milk, yogurt, onion and garlic, turmeric, etc. Bioactive components present in functional foods that possess these health benefits consist of essential fatty acids, carotenoids, dietary fiber, polyphenolic compounds, flavonoids, flavonols, flavanols, plant sterol and stanols, prebiotics and probiotics, isothiocyanates, minerals and vitamins, and other antioxidants. Further investigations on functional foods will reveal functional perspectives and core mechanisms of functional ingredients in prevention of diseases.

## KEYWORDS

- **antioxidants**
- **bioactive components**
- **flavonoids**
- **flavonols**
- **phytochemicals**
- **polyphenols**
- **vitamins**

## REFERENCES

1. Adelstein, B. A.; Macaskill, P.; Chan, S. F. Most Bowel Cancer Symptoms do not Indicate Colorectal Cancer and Polyps: A Systematic Review. *J. BMC Gastroenterol.* **2011,** *11* (9), 65–78.
2. Anderson, J. W. Whole Grains Protect Against Atherosclerotic Cardiovascular Disease. *J. Proc. Nutr. Soc.* **2003,** *62* (1), 135–142.

3. Aune, D. N.; Keum, E.; Giovannucci, L. T.; Fadnes, P.; Boffetta, D. C.; Greenwood, S.; Tonstad, L. J.; Vatten, E.; Riboli, T. Norat. Whole Grain Consumption and Risk of Cardiovascular Disease, Cancer, and All Cause and Cause Specific Mortality. *Syst. Rev. Dose-Response Meta-Anal. Prospect. Stud.* **2016,** *23* (56), 353–2716.
4. Bigliardi, B.; Galati, F. Innovation Trends in the Food Industry: The Case of Functional Foods. *Trends Food Sci. Technol.* **2013,** *32* (2), 118–129.
5. Chaturvedi, S. P. K.; Sharma, V. K.; Garg, Bansal, M. Role of Nutraceuticals in Health Promotion. *Int. J. Pharm. Technol. Res.* **2016,** *3* (23), 442–448.
6. Chen, P.; Zhang, X.; Wang, K.; Zhao, D. S.; Negi, L.; Zhuo, M.; Qi, X.; Wang, Zhang X. Lycopene and Risk of Prostate Cancer. *Syst. Rev. Meta-Anal. Med.* **2015,** *94* (10), 1260–1267.
7. Chowdhury, R.; Stevens, S.; Gorman, D.; Pan, A.; Warnakula, S.; Chowdhury, S.; Ward, H.; et al. Association Between Fish Consumption, Long Chain Omega 3 Fatty Acids, and Risk of Cerebrovascular Disease: Systematic Review and Meta-analysis. *BMJ* **2012,** *345*, e6698.
8. Corzo-Martínez, M.; Corzo, N.; Villamiel, M. Biological Properties of Onions and Garlic. *J. Trends Food Sci. Technol.* **2007,** *18* (64), 609–625.
9. Cramer, H.; Lauche, R.; Haller, H.; Steckhan, N.; Michalsen, A.; Dobos, G. Effects of High Phenolic Olive Oil on Cardiovascular Risk Factors: A Systematic Review and Meta-analysis. *J. Phytomed.* **2015,** *22* (20), 631–640.
10. De Brito, A.; de Sousa, M. P. C.; Neto, M.; Magnani, V.; de Andrade Braga, J. H.; da Costa-Silva, C. G.; Leandro, H.; Vidal, Pirola L. New Insights on the Use of Dietary Polyphenols or Probiotics for the Management of Arterial Hypertension. *J. Front. Physiol.* **2016,** *23* (7), 448–456.
11. Del-Gobbo, L. C.; Imamura, F.; Aslibekyan, S.; Marklund, M.; Virtanen, J. K.; Wennberg, M.; et al. Omega-3 Polyunsaturated Fatty Acid Biomarkers and Coronary Heart Disease: Pooling Project of 19 Cohort Studies. *J. Int. Med.* **2016,** *43* (17), 2925–2934.
12. Goldberg, I. *Functional Foods: Designer Foods, Pharmafoods, Nutraceuticals*; Springer Science & Business Media: New York, USA, 2012; pp 1123–1143.
13. Jooyandeh, H. Soy Products as Healthy and Functional Foods. *Middle-East J. Sci. Res.* **2015,** *42* (7), 71–80.
14. Jose, P. A.; Raj, D. Gut Microbiota in Hypertension. *J. Curr. Opin. Nephrol. Hypertens.* **2015,** *24* (14), 403–410.
15. Justina, G.; Justina, K.; Mindaugas, D. Functional, Organic or Conventional? Food Choices of Health Conscious and Skeptical Consumers. *Baltic J. Manag.* **2015,** *12* (2), 139–152.
16. Kahouli, I.; Tomaro-Duchesneau, C.; Prakash, S. Probiotics in Colorectal Cancer (CRC) with Emphasis on Mechanisms of Action and Current Perspectives. *J. Med. Microbiol.* **2015,** *62* (37), 1107–1123.
17. Kanza, A. A.; Butt, M. S.; Haq, I.; Suleria H. A. R. Investigating the Antioxidant Potential of Garlic (*Allium sativum*) Extracts Through Different Extraction Modes. *Curr. Bioact. Compd.* **2017,** *13*, 1–6.
18. Lai, L. R.; Hsieh, S. C.; Huang, H. Y.; Chou, C. C. Effect of Lactic Fermentation on the Total Phenolic, Saponin and Phytic Acid Contents as well as Anti-colon Cancer Cell Proliferation Activity of Soymilk. *J. Biosci. Bioeng.* **2015,** *115* (41), 552–556.
19. Lanzotti, V. The Analysis of Onion and Garlic. *J. Chromatogr.* **2015,** *11* (22), 13–22.

20. Liang, S.; Lv, G.; Chen, W.; Jiang, J.; Wang, J. Citrus Fruit Intake and Bladder Cancer Risk: A Meta-analysis of Observational Studies. *Int. J. Food Sci. Nutr.* **2015,** *65* (13), 893–898.
21. Liu, L.; Zubik, L.; Collins, F. W.; Marko, M.; Meydani, M. The Antiatherogenic Potential of Oat Phenolic Compounds. *J. Atheroscler.* **2015,** *175* (1), 39–49.
22. Mell, B.; Jala, V. R.; Mathew, A. V.; Byun, J.; Waghulde, H.; Zhang, Y.; Joe, B. Evidence for a Link Between Gut Microbiota and Hypertension in the Dahl Rat. *J. Physiol. Genom.* **2015,** *47* (7), 187–197.
23. Menrad, K. Market and Marketing of Functional Food in Europe. *J. Food Eng.* **2015,** *56* (11), 181–188.
24. Meydani, M. Potential Health Benefits of Avenanthramides of Oats. *J. Nutr. Rev.* **2015,** *67* (12), 731–735.
25. Micali, S.; Isgro, G.; Bianchi, G.; Miceli, N.; Calapai, G.; Navarra, M. Cranberry and Recurrent Cystitis: More Than Marketing. *Critic. Rev. Food Sci. Nutr.* **2015,** *54* (19), 1063–1075.
26. Mladenova, D.; Daniel, J. J.; Dahlstrom, J. E.; Bean, E.; Gupta, R.; Pickford, R.; Kohonen-Corish, M. R. The NSAID Sulindac is Chemopreventive in the Mouse Distal Colon but Carcinogenic in the Proximal Colon. *Int. J. Gut Health* **2015,** *60* (15), 350–360.
27. Radhika, P. R.; Singh, R. B. M.; Sivakumar, T. Nutraceuticals: An Era of Tremendous Scope. *Int. J. Res. Ayurved. Pharmacol.* **2015,** *2* (67), 410–415.
28. Rashid, S. K.; Idris-Khodja, N.; Auger, C.; Alhosin, M.; Boehm, N.; Oswald-Mammosser, M.; Schini-Kerth, V. B. Probiotics (VSL# 3) Prevent Endothelial Dysfunction in Rats with Portal Hypertension: Role of the Angiotensin System. *Int. J. Microbiol.* **2015,** *9* (97), 458–464.
29. Rezaei, M.; Sanagoo, A.; Jouybari, L.; Behnampoo, N.; Kavosi, A. The Effect of Probiotic Yogurt on Blood Glucose and Cardiovascular Biomarkers in Patients with Type II Diabetes. *Random. Controll. Trial. Evid. Based Care* **2015,** *6* (32), 26–35.
30. Saeed, F.; Arshad, M. U.; Pasha, I.; Suleria, H. A. R.; Arshad, M. S.; Qamar, A.; Sultan, S. Effect of Arabinoxylan and Arabinogalactan on Textural Attributes of Bread. *J. Food Process. Preserv.* **2015,** *39* (6), 1070–1088.
31. Saltzman, E.; Das, S. K.; Lichtenstein, A. H.; Dallal, G. E.; Corrales, A.; Schaefer, E. J.; Roberts, S. B. An Oat-Containing Hypocaloric Diet Reduces Systolic Blood Pressure And Improves Lipid Profile Beyond Effects of Weight Loss in Men and Women. *J. Nutrition* **2015,** *131* (5), 1465–1470.
32. Song, J. K.; Bae, J. M. Citrus Fruit Intake and Breast Cancer Risk: A Quantitative Systematic Review. *J. Breast Cancer* **2015,** *16* (25), 72–76.
33. Soofi, S.; Cousens, S.; Iqbal, S. P.; Akhund, T.; Khan, J.; Ahmed, I.; Zaidi, A. K.; Bhutta, Z. A. Effect of Provision of Daily Zinc and Iron with Several Micronutrients on Growth and Morbidity Among Young Children in Pakistan. *A Cluster-Randomised Trial. Lancet Lond. Engl.* **2015,** *382* (54), 29–40.
34. Statistics and Facts on the Functional Foods Market in the U.S. http://www.statista.com/topics/1321/functional-foods-market/ (accessed Aug 25, 2017).
35. Sultan, M. T.; Butt, M. S.; Ahmad, R. S.; Batool, R.; Naz, A.; Suleria, H. A. R. Supplementation of Powdered Black Cumin (*Nigella sativa*) Seeds Reduces the Risk of Hypercholesterolemia. *Funct. Foods Health Dis.* **2011,** *1* (12), 516–524.

36. Sultan, M. T.; Butt, M. S.; Karim, R.; Ahmad, A. N.; Suleria, H. A. R.; Saddique, M. S. Toxicological and Safety Evaluation of *Nigella sativa* Lipid and Volatile Fractions in Streptozotocin Induced Diabetes Mellitus. *Asian Pac. J. Trop. Dis.* **2014,** *4*, S693–S697.
37. Tamai, H.; Katoh, K.; Yamaguchi, T.; Hayakawa, H.; Kanmatsuse, K.; Haze, K.; Katoh, O. The Impact of Tranilast on Restenosis After Coronary Angioplasty: The Second Tranilast Restenos is Following Angioplasty Trial (TREAT-2). *Am. Heart J.* **2015,** *143* (3), 506–513.
38. Tokede, O. A., Onabanjo, T. A.; Yansane, A.; Gaziano, J. M.; Djousse L. Soya Products and Serum Lipids: A Meta-analysis of Randomised Controlled Trials. *Br. J. Nutr.* **2015,** *114* (34), 831–843.
39. U.S. Department of Health and Human Services (U.S. Food and Drug Administration), 2000. Structure/FunctionClaimshttp://www.fda.gov/Food/IngredientsPackaging Labeling/ LabelingNutrition/ucm2006881.htm (accessed Aug 25, 2017).
40. Uccello, M.; Malaguarnera, G.; Basile, F.; D'agata, V.; Malaguarnera, M.; Bertino, G.; Biondi, A. Potential Role of Probiotics on Colorectal Cancer Prevention. *BMC Surg.* **2012,** *12* (Suppl 1), S35.
41. Vasileiou, I.; Katsargyris, A.; Theocharis, S.; C. Giaginis, C. Current Clinical Status on the Preventive Effects of Cranberry Consumption Against Urinary Tract Infections. *J. Nutr. Res.* **2017,** *33* (58), 595–607.
42. Wang, A.; Zhu, C.; Fu, L.; Wan, X.; Yang, X.; Zhang, H.; Zhao, H. Citrus Fruit Intake Substantially Reduces the Risk of Esophageal Cancer: A Meta-analysis of Epidemiologic Studies. *Int. J. Oncol.* **2015,** *94* (27), 1390–1399.

## CHAPTER 2

# PLANT-BASED FUNCTIONAL FOODS FOR HUMAN NUTRITION: CURRENT TRENDS, FUTURE PROSPECTIVE, AND PHYTOCHEMICAL ATTRIBUTES

MUHAMMAD SOHAIB, AYESHA ZAFAR, and AZMAT ULLAH KHAN


## ABSTRACT

Globally, the consumption of therapeutic and functional foods is rising, as they are vital components of dietary interventions for health promotion and disease management. These foods may be raw or processed and contain functional ingredients with optimal bioavailability for optimal human health. They represent one of the intensively investigated focus areas in food and nutrition field. However, they are not "magic bullets" or universal panacea for poor dietary practices. Also, the scientific evidences have reported plant-based diet ameliorative potential against various ailments such as: hypercholesterolemia, hyperglycemia, obesity, and oncogenesis. The focus on plant-based dietary habits is driven by three factors: long-term health, daily health, and social reasons (including animal treatment and concerns about the environment). Various plants such as onion, garlic, flaxseed, and broccoli and other cruciferous vegetables, grapes, and tea are utilized to develop functional foods to improve human nutrition. Cereal products especially oats are widely used as a source of cholesterol-lowering dietary fiber β-glucan whose consumption can decrease cholesterol level, thus reducing coronary heart disorders. Soy-based foods have therapeutic role and have proven their significance against cardiovascular disease (CVD), osteoporosis, and mitigation of menopausal symptoms. Flaxseed being the rich source of linolenic acid can reduce cholesterol along with platelet aggregation. Similarly, vegetables such as tomatoes, garlic, onion, and other cruciferous have received significant attention because of their bioactive and polyphenolic compounds with long-term benefits. Lycopene, primary carotenoid found in fruits, can reduce risk of cancer. *Allium sativum* has chemo preventive, antihypertensive, and cholesterol-lowering potential. Moreover, epidemiological studies on broccoli and cruciferous vegetables have reported anticarcinoma properties due to high content of glucosinolates.

Tea is beneficial due to presence of polyphenolic constituents including epigallocatechin-3-gallate, epigallocatechin, epicatechin-3-gallate, and epicatechin effective against CVD. Recent trends have documented vegetable protein intake linked with 34% reduction in risk of fatal ischemic heart disease; higher vegetable protein and lower carbohydrate intake are linked to 20% reduction in CVD mortality; and a lower-carbohydrate, higher-vegetable diet is associated with 10% reduction in the risk of diabetes. Conclusively, functional foods consumption especially of plant origin is among leading foods for health benefits.

## 2.1 FUNCTIONAL FOODS: GENERAL OVERVIEW

The phrase "Let food be thy medicine and medicine be thy food" by Hippocrates almost 2500 years ago has received transformed interest by dieticians, nutritionists, and health conscious consumers. Above all, there has been an eruption of consumer attention in health improving role of functional foods,[41] which are raw, processed, whole, fortified, enriched, or enhanced foods with health benefits. Nutritionally, health claim of these functional foods is derived from scientific treaty standards by experts and by American Dietetic Association (ADA). Similarly, food and nutrition experts are working with health professionals, food traders, government agencies, and scientific community. Since foods are not assessed only in terms of micro and macronutrients evaluating the levels of physiologically active constituents and analyzing their health effects is becoming mandatory for the scientific community due to increasing consumer demand.[42]

The term functional food was used in Japan for first time in 1980 and it includes processed foods having functional ingredients to provide benefits beyond nutrition. Today, Japan is the only country where specific regulatory approval procedure exists for functional foods (Foods for Specified Health Use), by Japanese Ministry of Health and Welfare.[7] Blood pressure, blood cholesterol levels, and gastrointestinal health can be maintained by the use of functional foods.[6] Also, functional foods market in Japan is one of the most progressive markets in the world and more than 100 products have been licensed[118] Contrary to this in United States, there is no legal functional foods category.

Functional foods have become increasingly popular in United States due to health conscious baby boomers food sector estimates.[76] A recent study estimated the market value of functional foods (in Japan, United States, and Europe) at $130 billion.[74] Functional foods potential is to alleviate disease, promote health, and to reduce health care costs. According to ADA, all foods can impart benefits and can be considered functional at specific level considering physiological behavior of humans. They include all nutrients that help to nourish development, maintain essential processes, or furnish energy needs. Functional Foods can be classified into four groups: conventional, modified (i.e., fortified or improved), therapeutic, and foods for special dietary use.[43]

### 2.1.1 SIGNIFICANCE OF FUNCTIONAL FOODS

The word "functional foods" includes wide variety of foodstuffs, such as: yogurts containing probiotics, cholesterol-lowering food products, and breakfast cereals with added folic acid. It should be possible to consume a functional food as a part of a usual, balanced diet and still achieve proposed benefits. Although functional foods may provide health benefits for some persons, yet these products are not perceived as a substitute to a healthy, well-balanced diet and achieving a healthy lifestyle in other ways.[127] Moreover, researchers hold many promises regarding functional foods to increase the quality of life of consumers. However, to reach such a conclusion, scientific study must create the bioavailability and efficiency of these compounds at levels that are functionally attainable under particular dietetic patterns.[22]

The 2011 survey of functional foods/foods for the healthy consumers reported that about 73% consumers have faith in such foods as potential to improve their health. Around 70% believe that vegetables and fruits are most renowned functional foods for human health. According to this survey, 80% of the consumers approved that functional foods and beverages can help to improve or maintain health of individuals, comprising the immune health (79%), eye health (66%), bone health (81%), heart (79%) and circulatory health (74%), and digestive health (78%). And according to International Food Information Council (IFIC), 87% of Americans consider that certain foods have health advantages apart from basic nutrition.

Globally, the consumption of nutritional related food business is mounting and the driving factors of this market have increased the consumer knowledge, adaptability, and effectiveness of nutrition products. The side effects associated with drugs and costly treatments are also pushing the people toward healthy diets.[103] Also worldwide, functional foods can reduce the cost of health care by improving health and wellness, and generously consumers have better control over their health by giving an appropriate form of health-increasing constituents. The functional food market is being developed due to positive outcome of consumer trends and scientific improvements and is estimated to increase to $151 billion by 2020.

This chapter discusses current trends and future prospective of plant-based functional foods for human health.

## 2.2 TYPES OF FUNCTIONAL FOODS

Functional foods are categorized into four groups that are discussed in this section.[47]

### *2.2.1 CONVENTIONAL FOODS*

The conventional functional foods are present in their original form and/or are not modified, for example, fruits and vegetables (FAV). According to the 2009 survey conducted by the IFIC on "Consumer Attitude towards Functional Foods," vegetables and fruits were at the top of the list that was identified by the consumers. Examples may include: broccoli, raspberries, kale, and tomatoes with high levels of bioactive constituents such as: sulforaphane, ellagic acid, lutein, and lycopene. Many conventional foods may reduce cancer risk. There are different conventional foods, which can be used for the treatment of different types of cancers as discussed in this section.

Based on the epidemiologic and experimental studies, cruciferous vegetables reduce the risk of numerous types of cancer.[125] Citrus fruits reduce the risk of stomach cancer.[10] Tomato and its products reduce the risk of ovarian, prostate, pancreatic, and gastric cancer due to presence of high level of lycopene.[55] Peanuts and tree nuts reduce the risk of sudden cardiac death.[61] The consumption of dark chocolate can improve the endothelial function, which is good for functioning of heart.[32] Cranberry juice reduces bacteriuria for proper functioning of urinary tract. Probiotics (fermented dairy products) can improve the irritable bowel syndrome, which is used for the maintenance of intestinal health.[84]

### 2.2.2 MODIFIED FUNCTIONAL FOODS

Modified functional foods are enriched, enhanced, or fortified foods. Few of the drinks or beverages and some snacks have been criticized in the market due to the food safety issues and false medical claims because some products were modified by kava as mentioned in St John's wort and *Echinacea*. For this reason, such beverages and snacks were removed from the market at commercial level. However, using biotechnology techniques, different food products have been modified to improve the nutritional value and to maintain health attributes. Examples of modified food products include: breads enriched with folate for the development of proper fetal, orange juices fortified with calcium for improving bone health, food products enriched with the bioactive constituents as energy-stimulating constituents. Such drinks include: guarana, taurine, and ginseng. The level of cholesterol can be reduced using margarines because they are enriched with the plant stanols or sterol esters. The N-3 fatty acids are also modified by biotechnology technique.[49]

### 2.2.3 MEDICAL FUNCTIONAL FOODS

According to the Orphan Drug Act, medical food is "any food that is used intentionally for the particular nutritional management of the specific disease or illness; and which required distinctive dietary requirements; and the scientific based codes are set by medical assessment; and that is used enterally under the observation of a general practitioner." This category of medical foods include: phenylketonuria formulas that are free of phenylalanine, and renal/liver/diabetic formulations as oral supplements. The regulatory status of the food product can be determined by the assertions. For instance, the oral supplements of bottled or canned food are medical foods that can be used under the medical supervision, and when sold to the consumers it will become a food for the specific dietary use at the retail level [US FDA (Food and Drug Administration)].

### 2.2.4 FOODS FOR THE SPECIAL DIETARY USE

The US Federal Food, Drug, and Cosmetic Act [Section 411(c) (3)] explains the "Special Dietary Use" as a specific use for which a food is represented to be used, and comprises but is not restricted to the following:

- Providing a dietary need that exists by the purpose of a pathological, physical, physiological, or other ailment;
- Providing vitamins, minerals, and other constituents for the use of individuals to enhance the diet by increasing the overall nutritional intake; and
- Providing a specific dietary requirement by the purpose of being a food for use as the individual part of the diet. (FDA, 2009).

Hypoallergenic foods (lactose and gluten free) are provided to reduce weight and infant's foods are examples of such type of foods. Selected examples are mentioned in Table 2.1.

**TABLE 2.1** Types of Functional Food with Typical Examples.

| Types of functional food | Typical example |
|---|---|
| Conventional foods (unmodified foods) | Broccoli, tomatoes, nuts |
| Modified foods (enriched/ enhanced food or fortified food) | Iodized salt<br>Orange juice fortified with calcium |
| | Breads with folate (enriched) |
| | Energy bars, yogurts, snacks, bottled water, teas and other foods formulated with bioactive constituents such as lutein, St John's wort, ginkgo biloba, saw palmetto, fish oils, and/or classified amino acids (enhanced) |
| Medical foods | Phenylketonuria formulas that is free of phenylalanine |
| Foods for the special dietary use | Hypoallergenic foods (lactose-free foods and gluten-free foods); foods for weight loss and infant's food |

## 2.3 PLANT-BASED FUNCTIONAL FOODS

Functional foods of plant origin show a significant role in disease prevention and promoting or maintaining the health of the individuals. These are physiologically active ingredients, therefore only few of these have had functional health advantages. For the authorization of health claim, only few have been accepted for arduous standard of "Significant Scientific Agreement (SSA)." Plant-based functional foods that are currently eligible to have health claim of US FDA involve soluble fiber from psyllium seed husk, oat-soluble fiber (β-glucan), soy protein and stanol, and sterol-ester-fortified margarine. There are a few number of plant-based foods or components that are currently not approved by FDA considering their health claims. However, they have experimental exploration supporting their probable health benefits. Plant-based foods may involve garlic, grapes, cranberries, nuts, and chocolate.[40] Some plant-based functional foods (coriander, garlic, onion, flaxseed, fenugreek, FAV, and moringa) are described in this section.

### 2.3.1 PHYTOCHEMICAL PROPERTIES OF CORIANDER

Coriander *(Coriandrum sativum L.)* possesses medicinal value and is cultivated in India, Pakistan, Bangladesh, Russia, Central Europe, Morocco, and China where its seeds and leaves are used for various food applications and essential oil extraction.[20] Coriander leaves contain 84% water and are rich source of vitamin A (β-carotene), vitamin C, minerals, and iron. It contains vitamin C up to 160 mg/100 g and vitamin A up to 12 mg/100 g. It contains very less amount of saturated fat and cholesterol, and very high amount of thiamine, zinc, and dietary fiber. The two main essential fatty acids are linoleum and linolenic acid. Depending upon the origin, the coriander seeds contain up to 1.8% volatile oil and 5–7% of ash content.[14]

Coriander oil is especially useful for the control of mental exhaustion, tension, migraine, muscle spasms, and joint arthritis. It also increases appetite and improves gastric problems, relieves intestinal hampering, and has anti-inflammatory properties. Volatile constituents of oil have antioxidative activity that may limit lipid peroxidation and inhibit microorganism growth.[11] It possesses nutritional and medicinal properties and deployed in domestic formulations to cure seasonal cold, stomach disorders, and arthritis. It also has cholesterol-lowering properties and can be used as antioxidant, antidiabetic, antimicrobial, antimutagenic, anti anxiety, and analgesic.[97] Leaves of coriander have more antioxidants as compared to seeds due to presence of phenolics and flavonoids.[99,125]

Coriander has high value of coumarins that have been described to retard growth of breast, colon, lung, and prostate cancer. It hinders angiogenesis and attack of cancerous cells on body.[4] In a study, extract of roots, leaves, and stems of *C. sativum* was used to assess the anticancer antioxidant property by using hexane, dichloromethane, ethyl acetate, methanol, and water. The results showed that leaves of coriander had the highest antiproliferative activity and roots prevent DNA damage suggesting anticancer properties and inhibition of metastasis.[111] According to a research, coriander provided via feed (62.5 g/kg) and drinking water (2.5 g/L, prepared by 15 min decoction) was able to reduce hypoglycemia in streptozotocin(STZ)-induced diabetic mice.[38] The antihyperglycemic property of coriander leaves was evaluated in normoglycemic rats, which were fed with high-fat diet with additional cholesterol content. Results further documented that low-density lipoprotein (LDL) and very low-density lipoprotein (VLDL) cholesterol levels were decreased, whereas high-density lipoproteins (HDL) was increased.[26]

To evaluate antimicrobial properties of coriander oil, 12 essential oils were tested against several strains of *Clostridium jejuni*. The results indicate that coriander oil serves as natural antimicrobial in food against different microbes such as *C. jejuni*.[100] Food industry also faces challenges to protect food from foodborne pathogens, the results of different studies reveled the antimicrobial effect of coriander oil.[105] A study was done on histopathologically damaged rats to evaluate nephron-protective effect of coriander, the results of study show positive effect of coriander in nephron protection due to presence of flavonoids and polyphenols.[64]

Different animal models to assess anti-anxiety effect of coriander have shown that 100 and 200 mg/kg dose of coriander has similar effect as diazepam, a drug used for the treatment of anxiety.[71] Similarly, benzodiazepines, a drug used as anxiolytics, has unwanted adverse effects. To assess anti-anxiety effect of coriander, an aqueous extract of coriander was tested in albino mice, and it was found that coriander extract has potential muscle relaxant effect to show anxiolytic property.[31] Coriander has also a role in improving abdominal troubles (from indigestion to flatulence to diarrhea) and it is safe for infants to relieve abdominal discomforts. It also has a mild effect as eyewash to reduce irritation and burning sensation.[122] The coriander oil acts as a body cleanser to remove toxins and fluid waste. It also acts as a natural laxative.[11]

Considering the available data and reported studies, coriander oil has beneficial phytonutrients and seeds have health-associated behavior. It also possesses hypoglycemic, hypolipidemic, antibacterial, and antimutagenic activity showing a remarkable role of coriander in health industry. Due to its hypoglycemic and hypolipidemic activity, one should take precautions while using during surgery to avoid complications.

### *2.3.2 PHYTOCHEMICAL PROPERTIES OF GARLIC*

Garlic (*A. sativum*) belongs to a group of plants of nutritious, medicinal, and pro-health activities. Fresh garlic bulbs contain about 60% of water, 32% of carbohydrates, and 6.5% of fiber. Garlic also supplies vitamins, vitamin C is present in highest amount (about 31 mg in 100 g of product); and among mineral ingredients, potassium is present in the highest concentration, whereas iron, magnesium, and phosphorus in a slightly lower concentration. Therefore, garlic contains considerable amount of biologically active substances.[101] The US FDA conducted a survey of 900 people to determine the utilization of garlic supplement, and results showed almost 17% of population using garlic supplement in 12 months.[116] Allicin belongs to most important biologically active substance with characteristic odor and it is allyl cysteine sulfoxide, crystalline, colorless, odorless, and without antibiotic properties, but it is present in whole, unbruised garlic cloves.[69]

It is believed that allicin in garlic have antiparasitic, stabilizing intestinal flora, and hypotensive activity; and concurrently it may have irritating influence on skin and air passages; and it has been proved to inhibit blood platelet aggregation and lower the level of triacylglycerols in blood serum.[101] The sulfur compounds are most widely described group of bioactive substances contained in garlic bulbs, because they are responsible for its characteristic odor and flavor, in comparison with other plants belonging to the same group; garlic contains about five times more sulfur compounds as compared to onion and leek.[131]

Garlic inhibits adenosine deaminase which is an enzyme disintegrating adenosine that decreases cardiac muscle overload thus improving blood flow through coronary vessels. The adenosine increases glutathione peroxidase activity, which: decreases the activity of anion radical superoxide ($O_2^-$) and hydroxide, causing peroxidation of cell membrane lipids and unsaturated fatty acids; decreases adherence of incited granulocytes to endothelium of blood vessels and blood platelet aggregation; increases fibrynolitic activity of plasma; and lowers cholesterol and its fractional concentration without considerable influence on LDL/HDL plasma factor, and glycerol and phospholipid concentration in blood serum.[101]

Garlic exhibits properties as enhanced neuroprotection, natural NSAID, antioxidant, and memory enhancer. In Alzheimer's transgenic mice, aged garlic extract (AGE) prevented hippocampal memory tasks suggesting AGE potential against Alzheimer diseases (AD) progression.[19,90] Currently, garlic is being used as a single or in combination of other ingredients to produce synthetic pharmaceutical drugs for treatment of Alzheimer's disease with least adverse effects.[15]

Garlic also has antioxidant properties due to presence of allicin as active molecule. Allicin reacts with free thiol-containing enzymes and traps the free radicals thus serving as an effective antioxidant.[77] It has also found to forage hydroxyl radicals[94] and to hinder superoxide generation by phorbol ester-activated human granulocytes.[104] As an additional therapeutic property, allicin showed variation of SH-dependent activities.[129] Garlic also has anticancerous properties due to presence of allicin.[27] Allicin plays a major role as antiproliferative agent. Water-soluble garlic preparations increase the efficiency of allicin to reduce the intracellular gluthathione level. Moreover, reduced glutathione (GSH) level is correlated with reduced growth activity by allicin.[45] The therapeutic uses of garlic against cancer are well-documented as it inhibits the *nitrosamine* formation *and bioactivation*[8] Allicin is competent immune-modulator of cellular activities and macrophage secretory and showed distinct effects on production of cytotoxic molecules.[53] Garlic extract works as immune modulator thus, decreasing diseases mainly caused by dysfunction of immune system.[63] Garlic and its constituents help in production of cytokine as intermediary of inflammation.

A protein complex nuclear factor-KB (NF-KB) is a critical dimeric transcriptional factor, which plays an essential role in the activation of various genes that control immune reaction. NF-KB is implicated in activation and regulation of vascular inflammatory molecules associated with cancer cellular process such as apoptosis.[67] Additionally, garlic is indirectly involved in modulation of pro- and anti-inflammatory cytokine inhibition.[56] Diallyl sulfide and thiacremonone compounds of garlic possess antiarthritic properties as well as improve allergic airways inflammation (allergic rhinitis). Raw garlic juice is used to instantly stop burning due to rashes and bug bites. It also helps in treating throat irritations due to antimicrobial activity of garlic and can reduce frequency of cold by daily use of garlic. It may also reduce relentlessness of upper respiratory tract infections and its disorders such as asthma.[121]

According to reported studies, garlic and its bioactive components can affect many events associated with health. Its ability to reduce cholesterol, blood pressure, sterol synthesis, and thrombus formation makes it strong candidate to lower heart diseases. Garlic has very low side effects except its persistent odor. Precautions should be taken while using in pregnancy, bleeding disorders such as ulcers and gastrointestinal infections. Special concern is also needed during surgery due to its anticoagulant property. Long-term studies are still required to evaluate its performance in cancer and stroke. The strong smell of garlic is mostly due to sulfur-containing moieties while flavonoids and alk (en)-based cysteine sulfoxides (ACSOs) are two important chemical groups of garlic that have positive effects on human health.[110]

### *2.3.3 PHYTOCHEMICAL PROPERTIES OF ONION*

Onion (*Allium cepa* L.) contains approximately 90% water, dietary fiber, sugars, vitamins, and minerals. It contains low concentration of sodium and higher levels of calcium, magnesium, phosphorus, potassium, vitamin $B_6$, and folic acid. Naturally, it contains lower level of fats. However, amino acids especially arginine and glutamic acid are found in appreciable amounts. In all traditions and cultures, onions are often used as a raw ingredient. Onion is the most commonly traded vegetable owing to its storage and handling characteristics.

Health benefits of onion include: anticancer properties, antithrombotic activity, antiplatelet activity, antibiotic, and antiasthmatic effects. Nowadays, *Allium* vegetables including onion has significant role in prevention of two prevalent diseases, namely, cardiovascular disease (CVD) and cancer.[39] Studies have revealed that onion does not only prevent cancer development but also mitigates cancer especially rectal cancer. Various active compounds such as polyphenols have striking effect on cancer alleviation.

The consumption of polyphenol-enriched FAV has been used to reduce risk for CVD. Accordingly, double blind trails were conducted to determine effect of onion peel extract (OPE) on total antioxidant capacity, leukocyte DNA, and lipid peroxidation. For this study, healthy female participants were treated with either OPE or placebo for 14 days. The results revealed that total cholesterol (TC), LDL cholesterol, and atherogenic index were reduced. The study confirmed health benefits from OPE by reducing lipid level.[60] Similarly, another study was carried out to evaluate effects of onion and garlic juices on biochemical measurements, enzymes activity, and oxidation in diabetic rats. The rats were orally administrated with 1 mL of either garlic or onion juices per 100 g body weight for 1 month. Results indicated that glucose, urea, creatinine, and bilirubin were significantly enhanced in plasma of diabetic groups compared to the control. Other parameters such as aspartate aminotransferase (AST), alanine aminotransferase (ALT), lactate dehydrogenase (LDH), and alkaline and acid phosphatases performance were also predominantly improved in diabetic rats. Brain LDH was also significantly increased.[30]

A comparison study was carried out between onion and garlic to evaluate their antioxidant activities. The results revealed that onion had significantly high antioxidant activity than garlic. The red onion had higher activity as compared to yellow onion and OPE retained the highest scavenging activity.[85] A case–control study was conducted to examine role of allium in treating various neoplasms. The study analyzed the relationship between consumption of onion/garlic and cancer. The results showed the inverse link between consumption of onion and insurgence of cancer risk.[35] A study in China with 564 patients of stomach cancer and 1131 of gastric cancer also reported that higher consumption of onion is linked with significant reduction in gastric cancer risk.[132]

A study was done to address the immune-modulatory properties of onion lectin (*A. cepa agglutinin*: ACA). Purified ACA was obtained from the onion extract. The immune-modulatory activity of ACA showed a momentous increase in the production of nitric oxide at 24 h and the production of pro-inflammatory cytokines (tumor necrosis factor-alpha, TNF-α and interleukin-2, IL-12) at 24 h compared to the control. Furthermore, it significantly increased expression of cytokines compared to control.[95] Another study was done to study effects of *A. cepa* and quercetin to treat asthma. A study was also conducted to evaluate the effects of *A. cepa* L. extract and quercetinon cytokine and smooth muscle contraction for the therapeutic prospects. Findings reported reduced production of inflammatory cytokines that was used as antiasthmatic drug due to its immune-modulatory and bronchodilator properties.[86]

To evaluate antimicrobial properties, 50–100 mg of crushed fresh onions were used. The results showed significant inhibition of *Staphylococcus aureus* strain.[123] Another experimental study was done on broiler chicks to evaluate the effect of onion as antibiotic growth promoter, its immune response and serum biochemistry. Body weight of chicks was recorded on 1, 21, and 42nd day. The results revealed that increase in body weight and size of lymphoid organs of chicks at 21st; and increase in HDL and decrease in triglycerides (TG) concentration were observed on 42nd day compared to control.[37] These studies indicate role of onion in prevention of different diseases. It works as dietary fiber and can treat the gastrointestinal disorders and similarly can be used as antioxidant, antibiotic, and antifungal. It also lessens the burden of chronic diseases due to its hypolipidemic, hypoglycemic, antithrombotic, antiproliferative, and anticancerous properties.

### *2.3.4 HEALTH BENEFITS OF FLAXSEED*

Flax (*Linum usitatissimum*) has crispy texture and nutty taste.[102] Flaxseed is commonly known as linseed, and both terms are used synonymously in literature. The term "flaxseed" is often used when seeds are consumed by humans while the term "linseed" is used specifically for industrial purpose.[78] After refining, seed oil is widely utilized as edible oil. Flaxseeds possess substantial soluble and insoluble fiber.[117] The top five flaxseed producing countries are: Canada, China, India, United States, and Ethiopia. Canada is largest producer of linseed, producing 40% of the world's linseed.[106]

Recently, linseed has received much attention as functional food because of its high nutritional value and health benefits.[87] Nutritional composition of flaxseed depends upon cultivation conditions, environmental and physiological variation, and processing conditions. The total lipid content of linseed varies from 37 to 45 g/100 g of the seed. The oil of flaxseed constitutes 98% of phospholipids and triglycerides and 0.1% free fatty acids. It consists of 21% proteins, mainly albumin and globulin. Flaxseed is a good source of minerals especially potassium (5600–9200 mg/kg), phosphorus (650 mg/100 g), magnesium (350–431 mg/100 g), calcium (236–250 mg/100 g), and low content of sodium (27 mg/100 g). It comprises small quantity of water and fat-soluble vitamins. Vitamin E ($\gamma$-tocopherol) is available as 39.5 mg/100 g as antioxidant. It also contains antinutrients that may have adverse effects on health.[52]

There is no doubt that flaxseed has considerable potential as a functional food.[3] Dietitian and nutritionists highly recommend the incorporation of omega-3 fatty acid in the diet. Whole linseed, ground meal, and extracted oil are commonly used as nutritional additives in the development of a number of foods such as baked/ready to eat cereal products, fiber bars, salad toppings, meat extenders, bread, muffins, and spaghetti. In spite of various clinical evidences of linseeds, public awareness about its nutritional and therapeutic benefits is limited.[106]

For a research study to access the health effect of flax in CVD and in cancer, a randomized control trial study was conducted on 22 men and 7 postmenopausal women. They were fed with muffins and control with wheat bran having same micronutrient profiles. The results showed that flaxseed was effective in lowering LDL cholesterol, but further studies are needed for evaluating oxidative stress activity.[51] Secoisolariciresinol-di-glucoside (SDG) is a phytoestrogen present in flaxseed having the property such as platelet-activating factor—receptor antagonists and as an antioxidant. The study was conducted to evaluate SDG potential to effect blood lipid and aortic tissue oxidation and atherosclerosis progression in rabbits. The study suggested that SDG lessen hypercholesterolemia atherosclerosis that was linked with a reduction of serum cholesterol, lipid oxidation products, and increased HDL and antioxidant potential.[93] Another study was done to evaluate the antioxidant activity of flaxseed and its lignin SDG and metabolites enterodiol (ED) and enterolactone (EL). Results showed that decreased functionality of Glutathione disulfide (GSSG)-Red and unchanged glutathione (GSH) levels were related to antioxidant effect of ED and EL in muscles of flaxseed and SDG-fed rats.[133] Another research was done on postmenopausal women to evaluate effects of flaxseed supplement versus hormone replacement therapy; and it showed that 40 g of flaxseed is enough as oral estrogen—progesterone to restore mild menopausal symptoms and to decrease glucose and insulin concentrations.[65]

Research study was conducted to define the impact of dosage of 50 g/day of flaxseed for 4 weeks on healthy young adults. On ingesting, α-linoleate level was increased in adipose and n-3 polyunsaturated fatty acids (PUFAs) were enhanced in plasma lipids. Moreover, plasma LDL was also decreased up to 8%, however, total urinary lignin excretion was increased more than five-folds. The study concluded that flaxseed has positive effects on different parameters of nutritional status.[24] Further studies were done on healthy females who voluntarily consumed 50 g of ground raw flaxseed for 4 weeks that raised urinary thiocyanate excretion by 2.2-folds. It also lowered total cholesterol and LDL by 9% and 18%, respectively; and postprandial glucose was decreased by 27%. It can be concluded that consumption of 50 g flaxseed is safe and has nutritional benefits by increasing *n*-3 fatty acids in plasma and erythrocytes, and by reducing postprandial glucose response.[23]

Similarly, SDG was extracted from flaxseed to study effects on mammary tumorigenesis in rats. The consumption of SDG @ 1.5 mg/day for 20 weeks led to 37% reduction in tumor in treated rats compared to 46% reduction in the control. No enlargement of the major organs was seen in the SDG-treated rats. The data showed that SDG has an antitumor potential when consumed at the initial stage of tumorigenesis.[114] In a study on flaxseed, precursor (such as SDG and β-glucuronidase) activity was assessed to study their role as colon cancer protective factor in six groups of male rats fed for 100 days. Results indicated that urinary lignan excretion was significantly increased in the flaxseed and defatted flaxseed groups. It was concluded that flaxseed has a colon cancer protective effect due to SDG and β-glucuronidase activity.[50] The reported studies revealed the preventive and therapeutic effects of flax-seed as antioxidant, antitumerogenic, and anticancerous and in CVDs. More clinical trials are required to know its health benefits and minimum amount of flaxseed to be used so that problems posed by its overdose can be avoided. Development of new, rapid, reproducible, and economical techniques is needed for flaxseed nutraceuticals analysis.

### *2.3.5 HEALTH BENEFITS OF FENUGREEK*

Fenugreek (*Trigonella foenum-graecum*) seeds as whole or powder are used for medicine purpose. Fenugreek seeds have a slightly bitter taste comparable to burnt sugar, and are commonly used as flavor enhancers in teas and spices.

The leaves of fenugreek contain 26% starch, 25% proteins, 6.5% lipids, 13% fiber, 4% gum, and 11% ash. The leaves are also good source of calcium, iron, β-carotene, and some other vitamins. The seeds of fenugreek have 44–59% carbohydrates (galactomannans), 20–30% proteins (arginine, glycine, and alanine), and 6–10% lipids (mainly unsaturated). Fenugreek seeds contain high content of calcium, phosphorus, iron, zinc, and manganese. The seeds also found to have flavonoids, carotenoids, and coumarins.[126]

Most of health benefits of fenugreek seeds are due to high fiber content @ 65% of dietary fiber per 100 g of seeds. The health benefits include: antidiabetic, anticarcinogenic, remedy for hypocholesterolemia and hypoglycemia, antioxidant, antimicrobial agent, gastric stimulant, and anti-anorexia agent. It can be utilized as traditional, functional, or nutraceutical food.[80] Fenugreek seeds reduce the reabsorption of bile salts in the colon, and they also have a unique feature of galactose to mannose ratio (1.1) that is responsible to lower bad cholesterol levels in plasma. The potential health benefits of fenugreek seeds are due to presence of fibers along with saponins and flavonoids. Owing to these beneficial aspects, fenugreek seeds are recommended in daily dietary pattern.[58]

A study was conducted to evaluate the effects of fenugreek as antidiabetic. For this purpose, *T. foenum-graecum* was given to normal and alloxan-induced diabetic rats @ 2 and 8 g/kg dose orally. Results indicated a momentous decrease in blood glucose in normal and treated groups.[59] Another study identified the effects of fenugreek on adipocyte size and adipose tissue inflammation in diabetic obese rats. The rats were treated with high fat diet with 2% fenugreek. This study suggested that diosgenin in fenugreek helped to manage diabetes by triggering adipocyte differentiation and hindering inflammation in adipose tissues. It was concluded that fenugreek containing diosgenin may be useful for decreasing the burden of glucose-related metabolic disorders.[120]

In another study, antihyperglycemic effect of fenugreek seeds was observed in type-2 diabetic patients with add-on metformin therapy. For this purpose, group 1 and 2 were given 500 mg of metformin twice a day and 500 mg of metformin with 1 g fenugreek seed powder capsule thrice a day, respectively. Both groups showed decrease in fasting and postprandial blood glucose, however, group 2 showed more significant reduction in blood glucose as compared to group 1 after 12 weeks of treatment.[54]

A study was conducted to identify and extract antioxidant active ingredients in fenugreek seeds. Fenugreek seeds were subjected to active assessment using standard methods. High-antioxidant compounds were identified by UV scan, mass analysis nuclear magnetic resonance (NMR), and liquid chromatography mass spectrometry (LC–MS). The results showed that vitexin and isovitexin were major antioxidant compounds along with apigenin, kaempferol, and caffeic acid derivatives.[57] In a study, antioxidant and hypercholesteremic effects of fenugreek seeds were observed in cholesterol-fed rats by different extracts (water, methanol, ethyl acetate, hexane, and dichloro-methane). The results indicated that phenolics and flavonoid contents were highest in methanol and ethyl acetate extracts, showing hypercholesteremic and antioxidant effects of fenugreek seeds.[13]

The antinociceptive and anti-inflammatory effects of fenugreek seeds were evaluated by using liquid–liquid extraction method. Results of the study showed that both aqueous and acidified chloroform fractions could significantly inhibit paw edema at different dosages. It was concluded that alkaloid and flavonoid contents of fenugreek seeds may be responsible for antinociception and anti-inflammatory effects of the plant, respectively.[72]

The safety and efficacy of extract of fenugreek seeds were evaluated as nutritional adjuvant to levodopa in Parkinson's disease. Capsules of 300 mg extract were given twice daily to 50 patients with matching placebo for 6 months. The scores of Unified Parkinson's Disease Rating Scale (UPDRS), and Hoehn and Yahr (H&Y) staging at baseline were used to measure efficacy and outcome of treatments. Total UPDRS scores showed slower rise as compared to steep rise shown by placebo. Similar progress was indicated in H&Y staging as compared with placebo. The results concluded that IBHB (trigonelline based standardized fenugreek seed extract) was found to have exceptional safety and tolerability profile.[82]

Defensive potential of fenugreek seed extract was explored on experimental rat-induced gastric mucosal ulcer. For the purpose, 24 male rats were divided into 4 groups; and treatments included fenugreek seed extract and control were given normal saline and extract orally, respectively. At the termination of experiment, the glandular stomach and tissues were examined. The results documented that fenugreek seed extract momentously reduced erosion and ulcer in treated group compared to control.[9] Conclusively, fenugreek can be used in daily life that has potential health benefits on human health. Fenugreek seeds are used to treat hyperglycemia induced by obesity that can be used as nutraceutical for pharmacological purposes. Besides, it also works as antioxidant, anticancerous, antiulcers and in the treatment of neurological problems.

### *2.3.6 HEALTH BENEFITS OF MORINGA PLANT*

Moringa plant [*Moringa oleifera* (MOI)] is commonly known as "*Sohanjana*" in Asian countries. Nearly all parts of this miracle plant have a range of nutraceutical properties.[33] Nutritional profile indicated that it contains protein 9.4 g, iron 4 mg, potassium 337 mg, calcium 185 mg, and vitamin C 51.7 mg per 100 g of the composition. The leaves of moringa plant are rich in iron content about 100 g gives 31% of dietary iron that is useful against iron deficiency disorders in young children and females. Moringa plant has wide range of several antioxidants that fight against the damaging cells.[5]

Latest studies conclude that all the parts of this plant are useful. Some parts can be used to make lubricants to nourish the skin, whereas others can be used as a tea due to antioxidant properties. Moringa is effective against inflammation, type-2 diabetes, cancerous cells, and iron deficiency, pains associated with arthritis and some digestive track disorders such as peptic and duodenal ulcers, central nervous system (CNS) disorders such as prolonged headache, and epilepsy, etc. Moringa plant balances fluid and electrolyte content in our body thus maintaining kidney homeostasis, and prevents formation of kidney stones. It is also beneficial against certain viral and bacterial diseases. Presence of high percentage of protein, carotenoids, and ascorbic acid may delay the effects of aging. Moringa plant can fight against infections caused by the harmful bacteria and viruses. It can prevent skin infections, infections of urinary track, and digestive disorders. This plant acts as a clotting agent and prevents excessive blood loss. It also helps in healing of wounds and scratches. The moringa oil is applied on the skin and acts as bacteriostatic agent. Moringa has a characteristic odor and is used to impart pleasing smell in certain foods.

Moringa contains chlorogenic acid, which helps to maintain appropriate glucose levels in the blood. It also regulates release and storage of glucose (glycogenesis). Due to these activities, moringa plant is known as a natural remedy against diabetes because of presence of isothiocyanatesin leaves of this plant. Research study was done on albino rats to evaluate the effects of moringa plant for treatment of diabetes. Rats were treated with three doses of aqueous extract, tolbutamide, and normal saline. The results of study indicated that MOI leaves aqueous extract have capacity to decrease blood sugar level as documented by normoglycemic and alloxan-induced diabetic rat models.[29] Similarly, another study documented potential effect of moringa leaves for treatment of diabetes.[130] In another study, phenolic glycosides along with four different compounds were isolated from moringa seeds. The hypoglycemic effect of these compounds was checked on HepG2 cell and STZ-induced mice. The results showed remarkable reduction in blood glucose level in STZ-induced rats thus compounds may be used as safe hypoglycemic drug.[124]

Traditionally, MOI has been used to treat hypertension and intestinal disorders. For this, a study was conducted to observe hypertensive effect of moringa plant leaves on normotensive rats. The rats were fed with extract of moringa leaves @ 5–75 mg/kg body weight intravenously. The results indicated the effect of leaves as antihypertensive and antispasmodic on isolated duodenum smooth muscles therefore, it can be used to treat hypertension, diarrhea, and dysentery.[34] In another study, MOI seed powder showed to reduce cardiac diseases in spontaneous hypertensive rats (SHR). For this purpose, SHR were feed with 750 mg/day for 8 weeks with moringa seed powder and normal diet. The results suggested beneficial effect of powder on cardiac structure and function by upregulating peroxisome proliferator-activated receptor (PPAR)-α and δ-signaling. The results supported the use of moringa seed powder in treatment of cardiac diseases associated with hypertension.[98]

To check antioxidant activity of moringa leaves, in vitro models were used. The results showed strong antioxidant activity of moringa leaves due to presence of phenolic content.[108] A study was conducted to document active compound in MOI against cancer. The study showed that niazimicin is one of bioactive compounds found in MOI and it has potent antitumor activity.[89] In a study, anticancerous effect of moringa plant leaves extract was observed in cancerous esophageal cell line. The results suggested positive activity of moringa leaves extract on cancerous esophageal cell line by increasing lipid peroxidation, DNA fragmentation and initiation of apoptosis.[115] In another study, its effectiveness against breast cancer was observed.[36]

A research study showed significant antimicrobial activity of extract of moringa leaves against *Bacillus cereus* and *subtilis, S. aureus, Sarcinalutea, Escherichia coli,* and *Mycobacterium phlei.*[88] In vitro study to check antimicrobial effect of leaf juice and extract of MOI against Gram-positive and negative bacteria including *Shigella shiga, Pseudomonas aeruginosa, S. aureus, B. cereus* showed that extract can be used for pharmaceutical purposes to combat pathogenic bacteria.[96] In another study, different extracts were evaluated to check the activity of moringa crude extract against different Gram-positive and Gram-negative bacteria. The results showed that petroleum ether extract has the effective antibacterial property.[79]

A research study identified protein energy malnutrition in children who were fed with MOI powder @ 15 g twice daily and were reassessed after 2 months. The parameters of study showed 70% children with grade II protein energy malnutrition (PEM) were improved to grade I, whereas more than 60% children with grade I PEM also showed improvement in nutritional status suggesting its nutritional potential for alleviating malnutrition.[109] These studies concluded that MOI has the ability to retain maximum concentrations of minerals and vitamins and can effectively be used to treat malnutrition in children. Due to presence of many bioactive compounds, it can be considered as good antioxidant and can treat number of diseases due to its ability as antidiabetic, antihypertensive, and anticancerous and antimicrobial.

## 2.3.7 THERAPEUTIC ASPECTS OF FRUITS AND VEGETABLES

FAV are among the healthy food choices and their consumption on regular basis shows health benefits due to presence of specific nutrients and biochemical compounds. Overwhelming studies have documented therapeutic aspects of these foods to lower the risk of diseases. The antioxidants present in FAV include: ascorbic acid, carotenoids along with selenium and lycopene. Polyphenols represent a large group of dietary antioxidants that are found naturally in FAV; and these include: flavonoids including flavones, flavanols, flavanones, isoflavones, and anthocyanins, and non-flavonoid polyphenolics such as: lignans, phenolic acids, stilbenes, and soluble dietary fibers. The presence of vitamins can contribute to preservation of immune function, healthy vision, bone health, and cell integrity. Also these help to regulate calcium and phosphorus. Fruits are rich in antioxidant vitamins such as A, C, and E. Minerals play vital role in functioning of normal cells and tissues.[107]

This section describes functional importance of fruits. Banana fruit contains easily digestible, soft flesh made up of simple sugars (fructose and sucrose) that upon ingestion instantly replenishes energy and refreshes the body. It is also rich source of potassium (about 400 mg potassium in a single average size banana) that is essential for good nerve and muscle functioning and sustaining a healthy equilibrium of fluids in the body and prevents muscle cramps. The plantain is also low in fat and protein, high in fiber thus prevents diverticulosis. It is also rich in antioxidants and minerals that boosts immunity and promotes cell growth.[62]

Apples are rich source of iron, fiber, and phytochemicals that are considered instrumental for human health. Studies have shown that apple consumption is linked with decreased cancer risk, coronary heart diseases, diseases of lungs and bronchi, and type-2 diabetes.[2] Similarly, grapes contain polyphenolic compounds called resveratrol and anthocyanins that are said to improve memory and brain function.[128] Similarly, citrus fruits are excellent source of vitamin C and carotenes that promote healthy cell growth and reduce inflammation. Pineapple is a source of vitamins, fibers, and digestive enzyme bromelain, which helps in digestion. Also, bromelain reduces inflammation and joint pain. Peaches, plums, and nectarines contain a range of polyphenols and catechins that suppress growth and induce apoptosis in human prostate cancer, breast cancer, and endothelial cell proliferation. Passion fruit is rich in antioxidant providing great immunity. Pomegranate is also rich source of antioxidants and vitamins.[92] Diet rich in FAV reduces cancer risk. A study compared the relationship between risk of many types of cancers with consumption of FAV; and this study included about 10,000 cases of 14 different types of cancers and 17,000 controls. An inverse relationship was found between consumption of FAV with cancer especially intestinal cancer mainly due to high fiber content. This study also reported that Mediterranean diet has favorable role in prevention against not only CVDs but also on different kinds of cancers.[119]

The consumption of vegetables can promote health perspective due to high fiber content, levels of vitamins and minerals, polyphenols, and bioactive compounds. The presence of high fiber can help to reduce blood cholesterol and also protects against diabetes and prevents constipation. The peas are rich source of dietary fiber that enhances health of gastrointestinal track and high phosphorus promotes bone health. Spinach contains folate that helps cell growth and differentiation during periods of rapid growth such as pregnancy, infancy, and adolescence. Carrots are rich in vitamin A that promotes vision, bone health, reproduction, growth, and immunity. It also protects surface linings of mucous membrane in respiratory tract. Pumpkin and its seeds are rich source of magnesium that is essential for operation of kidney and heart. The magnesium also acts as cofactor in many enzyme systems and it helps to regulate calcium levels, and vitamin D, copper, potassium, zinc, and other important nutrients in the body. Peppers (both red and green) contain ample amount of ascorbic acid that not only prevents damage from free radicals but also promotes other antioxidants such as α-tocopherol. Similarly, tomatoes (*Lycopersicum esculentum*) contain phytochemical lycopene. Studies concluded that consumption of tomato can prevent CVD due to presence of lycopene.[18] A study was conducted to evaluate impact of lycopene, content of phenolic compounds, and antioxidant activity of nine different varieties of tomatoes (*L. esculentum*) and it suggested antioxidant property of extract depended on the variety and lycopene content.[73]

### 2.3.8 THERAPEUTIC ASPECTS OF GREEN TEA AND BLACK TEA

Tea belongs to family Theaceae (i.e., *Camellia sinensis* and *Camellia assamica*). It is available as white, black, green, and red tea containing different compounds having different physiological importance. Tea leaves contain many bioactive ingredients. For example, aqueous solution from dried leaves of *C. sinensis* contains a mixture of approximately 300 bioactive compounds, which give specific characteristics, properties, and taste. Tea leaves contain substances such as: tannin, flavonols, proteins, and amino acids, aroma-producing volatile substances, enzymes, vitamins, mineral compounds, and microelements and alkaloids. Tea beverages also have some importance in daily diet due to presence of several mineral ingredients (such as calcium, sodium, potassium, magnesium, and manganese) and high caffeine content.[25] Some other active compounds are flavanol monomers (catechins), including epicatechin and epigallocatechin. Among these, epigallocatechin-3-gallate is the most bioactive and the most studied one. Green tea extract can be used in different food items to prevent lipid oxidation and to increase the shelf-life of food products. They play major role as antioxidant, cancer prevention, to treat CVD by lowering cholesterol, neurological disorders, and arthritis.[81]

A research study compared the effect of antioxidants in black tea compared to green tea. Black tea contains theaflavins (TF) while green tea contains catechins. For this purpose, all TF and catechins tested inhibited copper-mediated LDL oxidation. The results suggested that TF in black tea showed almost same antioxidant property as catechins in green tea.[66] Pregnant female rats were used to evaluate the antioxidant activity of *C. sinensis* (green tea). The rats with mental deprivation had high level of free oxygen radicals, and they were supplemented with green tea. The results suggested that green tea had effective role in the prevention of mental deprivation.[75]

Epicatechin is natural flavonoid present in green tea. The major therapeutic role of epicatechin is in the control of diabetes and cancer.[1] Apoptotic role of epicatechin was studied to know the mechanism of human esophageal cancer. The results indicated that epigallocathingallate (EGCG) inhibits cell growth and increase apoptosis and it also increases anticancer drug concentration in cancerous cells.[68] The cancer patients showed high level of β-catenin. On analyzing, the results suggested that epigallocatechin gallate (EGCE) can inactivate the β-catenin singling pathway. Based on results, it may be concluded that EGCG may be a probable treatment for triple negative breast cancer.[46] The results suggested that both green tea and black tea extracts had role in lowering blood glucose and ameliorated glucose intolerance. The results also showed that green tea had more antihyperglycemic property as compared to black tea extract and the predominant mechanism for the antidiabetic effect of green tea extract was through insulin resistance, while for black tea extract it was through insulin secretion.[112]

According to a study, the soaked catechins (2–15% of intestinal level) bound (about 37%) to plasma HDL have a role in cholesterol metabolism. The results suggested tea and tea nutraceuticals are effective in regulation of plasma glucose and cholesterol.[113] For this purpose, male mice were fed high-fat/high-sucrose diet. The results indicated significant increase in gene expression of vascular endothelial growth factor A by green tea polyphenols and vascular endothelial growth factor receptor 2 by black tea polyphenols and a decrease in pigment epithelium-derived factor gene expression by oolong tea polyphenols and black tea polyphenols.[44] All reported information concluded that green tea and black tea show antioxidation property due to presence of multiple bioactive ingredients. Thus, both play role as anticancerous and in multiple neurological problems, in lowering cholesterol, antidiabetic and weight loss.

### *2.3.9 PHYTOCHEMICAL PROPERTIES OF CEREALS*

Cereals or whole grains are considered as functional foods containing many health-promoting compounds consisting of minerals and vitamins, phytochemicals including phenolic compounds. Phenolic compounds are more complex and diverse phytochemicals present in cereals are phenolic acids, flavonoids, condensed tannins, coumarins, and alkyl-resorcinols. Whole grain cereals lower the risk of gastrointestinal cancer, type-2 diabetes, metabolic syndrome, CVD, and ischemic heart diseases. Due to presence of phenolic acids, cereals have the antioxidant property. Cereals and whole grain foods also have dietary fiber that reduces the gastrointestinal problem such as gastroesophageal reflux disease, duodenal ulcer, diverticulitis, hemorrhoids, and constipation.[28] For example, oat (*Avena sativa*) cereal possesses higher nutritional profile that also makes it different from other cereals. It has ample amount of dietary fiber particularly β-glucan, vitamins notably B complex, minerals and protein, and low level of fat. The consumption of oat and its by-products showed positive influence in the management of diabetes and cardiovascular complications, and also facilitated the movement of bile acids toward intestinal tract and expedited the bile acid excretion thus, reduce the blood cholesterol and impart beneficial aspects to tackle coeliac disorder. Studies also documented that the incorporation of oat grains in combination of bran in the processed food-based products not only enhances nutritional profile but also can act as nutritional intervention to tackle various disease especially of metabolic concerns.[17] Barley (*Hordeum vulgare*) is rich in minerals and vitamins and show health benefits such as: weight loss, control blood glucose level, lower blood cholesterol, and improve digestion.[83]

Research study explained the role of oat and barley in lowering post-prandial glycemic and insulinemic and serum cholesterol level due to their (1→3) (1→4)-β-d-glucan (β-glucan) content.[48] In another study, β-glucan was purified from oat, barley, and wheat, to check its relationship between effects of regulation of cholesterol metabolism and antioxidant property on diabetic mice. Metformin was used and levels of blood glucose, serum lipid, liver lipid, superoxide dismutase (SOD), and malondialdehyde (MDA) were determined. Results suggested that average values of molecular weight (MW) of β-glucan from oat, wheat, and barley were 172 kDa, 635 kDa, and 743 kDa, respectively. Blood glucose differences were not significant due to high variability, however, lower levels of serum TC, TG, liver TC, TG, and LDL and higher levels of HDL were observed in β-glucan fed groups as compared to the control. The addition of β-glucan fed to diabetic mice considerably increased their liver SOD activities and reduced their MDA levels.[135]

A study was carried out to explain that if consumption of barley has any influence to decrease the incidence of cardiovascular disorder. For this purpose, slightly hypercholesterolemic patients (nine pre- and postmenopausal women and seven men) were provided with controlled diets for a period of 17 weeks followed by whole-grain foods containing 0, 3, or 6 g of β-glucan/day from barley after a washout period of 2 weeks. Diets were consumed for 5 weeks each and after that blood sampling (fasting) was done for analysis. The results showed that cholesterol was momentously decreased in patients on a diet containing 3 or 6 g β-glucan/d from barley compared to control in the men and postmenopausal women. LDL and VLDL fractions were significantly decreased, whereas HDL level was improved with the incorporation of whole grains in the diet. Findings of the study concluded that incorporation of barley in a regular diet can have a positive impact on health, especially reducing total and particularly LDL cholesterol.[12]

Oat (*A. sativa* L.) is a good source of antioxidants due to presence of many compounds including vitamin E, phytic acid, phenolics, and avenanthramides.[91] Two different species of barley (Falcon and Metcalfe) were evaluated for their antioxidant property. The results suggested that falcon had high antioxidant activity as compared to Metcalfe due to its high phenolic acid content. Barley (*H. vulgare*) fractions contained phenolic acids, namely, vanillic, caffeic, *p*-coumaric, ferulic, and sinapic acids.[70]

Antioxidant property of different cereals was evaluated mainly focusing on oat and barley. Diet containing barley, husked and naked oat, oat bran, and triticale were fed to rats for 4 weeks in order to assess the level of polyphenol, α-tocopherol, and total antioxidant capacity (TAC). The results suggested that barley was more potent antioxidant as compared to naked oat.[134] A study investigated the effect of 1–3, 1–4 oat β-glucan on normal, and cancerous cells. The study determined that low MW β-glucan from oat had no toxicity for normal cells.[21] The reported studies explained the role of cereals mainly oat and barley in therapeutic health. Due to presence of β-glucan and other phytochemicals mainly phenolic compounds make their role essential in weight loss, lowering serum cholesterol, and lowering blood glucose level. It also shows antioxidant property and can also play role as anticancerous.

## 2.4 NUTRITIONAL SIGNIFICANCE OF FUNCTIONAL FOODS

Health claims on the labels are mostly associated with functional foods showing their impact on health. Studies suggested high intake of fiber can lessen the chances of insurgence of various kind of cancer. Some countries have unique labeling system of food products such as the United States, Canada, European Union, and Sweden.

US FDA regulates and manages these kinds of claims. Steven Felice, Chairman of the Foundation of Innovation in Medicine in Cranford—New Jersey, thinks that more than 90% of functional foods are not tested clinically and are making unproven claims that are not supported through clinical trials or research studies. It is also important to know that processed functional food is good for health. For instance, a calcium fortified functional food with sugar and water is less nutritious because the additional nutrients found in orange juice are absent. During assessment of functional food, we have to check the facts that are mentioned on labels of food products, to find nutrient composition and its possible health benefits. The Daily Value for a single serving on a food label represents the percentage of nutrient such as saturated fat or carbohydrate that is suggested for an individual consuming 2000 Calories daily. We can choose the food product that contains less amount of cholesterol, high amount of protein, and no artificial colors. If the food is unprocessed, it must be in its raw state, having not been frozen or subjected to other forms of processing. The term fresh does not apply to food that is processed such as fresh milk or fresh bread. Nutritional content claims, structural and functional claims, or health claims can be cited on labels, as permitted by FDA. Nutritional content claims give information about the content of the foods and may include words such as "free," "low," and "reduced." Modern technologies of food processing have emerged along with traditional processing technologies of food items.[16] These include special coatings that are edible, dispersible films, microencapsulation, vacuum impregnation nanotechnology, etc. (Table 2.2).

**TABLE 2.2** Details of Functional Food Along with Potential Health Benefits and Claims.

| Functional food | Potential health benefits | Labeling claim |
|---|---|---|
| | **Whole foods** | |
| Oats | Decreases constipation, cholesterol, and coronary artery diseases | May reduce the risk of CVDs |
| Soy | Decreases LDL, bone disorders, and heart ailments | May decrease the chances of heart attack |
| Fruits and vegetables | Decreases risk of some cancers and elevated blood pressure | May lessen the chances of some cancers |
| Fish | Lessen levels of triglycerides and cholesterol | None |
| Garlic | Decreases risk of heart ailments, cholesterol | None |
| Grapes/grape juice | Lessen risk of heart malfunctions | Structure/function claim |
| Flaxseed | Decreases chances of dysfunctions of heart, elevated triglycerides, and increases blood glucose level | None |
| Nuts | Delay onset of heart ailments | None |
| | **Enriched foods** | |
| Grains | Diminish risk of heart disease, some cancers and fulfill nutrient insufficiencies | May lessen the risk of some cancers and heart ailments |
| | **Fortified foods** | |
| Juices with calcium | Decrease osteoporosis risk and hypertension | May prevent bone loss |
| Grains with folic acid | Delay onset of heart dysfunctions and birth defects (neural tube) | Might decrease brain and spinal cord birth defects |
| Iron-fortified formula for infants | Diminishes deficiency of iron | None |
| Fiber-enriched grains | Lessen chances of cancers, heart problems, enhance control of blood glucose | May delay the heart ailments and cancers |
| Vitamin D-fortified milk | Decreases osteomalacia chance along with osteoporosis | May prevent bone loss |
| Juices with added fiber | Lessen heart ailments, certain types of cancers, constipation, and hypertension | May delay onset of some cancers |

**TABLE 2.2** *(Continued)*

| Functional food | Potential health benefits | Labeling claim |
|---|---|---|
| | **Enhanced foods** | |
| Dairy products with probiotics | Decrease colon cancer and candida vaginitis, prevents inflammation | Structure/function claim |
| Salad dressings and beverages with antioxidants | Enhance overall health | |
| Foods and beverages containing herbal preparations | Differs with ingredients | |
| Sports bars | | |
| Spreads with stanol esters | Decrease levels of cholesterol | |
| Foods containing sugar alcohols in place of sugar | Decrease tooth decaying process | May diminish risk of tooth decay |
| Eggs with omega-3 fatty acids | Decrease risk of heart problems | Structure/function claim |

CVD, cardiovascular disease.

## 2.5 SUMMARY

Functional food either from plant or animal source contain bioactive compounds that may enhance health or reduce risk of diseases. These foods have an active role in healthy lifestyle and are beneficial to the public. Undoubtedly, nutrition is the major modifiable factor of disease, especially the burden of non-communicable diseases by modifying lifestyle behavior and by implementing effective preventive nutritional strategies. Such prevention strategies can eliminate or at least reduce the type-2 diabetes, heart diseases, stroke, and cancer to some extent. Enhancing conventional foods and offering functional foods have importance of translating the impact of nutrition on health and its control in reducing the public health burden. Functional foods have wide range of properties such as antioxidant, anti-inflammatory, and anticoagulant and lower the risk of CVD due to presence of bioactive compounds. Therefore, plant-based functional food should be encouraged in the daily diet such as wheat cereals and grains, FAV, green and black tea, moringa, etc. Also, consumer awareness is very critical on functional foods as they should be provided with knowledge and information and mechanisms must be developed to ensure to foster their growth and health claims for better acceptability.

## KEYWORDS

- **bioactive moieties**
- **functional foods**
- **phytochemicals**
- **phytonutrients**
- **therapeutic foods**

## REFERENCES


1. Abdulkhaleq, L. A.; Assi, M. A.; Noor, M. H. M.; Abdullah, R.; Saad, M. Z.; Taufiq-Yap, Y. H. Therapeutic Uses of Epicatechin in Diabetes and Cancer. *Vet. World*. **2017,** *10* (8), 869–872.
2. Adyanthaya, I.; Kwon, Y. I.; Apostolidis, E.; Shetty, K. Health Benefits of Apple Phenolics from Postharvest Stages for Potential Type 2 Diabetes Management Using In Vitro Models. *J. Food Biochem.* **2010,** *34* (1), 31–49.
3. Al-Okbi, S. Y. Highlights on Functional Foods, with Special Reference to Flaxseed. *J. Nat. Fibers* **2005,** *2* (3), 63–68.
4. Amri, E. The Role of Selected Plant Families with Dietary Ethnomedicinal Species Used as Anticancer. *J. Med. Plants*. **2014,** *2* (1), 28–39.
5. Anwar, F.; Latif, S.; Ashraf, M.; Gilani, A. H. *Moringa oleifera*, a Food Plant with Multiple Medicinal Uses. *Phytother. Res.* **2007,** *21* (1). 17–25.
6. Arai, S. Global View on Functional Foods: Asian Perspective. *Br. J. Nutr.* **2002,** *88* (S2), S139–S143.
7. Arai, S. Studies on Functional Foods in Japan: State of the Art. *Biosci. Biotechnol. Biochem.* **1996,** *60* (1), 9–15.
8. Atanasova-Goranova, V.; Dimova, P.; Pevicharova, G. Effect of Food Products on Endogenous Generation of N–nitrosamines in Rats. *Br. J. Nutr.* **1997,** *78* (2), 335–345.
9. Azari, O.; Kheirandish, R.; Shojaeepour, S. Protective Effect of Fenugreek Seeds (Trigonella foenum graecum) Extract Against Experimental Gastric Ulcer in Rats. *Compar. Clin. Pathol.* **2014,** *23* (6), 1743–1748.
10. Ashraf, H.; Butt, M. S.; Iqbal, M. J.; Suleria, H. A. R. Citrus Peel Extract and Powder Attenuate Hypercholesterolemia and Hyperglycemia Using Rodent Experimental Modeling. *Asian Pac. J. Trop. Biomed.* **2017,** *7* (10), 870–880.
11. Balasubramanian, S.; Roselin, P.; Singh, K. K.; Zachariah, J.; Saxena, S. N. Postharvest Processing and Benefits of Black Pepper, Coriander, Cinnamon, Fenugreek, and Turmeric Spices. *Critic. Rev. Food Sci. Nutr.* **2016,** *56* (10), 1585–1607.

12. Behall, K. M.; Scholfield, D. J.; Hallfrisch, J. Diets Containing Barley Significantly Reduce Lipids in Mildly Hypercholesterolemic Men and Women. *Am. J. Clin. Nutr.* **2004,** *80* (5), 1185–1193.
13. Belguith-Hadriche, O.; Bouaziz, M.; Jamoussi, K.; Simmonds, M. S.; El Feki, A.; Makni-Ayedi, F. Comparative Study on Hypocholesterolemic and Antioxidant Activities of Various Extracts of Fenugreek Seeds. *Food Chem.* **2013,** *138* (2), 1448–1453.
14. Bhat, S.; Kaushal, P.; Kaur, M.; Sharma, H. Coriander (*Coriandrum sativum* L.), Processing, Nutritional and Functional Aspects. *Afr. J. Plant Sci.* **2014,** *8* (1), 25–33.
15. Suleria, H. A. R.; Butt, M. S.; Anjum, F. M.; Sultan, S.; Khalid, N. Aqueous Garlic Extract; Natural Remedy to Improve Haematological, Renal and Liver Status. *J. Nutr. Food Sci.* **2014,** *4* (1), 252.
16. Budai, A.; Wang, L.; Gronli, M.; Strand, L. T.; Antal Jr, M. J.; Abiven, S.; Dieguez-Alonso, A.; Anca-Couce, A.; Rasse, D. P. Surface Properties and Chemical Composition of Corncob and Miscanthus Biochars, Effects of Production Temperature and Method. *J. Agric. Food Chem.* **2014,** *62* (17), 3791–3799.
17. Butt, M. S.; Tahir-Nadeem, M.; Khan, M. K. I.; Shabir, R.; Butt, M. S. Oat, Unique Among the Cereals. *Eur. J. Nutr.* **2008,** *47* (2), 68–79.
18. Canene-Adams, K.; Campbell, J. K.; Zaripheh, S.; Jeffery, E. H.; Erdman, J. W., Jr. The Tomato as a Functional Food. *J. Nutr.* **2005,** *135* (5), 1226–1230.
19. Chauhan, N. B.; Sandoval, J. Amelioration of Early Cognitive Deficits by Aged Garlic Extract in Alzheimer's Transgenic Mice. *Phytother. Res.* **2007,** *21* (7), 629–640.
20. Chawla, S.; Thakur, M. Coriandrum Sativum, A Promising Functional and Medicinal Food. *Med. Plants Int. J. Phytomed. Relat. Ind.* **2013,** *5* (2), 59–65.
21. Choromanska, A.; Kulbacka, J.; Rembialkowska, N.; Pilat, J.; Oledzki, R.; Harasym, J.; Saczko, J. Anticancer Properties of Low Molecular Weight Oat Beta-glucan—An In Vitro Study. *Int. J. Biol. Macromol.* **2015,** *80* (1), 23–8.
22. Crowe, K. M.; Francis, C. Position of the Academy of Nutrition and Dietetics, Functional Foods. *J. Acad. Nutr. Diet.* **2013,** *113* (8), 1096–1103.
23. Cunnane, S. C.; Ganguli, S.; Menard, C.; Liede, A. C.; Hamadeh, M. J.; Chen, Z.-Y.; Wolever, T. M. S.; Jenkins, D. J. A. High α-Linolenic Acid Flaxseed (*Linum usitatissimum*), Some Nutritional Properties in Humans. *Br. J. Nutr.* **2007,** *69* (2), 443–453.
24. Cunnane, S. C.; Hamadeh, M. J.; Liede, A. C.; Thompson, L. U.; Wolever, T. M.; Jenkins, D. J. Nutritional Attributes of Traditional Flaxseed in Healthy Young Adults. *Am. J. Clin. Nutr.* **1995,** *61* (1), 62–68.
25. Czernicka, M.; Zagula, G.; Bajcar, M.; Saletnik, B.; Puchalski, C. Study of Nutritional Value of Dried Tea Leaves and Infusions of Black, Green and White Teas from Chinese Plantations. *Roczniki Panstwowego Zakladu Higieny* **2017,** *68* (3), 237–245.
26. Dhanapakiam, P.; Joseph, J. M.; Ramaswamy, V.; Moorthi, M.; Kumar, A. S. The Cholesterol Lowering Property of Coriander Seeds (Coriandrum sativum), Mechanism of Action. *J. Environ. Biol.* **2007,** *29* (1), 53.
27. Suleria, H. A. R.; Butt, M. S.; Anjum, F. M.; Saeed, F.; Batool, R.; Ahmad, A. N. Aqueous Garlic Extract and Its Phytochemical Profile; Special Reference to Antioxidant Status. *Int. J. Food Sci. Nutr.* **2012,** *63* (4), 431–439.
28. Dykes, L.; Rooney, L. Phenolic Compounds in Cereal Grains and Their Health Benefits. *Cereal Foods World* **2007,** *52* (3), 105–111.
29. Edoga, C.; Njoku, O.; Amadi, E.; Okeke, J. Blood Sugar Lowering Effect of Moringa Oleifera Lam in Albino Rats. *Int. J. Sci. Technol.* **2013,** *3* (1), 88–90.

30. Suleria, H. A. R.; Butt, M. S.; Anjum, F. M.; Ashraf, M.; Qayyum, M. M. N.; Khalid, N.; Younis, M. S. Aqueous Garlic Extract Attenuates Hypercholesterolemic and Hyperglycemic Perspectives; Rabbit Experimental Modeling. *J. Med. Plants Res.* **2013,** *7* (23), 1709–1717.
31. Emamghoreishi, M.; Khasaki, M.; Aazam, M. F. Coriandrum Sativum, Evaluation of its Anxiolytic Effect in the Elevated Plus-maze. *J. Ethnopharmacol.* **2005,** *96* (3), 365–370.
32. Faridi, Z.; Njike, V. Y.; Dutta, S.; Ali, A.; Katz, D. L. Acute Dark Chocolate and Cocoa Ingestion and Endothelial Function, a Randomized Controlled Crossover Trial. *Am. J. Clin. Nutr.* **2008,** *88* (1), 58–63.
33. Farooq, F.; Rai, M.; Tiwari, A.; Khan, A. A.; Farooq, S. Medicinal Properties of Moringa Oleifera, An Overview of Promising Healer. *J. Med. Plants Res.* **2012,** *6* (27), 4368–4374.
34. Faso, B. Effects of Water Crude Leaf Extract of Moringa Oleifera Lam (Moringaceae) on Normotensive Rat Blood Pressure and Isolated Duodenum. *J. Physiol. Pharmacol. Adv.* **2015,** *5* (10), 771–778.
35. Galeone, C.; Pelucchi, C.; Levi, F.; Negri, E.; Franceschi, S.; Talamini, R.; Giacosa, A.; La Vecchia, C. Onion and Garlic Use and Human Cancer. *Am. J. Clin. Nutr.* **2006,** *84* (5), 1027–1032.
36. Ghosh, N. *Anticancer Effect of Moringa Oleifera Leaf Extract on Human Breast Cancer Cell.* M.Sc. Thesis, The University of Burdwan, India; 2014; p 110.
37. Goodarzi, M.; Landy, N.; Nanekarani, S. Effect of Onion (Allium cepa L.) as an Antibiotic Growth Promoter Substitution on Performance, Immune Responses and Serum Biochemical Parameters in Broiler Chicks. *Health* **2013,** *5* (08), 1210.
38. Gray, A. M.; Flatt, P. R. Insulin-releasing and Insulin-like Activity of the Traditional Anti-diabetic Plant Coriandrum Sativum (Coriander). *Br. J. Nutr.* **1999,** *81* (3), 203–209.
39. Griffiths, G.; Trueman, L.; Crowther, T.; Thomas, B.; Smith, B. Onions: A Global Benefit to Health. *Phytother. Res.* **2002,** *16* (7), 603–615.
40. Hasler, C. M. Functional Foods, Benefits, Concerns and Challenges: A Position Paper from the American Council on Science and Health. *J. Nutr.* **2002,** *132* (12), 3772–3781.
41. Hasler, C. M. Functional foods, Their Role in Disease Prevention and Health Promotion. *Food Technol. Champaign Then Chicago* **1998,** *52 (1)*, 63–147.
42. Hasler, C. M.; Brown, A. C. Position of the American Dietetic Association (ADA): Functional Foods. *J. Am. Diet. Assoc.* **2009,** *109* (4), 735–746.
43. Fawad, S. A.; Khalid, N.; Asghar, W.; Suleria, H. A. R. In Vitro Comparative Study of Bougainvillea Spectabilis "Stand" Leaves and Bougainvillea Variegata Leaves in Terms of Phytochemicals and Antimicrobial Activity. *Chin. J. Nat. Med.* **2012,** *10* (6), 441–447.
44. Heber, D.; Zhang, Y.; Yang, J.; Ma, J. E.; Henning, S. M.; Li, Z. Green Tea, Black Tea, and Oolong Tea Polyphenols Reduce Visceral Fat and Inflammation in Mice Fed High-fat, High-sucrose Obesogenic Diets. *J. Nutr.* **2014,** *144* (9), 1385–1393.
45. Hirsch, K.; Danilenko, M.; Giat, J.; Miron, T.; Rabinkov, A.; Wilchek, M.; Mirelman, D.; Levy, J.; Sharoni, Y. Effect of Purified Allicin, the Major Ingredient Offreshly Crushed Garlic, on Cancer Cell Proliferation. *Nutr. Cancer* **2000,** *38* (2), 245–254.
46. Hong, O. Y.; Noh, E. M.; Jang, H. Y.; Lee, Y. R.; Lee, B. K.; Jung, S. H.; Kim, J. S.; Youn, H. J. Epigallocatechin Gallate Inhibits the Growth of MDA-MB-231 Breast Cancer Cells Via Inactivation of the Beta-catenin Signaling Pathway. *Oncol. Lett.* **2017,** *14* (1), 441–446.
47. Hsieh, P. Y.-H.; Ofori, J. A. Innovations in Food Technology for Health. *Asia Pac. J. Clin. Nutr.* **2007,** *16* (S1), 65–73.
48. J. Wood, P. Cereal β-glucans in Diet and Health. *J. Cereal Sci.* **2007,** *46* (3), 230–238.

49. Jacobson, M. F.; Silverglade, B.; Heller, I. R. Functional Foods, Health Boon or Quackery? *West. J. Med.* **2000,** *172* (1), 8.
50. Jenab, M.; Thompson, L. U. The Influence of Flaxseed and Lignans on Colon Carcinogenesis and β-glucuronidase Activity. *Carcinogenesis*. **1996,** *17* (6), 1343–1348.
51. Jenkins, D. J., et al. Health Aspects of Partially Defatted Flaxseed, Including Effects on Serum Lipids, Oxidative Measures, and Ex Vivo Androgen and Progestin Activity, a Controlled Crossover Trial. *Am. J. Clin. Nutr.* **1999,** *69* (3), 395–402.
52. Kajla, P.; Sharma, A.; Sood, D. R. Flaxseed—A Potential Functional Food Source. *J. Food Sci. Technol.* **2015,** *52* (4), 1857–1871.
53. Kang, N.; Moon, E.; Cho, C.; Pyo, S. Immunomodulating Effect of Garlic Component, Allicin, on Murine Peritoneal Macrophages. *Nutr. Res.* **2001,** *21* (4), 617–626.
54. Kaur, M.; Singh, N.; Sharma, G.; Singh, D. To Study the Efficacy and Tolerability of Fenugreek Seed Powder as Add-on Therapy with Metformin in Patients of Type-2 Diabetes Mellitus. *Int. J. Basic Clin. Pharmacol.* **2016,** *5* (2), 378–383.
55. Perveen, R.; Suleria, H. A. R.; Anjum, F. M.; Butt, M. S.; Pasha, I.; Ahmad, S. Tomato (*Solanum lycopersicum*) Carotenoids and Lycopenes Chemistry; Metabolism, Absorption, Nutrition, and Allied Health Claims: A Comprehensive Review. *Crit. Rev. Food Sci. Nutr.* **2005,** *55* (7), 919–929.
56. Keiss, H.-P.; Dirsch, V. M.; Hartung, T.; Haffner, T.; Trueman, L.; Auger, J.; Kahane, R.; Vollmar, A. M. Garlic (Allium sativum L.) Modulates Cytokine Expression in Lipopolysaccharide-activated Human Blood Thereby Inhibiting NF-κB Activity. *J. Nutr.* **2003,** *133* (7), 2171–2175.
57. Khole, S.; Chatterjee, S.; Variyar, P.; Sharma, A.; Devasagayam, T.; Ghaskadbi, S. Bioactive Constituents of Germinated Fenugreek Seeds with Strong Antioxidant Potential. *J. Funct. Foods*. **2014,** *6* (1), 270–279.
58. Khorshidian, N.; Yousefi Asli, M.; Arab, M.; Adeli Mirzaie, A.; Mortazavian, A. M. Fenugreek, Potential Applications as a Functional Food and Nutraceutical. *Nutr. Food Sci. Res.* **2016,** *3* (1), 5–16.
59. Khosla, P.; Gupta, D.; Nagpal, R. Effect of Trigonella Foenum Graecum (Fenugreek) on Blood Glucose in Normal and Diabetic Rats. *Ind. J. Physiol. Pharmacol.* **1995,** *39* (2), 173–173.
60. Kim, J.; Cha, Y. J.; Lee, K. H.; Park, E. Effect of Onion Peel Extract Supplementation on the Lipid Profile and Antioxidative Status of Healthy Young Women, a Randomized, Placebo-controlled, Double-blind, Crossover Trial. *Nutr. Res. Pract.* **2013,** *7* (5), 373–9.
61. Akhtar, S.; Khalid, N.; Ahmed, I.; Shahzad, A.; Suleria, H. A. R. Physicochemical Characteristics, Functional Properties, and Nutritional Benefits of Peanut Oil, A Review. *Critic. Rev. Food Sci. Nutr.* **2014,** *54* (12), 1562–1575.
62. Kumar, K. S.; Bhowmik, D. Traditional and Medicinal Uses of Banana. *J. Pharmacogn. Phytochem.* **2012,** *1* (3), 51–63.
63. Kyo, E.; Uda, N.; Kasuga, S.; Itakura, Y. Immunomodulatory Effects of Aged Garlic Extract. *J. Nutr.* **2001,** *131* (3), 1075S–1079S.
64. Lakhera, A.; Ganeshpurkar, A.; Bansal, D.; Dubey, N. Chemopreventive Role of Coriandrum Sativum Against Gentamicin-induced Renal Histopathological Damage in Rats. *Interdiscip. Toxicol.* **2015,** *8* (2), 99–102.
65. Lemay, A.; Dodin, S.; Kadri, N.; Jacques, H.; Forest, J.-C. Flaxseed Dietary Supplement Versus Hormone Replacement Therapy in Hypercholesterolemic Menopausal Women. *Obstetrics Gynecol.* **2002,** *100* (3), 495–504.

66. Butt, M. S.; Imran, A.; Sharif, M. K., Ahmad, R. S.; Xiao, H.; Imran, M.; Rsool, H. A. Black Tea Polyphenols, a Mechanistic Treatise. *Critic. Rev. Food Sci. Nutr.* **2014,** *54* (8), 1002–1011.
67. Li, Q.; Verma, I. M. NF-[kappa] B Regulation in the Immune System. *Nat. Rev.* **2002,** *2* (10), 725.
68. Liu, L.; Ju, Y.; Wang, J.; Zhou, R. Epigallocatechin-3-gallate Promotes Apoptosis and Reversal of Multidrug Resistance in Esophageal Cancer Cells. *Pathology* **2017,** *213* (10), 1242–1250
69. Lutomski, J. Garlic Fascination–Yesterday and Today. *Postępy Fitoterapii* **2001,** *1* (1), 7–14.
70. Madhujith, T.; Izydorczyk, M.; Shahidi, F. Antioxidant Properties of Pearled Barley Fractions. *J. Agric. Food Chem.* **2006,** *54* (9), 3283–3289.
71. Mahendra, P.; Bisht, S. Anti-anxiety Activity of Coriandrum Sativum Assessed Using Different Experimental Anxiety Models. *Ind. J. Pharmacol.* **2011,** *43* (5), 574–7.
72. Mandegary, A.; Pournamdari, M.; Sharififar, F.; Pournourmohammadi, S.; Fardiar, R.; Shooli, S. Alkaloid and Flavonoid Rich Fractions of Fenugreek Seeds (Trigonella foenum-graecum L.) with Antinociceptive and Anti-inflammatory Effects. *Food Chem. Toxicol.* **2012,** *50* (7), 2503–2507.
73. Martínez-Valverde, I.; Periago, M. J.; Provan, G.; Chesson, A. Phenolic Compounds, Lycopene and Antioxidant Activity in Commercial Varieties of Tomato (Lycopersicum esculentum). *J. Sci. Food Agric.* **2002,** *82* (3), 323–330.
74. Martirosyan, D. M.; Singh, J. A New Definition of Functional Food by FFC, What Makes a New Definition Unique? *Funct. Foods Health Dis.* **2015,** *5* (6), 209–223.
75. Menezes, J.; Neves, B. H.; Souza, M.; Mello-Carpes, P. B. Green Tea Protects Against Memory Deficits Related to Maternal Deprivation. *Physiol. Behav.* **2017,** *182* (1), 121–127.
76. Meyer, A. The 1998 Top 100® R&D Survey. *Food Process.* **1998,** *58* (8), 32–40.
77. Miron, T.; Rabinkov, A.; Mirelman, D.; Wilchek, M.; Weiner, L. The Mode of Action of Allicin, its Ready Permeability Through Phospholipid Membranes may Contribute to its Biological Activity. *Biochimica et Biophysica Acta (BBA).* **2000,** *1463* (1), 20–30.
78. Morris, D. H.; Canada, F. C. *Flax, a Health and Nutrition Primer*. Flax Council of Canada; 2007; p 234.
79. Mukherjee, D.; Bhattacharya, K.; Chandra, G. Extracts of Edible Pods of Moringa Oleifera Lam.(Moringaceae) as Novel Antibacterial Agent Against Some Pathogenic Bacteria. *Int. J. Pharma Biosci.* **2015,** *6* (3), 513–520.
80. Murlidhar, M.; Goswami, T. A Review on the Functional Properties, Nutritional Content, Medicinal Utilization and Potential Application of Fenugreek. *J. Food Process. Technol.* **2012**, *3* (9), 1–10.
81. Namal Senanayake, S. P. J. Green Tea Extract, Chemistry, Antioxidant Properties and Food Applications: A Review. *J. Funct. Foods* **2013,** *5* (4), 1529–1541.
82. Nathan, J.; Panjwani, S.; Mohan, V.; Joshi, V.; Thakurdesai, P. A. Efficacy and Safety of Standardized Extract of Trigonella foenum-graecum L Seeds as an Adjuvant to L-dopa in the Management of Patients with Parkinson's Disease. *Phytother. Res.* **2014,** *28* (2), 172–178.
83. Newman, R. K.; Newman, C. W. *Barley for Food and Health, Science, Technology, and Products*; John &Wiley: New York; 2008; pp 281.
84. Nowack, R. Cranberry Juice–A Well-characterized Folk-remedy Against Bacterial Urinary Tract Infection. *WMW Wiener Medizinische Wochenschrift* **2007,** *157* (13), 325–330.

85. Nuutila, A. M.; Puupponen-Pimiä, R.; Aarni, M.; Oksman-Caldentey, K.-M. Comparison of Antioxidant Activities of Onion and Garlic Extracts by Inhibition of Lipid Peroxidation and Radical Scavenging Activity. *Food Chem.* **2003,** *81* (4), 485–493.
86. Oliveira, T. T.; Campos, K. M.; Cerqueira-Lima, A. T.; Cana Brasil Carneiro, T.; da Silva Velozo, E.; Ribeiro Melo, I. C. A.; Figueiredo, C. A. Potential Therapeutic Effect of Allium cepa L. and Quercetin in a Murine Model of Blomia Tropicalis Induced Asthma. *DARU J. Pharmaceut. Sci.* **2015,** *23* (1), 1–18.
87. Oomah, B. D. Flaxseed as a Functional Food Source. *J. Sci. Food Agric.* **2001,** *81* (9), 889–894.
88. Pal, S. K.; Mukherjee, P. K.; Saha, K.; Pal, M.; Saha, B. P. Antimicrobial Action of the Leaf Extract of Moringa Oleifera Lam. *Ancient Sci. Life* **1995,** *14* (3), 197–199.
89. Pangastuti, A.; Amin, I. F.; Amin, A. Z.; Amin, M. Natural Bioactive Compound From Moringa Oleifera against Cancer Based On In Silico Screening. *J. Tecnol. Sci. Eng.* **2016,** *78*, 5.
90. Peng, C. Y. J.; Lee, K. L.; Ingersoll, G. M. An Introduction to Logistic Regression Analysis and Reporting. *J. Educ. Res.* **2002,** *96* (1), 3–14.
91. Peterson, D. M. Oat Antioxidants. *J. Cereal Sci.* **2001,** *33* (2), 115–129.
92. Khalid, N.; Suleria, H. A. R.; Ahmed, I. Pineapple Juice. In *Handbook of Functional Beverages and Human Health*; CRC Press: Boca Raton; 2018; pp 489–500.
93. Prasad, K. Reduction of Serum Cholesterol and Hypercholesterolemic Atherosclerosis in Rabbits by Secoisolariciresinol Diglucoside Isolated from Flaxseed. *Circulation* **1999,** *99* (10), 1355–1362.
94. Prasad, K.; Laxdal, V. A.; Yu, M.; Raney, B. L. Antioxidant Activity of Allicin, an Active Principle in Garlic. *Mol. Cell. Biochem.* **1995,** *148* (2), 183–189.
95. Sultan, M. T.; Buttxs, M. S.; Qayyum, M. M. N.; Suleria, H. A. R. Immunity: Plants as Effective Mediators. *Critic. Rev. Food Sci. Nutr.* **2014,** *54* (10), 1298–1308.
96. Rahman, M. M.; Sheikh, M. M. I.; Sharmin, S. A.; Islam, M. S.; Rahman, M. A.; Rahman, M. M.; Alam, M. Antibacterial Activity of Leaf Juice and Extracts of *Moringa Oleifera Lam*. Against Some Human Pathogenic Bacteria. *CMU J. Nat. Sci.* **2009,** *8* (2), 219.
97. Rajeshwari, U.; Andallu, B. Medicinal Benefits of Coriander (*Coriandrum Sativum* L). *Spatula DD* **2011,** *1* (1), 51–58.
98. Randriamboavonjy, J. I.; Loirand, G.; Vaillant, N.; Lauzier, B.; Derbré, S.; Michalet, S.; Pacaud, P.; Tesse, A. Cardiac Protective Effects of *Moringa Oleifera* Seeds in Spontaneous Hypertensive Rats. *Am. J. Hypertens.* **2016,** *29* (7), 873–881.
99. Rathore, S.; Saxena, S.; Singh, B. Potential Health Benefits of Major Seed Spices. *Int. J. Seed Spices* **2013,** *3* (2), 1–12.
100. Rattanachaikunsopon, P.; Phumkhachorn, P. Potential of Coriander (*Coriandrum Sativum*) Oil as a Natural Antimicrobial Compound in Controlling Campylobacter Jejuni in Raw Meat. *Biosci. Biotechnol. Biochem.* **2010,** *74* (1), 31–35.
101. Rola, S. B. C. I. I.; Dietoprofilaktyce, W.; Dietoterapii, I. Bioactive Substances of Garlic and Their Role in Dietoprophylaxis and Dietotherapy. *Roczniki Państwowego Zakładu* **2007,** *58* (1), 41.
102. Rubilar, M.; Gutiérrez, C.; Verdugo, M.; Shene, C.; Sineiro, J. Flaxseed as a Source of Functional Ingredients. *J. Soil Sci. Plant Nutr.* **2010,** *10* (3), 373–377.
103. Schieber, A. Functional Foods and Nutraceuticals. *Food Res. Int.* **2012,** *46* (2), 437–572.

104. Siegers, C.P.; Steffen, B.; Röbke, A.; Pentz, R. The Effects of Garlic Preparations Against Human Tumor Cell Proliferation. *Phytomedicine* **1999,** *6* (1), 7–11.
105. Silva, F.; Domingues, F. C. Antimicrobial Activity of Coriander Oil and its Effectiveness as Food Preservative. *Critic. Rev. Food Sci. Nutr.* **2017,** *57* (1), 35–47.
106. Singh, K. K.; Mridula, D.; Rehal, J.; Barnwal, P. Flaxseed: A Potential Source of Food, Feed and Fiber. *Critic. Rev. Food Sci. Nutr.* **2011,** *51* (3), 210–222.
107. Slavin, J. L.; Lloyd, B. Health Benefits of Fruits and Vegetables. *Adv. Nutr. Int. Rev. J.* **2012,** *3* (4), 506–516.
108. Sreelatha, S.; Padma, P. R. Antioxidant Activity and Total Phenolic Content of *Moringa Oleifera* Leaves in Two Stages of Maturity. *Plant Foods Hum. Nutr.* **2009,** *64* (4), 303.
109. Srikanth, V.; Mangala, S.; Subrahmanyam, G. Improvement of Protein Energy Malnutrition by Nutritional Intervention with *Moringa Oleifera* Among Anganwadi Children in Rural Area in Bangalore, India. *Int. J. Sci. Stud.* **2014,** *2* (1), 32–35.
110. Suleria, H. A.; Butt, M. S.; Anjum, F. M.; Saeed, F.; Khalid, N. Onion: Nature Protection Against Physiological Threats. *Critic. Rev. Food Sci. Nutr.* **2015,** *55* (1), 50–66.
111. Tang, E. L. H.; Rajarajeswaran, J.; Fung, S. Y.; Kanthimathi, M. S. Antioxidant Activity of Coriandrum Sativum and Protection Against DNA Damage and Cancer Cell Migration. *BMC Complement. Alternat. Med.* **2013,** *13* (1), 347.
112. Yousaf, S.; Butt, M. S.; Suleria, H. A. R.; Iqbal, M. J. The Role of Green Tea Extract and Powder in Mitigating Metabolic Syndromes with Special Reference to Hyperglycemia and Hypercholesterolemia. *Food Funct.* **2014,** *5* (3), 545–556.
113. Tenore, G. C.; Campiglia, P.; Giannetti, D.; Novellino, E. Simulated Gastrointestinal Digestion, Intestinal Permeation and Plasma Protein Interaction of White, Green, and Black Tea Polyphenols. *Food Chem.* **2015,** *169* (1), 320–326.
114. Thompson, L. U.; Seidl, M. M.; Rickard, S. E.; Orcheson, L. J.; Fong, H. H. S. Antitumorigenic Effect of a Mammalian Lignan Precursor from Flaxseed. *Nutr. Cancer* **1996,** *26* (2), 159–165.
115. Tiloke, C.; Phulukdaree, A.; Chuturgoon, A. A. The Antiproliferative Effect of Moringa oleifera Crude Aqueous Leaf Extract on Human Esophageal Cancer Cells. *J. Med. Food* **2016,** *19* (4), 398–403.
116. Timbo, B. B.; Ross, M. P.; McCarthy, P. V.; Lin, C. T. J. Dietary Supplements in a National Survey, Prevalence of Use and Reports of Adverse Events. *J. Am. Diet. Assoc.* **2006,** *106* (12), 1966–1974.
117. Tolkachev, O. N.; Zhuchenko, A. A. Biologically Active Substances of Flax, Medicinal and Nutritional Properties (A Review). *Pharmaceut. Chem. J.* **2004,** *34* (7), 360–367.
118. Toshinai, K., et al. Des-acyl Ghrelin Induces Food Intake by a Mechanism Independent of the Growth Hormone Secretagogue Receptor. *Endocrinology* **2006,** *147* (5), 2306–2314.
119. Turati, F.; Rossi, M.; Pelucchi, C.; Levi, F.; La Vecchia, C. Fruits and Vegetables and Cancer Risk, a Review of Southern European Studies. *Br. J. Nutr.* **2015,** *113* (S2), S102–S110.
120. Uemura, T., et al. Diosgenin Present in Fenugreek Improves Glucose Metabolism by Promoting Adipocyte Differentiation and Inhibiting Inflammation in Adipose Tissues. *Mol. Nutr. Food Res.* **2010,** *54* (11), 1596–1608.
121. Suleria, H. A. R.; Butt, M. S.; Khalid, N.; Sultan, S.; Raza, A.; Aleem, M.; Abbas, M. Garlic (Alium sativum) Diet Based Therapy of 21st Century: A Review. *Asian Pac. J. Trop. Dis.* **2015,** *5* (4), 271–278.
122. Iqbal, M. J.; Butt, M. S.; Suleria, H. A. R. Coriander (*Coriandrum Sativum* L.): Bioactive Molecules and Health Effects. In: *Bioactive Molecules in Food*; Mérillon, J. M., Ramawat, K. G. Eds.; Springer International Publishing: New York, 2017; pp 1–37.

123. Virtanen, A. I.; Matikkala, E. The Isolation of S-methyl-L-cysteine Sulphoxide and Sn-propyl-L-cysteine Sulphoxide from Onion (*Allium cepa*) and the Antibiotic Activity of Crushed Onion. *Acta Chem. Scand.* **1959,** *13* (9), 1898–1900.
124. Wang, F.; Zhong, H.-H.; Chen, W.-K.; Liu, Q.-P.; Li, C.-Y.; Zheng, Y.-F.; Peng, G.-P. Potential Hypoglycaemic Activity Phenolic Glycosides from Moringa Oleifera Seeds. *Nat. Prod. Res.* **2017,** *31* (16), 1869–1874.
125. Wangensteen, H.; Samuelsen, A. B.; Malterud, K. E. Antioxidant Activity in Extracts from Coriander. *Food Chem.* **2004,** *88* (2), 293–297.
126. 126.Watson, R. R.; Preedy, V. R. *Bioactive Foods in Promoting Health, Fruits and Vegetables*; Elsevier Science: New York, 2014; pp 312.
127. Williamson, C. Functional Foods: What are the Benefits? *Br. J. Commun. Nurs.* **2009,** *14* (6), 230–236.
128. Abbas, M.; Saeed, F.; Anjum, F. M.; Afzaal, M.; Tufail, T.; Bashir, M. S.; Suleria, H. A. R. Natural Polyphenols: An Overview. *Int. J. Food Prop.* **2017,** *20* (8), 1689–1699.
129. Wills, E. Enzyme Inhibition by Allicin, the Active Principle of Garlic. *Biochem. J.* **1956,** *63* (3), 514.
130. Yassa, H. D.; Tohamy, A. F. Extract o*f Moringa Oleifera* Leaves Ameliorates Streptozotocin-induced *Diabetes Mellitus* in Adult Rats. *Acta Histochem.* **2014,** *116* (5), 844–854.
131. Yoo, K. S.; Pike, L. M. Determination of Flavor Precursor Compound S-alk (en) yl-L-cysteine Sulfoxides by an HPLC Method and Their Distribution in Allium Species. *Scientia Horticult.* **1998,** *75* (1), 1–10.
132. You, W. C.; Blot, W. J.; Chang, Y. S.; Ershow, A.; Yang, Z. T.; An, Q.; Henderson, B. E.; Fraumeni, J. J. F.; Wang, T. G. Allium Vegetables and Reduced Risk of Stomach Cancer. *JNCI* **1989,** *81* (2), 162–164.
133. Yuan, Y. V.; Rickard, S. E.; Thompson, L. U. Short-term Feeding of Flaxseed or its Lignan has Minor Influence on In Vivo Hepatic Antioxidant Status in Young Rats. *Nutr. Res.* **1999,** *19* (8), 1233–1243.
134. Zduńczyk, Z.; Flis, M.; Zieliński, H.; Wróblewska, M.; Antoszkiewicz, Z.; Juśkiewicz, J. In Vitro Antioxidant Activities of Barley, Husked Oat, Naked Oat, Triticale, and Buckwheat Wastes and Their Influence on the Growth and Biomarkers of Antioxidant Status in Rats. *J. Agric. Food Chem.* **2006,** *54* (12), 4168–4175.
135. Zhao, Q.; Hu, X.; Guo, Q.; Cui, S. W.; Xian, Y.; You, S.; Chen, X.; Xu, C.; Gao, X. Physicochemical Properties and Regulatory Effects on db/db Diabetic Mice of β-glucans Extracted from Oat, Wheat and Barley. *Food Hydrocolloids* **2014,** *37* (1), 60–68.

# CHAPTER 3

# NATURAL OR SYNTHETIC ANTIOXIDANTS IN FOODS

TOSIN MICHAEL OLAJIDE, MD. RAMIM TANVER RAHMAN, ZAIXIANG LOU, and JING QI

## ABSTRACT

The antioxidants are bioactive compounds in foods that help to neutralize free radicals, which attack fats, protein, and DNA in the cells, causing various types of diseases and accelerating the aging process. Concurrently, autoxidation causes rancidity of fats and oils leading to deterioration of fatty foods. The best natural antioxidant sources are plant materials such as fruits, vegetables, and herbs. However, synthetic antioxidants are also available to the food industry and are further improvised for better effectiveness in delaying lipid oxidation thus preserving healthy aspects of food. This chapter reviews impact, sources, and suspected adverse health effects of natural and synthetic antioxidants.

## 3.1 INTRODUCTION

Foods such as dairy products, meats, nuts, and poultry contain fats and oil, which are nutritious components from which humans derive energy. Fats contribute to the overall palatability of these foods, but also make them rancid due to oxidation (a reaction process that generates free radicals capable of progressively causing cell damage). This rancidity, which is often spontaneous is one of the main factors that worsen food conditions during processing and storage.[1] In addition, oxidation of lipids in the body system gives rise to heart diseases, carcinogenesis, and aging.[5,26] Although modification of food preparation, refrigeration, nitrogen blanketing, and packaging include various attempts previously employed to protect against deterioration, these methods, however, may not be enough to prevent oxidation. Hence, antioxidants (an organic substance that prevents other biologically useful substances from becoming oxidized), that is, natural or synthetic, are often used in oils and fatty foods to retard autoxidation and enhance their shelf life. This concept of incorporating antioxidant into food to prevent food deterioration remains the most operative, resourceful, and cost-effective method.[27,31]

This chapter reviews impact, sources, and suspected adverse health effects of natural and synthetic antioxidants.

Antioxidants used in edible oils and fatty acids usually satisfy the following requirements[18,31]:

- Able to retard lipid autoxidation at small amounts
- Lipophilic
- Nontoxic to humans
- Nonvolatile and chemically stable at cooking, hot-oven and ultrahigh temperatures processing
- Not too expensive
- Odorless, tasteless, and colorless

Antioxidants can be synthetic or natural. But, based on their antioxidative mechanisms, they are classified into primary and secondary antioxidants. Primary antioxidants are free-radical scavengers. On the other hand, secondary antioxidants are also called synergists, and help regenerate or sustain primary antioxidants by chelating metal ions, reduce resulting peroxides, or scavenge oxygen. Structurally, the phenolic antioxidants, which possess at least one phenol ring, are most commonly utilized antioxidants in oil- and fat-based foods. They include: tocopherols, butylated hydroxyanisole (BHA), propyl gallate (PG), *tert*-butylhydroquinone (TBHQ), and butylated hydroxytoluene (BHT).[25]

## 3.2 NATURAL ANTIOXIDANTS

Natural antioxidants are compounds found in nature that possess antioxidative activities. Many fruits and vegetables, spices, and herbs have potent antioxidant activity and are rich sources of natural antioxidants.[32] The common examples of this class of antioxidants are: vitamins, polyphenolic compounds, tocopherols, carotenoids, ascorbic acid, and polyphenols (flavanoids and hydroxycinnamic acids).

Spice extracts, tocopherols, and the salts of citric and ascorbic acids are commonly used commercial natural antioxidants.[30] The tocopherols (α, β, γ, and δ: Fig. 3.1) predominate in fats, and are found in combination in most vegetable oils[7]; while the most available and the most biologically potent one is α-tocopherol.[23,33]

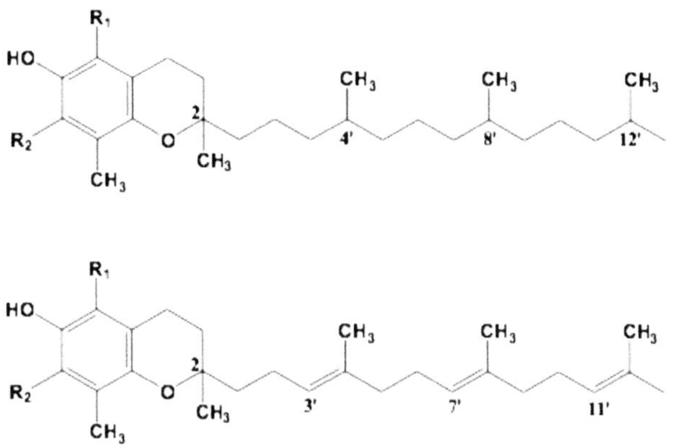


| $R_1$ | $R_2$ | |
|---|---|---|
| $CH_3$ | $CH_3$ | α-Tocopherol |
| $CH_3$ | H | β-Tocopherol |
| H | $CH_3$ | γ-Tocopherol |
| H | H | δ-Tocopherol |

| $R_1$ | $R_2$ | |
|---|---|---|
| $CH_3$ | $CH_3$ | α-Tocotrienol |
| $CH_3$ | H | β-Tocotrienol |
| H | $CH_3$ | γ-Tocotrienol |
| H | H | δ-Tocotrienol |

**FIGURE 3.1** Structures of α-, β-, γ-, and δ-tocopherol and tocotrienol.

Ascorbyl palmitate is an ester of ascorbic acid soluble in oil and is commonly used for food applications. In tissues of plants and animals, citric acid can be found, which is capable of chelating metal ions by bonding its carboxyl or hydroxyl groups with them,[2] thereby effectively preventing the spontaneous rancidity of food lipids and vegetable oils.[8] Importantly, there are varying methods by which antioxidants control the autoxidation process such as citric acid chelates; while both tocopherol and rosemary extracts and ascorbyl palmitate scavenge radicals and oxygen, respectively.[15]

### *3.2.1 SOURCES OF NATURAL ANTIOXIDANTS*

Vegetable oils contain considerable amounts of tocopherols (Table 3.1) including oilseeds, leaves, and other green parts of higher plants.[14] Tocopherols present in unrefined vegetable oils usually decrease after processing. For instance, after refining soybean oil, there was a 4.3% loss of α-tocopherols, 15% after bleaching, and 20–51% after deodorization.[3]

**TABLE 3.1** Selected Vegetable Oils with Approximate Tocopherol Contents.

| Tocopherols (mg/kg) | Olive oil | Palm oil | Soybean oil | Sunflower oil | Rapeseed/ canola oil |
|---|---|---|---|---|---|
| α-Tocopherol | 1–240 | 180–260 | 30–120 | 350–700 | 180–280 |
| β-Tocopherol | 0 | Trace | 0–20 | 20–40 | – |
| γ-Tocopherol | 0 | 320 | 250–930 | 10–50 | 380–590 |
| δ-Tocopherol | 0 | 70 | 50–450 | 1–10 | 10–20 |

**Source:** Adapted from Ref. [24].

## 3.3 SYNTHETIC ANTIOXIDANTS

These are antioxidants synthesized in the laboratory. They are colorless, odorless, tasteless, and cheaper and are used frequently compared to natural antioxidants that are more expensive. The preservation of food quality by protecting against lipid oxidation using synthetic antioxidants remains the most resourceful and effective method.[31,20] Common antioxidants of synthetic origin generally are effective only at not too high temperature, but lose their effectiveness at temperature above 150°C.[20] Synthetic antioxidants (Fig. 3.2) that are frequently used for oil and fatty foods include: BHA, BHT, TBHQ, and PG.[21]

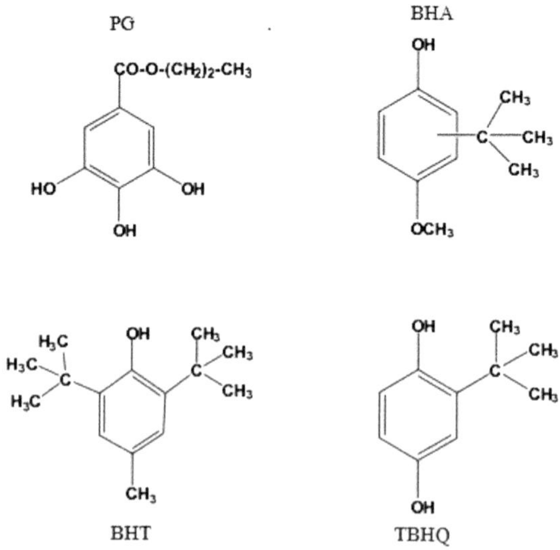


**FIGURE 3.2** Chemical structures of common synthetic antioxidants.

**Source:** Adapted from Ref. [21].

**TABLE 3.2** Flavonoids (mg/kg Eatable Portion), Hydroxycinnamic Acids (mg/kg per Eatable Portion), Carotenoids (g/100 g Eatable Portion) and Total Ascorbic Acid (mg/100 g Eatable Portion) in Commercial Fresh Tomato, Lettuce, Onion, and Kale.

| Vegetable | Flavonoids | | Hydroxycinnamic acids | | | Carotenoids | | | Total ascorbic acid |
|---|---|---|---|---|---|---|---|---|---|
| | Quercetin | Kaempferol | Caffeic acid | Coumaric acid | Ferulic acid | Lutein | Lycopene | β-carotene | |
| Tomato | 9.7 ± 3.3 | 18.0 ± 0.4 | 46.7 ± 1.1 | 13.0 ± 1.2 | 4.0 ± 0.3 | 58.7 ± 11.6 | 2205.0 ± 997.0 | 496.0 ± 100.0 | 10.70 ± 1.89 |
| Lettuce | 24.2 ± 2.1 | <2.0[a] | 12.6 ± 1.9 | n.d. | n.d. | 880.0 ± 321.0 | n.d. | 556.0 ± 178.0 | 10.74 ± 2.14 |
| Onion | 417.0 ± 10.4 | 23.9 ± 8.9 | 33.3 ± 1.2 | 30.0 ± 1.0 | 39.0 ± 1.9 | – | – | – | 5.15 ± 1.18 |
| Kale | 14.2 ± 27.0 | 35.2 ± 7.3 | 59.4 ± 0.2 | 40.1 ± 0.9 | n.d | 1056.0 ± 142.0 | <2.0[a] | n.d. | [b] |

n.d. = not detectable.

[a]Below limit of detection.

[b]Not evaluated.

Values are expressed in mean ± SD.

When selecting antioxidants their synergistic capacity, optimal concentration, and stability at high temperatures are usually considered[10] BHA and BHT are good primary antioxidants, having phenolic rings that contain *tert*-butyl groups.[22] Some confectioneries and cereals containing vegetable oils are frequently preserved by putting BHA into them. It remains unchanged under heat and moderate alkaline conditions, thus is good for baking, but limited due to its volatility. In addition, it has been shown that BHA has antimicrobial properties.[6] On the contrary, although BHT is not as stable as BHA, it is currently favored for high-temperature processing in the food industry.[22] Also, it can be solely used or combined with other compounds that prevent oxidation reactions such as citric acid, PG, and BHA to stabilize low or high lipid-containing foods and dispensers.[17] TBHQ, however, is effective in stabilizing polyunsaturated vegetable oils and PG is a good antioxidant in soybean oil, but its usage is hindered by low oil solubility and tendency to react with metals to cause discoloration.

### *Protective factor*

The activities of antioxidant compounds can easily be tested using Rancimat to evaluate "protective factor" (PF). As indicated in eq 3.1, an antioxidant's effectiveness is proportional to its PF value (i.e., a higher PF value means a stronger antioxidant activity of compound).[28]

$$PF = I_A / I_O \tag{3.1}$$

where $I_A$ is the induction time of oil samples with added antioxidants, and $I_O$ is the induction time of oil samples only.

### *Recent developments*

The ineffectiveness of common synthetic antioxidants at high temperatures, especially because of their low molecular weights, has continuously prompted food scientists to synthesize new ones of improved antioxidant efficiencies. This section presents novel antioxidants, which are built upon the structures of previously existing ones (i.e., mother compounds) or sometimes from their breakdown.

An acid-catalyzed condensation reaction involving glyoxal and toluhydroquinone (THQ) or TBHQ, respectively,[13] was used to synthesize two novel antioxidant compounds with four hydroxyphenyl groups (Fig. 3.3). These newly synthesized compounds were observed to have PF superior to TBHQ at 140°C, and can improve the oxidative stability of lipid products during high-temperature processing.

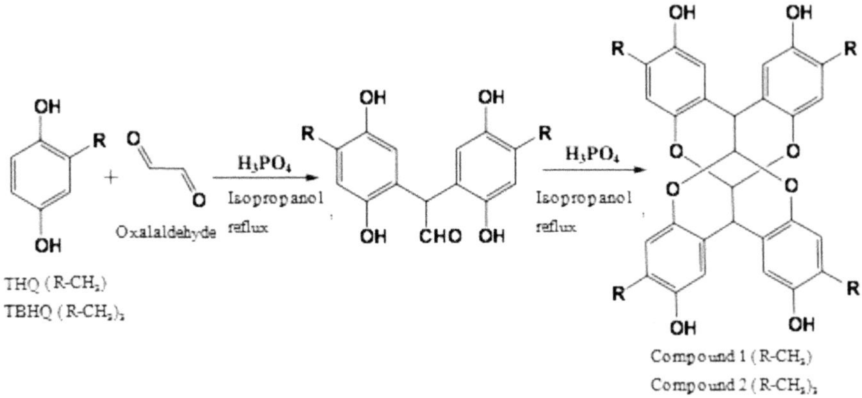


**FIGURE 3.3** Synthesis of two novel tetraphenolic compounds derived from THQ and TBHQ.
**Source:** Adapted from Ref. [20].

Other research studies[16] involving TBHQ as mother compound have also been reported. Lauryl *tert*-butylhydroquinone (LTBHQ) and lauryl *tert*-butylquinone (LTBQ) were derived by reacting TBHQ with lauryl alcohol.[35] While 5,5'-(ethane-1,1-diyl)bis(2-(tert-butyl)benzene-1,4-diol) (DHPE) was synthesized through condensation reaction between TBHQ and acetaldehyde (Fig. 3.4). These new antioxidants all have stronger radical scavenging activities at high temperatures (>140°C) than their common synthetic counterparts such as BHA, BHT, TBHQ, and PG.

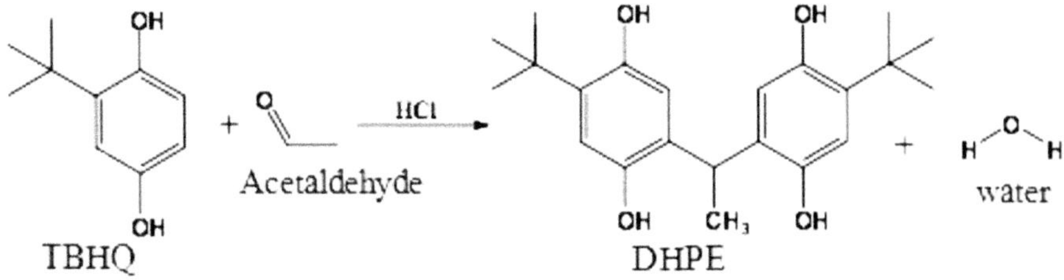


**FIGURE 3.4** Synthesis of DHPE by condensation reaction between TBHQ and acetaldehyde.
**Source:** Adapted from Ref. [27].

In addition using tertiary butanol, the 3-(tert-butyl)-5-methylbenzene-1,2-diol (TBHPC) was produced through alkylation.[12] TBHPC showed most potent antioxidant activity and formed a more stable free radical intramolecularly. This synergistic mechanism leading to a more stable intermediate by TBHPC is elucidated more clearly by Mukhopadhyay.[13]

## 3.4 HEALTH CONTROVERSIES

Many common substances with antioxidative properties have been studied to negatively affect the human health.[4] However, sufficient data, especially on their long-term (over 6 months) consequences on health are lacking, making the characterization of this subject matter inconclusive. Nonetheless based on as much known, different countries vary the permissible amounts of antioxidants usable to avoid health risks.[22,23] As a result, there are systems in place that monitor and implement the safe use of these man-made substances in food products. For instance, the US and Canadian food authorities only allow a single antioxidant in food at not more than 100 ppm, and at 200 ppm when combined with others.[19]

It is noteworthy that the adverse health effects of antioxidants (artificial or natural) arise partly due to their ability to become pro-oxidant at high concentrations, thereby forming unstable free radicals in the biological systems. Ascorbate, for example, at high concentration of iron may display pro-oxidative properties, which is often seen in hemochromatic patients.[29] It was observed[9] that sodium ascorbate enhances the adverse effects of known cancer-inducing substance in rats. However, there have been no such studies on humans.

The safety consumption of artificial antioxidants has long been a subject of expansive debate prompting further research. However, naturally occurring antioxidants in food on the other hand have mostly been free of controversies, and in fact, known to have anticarcinogenic properties.[19] For instance, the use of some natural antioxidants such as β-carotene to supplement diets has been suspected to increase the chance of developing chronic heart problems.[20] While most of studies on adverse reactions to BHA, BHT, and/or TBHQ have been done on animals, there is a considerable research on human that identified problems with these for humans, such as allergy, joint pains, asthma, dermatitis, headache, rhinitis, abdominal problems, sleepiness, and eye problems.[19]

### *3.4.1 ROLE IN CANCER*

BHA has been studied, in high doses, to promote forestomach cancer in rats and hamsters,[29] while BHT caused lung damage and increased the incidence of liver tumors in mice. However, there has been no evidence of these in humans. TBHQ, on the other hand, has been shown to induce carcinogenicity at chronic exposure.[6] All in all, there are no cancer-causing activities by BHT and BHA. In fact, they are likely to be anticarcinogenic at their present permissible usage limit.[34]

## 3.5 SUMMARY

Vegetable oils, oilseeds, nuts, and some other foods in nature have naturally occurring bioactive compounds with antioxidant activities that keep them from becoming rancid; and these substances, however, cannot guarantee a longer shelf life. Hence, the need for stronger, cheap, and more easily accessible antioxidants has brought production and release of synthetic alternatives to the food industries. However, despite acceptable daily intake (ADI) that were recommended by regulatory bodies for appropriate usage of these antioxidants (BHA, TBHQ, BHT, and PG) in oil- and fat-based foods, have been a subject of expansive debate due to some suspected adverse health effects which are yet to be fully characterized in human.

## KEYWORDS

- **antioxidant**
- **free radical**
- **primary antioxidants**
- **secondary antioxidants**
- **synthetic antioxidants**
- **natural antioxidants**

## REFERENCES

1. Anonymous. *Ullmann's Encyclopedia of Industrial Chemistry*. Wiley-VCH Verlag: Weinheim, Germany, 1985; pp 103–108.
2. Chattopadhyay, M. Adverse Effects of Antioxidants. *Curr. Sci.* **2003,** *85* (2), 121.
3. Frankel, E. N.; Huang, S. W.; Aeschbach, R.; Prior, E. Antioxidant Activity of a Rosemary Extract and Its Constituents, Carnosic Acid, Carnosol, and Rosmarinic Acid, in Bulk Oil and Oil-in-Water Emulsion. *J. Agric. Food Chem.* **1996,** *44* (1), 131–135.
4. Gharavi, N.; El-Kadi, A. O. S. *tert*-Butylhydroquinone is a Novel Aryl Hydrocarbon Receptor Ligand. *Drug Metab. Dispos.* **2004,** *33* (3), 365–372.
5. Gordon, M. H. Oils and Fats: Taint or Flavor? *Chemistry in Britain*, 1991; pp 1020–1022.

6. Hilton, J. Antioxidants: Function, Types and Necessity of Inclusion in Pet Foods. *Can. Veter. J.* **1989,** *30* (8), 682–684.
7. Houlihan, C. M; Ho, C. T; Chang, S. S. Elucidation of the Chemical Structure of a Novel Antioxidant, Rosmaridiphenol, Isolated from Rosemary. *J. Am. Oil Chem. Soc.* **1984,** *61* (6), 1036–1039.
8. Huang, Y.; Jiang, Z.; Liao, X.; Hou, J.; Weng, X. C. Antioxidant Activities of Two Novel Synthetic Methylbenzenediol Derivatives. *Czech J. Food Sci.* **2014,** *32* (4), 348–353.
9. Jiang, Z. W.; Weng, X. C; Huang, Y.; Hou, J. P.; Liao, X. Y. Synthesis and Antioxidant Activity of Two Novel Tetraphenolic Compounds Derived from Toluhydroquinone and Tertiary Butylhydroquinone. *Grasas Aceites* **2014,** *65* (2), e016.
10. Kamal-Eldin, A.; Appelqvist, L. The Chemistry and Anti-oxidant Properties of Tocopherols and Tocotrienols. *Lipids* **1996,** *31* (7), 671–701.
11. Kochhar, S. P.; Rossell, J. B. Detection, Estimation and Evaluation of Antioxidants in Food Systems. In *Food Antioxidants*; Hudson, B. J., Ed.; Springer: London, 1990; pp 19–64.
12. Li, J. Y.; Wang, T.; Wu, H.; Ho, C. T.; Weng, X. C. 1,1-Di-(2,5-dihydroxy-4 tertbutylphenyl) Ethane: A Novel Antioxidant. *J. Food Lipids* **2006,** *13* (3), 331–340.
13. Mukhopadhyay, A. *Antioxidants: Natural and Synthetic;* Amani International Publishers Kiel: Germany, 2007; pp 213.
14. Nakatani, N.; Inatani, R. Two Antioxidant Diterpenes from Rosemary (*Rosmarinus officinalis* L.) and a New Revised Structure for Rosmanol. *Agric. Biol. Chem.* **1984,** *48* (8), 2081–2085.
15. Race, S. *Antioxidants: The Truth About BHA, BHT, TBHQ and Other Antioxidants Used as Food Additives*; Tigmor Books, 2009; pp 6–11. http://www.foodcanmakeyouill.co.uk/uploads/1/2/7/4/12746572/antioxidants.pdf (accessed 15 July, 2017).
16. Rapola, J. M.; Virtamo, J.; Ripatti, S.; Huttunen, J. K.; Albanes, D.; Taylor, P. R.; Heinonen, O. P. Randomized Trial of α-Tocopherol and β-Carotene Supplements on Incidence of Major Coronary Events in Men with Previous Myocardial Infarction. *Lancet* **1997,** *349* (9067), 1715–1720.
17. Raza, A.; Shabbir, M. A.; Khan, M. I.; Suleria, H. A. R.; Sultan, S. Effect of Thermal Treatments on the Formation of Heterocyclic Aromatic Amines in Various Meats. *J. Food Process. Preserv.* **2015,** *39* (4), 376–383.
18. Reische, D. W.; Llillard, D. A.; Eitenmiller, R. R. Antioxidants. In *Food Lipids: Chemistry, Nutrition and Biotechnology*; Akoh, C. C., Min, D. M., Eds.; CRC Press: Boca Raton, FL, 2002; pp 489–516.
19. Rice-Evans, C. A.; Miller, N. J.; Paganga, G. Antioxidant Properties of Phenolic Compounds. *Trends Plant Sci.* **1997,** *2* (4), 152–159.
20. Saad, B.; Sing, Y. Y.; Nawi, M. A. Determination of Synthetic Phenolic Antioxidants in Food Items Using Reversed-phase HPLC. *Food Chem.* **2014,** *105* (1), 389–394.
21. Schuler, P. Natural Antioxidants Exploited Commercially. In *Food Antioxidants*; Hudson, B. J., Ed.; Springer: London, 1990; pp 99.
22. Shabbir, M. A.; Dilber, A.; Raza, A.; Suleria, H. A. R.; Saeed, M.; Sultan, S. Influence of Thermal Processing on the Formation of Trans Fats in Various Edible oils. *J. Food Process. Preserv.* **2015,** *39* (6), 1475–1484.
23. Shabbir, M. A.; Raza, A.; Anjum, F. M.; Khan, M. R.; Suleria, H. A. R. Effect of Thermal Treatment on Meat Proteins with Special Reference to Heterocyclic Aromatic Amines (HAAs). *Critic. Rev. Food Sci. Nutr.* **2015,** *55* (1), 82–93.

24. Shahidi, F.; Zhong, Y. Antioxidants: Regulatory Status. In *Industrial Oil and Fat Products*; John Wiley & Sons Inc.: Hoboken, NJ, USA, 2005; pp 491–512.
25. Tsuda, T.; Osawa, T.; Ohshima, K.; Kawakishi, S. Antioxidative Pigments Isolated from the Seeds of *Phaseolus vulgaris* L. *Food Chem.* **1994,** *42* (2), 248–251.
26. Wanasundara, P. K. J. P. D.; Shahidi, F. Antioxidants: Science, Technology and Applications. In *Bailey's Industrial Oil & Fat Products*; Wiley InterScience: Hoboken, NJ, 2005; pp 431–489.
27. Wang, W.; Weng, X. C.; Cheng, D. L. Antioxidant Activities of Natural Phenolic Components from *Dalbergia odorifera* T. *Food Chem.* **2000,** *71* (1), 45–49.
28. Weber, R. W. Adverse Reactions to Butylated Hydroxytoluene and Butylated Hydroxyanisole (BHA and BHT). *Toxicol. Sci.* **1999,** 51 (2), 202–210.
29. Weinreich, B. Frischeschutz Mit Weisser Weste. *Die Zeitschrift fuÈr die Lebensmittelwirtschaft* **1998,** *49*, 24–26.
30. Weng, X. C.; Ren, G. P.; Duan, S.; Dong, X. W.; Jiang, A. L. Screening of Natural Antioxidants from Chinese Medicines, Herbs and Spices. *J. Chin. Cereal Oils Assoc.* **1998,** *13* (4), 46–48.
31. Weng, X. C. Antioxidants and Their Antioxidation Mechanism. *J. Zhengzhou Grain Coll.* **1993,** *3*, 20–28.
32. White, P. J.; Xing, Y. Antioxidants from Cereals and Legumes. In *Natural Antioxidants;* AOCS Press: Champaign, IL, 1997; pp 25–63.
33. Williams, G. M.; Iatropoulos, M. J.; Whysner, J. Safety Assessment of Butylated Hydroxyanisole and Butylated Hydroxytoluene as Food Additives. *Food Chem. Toxicol.* **1999,** *37* (9), 1027–1038.
34. Younis, M. S.; Butt, M. S.; Sharif, M. K.; Sulerai, H. A. R.; Hameed, F. Effect of Preservatives on Physicochemical, Microbial and Sensory Attributes of Mangoes. *Internet J. Food Saf.* **2011,** *13*, 246–263.
35. Zhang, C. X., Wu, H.; Weng, X. C. Two Novel Synthetic Antioxidants for Deep Frying Oils. *Food Chem.* **2004,** *84* (2), 219–222.

# PART II

# Pharmacological Aspects of Fruits and Vegetables

## CHAPTER 4

# HEALTH BENEFITS OF ANTHOCYANINS IN BLACK CARROT *(Daucus carota)*

IAHTISHAM-UL-HAQ, MASOOD SADIQ BUTT,
ANEELA SHAMSHAD, and HAFIZ ANSAR RASUL SULERIA

### ABSTRACT

Since dawn of the civilization, diet has played a basic role in maintenance of human health. Due to changing lifestyles and consumption of energy foods alongside sedentary living practices, several life-threatening diseases have prevailed across the globe, which are escalating rate of mortality and morbidity in humans. Nowadays, nutritionists and dietary experts are emphasizing to use functional foods and nutraceuticals in routine meals to manage lifestyle-related disorders. In this milieu, black carrot (*Daucus carota*) holds a firm position among vegetables due to presence of high densities of anthocyanins. Black or purple carrots acquire several functional compounds that may serve as a shield against various degenerative diseases. However, anthocyanins are principal biocomponents among black carrot bioactives. These functional constituents are of primitive significance in medicinal and therapeutic use in various traditional and modern dietary therapies. Anthocyanins and its derivatives have shown favorable health benefits in alleviating oxidative stress, cardiovascular complications, diabetes mellitus, obesity, bone degeneration, and cancer, and various inflammatory conditions. This chapter focuses on the importance of black carrot, its bioactive compounds with special reference to anthocyanins, and disease management.

### 4.1 INTRODUCTION

Plant-based foods loaded with plenty of phytochemical compounds are considerably important to prevent onset of various degenerative disorders.[80]

In this respect, functional and nutraceutical foods not only fulfill nutritional requirements but also provide medicinal benefits. Therapeutic foods and medicine are consumed for similar functions but medicines are not as important as food in healthy persons on daily basis.[84] Changing lifestyles and sedentary living practices, poor cultural habits, and higher medicinal costs have enforced researchers to find out diet-based therapies that are cost effective and safe.[11] Nowadays, functional and nutraceutical foods have become an integral component of our daily diet due to their therapeutic effects against various lifestyle-related disorders such as cardiovascular diseases, inflammatory disorders, hypercholesterolemia, hyperglycemia, cancer, etc.[74]

Carrots (*Daucus carota*) contain plethora of nutritionally imperative biocompounds responsible for various physiological benefits. Among different types of carrots, black carrot holds firm position due to presence of highly important antioxidant moieties, that is, anthocyanins.[5,7] Carrots are among commonly consumed vegetables around the globe. However, the use of black carrot is not as wide as other carrot varieties. Nevertheless, it finds its nutritional and medicinal worth in folk therapies.[41] Carrot is a biennial plant which produces high sugar contents in first year that are stored in the taproots for plant nourishment.

This chapter reviews (1) concepts of functional foods and nutraceuticals, (2) nutritional significance of black carrots, (3) bioactive compounds with special reference to anthocyanins, and (4) disease-preventing perspectives of anthocyanins.

## 4.2 FUNCTIONAL FOODS AND NUTRACEUTICALS

From a last few years, concepts about role of food have been modified from substances that are consumed to mitigate hunger or provide basic nutrition to body cells for growth and nourishment to the substances that boost up health and prevent diseases.[11] In this era of climate change with increasing inhabitants and relative depletion of resources for sustenance, it is important to take actions to improve agricultural techniques to ensure food security.[76] Furthermore, changing lifestyle has modified basic food habits of people around the globe. Escalated utilization of snack foods is a major contributory factor for improper/unbalanced nutrition leading toward several diseases. Among these, heart-related diseases are leading cause of mortality in developing countries followed by arthritis, hypercholesterolemia, cancer, and diabetes.[14] Keeping in view the present conditions, new health strategies are emerging to manage this nutritional dilemma. A quality food has better impact on consumer's health as there is a direct relationship between food and health. In this context, functional foods and nutraceuticals have prospects for food industries not only to develop innovative food products but also to reduce medical care costs and improve revenue for population.[27,83]

Functional foods refer to the foods that help in improving specific body functions in addition to supplying basic nutrients for sustenance,[89] whereas nutraceuticals are isolated/purified compounds from foods that help to maintain normal physiological functions and prevent diseases. Fundamentally, nutraceutical is an amalgam of "nutrition" and "pharmaceuticals." Hence, functional foods and nutraceuticals are foods or component(s) of food having health-promoting and disease-curing perspectives beyond providing basic nutrition to the body. Commonly, functional foods and nutraceuticals are being consumed as whole or in different products including dietary fiber, antioxidants, probiotics, polyunsaturated fatty acids, bioactive components, and numerous other varieties of herbal or natural foods.[81] These have made food industry a research oriented to contemporary epoch of natural medicine with positive health effects derived from edibles. For this purpose, plant-based functional foods are gaining more popularity owing to their larger spectra of therapeutic effects with safer nature.[27]

Globally, increased trend for junk food items has created numerous health problems. Especially, heart diseases are leading cause of mortality among various age groups worldwide followed by arthritis, diabetes, cancer, and osteoporosis that are severely affecting quality of life. Advances in medical science and emerging technologies have introduced latest treatments for these disorders in developed countries. But, these treatments are rather expensive and are not affordable to every person.[101]

Recent scientific efforts have linked food and health as a vital approach to prevent lifestyle-related disorders. Regardless of latest medical technologies, dietary interventions are gaining more attraction owing to their long-term administration consistency and counteracting effects of various pharmaceuticals.[82] In this milieu, functional and nutraceutical foods are considered as more effective and sustainable approach for health management.[27] Although use of food as remedy against various diseases is ancient; however, the conceptual approach regarding functional foods was first used in Japan in 1980s for the foods that provide physiological benefits. In the United States, functional foods have no exact definition because of the complexity of the word "functionality" but it is defined by various organizations.[36] According to Institute of Food Technologists (IFT), "functional foods are foods and their components that provide health benefits for the intended population beyond basic nutrition." Academy of Nutrition and Dietetics explains it as "food that is enriched, fortified or enhanced and provides physiological health benefits when consumed as conventional food at effective level." Moreover, functional foods are available as conventional foods not in the form of dietary supplements, pills, or capsules.[11]

As previously mentioned, nutraceutical is an amalgam of "pharmaceutical" and "nutrition" and was coined in 1989 by the founder of the Foundation for Innovation in Medicine, Stephen De Felice. Nutraceuticals are dietary supplements that provide a specific amount of bioactive ingredient from a food and are available in the form of pills or capsules, having capability to enhance health more than normal foods. These have therapeutic impact and are effective against osteoporosis, arthritis, sleeping disorders, altered cholesterol profile, diabetes, cancer, cold and cough, cardiovascular diseases, and blood pressure, etc.[27]

Plant-based foods have capacity to lower the risk of chronic diseases owing to the presence of biologically active secondary metabolites. These phytochemicals have low potency as compared to pharmaceuticals, though they are regularly ingested in significant amounts that have long-lasting physiological influences. Pharmaceutical drugs are expensive and possess certain side effects while nutraceuticals are safer and less likely to cause harmful effects. However, the problem associated with the production and consumption of nutraceuticals is that natural plant-based active ingredients vary in content and composition as their composition and concentration depend on soil, temperature, season, humidity, climate, and various other factors involved in farming practices of these plants. Thus, use of innovative analytical techniques is mandatory in nutraceuticals research to ensure the critical aspects such as identification, collection, and standardization for uniform quality.[62] These techniques may involve different extraction procedures, such as: capillary electrophoresis (CE), nuclear magnetic resonance (NMR), high-performance liquid chromatography (HPLC), HPLC–NMR, gas chromatography (GC), mass spectrometry (MS), GC-MS, HPLC-MS, and mass chromatography.[18]

In short, functional foods and nutraceuticals are providing a new era of health and medicine-related research consequently, food industry is flourishing in an imperative manner. The main difference between functional foods and nutraceuticals is the format in which they are consumed: functional foods are consumed as conventional or ordinary food items, whereas nutraceuticals are consumed as tablet, capsules, or pills. If a phytochemical or bioactive ingredient is added in a food formulation then it is a functional food but if the same active ingredient or phytochemical is added in a pill or capsule it is considered as nutraceutical.[18] This approach also favors the development of designer foods, which are tailored in such a way that these may supply active ingredients in sufficient quantities to address target disease in an individual. Nevertheless, the spectrum of research in this area is as broader as one can consider that allows food technologists, researchers, and nutritionists to further develop and investigate novel products.

## 4.3 CHEMICAL COMPOSITION OF BLACK CARROTS

Carrots (*D. carota*) belong to the family Apiaceae and is an important food source for human beings.[2] Nature has blessed Pakistan with versatile environment and land for the cultivation of an array of fruits and vegetables throughout the year. Vegetables production in Pakistan is well diversified in terms of range of vegetable varieties grown.[16,40] More than 36 vegetables are grown in various parts of the country both in winter and summer seasons including potato, carrot, chili, onion, and tomato as the major vegetables. In Pakistan, production of fruits and vegetables is expanding due to good economic return received by growers. In this context, the export is also increasing on the yearly basis.[24]

In this context, carrots are fourth major vegetable crop in terms of production in Pakistan. In 2008, its yield was estimated as 236,869 t with a cultivation area of 13,248 ha. In Pakistan, its cultivation is limited in Sheikhupura, Hyderabad, Lahore, and Kasur. The United States, China, and Russia are top carrot-producing countries and contribute about 50% of carrot production in the world.[9] According to Arscott and Tanumihardjo,[9] annual consumption of carrots in the United States was 11.7 pounds per individual in 2005. Prepackaged carrots are also available in market having huge consumption and economic advantages. Carrots are graded sixth among 22 vegetables that are usually consumed in the United States.[44,85]

It has been established that vegetables have capability to reduce the occurrence of cancer, macular degeneration, and cardiovascular diseases owing to the presence of antioxidants especially polyphenols.[1] These antioxidants act as scavenger of free radicals that are produced in human body, having damaging effect on cells and tissues. Plant-based phenolic compounds have antioxidant potential and various vegetables such as carrots, red cabbage, spinach, etc. have phenolic compounds capable of reducing oxidative stress.[85]

Carrots are important to human nutrition and have valuable health-enhancing pigments such as anthocyanins and carotenes that have antioxidant properties to fight against various degenerative disorders. Different carrot varieties have different colors owing to their bioactive pigments. Lycopene is responsible for red carrot color and lutein produces yellow-colored carrots, whereas black or purple carrots have anthocyanins. All these carrot pigments have been endorsed for having disease-preventive perspectives.[9]

Black or purple carrot (*D. carota* ssp. *sativus* var. *atrorubens Alef.*) holds imperative nutritional profile. Carrot cultivars are subdivided into two categories: carotene group or Western carrots (*D. carota* ssp. *sativus* var. *sativus*) and anthocyanin group or Eastern carrots (*D. carota* ssp. *sativus* var. *atrorubens Alef.*). Orange-colored or Western carrots are well known in Western world as compared to black carrots but have lesser stability toward increased pH, light, and heat. Recently, black carrots are gaining popularity owing to their high anthocyanin contents and used as a source of natural colorant in food industries. Because of consumer demand and intensified legal limitations, synthetic or artificial additives are progressively replaced by plant-based natural pigments such as: anthocyanins, carotenes, betalains, and lycopene.[44,57] Black or purple carrot extracts are commonly used in ice-cream, jellies, preserves, nectars, juices, soft drinks, candies, fermented beverages, and confectionery as a healthier substitute to synthetic colors. Because of natural color, it does not need any regulatory abuse.

Black carrot contains 3% fiber. Its root is composed of 88% water, 7% carbohydrates, 3% fiber, 1% protein, and 0.2% fat. The carbohydrate proportion of carrot includes simple sugars: mainly sucrose, fructose, and glucose, and a small amount of starch.[9] Black carrots also have small quantities of nonanthocyanin contents such as phenolic compounds, which produce haze and precipitation in fruit juices. Additionally, it has high antioxidant activity that provides health benefits and prevents chronic diseases.[46,57]

Black carrots have appealing purple color with significant quantity of anthocyanins and are more resistant to pH, heat, and light than orange carrots.[44] Presence of intense purple color due to anthocyanins in outer parts of this vegetable is the main attracting feature of this commodity. Owing to phenolic compounds contributing to high antioxidant potential, black carrots have attracted the attention of scientific community.[7] In South Asian countries, black carrot is used to prepare *Kanji*, which is a famous fermented drink with high acceptability among population. This drink is popular due to its various medicinal benefits.[46] Black carrots also have aromatic volatile compounds in low concentrations, which give it or its products a specific feature. Presence of volatile compounds mainly depends on harvesting, processing, and storage conditions. However, it also correlates with various sensory features such as odor, flavor, and taste.[31] Black carrots do not provide a sufficient number of calories but a significant number of phytochemicals such as anthocyanins and phenolic components. These constituents are promising health improvers in humans *via* dietary intakes. Black carrot anthocyanins are chiefly composed of cyaniding-based pigments. Anthocyanin content of black carrots is acylated with *p*-hydroxybenzoic, *p*-coumaric, ferulic, and sinapic acids; therefore, these show more resistance to pH, light, heat, and hydration. This improved stability, in comparison to other anthocyanin extracts such as grape pomace having low acylation, is dependent on their structural features. Anthocyanin contents ranged up to 1750 mg/kg on fresh weight (FW) basis of black carrots [46,57]. The mineral profile of black carrots also has significance as they contain potassium in abundance (443–758 mg/100 g FW).[9] Furthermore, it also possesses vitamins such as tocopherols and ascorbic acid, which additionally function as antioxidants.[7]

Anthocyanin content of black carrots has been identified by several researchers who have found that anthocyanins of black carrot have cyanidin as an aglycon and the major pigments in it are 3-*O*-(6-*O*-acyl-*â*-D-glucopyranosyl)-(1f6)-[*â*-Dxylopyranosyl-(1f2)]-*â*-D-galactopyranosyl cyanidins acylated with *p*-coumaric, sinapic, or ferulic acid.[72] According to Özen et al.,[59] four main anthocyanins exist in black or purple carrot extracts: 41% of anthocyanins were acylated that acquire cyanidin 3-feruloyl-xylosyl-glucosylgalactoside (13.5%) and cyanidin 3-sinapoyl-xylosylglucosyl-galactoside (27.5%). While Schwarz et al.[72] identified that noncolored phenolic compounds mainly cinnamic acid are also found in black carrots. Among phenolic acids, chlorogenic or 5-caffeoylquinic acid were present in higher concentrations, whereas coumaric, ferulic, and caffeic acids were found in small quantities. As compared to yellow and orange varieties, black carrots have five times more phenolic contents. All these chemical constituents make black carrot nutritionally imperative vegetable having plethora of bioactive moieties capable of improving overall health.

## 4.4 BIOACTIVE COMPOUNDS FROM BLACK CARROTS

Plants provide beneficial components that positively influence human health and reduce the risk of lifestyle-related maladies.[3,10] Based on their abundance in plants, phytonutrients are divided into two groups: major and minor constituents. The major compounds include carbohydrates, proteins, and lipids while minor components are vitamins, minerals, and health-enhancing secondary metabolites.[34] Plant's nonnutritive compounds are categorized into three large groups: phenolics, steroids, and terpenes depending upon their biosynthesis pathways. Various technologies have been developed for the identification and isolation of active components to improve efficacy of food regarding functional and physiological properties.[74]

Naturally occurring phenolics are split up into 15 main structural groups. The main classes of plant-based phenolic compounds involve (Fig. 4.1): C6, resorcinol (simple phenols); C6–C1, *p*-hydroxybenzoic acid (phenolic acid); C6–C2, phenylacetic acids and acetophenones; C6–C3, hydroxycinnamic acid (caffeic acid); C6–C4, hydroxyanthraquinones; C6–C2–C6, resveratrol (stilbenes); C6–C3–C6, flavonoids (quercetin); (C6–C3–C6)2, agathisflavone (biflavonoids); (C6–C3)2, matairesinol (lignans); (C6–C3)n, lignins and (C6–C3–C6)n, and procyanidin (condensed tannins).[8]

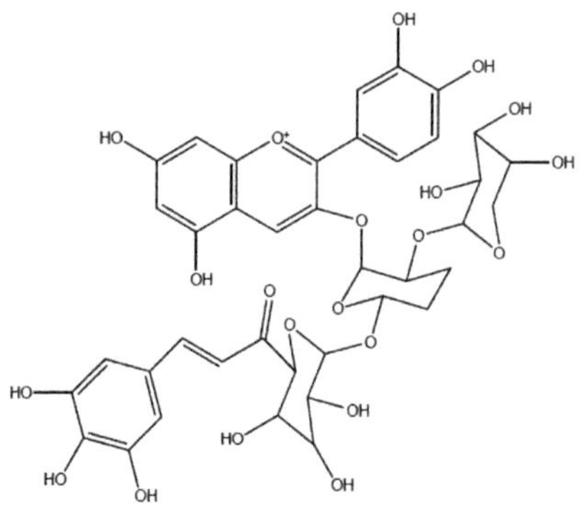

Ferulic acid derivative of pelargonidin 3 -xylosylglucosylgalactoside

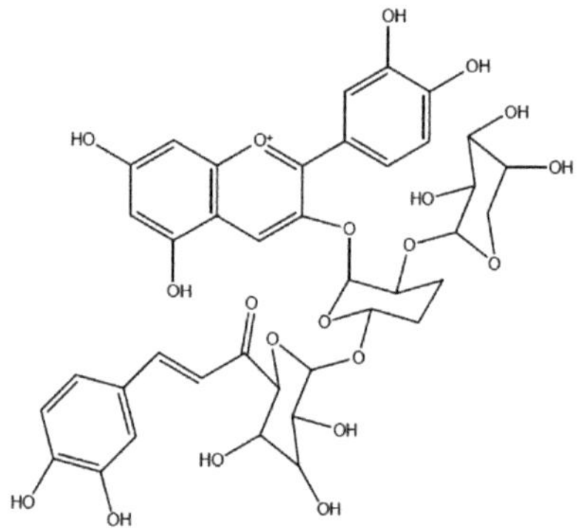

Ferulic acid derivative of peonidin 3-xylosylglucosyl galactoside

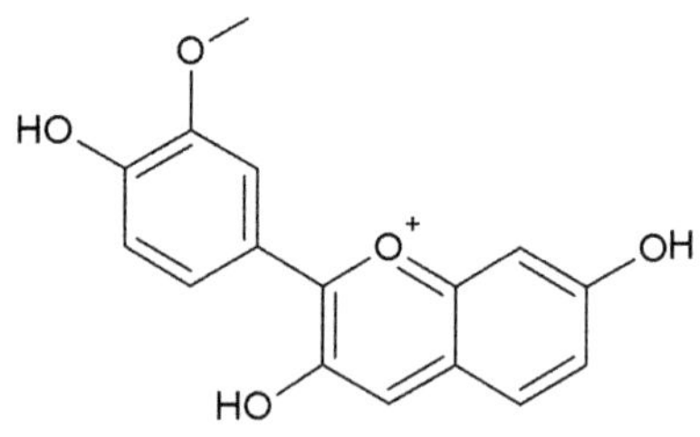

Peonidin

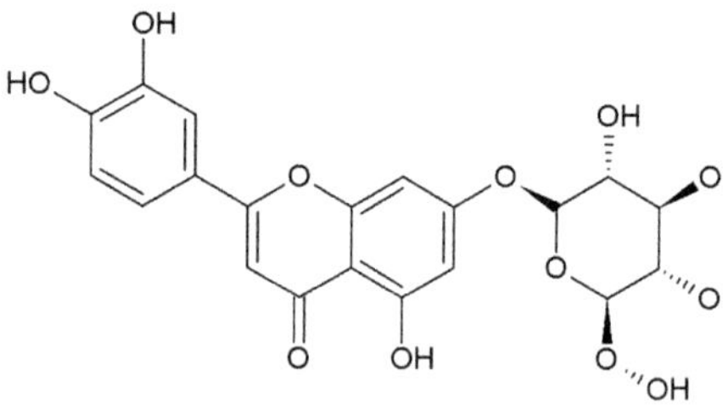

Lutilin

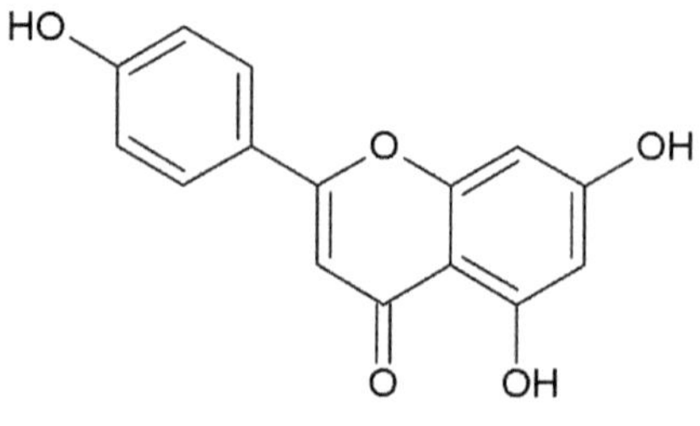

Epigen

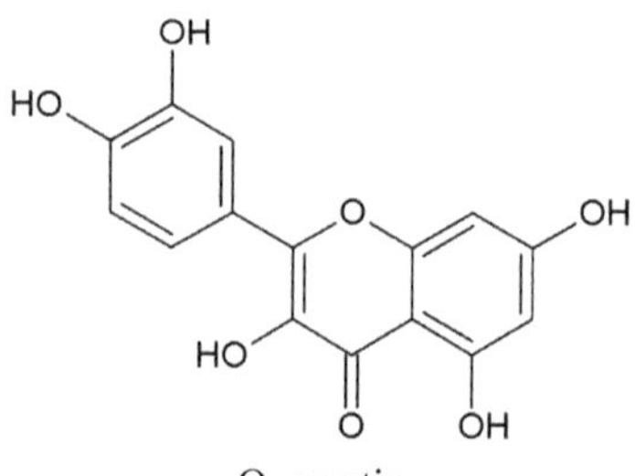

Quercetin

**FIGURE 4.1** *(Continued)*

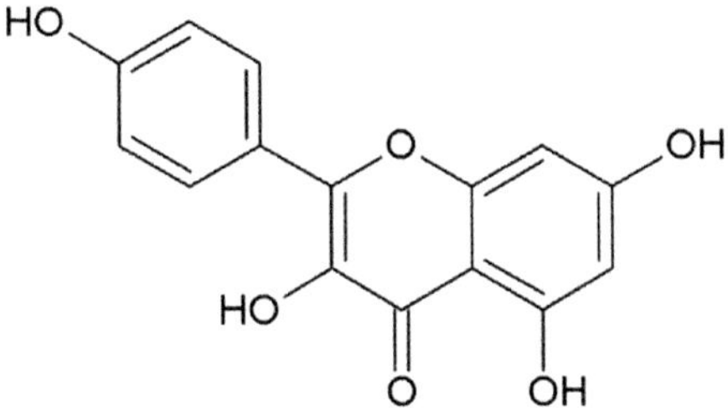

Kaempferol

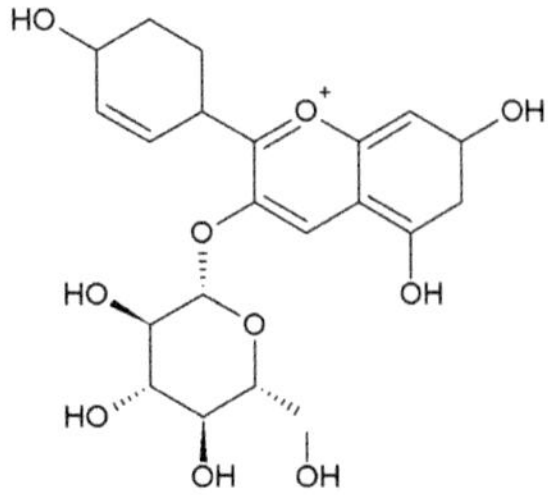

Chlorogenic acid

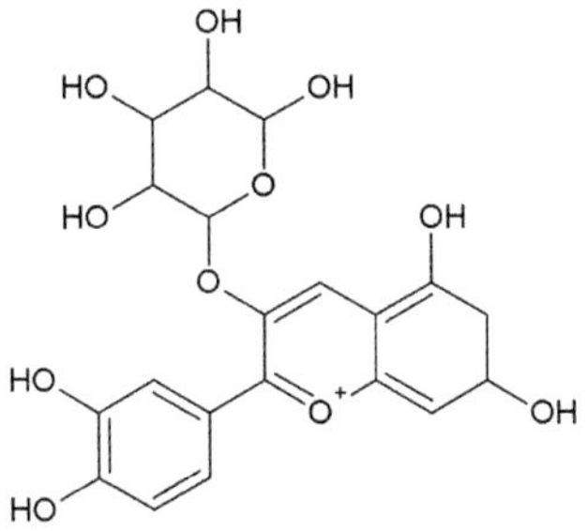

Cyanidin 3-glucoside

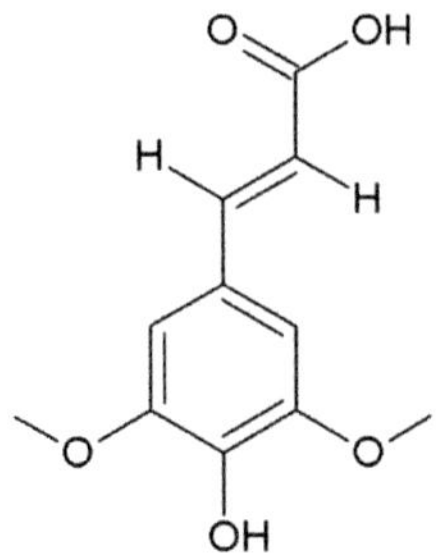

Sinapic acid

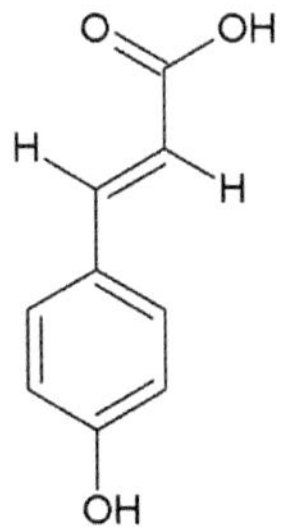

*p*-coumaric acid

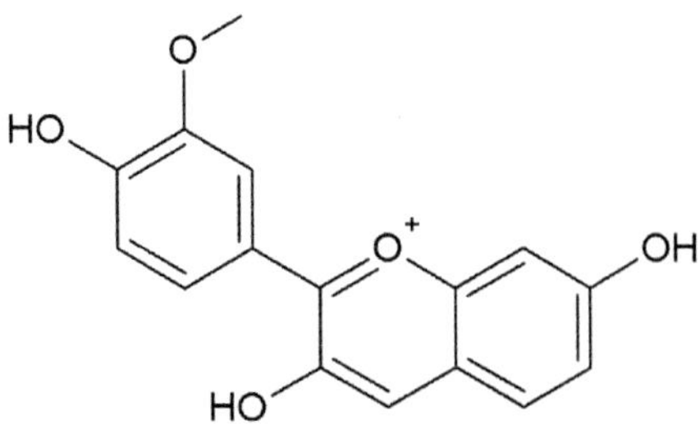

Ferulic acid

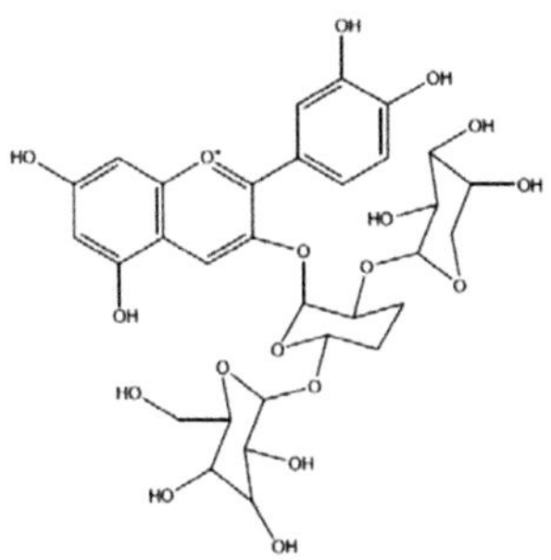

Cyanidin 3-xylosylglucosyl galactoside

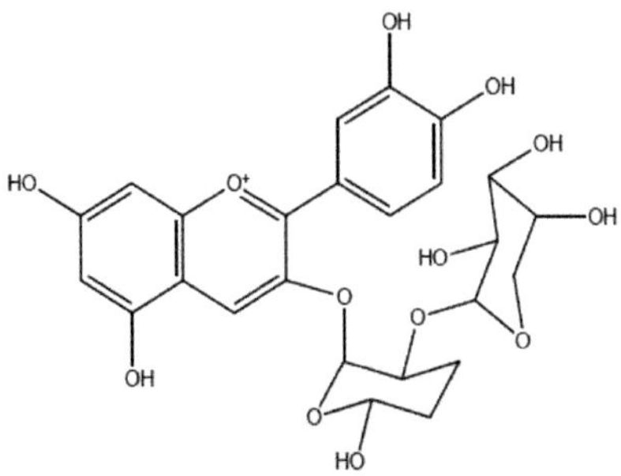

Cyanidin 3-xylosylgalactoside

**FIGURE 4.1** Functional ingredients in black carrot.

**Source:** Adapted from Ref. [3].

The various forms of phenolics include: free, bound, and soluble esters. Mostly, phytochemicals in plant-based sources are available in soluble or liberated forms. The bound forms need to be hydrolyzed in gastrointestinal tract and colon to provide healthier effects. Quantity and quality of phenolic compounds vary in plants due to multiple factors such as agronomic practices, genetics, and environmental and postharvest conditions.[74]

Considering the phytochemical diversity of the black carrots, these have been found quite valuable for medicinal usage. In routine cases, the carrot color indicates the type of chemical compounds it may possess.[3] Investigations have shown that black carrots acquire higher concentrations of phenolics and flavonoids than its red and orange counterparts. Major flavonoids in carrots include: quercetin, kaempferol, luteolin, and myricetin.[15,54] Moreover, every colored carrot predominately contains chlorogenic acid as major compound. However, purple carrots acquire highest chlorogenic acid concentrations of 54.1 mg/100 g.[6,7]

Carotenoids are phytochemicals that consist of 700 compounds and accountable for color of many fruits and vegetables. They are produced in plants and present in the chloroplast of photosynthetic while in the chromoplast of nonphotosynthetic tissues such as fruits and carrot roots. Carotenoids provide photoprotection to plants by reducing energy of singlet oxygen or excited chlorophyll and also attract animals for pollination in flowers. Carrots are rich source of carotenoids and its concentration in dark red carrots is found to be 500 ppm is variable depending on the cultivar, genetic factors, growing season, maturity, and soil. Carotenoids are variably dispersed in carrot roots and highest in phloem or outer root while lower in xylem or core. Carotenoids provide protection from macular degeneration, cancer, and eye diseases. Yellow and white carrots contain lowest level of carotenoids (2–6 ppm FW), orange and dark orange carrots contain 98–160 ppm, whereas purple carrots contain 73–92 ppm on FW basis. Black carrots with white xylem have little amounts of carotenes (4–38 ppm) whereas purple orange carrots contain more concentrations (130 ppm). Researchers have produced a novel variety of carrots, which is purple–orange–red carrot and it contains almost 50 ppm of carotenes and 63 ppm of lycopene. Purple carrots with diversity of pigments contribute to diet and are responsible for various health benefits.[9]

Anthocyanins are soluble in water and provide red, purple, and blue color to numerous fruits and vegetables. These belong to a class of flavonoids and are derivatives of anthocyanidins and flavylium (glycosylated salts of polyhydroxy or polyethoxylated 2-phenylbenzopyrolium, respectively). Naturally, they occur as glycosides of anthocyanidin acylated with aromatic or aliphatic acids. Researchers have indicated that acylated anthocyanins have more stability to heat, light, high pH and hydration, and recommended coloring agents. Acylated anthocyanins impart more appealing color and improved stability in foods as compared to the nonacylated anthocyanins.[93]

Black carrots are rich in five major anthocyanin pigments: (1) comprising three cyanidins acylated with sinapic acid [cyanidin 3-sinapoylxylosylglucosylgalactoside, cyaniding 3-feruloylxylosylglucosylgalactoside (ferulic acid), and cyaniding 3-*p*-coumaroylxylosylglucosylgalactoside (*p*-coumaric acid)] and (2) two nonacylated cyanidins (cyanidin 3-xylosylglucosylgalactoside and cyanidin 3-xylosylgalactoside).[6,7] Additionally, black carrots also contain varying concentrations of acylated anthocyanins ranging between 55% and 99% [44]. Further, acylated anthocyanins have been reported to have 25–50% of the total phenolic compounds in certain varieties of black carrot. Typical to black carrots, cyanidin 3-O-feruloyl-(xylosylglucosylgalactoside) shows 40–80% of the total anthocyanin moieties in purple haze and Antonina varieties of black carrots.[3,7] Certain phytochemical compounds and their concentrations in black carrot are presented in Table 4.1.

**TABLE 4.1** Phytochemical Constituents in Black Carrot.

| Compounds | Concentration (mg/100 g) | References |
|---|---|---|
| Anthocyanins | 64.9–837.9 | [52,95] |
| Bound phenolics | 0.50–13.41 | [47] |
| Falcarindiol 3-acetate | 0.97 | [56] |
| Falcarinol | 1.55 | [56] |
| Flavonols | 51.6 | [52] |
| Flavonoids | 3–111.7 | [47] |
| Free phenolics | 31.95–290.03 | [47] |
| Polyacetylenes | 3.02 | [56] |
| Total phenolics | 7.98–291.48 | [47] |
| Total phenols | 187.8–492 | [7] |

## 4.5 ANTHOCYANINS IN BLACK CARROTS

Anthocyanins are imperative plant-based pigments and observable to unaided human eye. They are water-soluble and belong to a class of phenolic compounds known as flavonoids. Anthocyanins include over 600 compounds that provides dark purple, blue, red, and orange colors to numerous fruits, vegetables, plants, cereals, and flowers. Anthocyanins are the glycosides of polymethoxy and polyhydroxy derivatives of flavylium and 2-phenlybenzopyrylium salts. Anthocyanins exhibit various structures depending on the number of hydroxyl groups, number, nature, and position of sugars attached to the molecule and number, and nature of the acids (aromatic or aliphatic) attached to sugars in the molecule. The sugar moieties attached to molecule are usually mono- and disaccharides, normally glucose, xylose, arabinose, galactose, rutinose, and rhamnose. Anthocyanins can be glycosylated in different positions such as 3-OH, 5-OH, and 7-OH and less commonly in both positions (3,5-*O*-diglycosides). To date, 600 anthocyanins or aglycones have been isolated from plant-based sources; however, they differ in methoxylation and hydroxylation configuration of aromatic ring.[7,9,49]

Anthocyanidins or aglycones are basic structures of anthocyanins and are composed of an aromatic ring (A), which is bound to heterocyclic ring (C) having oxygen, and this C ring is also bound to third aromatic ring (B) by a carbon–carbon bond. When these anthocyanidins are bound to sugar by glycosylation, they are called as anthocyanins.[22] Anthocyanins are polyphenolic compounds and encompass an anthocyanidin backbone that is 2-phenylbenzopyrylium also expressed as flavylium cation. This backbone may vary because of variation in number and position of hydroxyl and methoxyl groups and sugar molecule (which possibly is acylated by aromatic or aliphatic acids).

In higher plants, six anthocyanidins are most common: cyanidin (Cy), delphinidin (Dp), petunidin (Pt), malvidin (Mv), peonidin (Pn), and pelargonidin (Pg). Their distribution in comestible parts of plants includes: cyanidin (50%), peonidin (12%), delphinidin (12%), pelargonidin (12%), malvidin (7%), and petunidin (7%). From these, cyanidin, pelargonidin, and delphinidin having nonmethylated anthocyanidins are well known in nature, as they are present in 50% of flowers, 80% of leaves, and 69% of fruits.[9] Moreover, there are four common classes of anthocyanidin glycosides: 3-monosides, 3-biosides, 3,5-diglycosides, and 3,7-diglycosides. Moreover, 3,5-diglycosides are two and half time less frequent than 3-glycosides. Hence, the most common anthocyanin in nature is cyanidin 3-glucoside.[49]

Black carrots are a good source of anthocyanins chiefly present in acylated form. According to Kammerer et al.,[44] 55–99% of anthocyanins are acylated. Ordinarily, black carrots encompass cyaniding-based anthocyanins containing distinct sugar moieties, which may be nonacylated or acylated with ferulic acid, sinapic acid, or coumaric acid. Black carrots also possess peonidin and pelargonidin glycoside-based anthocyanins.[7] Arscott and Tanumihardjo[9] identified five main anthocyanin derivatives in black carrots: cyanidin-3-(2-xylose-galactoside) (Cy3XG), cyanidin-3-(2-xylose-6-glucose-galactoside) (Cy3XGG), cyanidin-3-(2-xylose-6-feruloyl-glucose-galactoside) (Cy3XFGG), cyanidin-3-(2- xylose-6-sinapoyl-glucose-galactoside) (Cy3XSGG), and cyanidin-3-(2-xylose-6-(4-coumaroyl)glucose-galactoside) (Cy3XCGG). In a study, Algarra et al.[7] detected three other anthocyanins in black carrots in small quantities, which were ferulic acid derivative of peonidin 3-xylosylglucosylgalactoside, peonidin 3-xylosylglucosylgalactoside, and ferulic acid derivative of pelargonidin 3-xylosylglucosylgalactoside. From this exploration, it can be concluded that acylation of anthocyanidin-O-glycosides with aromatic acid exhibits improved color stability. Although black carrots are a good source of anthocyanins, its concentration depends on cultivar, growing conditions, growing season, root-coloring, genotype, and maturity. According to Sun et al.,[85] total anthocyanin content ranged from 0 mg/100 g FW in orange carrots to 350 mg/100 g FW in black carrots. Kammerer et al.[44] narrated that black carrots encompass anthocyanins from 17.4 g/kg to 45.4 mg/kg on dry matter basis. Black carrots are a superior source of anthocyanins as compared to others such as blueberries containing 113 g/100 g; cherries having 117 mg/100 g while raspberries acquiring 48 mg/100 g anthocyanins.[20] Due to diversity of these pigments in various fruits and vegetables, anthocyanins are claimed to be most consumed plant pigment on daily basis in the United States.[96]

Conclusively, anthocyanins are a source of natural colorant and are used in food, beverage, and confectionary industries. Their role in plants is to attract animals for pollination, provide photoprotection, and scavenge free radicals. Anthocyanins pigments have significance as medicinal agents and are exploited in folk medicine throughout the world. Anthocyanins from different sources have been used for alleviation of hypertension, vision disorders, microbial infections, liver dysfunctions, diarrhea, and various other diseases. They play imperative role in health enhancement, prevention of cardiovascular disorders and cancer, boost memory, reduce age-related neural disorders, and act as dietary antioxidants to reduce the inflammation, lipid oxidation, and apoptosis.[9,60]

## 4.6 POTENTIAL HEALTH BENEFITS OF ANTHOCYANINS

Currently, the main focus of food technologists and nutritionists is to classify the essential food nutrients and to explore their health-enhancing potential.[21,53,77] Meanwhile, China and Mediterranean countries are focusing on the exploration of nonnutritive phytochemicals and bioactive molecules for their disease-fighting potentials.[87] People consuming functional and nutraceutical diets have lower risk of illness so they enjoy healthy life for long period of time.[42] Among new generation, cardiovascular complications (i.e., blood pressure, heart attack, and other such ailments) are spreading at alarming rates. The major causes of these life threatening ailments include unhealthy lifestyle, poor eating habits, and higher consumption of junk foods.[66]

In recent era, anthocyanins have grabbed significant attention among the scientific communities as a promising agent against various disorders involving cardiovascular complications, obesity, diabetes, inflammatory and oncogenic events, bone degeneration alongside allergic, as well as mutagenic conditions.[32,65,67] Anthocyanins from various plant products are consumed in fairly large quantities as compared to other phytochemicals. These are then absorbed and metabolized to obtain several health effects. Furthermore, the concentration and individual component of anthocyanins are variable in different fruits. The anthocyanins are absorbed in the body in intact form and then metabolized. The information is limited in humans regarding the metabolites of these compounds; however, various trials have been conducted using animal models. The ameliorative potential of anthocyanins has been devised in response to lowered oxidative stress and its biomarkers in animal-based trials. It has been anticipated that the beneficial health effects are observed while consuming relatively higher concentration (2–400 mg/Kg BW) of anthocyanins.[63] It has been estimated that average anthocyanin consumption in the United States is 215 mg on daily basis during summer compared to 180 mg in winter.[25] The potential health benefits against various physiological dysfunctions have briefly been discussed in this section.

### 4.6.1 OXIDATIVE STRESS

Antioxidant potential is defined as an ability to scavenge free radicals, chelate metal ions and delay action of enzymes involved in oxidation. The antioxidants are categorized into two classes such as primary and secondary antioxidants. The primary antioxidants break the initiation and propagation of free radicals' chain. In contrast, secondary antioxidants slow down the oxidation reactions. Resultantly, antioxidant potential of any substance is explained by its capability to suppress, sequester, and retard the free radicals, metal ions, and enzymatic reactions, respectively, involved in the process of oxidation.

The antioxidant potential to scavenge free radicals is effected by two factors: concentration and reactivity. Previous studies have revealed that foods rich in polyphenols exhibit strong antioxidant potential both in vivo and in vitro. In this respect, functional and nutraceutical ingredients present in black carrots especially anthocyanins and other phytochemicals possess promising feature in alleviating in vitro and in vivo oxidative stress.[7,23] The antioxidant capacity of black carrots has been identified as 182.0 μM TE/100 g and 240 μM TE/100 g FW for ferric-ion-reducing antioxidant capacity (FRAP) and 1, 1-diphenyl-2-picrylhydrazyl (DPPH) assays, respectively. Furthermore, 2, 2'-azino-bis (3-ethylbenzothiazoline-6-sulphonic acid (ABTS) assay revealed 73–285 μM TE/g dry weight radicals scavenging activities by black carrot.[85]

Hypergeneration of reactive species involving oxidizing agents from various sources may interact with routine physiological processes and cause oxidative stress that must be managed to acquire normal health of an individual.[38] In this respect, anthocyanins demonstrated significant antioxidant abilities that control oxidative damage caused by reactive species.[69] The antioxidative benefits of anthocyanins may be received either directly or indirectly. In case of direct antioxidant effect, the anthocyanins may show their positive feature by scavenging free radicals due to capacity of anthocyanins structure to donate electron/hydrogen, whereas, indirect effects may involve restoration of superoxide dismutase and glutathione peroxidase activities,[90] upregulation of the genes for these enzymes[75] and decreasing the DNA abducts, hence lowering the production of reactive species during metabolism.[78] However, it is not necessary that all the anthocyanins show similar degree of antioxidant abilities.

The antioxidant power of anthocyanins depends on their chemical structure.[30,69] The phenol group in anthocyanins structure is believed to be responsible for its antioxidant effects, whereas, the number and position of hydroxyl and methyl groups on B ring is considered to be associated with antiproliferative perspectives.[97] The anthocyanins and their derivatives are promising agents for neutralizing free radicals via hydrogen donation and these have been found to acquire higher antioxidant abilities than ascorbic acid, butylated hydroxyl toluene (BHT) and butylated hydroxyl anisole (BHA), and vitamin E.[69] The potential H-donating abilities of anthocyanins are linked with positively charged O-atoms.[49] Such a high antioxidant abilities of anthocyanins may protect a number of degenerative disorders arising from oxidative damages.[88]

### 4.6.2 CARDIOVASCULAR PROTECTION

Primarily, low-density lipoproteins (LDL) trigger the accretion of white blood cell macrophages in blood vessels upon oxidation. The oxidized cholesterol deposits in the form of plague in arteries leading to cardiovascular diseases due to atherosclerosis.[12,13] Other possible factors that may compromise cardiovascular health include hypertension, vascular endothelium dysfunction, and platelet aggregation.[69] It has been noted that dietary antioxidants increase antioxidant potential in serum, hence avoid LDL oxidation thereby reducing the risk of cardiovascular diseases.[38]

Ziberna et al.[103] noticed that anthocyanins lower oxidative stress in vascular endothelium alongside protecting cardiac health under ischemia-perfusion conditions. Earlier, it was proved that anthocyanins protect against vascular endothelium degeneration due to oxidative stress. These moieties are incorporated in the cell cytosol thereby restrict the damages arising from free radical moieties hence, halting the risk of cardiac ailments.[99] The radical scavenging abilities of anthocyanins may be affected by anaerobic metabolism that substantially decrease the intercellular pH under reperfusion conditions.[26]

In various animal model studies, it has been seen that anthocyanins from fruits positively manage the cardiac damages that signifies their heart-protecting perspectives.[19,35,90] As anthocyanins are most widely consumed flavonoid entities, their significance in dietary regimen is of primitive importance in controlling cardiac complications and diseases.[69] Furthermore, cardio-protective effects of anthocyanins have also been endorsed in various investigations.[28,37,48,55]

### 4.6.3 OBESITY MANAGEMENT

Anthocyanins-rich foods have been observed to tackle the dilemma of obesity. Mechanistically, anthocyanins upregulate adiponectin which is an important adipocytokines indicative of lowered risk of obesity as its concentration have been found to be lowered in case of obesity prevalence.[70]

Tsuda[91] reported that animals reared on anthocyanins-containing corn showed prevention to obesity. Similarly, black soybean and blueberry anthocyanins also depicted antiobesity effects. Alongside, this treatment prevented from increased cholesterol and triglycerides while, improved high-density lipoprotein (HDL) levels.[64] Tsuda et al.[92] reported that purified anthocyanins showed antiobesity response while whole blueberry consumption promoted obesity biomarkers. Cyanidin-3-glucoside upregulated 633, whereas cyanidin showed upregulation of 427 genes in the total RNA recovered from adipocytes which demonstrated enhancement of hormone-sensitive lipase alongside lipolytic activities as confirmed from their gene expression assays.[92] Furthermore, anthocyanins have also been observed to attenuate insulin resistance conditions arising due to high-fat diet-induced obesity.[29] Another investigation showed that anthocyanin treatment retarded obesity development by lowering serum leptin concentrations.[64] Hence, dietary regimen can be devised using anthocyanins-rich foods such as black carrots to manage obesity biomarkers.

### 4.6.4 *DIABETES MELLITUS*

Diabetes mellitus must be managed through dietary regimen and lifestyle practices to overcome this worse scenario.[70] Anthocyanins have been endorsed by various researchers to prevent hyperglycemia by improving glucose metabolism and insulin sensitivity, lowering blood glucose levels, and hindering production of excessive free radicals.[33,86] Interestingly, healthy humans showed no change in the blood glucose and serum insulin status in this respect.[17] Additionally, anthocyanins showed antioxidant abilities which is important in prevention of glucose-induced oxidative stress in β cells under type-2 diabetes conditions.[4]

Sugimoto et al.[79] stated that anthocyanins represent significant antioxidant potential that is vital in controlling oxidative damage in streptozotocin-induced diabetic conditions in rats. Similarly, Grace et al.[33] examined that anthocyanin fractions are more active in reducing type 2 diabetes than consumption of whole fruit. This study implies that hypoglycemic effect of anthocyanins may be attributed to specific fractions of these moieties.[33] However, the consumption of anthocyanins-rich fruits and vegetables may reduce the risk to develop diabetes and its associated malfunctions.[70]

### 4.6.5 *BONE HEALTH*

In elderly people, the osteoporosis is characterized by decreased mineral content of the bones especially calcium. This condition may arise due to hormonal imbalances, oxidative stress, and chronic inflammatory conditions in bones. In this respect, fruits and vegetable with significant antioxidant abilities may reduce oxidative stress-mediated bone damages.[39] Anthocyanins have been thought to possess significant impact on bone health due to its vital antioxidant abilities. It has been seen that consumption of anthocyanins-rich food counteracts oxidative damage caused by free radical species ultimately decreasing bone resorption.[69]

Scientific reports depicted that women who consumed larger portions of fruit in their diet during their childhood possessed denser bones compared to normal. Alongside, individuals of either gender having higher quantities of fruits, vegetables, and grains in their diet represented high bone mineral densities than normal subjects.[39,58] High bone mineral densities have also been observed in women habitually consuming flavonoids and anthocyanins.[94] Besides, frequent grains, fruits, and vegetables consumption has also linked with decreased risk of bone fracture in a study conducted by Langsetmo et al.[51] All these studies imperatively support the positive impact of anthocyanins and other phytochemicals from fruits and vegetables on bone health.

### *4.6.6 CANCER AND INFLAMMATORY CONDITIONS*

In recent era, cancer and inflammatory conditions in various organs of the body are prevailing at larger extents due to poor dietary patterns and unbalanced lifestyle issues. Additionally, genetic factors and metabolic disorders also play vital role in prevalence of inflammatory diseases. Furthermore, cancer is among principal reasons of mortality around the globe, presently. Demographical data showed that one-third of global population prevails cancer, whereas 20% deaths are due to this serious medical problem nowadays.[68] Since dawn of civilization, dietary therapies have been used to avoid various maladies including cancer. In this respect, fruits and vegetables rich in bioactive compounds have been endorsed to avoid onset of cancer issues. Hence, dietary regimen has grabbed global attention to address these metabolic malfunctions by reducing the chances of their occurrence.[100]

Cancer prevention through the use of anthocyanins takes place by various biological functions. Anthocyanins may exert beneficial effects either by (1) arresting cell cycle, (2) induction of apoptosis and antiangiogenesis, (3) preventing oxidative damage to DNA, (4) upregulation of phase II detoxification enzymes, and (5) restricting cyclooxygenase-2 (COX-2) enzymes.[68] Anthocyanins are better anticarcinogenic agents than flavonoids while anthocyanidins are more potent compounds for inhibition of cell proliferation.[102] Furthermore, anthocyanins fractions have also exhibited better antiproliferative activities than crude phenolic extracts obtained from grapes.[98] It has further been seen that anthocyanins exhibit structure-dependent antiproliferative effect.[43] Alongside, they also represented chemoprotective effects in colon cancer cells.[50] Similarly, antiproliferative influences of anthocyanins have been documented in human cancer cells from malignant tissues[102] and leukemia cells.[45]

Inflammation is involved in progression and instigation of different degenerative disorders. It is a biological response caused by injuries, irritants and other stimulants in vascular tissues. It may be associated with other diseases such as diabetes, cardiovascular ailments, and certain cancers. In this respect, the agents having anti-inflammatory activities may also lessen the risk to develop diseases. A number of scientific reports have confirmed anti-inflammatory effects of anthocyanins and its fractions.[68] Seeram et al.[73] reported that anthocyanins fractions from cherries and berries acquire potent anti-inflammatory activities attributed to cyanidin glycosides. In another investigation, dose-dependent reduction in inflammatory biomarkers was observed in rats under lung inflammation.[71]

Like anticancer activities, anthocyanins also depict structure-dependent anti-inflammatory responses. These compounds suppress nuclear factor-κB by down regulating mitogen-activated protein kinase pathways which are promising in reducing pro-inflammatory cytokines.[61] Hence, anthocyanins-rich fruits and vegetables are quite helpful in reducing the chances of carcinogenesis and inflammatory conditions arising from poor dietary habits and lifestyle practices.

## 4.7 SUMMARY

Black carrots are promising source of dietary antioxidants especially anthocyanins. These bioactive moieties possess significant antioxidant abilities and disease-preventing perspectives. The dietary regimen using anthocyanins-rich fruits and vegetables should be encouraged in routine diet to manage the menace of oxidative stress-induced malfunctions. However, the dietary therapies should be critically designed and conducted as anthocyanins and their fractions are structure dependent in their specific actions. In this respect, precisely modeled trials with clinical setups should be implanted to design and develop novel functional foods and nutraceutics. In future, these designer foods and nutraceutical supplements are expected to capture a huge market share due to consumer's concerns about health and relative medical costs. Nevertheless, further research in functional and nutraceutical foods will help in alleviating overall mortality and morbidity in global communities.

## KEYWORDS


- **black carrot**
- **functional foods**
- **antioxidants**
- **anthocyanins**
- **carotenoids**
- **cyanidin 3-glycoside**
- **metabolic disorders**


## REFERENCES


1. Abbas, M.; Saeed, F.; Anjum, F. M.; Afzaal, M.; Tufail, T.; Bashir, M. S.; Ishtiaq, A.; Hussain, S.; Suleria, H. A. R. Natural Polyphenols: An Overview. *Int. J. Food Prop.* **2017,** *20* (8), 1689–1699.
2. Ahmad, T.; Amjad, M.; Nawaz, A.; Iqbal, Q.; Iqbal, J. Socio-economic Study of Carrot Cultivation at Farm Level in the Punjab Province of Pakistan. *African J. Agr. Res.* **2012,** *7* (6), 867–875.
3. Akhtar, S.; Rauf, A.; Imran, M.; Saleem, Q.; Riaz, M.; Mubarak, M. S. Black Carrot (*Daucus carota* L.), Dietary and Health Promoting Perspectives of Its Polyphenols: A Review. *Trends Food Sci. Technol.* **2017,** *66,* 36–47
4. Al-Awwadi, N. A.; Araiz, C.; Bornet, A.; Delbosc, S.; Cristol, J. P.; Linck, N.; Azay, J.; Teissedre, P. L.; Cros, G. Extracts Enriched in Different Polyphenolic Families Normalize Increased Cardiac NADPH Oxidase Expression While Having Differential Effects on Insulin Resistance, Hypertension, and Cardiac Hypertrophy in High-fructose-fed Rats. *J. Agric. Food Chem.* **2005,** *53* (1), 151–157.
5. Al-Dabbas, M. M.; Saleh, M. I.; Al-Ismail, K. Preservation Methods Impacted Phenolic, Flavonoid and Carotenoid Contents and Antioxidant Activities of Carrots (*Daucus carota* L.). *J. Food Process. Preserv.* **2015,** *39* (6), 1618–1625.
6. Alasalvar, C.; Grigor, J. M.; Zhang, D.; Quantick, P. C.; Shahidi, F. Comparison of Volatiles, Phenolics, Sugars, Antioxidant Vitamins, and Sensory Quality of Different Colored Carrot Varieties. *J. Agr. Food Chem.* **2001,** *49* (3), 1410–1416.
7. Algarra, M.; Fernandes, A.; Mateus, N.; De Freitas, V.; Da Silva, J. C. E.; Casado, J. Anthocyanin Profile and Antioxidant Capacity of Black Carrots (*Daucus carota* L. Ssp. *Sativus* Var. *Atrorubens Alef.*) from Cuevas Bajas, Spain. *J. Food Compos. Anal.* **2014,** *33* (1), 71–76.
8. Apak, R.; Güçlü, K.; Demirata, B.; Özyürek, M.; Çelik, S. E.; Bektaşoğlu, B.; Berker, K. I.; Özyurt, D. Comparative Evaluation of Various Total Antioxidant Capacity Assays Applied to Phenolic Compounds with the CUPRAC assay. *Molecules* **2007,** *12* (7), 1496–1547.

9. Arscott, S. A.; Tanumihardjo, S. A. Carrots of Many Colors Provide Basic Nutrition and Bioavailable Phytochemicals Acting as a Functional Food. *Compr. Rev. Food Sci. Food Saf.* **2010,** *9* (2), 223–239.
10. Aruoma, O. I. The impact of Food Regulation on the Food Supply Chain. *Toxicology* **2006,** *221* (1), 119–127.
11. Aryee, A. N.; Boye, J. I. Current and Emerging Trends in the Formulation and Manufacture of Nutraceuticals and Functional Food Products. *Nutraceut. Function. Food Process. Technol.* **2015,** 1–52.
12. Aviram, M. Review of Human Studies on Oxidative Damage and Antioxidant Protection Related to Cardiovascular Diseases. *Free Rad. Res.* **2000,** *33*, 85–97.
13. Aviram, M.; Kaplan, M.; Rosenblat, M.; Fuhrman, B. Dietary Antioxidants and Paraoxonases Against LDL Oxidation and Atherosclerosis Development. *Atheroscler. Diet Drugs* Springer; 2005; pp 263–300
14. Awan, K. A.; Butt, M. S.; Iahtisham-Ul-Haq; Ashfaq, F.; Suleria, H. A. R. Storage Stability of Garlic Fortified Chicken Bites. *J. Food Chem. Nanotechnol.* **2017,** *3* (2), 80–85.
15. Bahorun, T.; Luximon-Ramma, A.; Crozier, A.; Aruoma, O. I. Total Phenol, Flavonoid, Proanthocyanidin and Vitamin C Levels and Antioxidant Activities of Mauritian Vegetables. *J. Sci. Food Agric.* **2004,** *84* (12), 1553–1561.
16. Bakhsh, K.; Ashfaq, M.; Alam, M. W. Effects of Poor Quality of Ground Water on Carrot Production: A Comparative Study. *J. Agric. Soc. Sci.* **2005,** *8* (11), 15–78.
17. Basu, A.; Penugonda, K. Pomegranate Juice: A Heart-healthy Fruit Juice. *Nutr. Rev.* **2009,** *67* (1), 49–56.
18. Bernal, J.; Mendiola, J.; Ibáñez, E.; Cifuentes, A. Advanced Analysis of Nutraceuticals. *J. Pharmaceut. Biomed. Anal.* **2011,** *55* (4), 758–774.
19. Bertuglia, S.; Malandrino, S.; Colantuoni, A. Effect of Vaccinium Myrtillus Anthocyanosides on Ischaemia Reperfusion Injury in Hamster Cheek Pouch Microcirculation. *Pharmacol. Res.* **1995,** *31* (3), 183–187.
20. Bhagwat, S.; Haytowitz, D. B.; Holden, J. M. *USDA Database for the Flavonoid Content of Selected Foods, Release 3.1.* US Department of Agriculture; Beltsville, MD; 2011 (Part 3–1).
21. Butt, M.; Sultan, M.; Anjum, F.; Sharif, M.; Khan, S. Functional Foods: Emerging Trend in Nutritional Support Programs. *International Symposium on Emerging Trends in Food Science and Technology at National Institute of Food Science and Technology.* University of Agriculture, Faisalabad, Pakistan; November 5–8, 2007; pp 5–8.
22. Castañeda-Ovando, A.; De Lourdes Pacheco-Hernández, M.; Páez-Hernández, M. E.; Rodríguez, J. A.; Galán-Vidal, C. A. Chemical Studies of Anthocyanins: A Review. *Food Chem.* **2009,** *113* (4), 859–871.
23. Chatatikun, M.; Chiabchalard, A. Phytochemical Screening and Free Radical Scavenging Activities of Orange Baby Carrot and Carrot (*Daucus carota* Linn.) Root Crude Extracts. *J. Chem. Pharm. Res.* **2013,** *5* (4), 97–102.
24. Chaudhry, M.; Ahmad, B. *Dynamics of Vegetable Production, Distribution and Consumption in Pakistan.* Asian Vegetable Research and Development Center, Tainan, Taiwan; 2000; p 112.
25. Clifford, M. N. Anthocyanins–nature, Occurrence and Dietary Burden. *J. Sci. Food Agric.* **2000,** *80* (7), 1063–1072.
26. Cvorovic, J.; Tramer, F.; Granzotto, M.; Candussio, L.; Decorti, G.; Passamonti, S. Oxidative Stress-based Cytotoxicity of Delphinidin and Cyanidin in Colon Cancer Cells. *Arch. Biochem. Biophys.* **2010,** *501* (1), 151–157.

27. Das, L.; Bhaumik, E.; Raychaudhuri, U.; Chakraborty, R. Role of Nutraceuticals in Human Health. *J. Food Sci. Technol.* **2012,** *49* (2), 173–183.
28. De Pascual-Teresa, S.; Moreno, D. A.; García-Viguera, C. Flavanols and Anthocyanins in Cardiovascular Health: A Review of Current Evidence. *Int. J. Mol. Sci.* **2010,** *11* (4), 1679–1703.
29. Defuria, J.; Bennett, G.; Strissel, K. J.; Perfield, J. W.; Milbury, P. E.; Greenberg, A. S.; Obin, M. S. Dietary Blueberry Attenuates Whole-body Insulin Resistance in High Fat-fed Mice by Reducing Adipocyte Death and Its Inflammatory Sequelae. *J. Nutr.* **2009,** *139* (8), 1510–1516.
30. Ferrars, R.; Czank, C.; Zhang, Q.; Botting, N.; Kroon, P.; Cassidy, A.; Kay, C. The Pharmacokinetics of Anthocyanins and Their Metabolites in Humans. *Br. J. Pharmacol.* **2014,** *171* (13), 3268–3282.
31. Gamboa-Santos, J.; Soria, A. C.; Pérez-Mateos, M.; Carrasco, J. A.; Montilla, A.; Villamiel, M. Vitamin C Content and Sensorial Properties of Dehydrated Carrots Blanched Conventionally or by Ultrasound. *Food Chem.* **2013,** *136* (2), 782–788.
32. Ghosh, D.; Konishi, T. Anthocyanins and Anthocyanin-rich Extracts: Role in Diabetes and Eye Function. *Asia Pacific J. Clin. Nutr.* **2007,** *16* (2), 200–208.
33. Grace, M. H.; Ribnicky, D. M.; Kuhn, P.; Poulev, A.; Logendra, S.; Yousef, G. G.; Raskin, I.; Lila, M. A. Hypoglycemic Activity of a Novel Anthocyanin-rich Formulation from Lowbush Blueberry (*Vaccinium angustifolium*) Aiton. *Phytomedicine* **2009,** *16* (5), 406–415.
34. Grusak, M. A. Phytochemicals in Plants: Genomics-assisted Plant Improvement for Nutritional and Health Benefits. *Curr. Opin. Biotechnol.* **2002,** *13* (5), 508–511.
35. Halliwell, B. Are Polyphenols Antioxidants or Pro-oxidants? What Do We Learn from Cell Culture and In Vivo Studies? *Arch. Biochem. Biophys.* **2008,** *476* (2), 107–112.
36. Hasler, C. M. Functional Foods: Benefits, Concerns and Challenges: A Position Paper from the American Council on Science and Health. *J. Nutr.* **2002,** *132* (12), 3772–3781.
37. Hassellund, S.; Flaa, A.; Kjeldsen, S.; Seljeflot, I.; Karlsen, A.; Erlund, I.; Rostrup, M. Effects of Anthocyanins on Cardiovascular Risk Factors and Inflammation in Pre-hypertensive Men: A Double-blind Randomized Placebo-controlled Crossover Study. *J. Hum. Hypertens.* **2013,** *27* (2), 100.
38. He, J.; Giusti, M. M. Anthocyanins: Natural Colorants with Health-promoting Properties. *Ann. Rev. Food Sci. Technol.* **2010,** *1*, 163–187.
39. Hubert, P. A.; Lee, S. G.; Lee, S.-K.; Chun, O. K. Dietary Polyphenols, Berries, and Age-related Bone Loss: A Review Based on Human, Aanimal, and Cell Studies. *Antioxidants* **2014,** *3* (1), 144–158.
40. Iahtisham-Ul-Haq; Butt, M. S.; Suleria, H. A. R.; Ahmed, W. Physicochemical Behavior of Zinc-fortified, Sodium Caseinate-based, Edible-coated Apricots During Storage in Controlled Atmosphere. *J. Food Process. Preserv.* **2015,** *39* (6), 2431–2441.
41. Iorizzo, M.; Senalik, D. A.; Ellison, S. L.; Grzebelus, D.; Cavagnaro, P. F.; Allender, C.; Brunet, J.; Spooner, D. M.; Van Deynze, A.; Simon, P. W. Genetic Structure and Domestication of Carrot (*Daucus carota* Subsp. *Sativus*). *Am. J. Bot.* **2013,** *100* (5), 930–938.
42. Jenkins, D. J.; Kendall, C. W.; Nguyen, T. H.; Marchie, A.; Faulkner, D. A.; Ireland, C.; Josse, A. R.; Vidgen, E.; Trautwein, E. A.; Lapsley, K. G. Effect of Plant Sterols in Combination with Other Cholesterol-lowering Foods. *Metabolism* **2008,** *57* (1), 130–139.

43. Jing, P.; Bomser, J. A.; Schwartz, S. J.; He, J.; Magnuson, B. A.; Giusti, M. M. N. Structure-function Relationships of Anthocyanins from Various Anthocyanin-rich Extracts on the Inhibition of Colon Cancer Cell Growth. *J. Agric. Food Chem.* **2008,** *56* (20), 9391–9398.
44. Kammerer, D.; Carle, R.; Schieber, A. Quantification of Anthocyanins in Black Carrot (*Daucus carota* Ssp. *Sativus* Var. *Atrorubens Alef.*) Extracts and Evaluation of Their Color Properties. *Eur. Food Res. Technol.* **2004,** *219* (5), 479–486.
45. Katsube, N.; Iwashita, K.; Tsushida, T.; Yamaki, K.; Kobori, M. Induction of Apoptosis in Cancer Cells by Bilberry (*Vaccinium myrtillus*) and the Anthocyanins. *J. Agric. Food Chem.* **2003,** *51* (1), 68–75.
46. Khandare, V.; Walia, S.; Singh, M.; Kaur, C. Black Carrot (*Daucus carota* Ssp. *Sativus*) Juice: Processing Effects on Antioxidant Composition and Color. *Food Bioproducts Process.* **2011,** *89* (4), 482–486.
47. Koley, T. K.; Singh, S.; Khemariya, P.; Sarkar, A.; Kaur, C.; Chaurasia, S.; Naik, P. Evaluation of Bioactive Properties of Indian Carrot (*Daucus carota* L.): Chemometric approach. *Food Res. Int.* **2014,** *60,* 76–85.
48. Konczak, I.; Zhang, W. Anthocyanins - More Than Nature's Colors. *Biomed Res. Int.* **2004,** *2004* (5), 239–240.
49. Kong, J. M.; Chia, L. S.; Goh, N. K.; Chia, T. F.; Brouillard, R. Analysis and Biological Activities of Anthocyanins. *Phytochemistry* **2003,** *64* (5), 923–933.
50. Lala, G.; Malik, M.; Zhao, C.; He, J.; Kwon, Y.; Giusti, M. M.; Magnuson, B. A. Anthocyanin-rich Extracts Inhibit Multiple Biomarkers of Colon Cancer in Rats. *Nutrition Cancer* **2006,** *54* (1), 84–93.
51. Langsetmo, L.; Hanley, D. A.; Prior, J. C.; Barr, S. I.; Anastassiades, T.; Towheed, T.; Goltzman, D.; Morin, S.; Poliquin, S.; Kreiger, N. Dietary Patterns and Incident Low-trauma Fractures in Postmenopausal Women and Men Aged ≥ 50 Y: A Population-based Cohort Study. *Am. J. Clin. Nutr.* **2011,** *93* (1), 192–199.
52. Leja, M.; Kamińska, I.; Kramer, M.; Maksylewicz-Kaul, A.; Kammerer, D.; Carle, R.; Baranski, R. The Content of Phenolic Compounds and Radical Scavenging Activity Varies with Carrot Origin and Root Color. *Plant Foods Hum. Nutr. (Dordrecht, Netherlands)* **2013,** *68* (2), 163.
53. Liu, H.; Xu, X.; Guo, S. D. Rheological, Texture and Sensory Properties of Low-fat Mayonnaise with Different Fat Mimetics. *LWT-Food Sci. Technol.* **2007,** *40* (6), 946–954.
54. Marinova, D.; Ribarova, F.; Atanassova, M. Total Phenolics and Total Flavonoids in Bulgarian Fruits and Vegetables. *J. Univ. Chem. Technol. Metall.* **2005,** *40* (3), 255–260.
55. Mazza, G. Anthocyanins and Heart Health. *Annali-Istituto Superiore di Sanita* **2007,** *43* (4), 369.
56. Metzger, B. T.; Barnes, D. M. Polyacetylene Diversity and Bioactivity in Orange Market and Locally Grown Colored Carrots (*Daucus carota* L.). *J. Agric. Food Chem.* **2009,** *57* (23), 11134–11139.
57. Montilla, E. C.; Arzaba, M. R.; Hillebrand, S.; Winterhalter, P. Anthocyanin Composition of Black Carrot (*Daucus carota* Ssp. *Sativus* Var. *Atrorubens alef.*) Cultivars Antonina, Beta Sweet, Deep Purple, and Purple Haze. *J. Agric. Food Chem.* **2011,** *59* (7), 3385–3390.
58. New, S. A.; Robins, S. P.; Campbell, M. K.; Martin, J. C.; Garton, M. J.; Bolton-Smith, C.; Grubb, D. A.; Lee, S. J.; Reid, D. M. Dietary Influences on Bone Mass and Bone Metabolism: Further Evidence of a Positive Link Between Fruit and Vegetable Consumption and Bone Health? *Am. J. Clin. Nutr.* **2000,** *71* (1), 142–151.

59. Özen, G.; Akbulut, M.; Artik, N. Stability of Black Carrot Anthocyanins in the Turkish Delight (Lokum) During Storage. *J. Food Process Eng.* **2011,** *34* (4), 1282–1297.
60. Ozgen, M.; Reese, R. N.; Tulio, A. Z.; Scheerens, J. C.; Miller, A. R. Modified 2, 2-Azino-Bis-3-Ethylbenzothiazoline-6-Sulfonic Acid (ABTS) Method to Measure Antioxidant Capacity of Selected Small Fruits and Comparison to Ferric Reducing Antioxidant Power (FRAP) and 2, 2 '-Diphenyl-1-Picrylhydrazyl (DPPH) Methods. *J. Agric. Food Chem.* **2006,** *54* (4), 1151–1157.
61. Pergola, G.; Maldera, S.; Tartagni, M.; Pannacciulli, N.; Loverro, G.; Giorgino, R. Inhibitory Effect of Obesity on Gonadotropin, Estradiol and Inhibin B Levels in Fertile Women. *Obesity* **2006,** *14* (11), 1954–1960.
62. Perveen, R.; Suleria, H. A. R.; Anjum, F. M.; Butt, M. S.; Pasha, I.; Ahmad, S. Tomato (*Solanum lycopersicum*) Carotenoids and Lycopenes Chemistry; Metabolism, Absorption, Nutrition, and Allied Health Claims—A Comprehensive Review. *Crit. Rev. Food Sci. Nutr.* **2015,** *55* (7), 919–929.
63. Prior, R. *Absorption and Metabolism of Anthocyanins: Potential Health Effects*. CRC Press: Boca Raton, FL; 2004; p 237.
64. Prior, R. L.; E. Wilkes, S.; R. Rogers, T.; Khanal, R. C.; Wu, X.; Howard, L. R. Purified Blueberry Anthocyanins and Blueberry Juice Alter Development of Obesity in Mice Fed an Obesogenic High-fat Diet. *J. Agric. Food Chem.* **2010,** *58* (7), 3970–3976.
65. Prior, R. L.; Wu, X. Anthocyanins: Structural Characteristics That Result in Unique Metabolic Patterns and Biological Activities. *Free Rad. Res.* **2006,** *40* (10), 1014–1028.
66. Retelny, V. S.; Neuendorf, A.; Roth, J. L. Nutrition Protocols for the Prevention of Cardiovascular Disease. *Nutr. Clin. Pract.* **2008,** *23* (5), 468–476.
67. Riaz, M.; Zia-Ul-Haq, M.; Saad, B. *Anthocyanins and Human Health: Biomolecular and Therapeutic Aspects*. Http://Www.Springer.Com/Series/10203; Springer: New York; 2016; p 138.
68. Riaz, M.; Zia-Ul-Haq, M.; Saad, B. Anthocyanins Effects on Carcinogenesis, Immune System and the Central Nervous System. In *Anthocyanins and Human Health: Biomolecular and Therapeutic Aspects*; Springer: New York; 2016; pp 125–138.
69. Riaz, M.; Zia-Ul-Haq, M.; Saad, B. The Role of Anthocyanins in Health as Antioxidant, in Bone Health and as Heart Protecting Agents. In: *Anthocyanins and Human Health: Biomolecular and Therapeutic Aspects*; Springer: New York; 2016; pp 87–107.
70. Riaz, M.; Zia-Ul-Haq, M.; Saad, B. The Role of Anthocyanins in Obesity and Diabetes. *Anthocyanins and Human Health: Biomolecular and Therapeutic Aspects*; Springer; 2016; pp 109–123.
71. Rossi, A.; Serraino, I.; Dugo, P.; Di Paola, R.; Mondello, L.; Genovese, T.; Morabito, D.; Dugo, G.; Sautebin, L.; Caputi, A. P. Protective Effects of Anthocyanins from Blackberry in a Rat Model of Acute Lung Inflammation. *Free Rad. Res.* **2003,** *37* (8), 891–900.
72. Schwarz, M.; Wray, V.; Winterhalter, P. Isolation and Identification of Novel Pyranoanthocyanins from Black Carrot (*Daucus carota* L.) Juice. *J. Agric. Food Chem.* **2004,** *52* (16), 5095–5101.
73. Seeram, N.; Momin, R.; Nair, M.; Bourquin, L. Cyclooxygenase Inhibitory and Antioxidant Cyanidin Glycosides in Cherries and Berries. *Phytomedicine* **2001,** *8* (5), 362–369.
74. Shahidi, F. Nutraceuticals and Functional Foods: Whole Versus Processed Foods. *Trends Food Sci. Technol.* **2009,** *20* (9), 376–387.
75. Shih, P. H.; Yeh, C. T.; Yen, G. C. Effects of Anthocyanidin on the Inhibition of Proliferation and Induction of Apoptosis in Human Gastric Adenocarcinoma Cells. *Food Chem. Toxicol.* **2005,** *43* (10), 1557–1566.

76. Sibbel, A. The Sustainability of Functional Foods. *Soc. Sci. Med.* **2007,** *64* (3), 554–561.
77. Siro, I.; Kápolna, E.; Kápolna, B.; Lugasi, A. Functional Food. Product Development, Marketing and Consumer Acceptance: A Review. *Appetite* **2008,** *51* (3), 456–467.
78. Steffen, Y.; Gruber, C.; Schewe, T.; Sies, H. Mono-O-methylated Flavanols and Other Flavonoids as Inhibitors of Endothelial NADPH Oxidase. *Arch. Biochem. Biophys.* **2008,** *469* (2), 209–219.
79. Sugimoto, M.; Kuo, M.-L.; Roussel, M. F.; Sherr, C. J. Nucleolar Arf Tumor Suppressor Inhibits Ribosomal RNA Processing. *Mol. Cell* **2003,** *11* (2), 415–424.
80. Suleria, H. A. R.; Butt, M. S.; Anjum, F. M.; Saeed, F.; Khalid, N. Onion: Nature Protection Against Physiological Threats. *Crit. Rev. Food Sci. Nutr.* **2015,** *55* (1), 50–66.
81. Suleria, H. A. R.; Butt, M. S.; Khalid, N.; Sultan, S.; Raza, A.; Aleem, M.; Abbas, M. Garlic (*Allium sativum*): Diet Based Therapy of 21st Century: A Review. *Asian Pacific J. Trop. Dis.* **2015,** *5* (4), 271–278.
82. Suleria, H. A. R.; Gobe, G.; Masci, P.; Osborne, S. A. Marine Bioactive Compounds and Health Promoting Perspectives; Innovation Pathways for Drug Discovery. *Trends Food Sci. Technol.* **2016,** *50*, 44–55.
83. Suleria, H. A. R.; Osborne, S.; Masci, P.; Gobe, G. Marine-Based Nutraceuticals: an Innovative Trend in the Food and Supplement Industries. *Marine Drugs* **2015,** *13* (10), 6336–6351.
84. Sultan, M. T.; Buttxs, M. S.; Qayyum, M. M. N.; Suleria, H. A. R. Immunity: Plants as Effective Mediators. *Crit. Rev. Food Sci. Nutr.* **2014,** *54* (10), 1298–1308.
85. Sun, T.; Simon, P. W.; Tanumihardjo, S. A. Antioxidant Phytochemicals and Antioxidant Capacity of Biofortified Carrots (*Daucus carota* L.) of Various Colors. *J. Agric. Food Chem.* **2009,** *57* (10), 4142–4147.
86. Takikawa, M.; Inoue, S.; Horio, F.; Tsuda, T. Dietary Anthocyanin-rich Bilberry Extract Ameliorates Hyperglycemia and Insulin Sensitivity Via Activation of AMP-activated Protein Kinase inDiabetic Mice. *J. Nutr.* **2010,** *140* (3), 527–533.
87. Tanaka, T.; Kohno, H.; Mori, H. Chemoprevention of Colon Carcinogenesis by Dietary Non-nutritive Compounds. *Asian Pacific J. Cancer Prevent.* **2001,** *2* (3), 165–177.
88. Tedesco, I.; Russo, G. L.; Nazzaro, F.; Russo, M.; Palumbo, R. Antioxidant Effect of Red Wine Anthocyanins in Normal and Catalase Inactive Human Erythrocytes. *J. Nutr. Biochem.* **2001,** *12* (9), 505–511.
89. Tiwari, P.; Kumar, B.; Kaur, M.; Kaur, G.; Kaur, H. Phytochemical Screening and Extraction: A Review. *Int. Pharmaceut. Sci.* **2011,** *1* (1), 98–106.
90. Toufektsian, M.-C.; De Lorgeril, M.; Nagy, N.; Salen, P.; Donati, M. B.; Giordano, L.; Mock, H.-P.; Peterek, S.; Matros, A.; Petroni, K. Chronic dietary Intake of Plant-derived Anthocyanins Protects the Rat Heart Against Is Chemia-reperfusion Injury. *J. Nutr.* **2008,** *138* (4), 747–752.
91. Tsuda, T. Regulation of Adipocyte Function by Anthocyanins; Possibility of Preventing the Metabolic Syndrome. *J. Agric. Food Chem.* **2008,** *56* (3), 642–646.
92. Tsuda, T.; Ueno, Y.; Kojo, H.; Yoshikawa, T.; Osawa, T. Gene Expression Profile of Isolated Rat Adipocytes Treated with Anthocyanins. *Biochimica Et Biophysica Acta (BBA)-Molecular Cell Biol. Lipids* **2005,** *1733* (2), 137–147.
93. Türker, N.; ErdogˇDu, F. Effects of pH and Temperature of Extraction Medium on Effective Diffusion Coefficient of Anthocynanin Pigments of Black Carrot (*Daucus carota* Var. L.). *J. Food Eng.* **2006,** *76* (4), 579–583.

94. Welch, A.; Macgregor, A.; Jennings, A.; Fairweather-Tait, S.; Spector, T.; Cassidy, A. Habitual Flavonoid Intakes Are Positively Associated with Bone Mineral Density in Women. *J. Bone Mineral Res.* **2012,** *27* (9), 1872–1878.
95. Witrowa-Rajchert, D.; Bawoł, A.; Czapski, J.; Kidoń, M. Studies on Drying of Purple Carrot Roots. *Drying Technol.* **2009,** *27* (12), 1325–1331.
96. Wu, X.; Beecher, G. R.; Holden, J. M.; Haytowitz, D. B.; Gebhardt, S. E.; Prior, R. L. Concentrations of Anthocyanins in Common Foods in the United States and Estimation of Normal Consumption. *J. Agric. Food Chem.* **2006,** *54* (11), 4069–4075.
97. Yang, M.; I Koo, S.; O Song, W.; K Chun, O. Food Matrix Affecting Anthocyanin Bioavailability. *Curr. Med. Chem.* **2011,** *18* (2), 291–300.
98. Yi, W.; Fischer, J.; Akoh, C. C. Study of Anticancer Activities of Muscadine Grape Phenolics *In Vitro*. *J. Agric. Food Chem.* **2005,** *53* (22), 8804–8812.
99. Youdim, K. A.; Martin, A.; Joseph, J. A. Incorporation of the Elderberry Anthocyanins by Endothelial Cells Increases Protection Against Oxidative Stress. *Free Rad. Biol. Med.* **2000,** *29* (1), 51–60.
100. Zaid, H.; Rayan, A.; Said, O.; Saad, B. Cancer Treatment by Greco-Arab and Islamic Herbal Medicine. *Open Nutraceut. J.* **2010,** *3* (1), 203–212.
101. Zhang, F.; Aravanis, A. M.; Adamantidis, A.; De Lecea, L.; Deisseroth, K. Circuit-breakers: Optical Technologies for Probing Neural Signals and Systems. *Nat. Rev. Neurosci.* **2007,** *8* (8), 577–581.
102. Zhang, Y.; Vareed, S. K.; Nair, M. G. Human Tumor Cell Growth Inhibition by Nontoxic Anthocyanidins, the Pigments in Fruits and Vegetables. *Life Sci.* **2005,** *76* (13), 1465–1472.
103. Ziberna, L.; Tramer, F.; Moze, S.; Vrhovsek, U.; Mattivi, F.; Passamonti, S. Transport and Bioactivity of Cyanidin 3-glucoside into the Vascular Endothelium. *Free Rad. Biol. Med.* **2012,** *52* (9), 1750–1759.

# CHAPTER 5

# OLIVE OIL PHENOLS: CHEMISTRY, SYNTHESIS, METABOLISM, FATE, AND THEIR ALLIED HEALTH CLAIMS

MUHAMMAD IMRAN, FAQIR MUHAMMAD ANJUM, SHINAWAR WASEEM ALI, IAHTISHAM-UL-HAQ, and HAFIZ ANSAR RASUL SULERIA

## ABSTARCT

Oleuropein, 2-(3,4-dihydroxyphenyl) ethanol, amounts to 60–140 mg/g of dry weight of olive leaves and is the most prevalent nontoxic secoiridoids found in olive cultivars. Oleuropein and its derivative hydroxytyrosol are structural components of olive stone. Likewise, hydroxytyrosol 4-(2-hydroxyethyl-1, 2-benzenediol), is quantitatively the major phenolic constituents in olive oil. Both phenolic compounds have an inductive effect on liver glutathione S-transferases owing to high antioxidant activity and reveal several pharmacological characteristics such as: preventing cancer, lowering cardiovascular disorders, and inhibiting platelet aggregation. Oleuropein and hydroxytyrosol exhibited anticancer potential on HT-29 and leukemia HL-60 cancer cells; and suppressed the cell growth of 786-O, LN-18, renal cell adenocarcinoma, malignant melanoma of the skin–lymph node metastasis, RPMI-7951, colorectal adenocarcinoma cells, TF-1a, T-47D, and erythroleukemia. Moreover, these phenolics possess protective effect on arrhythmia, adrenaline, and intestinal muscle spasms in cardiovascular diseases. Oleuropein also exhibits neuroprotective activity through developing noncovalent bonding with the Aβ peptide. Furthermore, hydroxytyrosol and oleuropein have antimicrobial potential that inhibits the growth rate of numerous human intestinal pathogens such as *Moraxella catarrhalis* and *Staphylococcus aureus*. This chapter reviews research on potential benefits of consumption of hydroxytyrosol and oleuropein and their pharmacological use to prevent cardiovascular and neurodegenerative ailments.

## 5.1 INTRODUCTION

Olive is widely distributed over the Indian subcontinent including Northwest Himalayas, adjoining mountains, and mountainous areas of Pakistan. Olive (*Olea europaea*) belongs to family *Oleaceae*, which comprises 30 genera and 600 species. In various countries around the world, it is an essential dietary constituent and is eaten in both unripe green fruit and fully ripe black fruit.[124] Structurally, the olive fruit can be divided into three different portions: (1) the skin, also called epicarp (1.0–3.0% of drupe weight) that encompasses anthocyanins, carotenoids, and chlorophyll giving coloration to the fruit; (2) the inner part is pulp, called mesocarp, comprising 70–80% of the fruit; and (3) the stone, called the woody endocarp, composing 18–22% of the whole seed.[116]

The chemical composition of olive oil depends on variety of olive, location, and growing conditions.[43] Virgin olive oil is distinctive oil exhibiting high levels of unsaturated fatty acids, oleic acid, and phenolic compounds as compared to other vegetable oils.[92] The nutritional profile of olive oil is unique, because of the presence of appreciable oleic acid content (80%), hydrocarbons and oxygenated compounds, and low levels of pigments alongside free fatty acids. Chemically, olive fruit comprises 50% moisture, 20% carbohydrates, and 20% oil. The major part (96–98%) of the oil is found in flesh and skin and only 2–4% oil is contained in the pit of olive fruit.[72] Olive oil contains only 2% of the total phenolics, whereas remaining 98% are lost in olive mill waste.[32]

During ripening, oleuropein and ligstroside-aglycones are produced through elimination of glucose from oleuropein and ligstroside-glycoside with the application of glucosidase enzyme. These aglycones and their allied derivatives are present abundantly as phenols in olive oil.[96] The major phenolic glycosides (such as oleuropein, ligstroside, dimethyl oleuropein, and derivative of aglycones) are formed upon the enzymatic hydrolysis. The ligstroside aglycones, dialdehydic forms of decarboxy methyl oleuropein, and aldehydic forms of oleuropein are commonly known as secoiridoid compounds. During olive oil extraction, enzymatic activity of β-glucosidase controls the hydrolysis of phenolic glycosides.[117] The polyphenols commonly present in olive are used in defensive mechanisms of several plant species, where they penetrate the food system and aid in neutralizing free radicals.[127,136] In addition, these are also characterized as natural anti-inflammatory and preventive agents from the deleterious effects of the synthetic drugs.[143]

The phenolic content of olive fruit is affected by climate, production process, and degree of ripening.[21,133] Likewise, lutein and β-carotene are main yellow pigments of virgin olive oils @ 1.4 ppm and 0.3–4.4 ppm, respectively. The carotenoids concentrations in Spanish olive oils varies from 3.1 to 9.2 mg/kg.[55] Tyrosol 4-(2-hydroxyphenyl) ethanol is amongst chief phenolic moieties contained in olive oil as conjugated or free forms (as aglycones or secoroids) and exhibits high antioxidant potential.[90] The contents of tyrosol and hydroxytyrosol are influenced by cultivar and harvest time. Furthermore, both phenols and their secoroids derivatives are comprised up to 30% of virgin olive oil phenolics, whereas conjugated forms (like oleuropein and ligstroside aglycones) share half of the total phenolic contents of virgin olive oil.[96] In a research study, it protected Caco-2 intestinal mucosa cells and inhibited leukocyte 5-lipooxygenase against cytotoxic and cytostatic effects produced by oxidized low-density lipoproteins (LDL).[28]

## 5.2 PHENOLIC COMPOUNDS IN OLIVE FRUITS

### *5.2.1 OLEUROPEIN*

Oleuropein is the major active polyphenolic compound of olive fruit and was discovered in 1908 by Bourquelot and Vintilesco.[15] Its structure was first illustrated by Panizzi and Scarpati[101] from fresh leaves. It is also present in numerous other genera from family Oleaceae like *Fraxinus angustifolia*, *Fraxinus chinensis*, *Fraxinus excelsior*, *Ligustrum vulgare*, *Ligustrum ovalifolium*, *Syringa vulgaris*, *Syringa josikaea,* and *Phillyrea latifolia.*[126] Oleuropein is a bitter secoiridoid glycoside that is present in diverse parts of olive fruit.[127] It is an ester of 2-(3, 4-dihydroxyphenyl) ethanol having oleosidic structure.[126]

Oleuropein exhibits health-promoting properties such as: anti-arrhythmic, cardioprotective, spasmolytic, hypotensive, immune-stimulant, anti-inflammatory, hypoglycemic, cytostatic, and antimicrobial. In the presence of β-glucosidase, it is converted to glucose and oleuropein aglycone (Table 5.1). Dual mechanism of oleuropein prevents formation of free radicals due to chelating ability and may also directly neutralize these radicals through hydroxyl groups.[108] Oleuropein and its metabolite hydroxytyrosol linked with catechol groups are essential for high antioxidant activity. Both bioactive compounds have been used to scavenge superoxide anions and inhibit the hypochlorous acid-derived radicals and neutrophils. Moreover, it inhibits the growth rate of human intestinal or respiratory pathogens such as: *M. catarrhalis*, *Salmonella typhimurium,* and *Vibrio parahaemoyticus*.[102]

**TABLE 5.1** Functions of Phenolic Compounds in Olive Fruits.

| Compounds | Functions |
|---|---|
| (-)-oleocanthal | Constrains cyclooxygenases (COX-1 and COX-2) activities |
| Caffeic acid | Excites central nervous system (CNS) and deactivates reactive oxygen species (ROS) |
| Dimethyl-oleuropein | Inhibits platelet aggregation and neutralizes ROS |
| Hydroxytyrosol | Demonstrates antimicrobial, anticancer, and antidiabetic activities and suppresses platelet aggregation |
| Lingstroside | Exhibits anticancer and antihyperlipidemic properties and scavenges ROS |
| Oleic acid | Lowers liposomal oxidation, reduces endothelial cell sensitivity to oxidation, and decreases low-density lipoproteins (LDL) cholesterol |
| Oleuropein | Demonstrates antihypertensive, antispasmodic, and antiarrhythmic activities |
| *p*-coumaric acid | Possesses antiaging and antihyperlipidemic effects and scavenges ROS |
| Pinoresinol | Neutralizes ROS and suppresses cAMP phosphodiesterase |
| Sterols | Suppress cholesterol acyl-transferase and inhibit acyl coenzyme A |
| Triterpenes | Exhibit anti-inflammatory, cardiotonic, and antioxidant effects |
| Verbascoside | Interferes with cyclooxygenase, possesses anti-inflammatory properties, modulates eicosanoid metabolism, and reduces DNA damage |

**Source:** Adapted from Refs. [153, 154].

#### 5.2.1.1 OLEUROPEIN: BIOSYNTHESIS, FATE, AND BIOAVAILABILITY

The biosynthesis of oleuropein is a complex biochemical process that involves a series of reactions and intermediate precursors. In olive fruit, it involves development of oleosides through mevalonic acid pathway and production of cologanin-derived secoiridoids.[63] Furthermore, geraniol, 10-hydroxy geraniol, iridodial, and 10-hydroxynerol are found to be precursors of loganin that further act as precursor of oleuropein. Subsequently, the precursors of oleuropein (including loganic acid, 7-epiloganic acid, and deoxyloganic acid) penetrate into ligstroside through 7-ketologanic acid intermediate.[106] At initiation of green maturation of olive fruit, the concentration of oleuropein is decreased and glycosylated derivatives of oleuropein such as elenolic acid, glucoside, and dimethyl-oleuropein are produced. In black maturation stage, these derivatives accumulate and attain maximum level until dimethyl-oleuropein becomes major component of black olives through the action of esterases enzymes.[17] Alongside, reduction in other oleosides such as ligstrosides is also observable. However, antioxidant compounds such as verbascoside and flavonoids increase simultaneously.

In small young olives, cornoside and ligstroside are abundantly present while the verbascoside exhibits in relatively small amounts. When green olives acquire appropriate size, the cornoside and ligstroside disappears following same fashion as other bioactive molecules and are easily converted into halleridone.[95] The hydroxytyrosol, oleuropein, and elenolic acid glucoside promote browning and maturation of olives due to associated action of diphenol oxidase.[61]

Considering the acid conditions, it is neither degraded nor absorbed in the small intestine. However, oleuropein is rapidly degraded into hydroxytyrosol through the action of the colonic microflora as soon as it reaches the large intestine.[35] In a scientific exploration, Vissers et al.[144] observed the absorption of supplemented hydroxytyrosol, oleuropein-aglycone, tyrosol, and ligstroside-aglycone (55–60%) in humans and animals. They proposed metabolism of oleuropein-glycoside to oleuropein and ligstroside-aglycones, which then is converted to hydroxytyrosol.[144] Likewise, another study revealed absorption of oleuropein with a maximum plasma concentration attaining after 2 h. It was further noticed that oleuropein in the form of glucuronides is rapidly distributed and excreted through urine.[130]

### *5.2.1.2 CHEMISTRY OF OLEUROPEIN*

The chemical structure of oleuropein was first reported by Panizzi in 1958.[54] The endogenous or exogenous enzymes play a significant role, where α-glucosidase transforms oleuropein into glucose and oleuropein aglycone. In addition, hydroxytyrosol is also a valuable bioactive phenol obtained by subsequent esterolysis. The decarboxylation, methylation, and oxidation of glycone produce new phenol aglycone structures.[26] Oleuropein being olive's major constituent is primarily elenolic acid ester with 3,4-dihydroxyphenyl ethanol produced by the removal of glucose moiety by the action of β-glucosidase enzyme during ripening.[96] It has been found that ortho-diphenols (such as hydroxytyrosol and oleuropein) contain hydroxyl groups, which are vital for antioxidant abilities. In this context, phenols and ortho-diphenols considerably contribute in oxidation stability of the oil.[137]

### 5.2.2 HYDROXYTYROSOL AND ITS SYNTHESIS

Amongst several bioactive polyphenol compounds present in olive tree, hydroxytyrosol, 4-(2-hydroxyethyl-1,2-benzenediol, has shown strong scavenging effects against reactive oxygen species (ROS). Chemically, it is amongst bioactive alcoholic ortho-diphenol components of secoiridoids. Hydroxytyrosol, a metabolite of oleuropein, is also found in olive and extra virgin olive oil. Furthermore, olive oil mill wastewater is also considered a potential source of this compound.[121] Hydroxytyrosol exhibits health-promoting perspectives such as: anticancer and antidiabetic, protection against LDL oxidation, oxidative deoxyribose nucleic acid (DNA) damage, inhibition of 5- and 12-lipoxygenases, and prevention of platelet aggregation due to catechol quinine.[114]

Hydroxytyrosol is generally isolated from olive tree through hydrolysis of oleuropein or utilizing 3,4-dihydroxy phenyl acetic acid as precursor[82] as it is not commercially available. Mosca et al.[91] used spectrophotometric assay method to determine phenol contents in olive oil. In this method, tyrosinase and nicotinamide adenine dinucleotide (NADH) were used as enzyme and reducing agent, respectively. Tyrosinase acts as a polyphenol oxidase enzyme that oxidizes monophenols into *O*-diphenols and subsequently converting it to *O*-quinones. The *O*-quinones bind proteins through covalent bonds and produce brown pigments. Nevertheless, under reducing conditions (either excess of ascorbic acid), O-quinones are converted to their O-diphenol precursors.[68]

#### 5.2.2.1 HYDROXYTYROSOL: ABSORPTION AND BIOAVAILABILITY

Hydroxytyrosol is normally absorbed in intestine through bidirectional passive diffusion mechanism.[82] Past studies have revealed hydroxytyrosol to be excreted through the kidney and also oxidized to 3,4-dihydroxyphenyl acetaldehyde sulfate conjugate, 3, 4-dihydroxyphenyl acetic acid, glucuronide conjugate, homovanillyl alcohol, and homovanillic acid.[144] The bioavailability and excretion of hydroxytyrosol were first determined in human by Visioli et al.[140] The researchers suggested that hydroxytyrosol is excreted as glucuronide conjugates via urine in humans after oral administration. Therefore, higher supplementation of hydroxytyrosol enhances the excretion quantity of conjugated glucuronide.[140]

The lethal dose ($LD_{50}$) of hydroxytyrosol has been found to be greater than 5000 mg/kg as represented by oral supplementation of this compound in a rat model with various concentrations. The toxic effects of hydroxytyrosol during pregnancy and reproduction of rats have been observed. In addition, genetic effects of this compound have also been reported in rats. Nonetheless, it does not exhibit any deleterious effects at concentration as high as 2000 mg/kg/day.[31]

### *5.2.3 ANTIOXIDANT STATUS OF OLEUROPEIN AND HYDROXYTYROSOL*

During metabolic processes, free radicals are produced that contribute to the development of cancer, diabetes, cardiovascular, atherosclerosis, and rheumatoid arthritis (RA).[11] Oleuropein and hydroxytyrosol contained in olive oil possess significant antioxidant abilities owing to presence of an ortho-diphenolic (catecholic) structure.[42] Hydroxytyrosol represents higher antioxidant potential due to ortho-dihydroxy structure as compared to other phenolic compounds such as ascorbic acid or α-tocopherol followed by 4-*O*-monohydroxy moieties (ligstroside and tyrosol) and 3-*O*-hydroxy-substituted catechols.[87] The hydroxytyrosol prevents production of ROS catalyzed by metal ions and breaking of preoxidative chain reactions in radical generation by scavenging free radicals.[33]

Clinical trials have shown that hydroxytyrosol provides an indirect protection by improving endogenous defense mechanism.[79] Furthermore, it has numerous pharmacological effects including formation and maintenance of bone, stimulation of calcium deposition by prevention from osteoporosis, and inhibition of multinucleated osteoclasts.[65] During in vitro studies, oleuropein has exhibited higher antioxidant activity than hydrosoluble analogue of tocopherol. Further, it has been seen that it inhibited hypochlorous acid-derived radicals and respiratory burst of neutrophils.[27]

### *5.2.4 PHARMACOKINETICS STUDY OF OLEUROPEIN AND HYDROXYTYROSOL*

To verify and correlate in vivo effects of phenolic compounds as depicted during in vitro trials, it is imperative to determine the absorption mechanism of these compounds in intestine followed by their distribution. In this context, rate of oleuropein absorption has been determined using an intestinal perfusion method.[47] The permeability coefficient of absorption of oleuropein was found to be $1.47 \pm 0.13 \times 10^{-6}$ cm/s in isoosmotic conditions, while significantly higher value of $5.92 \pm 0.49 \times 10^{-6}$ cm/s in hypotonic conditions due to an increase in the opening of paracellular junctions in response to hypotonicity. The constant rate of degradation of oleuropein at pH 7 was −0.023/min ($r^2$ = 0.962) in an isoosmotic solution. The pH is a critical determinant in controlling its stability.

Owing to polar nature, oleuropein passes through lipid bilayer of the epithelial cell membrane by transcellular pathway.[47] With respect to glycoside property, oleuropein certainly uses a glucose carrier for transportation of molecules. Carriers are used for absorption of oleuropein in epithelial cells of small intestine. The Glut-2 and Glut-5 act as carrier for glucose through diffusion. Glut-5 participates in the transportation and absorption of oleuropein and fructose molecules. It has been noticed that these molecules were present on apical region of epithelial cells in intestine,[75] whereas Glut-2 mediates the route of glucose into the circulation from epithelial cells and locates on the baso-lateral region of epithelial cells.[93]

Bai et al.[10] reported that the absorption of hydroxytyrosol gave highest plasma concentration in 5–10 min, while the quantity of hydroxytyrosol was smaller than the amount administered in the rat plasma after 60 min. In vivo metabolic fate of radiolabeled hydroxytyrosol has been estimated in rats by its supplementation. After oral administration, its concentration appeared to be maximum in plasma at 10 min that was ultimately excreted through urine. It was also bound to glucuronic acid (5%) and further eliminated in the form of waste products.[40]

Visioli et al.[140] illustrated that absorption and urinary excretion of hydroxytyrosol was dose dependent and regularly controlled through conjugation with glucuronic acid. Urinary excretion of this compound was much higher in the first 4 h due to increased rate of conjugation of hydroxytyrosol with glucuronide. Particularly, the excretion of hydroxytyrosol from human occurred at 2.43 h and its free form was not detected in plasma because of short half life.[29] Moreover, it is further metabolized into 3,4-dihydroxyphenyl acetic acid, homovanillic alcohol, 3,4-dihydroxyphenyl acetaldehyde, and sulfate conjugate through enzymatic actions. After administration of hydroxytyrosol to healthy subjects, a considerable increment was reported in homovanillic alcohol and acid urinary excretion after 24 h.[134]

## 5.3 HEALTH PERSPECTIVES OF OLIVE'S BIOACTIVES

### 5.3.1 CANCER-PREVENTIVE ROLE

Olive's bioactives have been shown to possess wide range of health perspectives (Table 5.2). In papillary and follicular thyroid cancer cell lines, hydroxytyrosol abridged viability of cancer cells concomitantly with a decrease of cyclin D1 expression and an up-regulation of cell cycle key modulator p21 levels. It also exerts proapoptotic effects on follicular and papillary cancer cells.

In thyroid cancer cells, hydroxytyrosol promoted apoptosis at higher doses with respect to other cancer cells line.[131,132] It inhibits mouse embryonic fibroblast cell line (3T3-L1) cell differentiation by down-regulating cannabinoid receptor type 1 (CB1) receptor gene expression and cell proliferation. It showed anti-adipogenic perspectives and the expression of lipoprotein lipase (LPL) and fatty acid synthase (FAS) genes was amplified after treating with hydroxytyrosol.[135] In cholangiocarcinoma human cell lines derived from extrahepatic bile duct carcinoma (TFK-1), human extrahepatic bile duct carcinoma cell line (KMBC), and gallbladder series cells (GBS-SD), the concentration of 75 μM hydroxytyrosol inhibited proliferation. Nevertheless, 200 μM hydroxytyrosol treatments (250 and 500 mg/kg d) noticeably repressed the CCA xenografts growth in mice. Additionally, it induces apoptosis and cell cycle arrest in vitro and in vivo.[24]

Oleuropein provides protection against azoxymethane through multiple mechanisms, such as: (1) reduction in intestinal interleukin-6 (IL-6),tumor necrosis factor-alpha (TNF-α), interleukin-17A (IL-17A), and interferon gamma (IFN-γ) concentration; (2) reduction of cyclooxygenase (COX)-2, Bax, and proliferating cell nuclear antigen protein expression; (3) down-regulation of Wnt/β-catenin, nuclear factor-κB (NF-κB), signal transducer and activators of transcription (STAT)3, and phosphatidylinositol-3-kinase (PI3K)/Akt; (4) inhibition of Th17 response; and (5) reduction of IL-17A and IFN-γ expression.[56,149]

In another study reported by Hassan et al.,[69] oleuropein represented potential to induce apoptotic cell death in luminal human breast cancer (MCF-7) cell lines via up-regulating p53 and Bax gene expression levels and down-regulating B-cell lymphoma 2 (Bcl2). Similarly, Elamin et al.[48] investigated that oleuropein has cytotoxic effect against breast cancer cell lines (MDA-MB-231) and luminal (MCF-7) cells through induction of apoptosis *via* mitochondrial pathway and suppressing cell proliferation.

Likewise, hydroxytyrosol also inhibits proliferation of HT29, human colon tumor cells (HT29 clone 19A), and HL60 cells. In fact, it showed antiproliferative effect by inhibiting cyclin-dependent kinase inhibitors.[50] Hepatocellular carcinoma (HCC) is linked with increased level of several angiogenic factors such as: vascular endothelial growth factor (VEGF) and interleukin (IL)-8 in patients.[100,119] There are two activators, that is, hypoxia-inducible factor (HIF)-1R and basic fibroblast growth factor (bFGF), which are associated with cancer proliferation. Among these activators, HIF-1R mainly promotes angiogenesis that controls critical adaptive responses of cancer cell lines to hypoxia, whereas bFGF mediates extracellular matrix degradation,[107] Additionally, vascular permeability enhances endothelial cell migration and proliferation of VEGF that are necessary for sprouting development and vascular remodeling of new blood vessels,[52] Similarly, production of free radicals has shown to increased oxidative stress and stimulate VEGF production that leads to cancer development.[147]

The progression of human breast cancer cells is stimulated through the binding of 17b-estradiol (E2) to the different estrogen receptor (ER) isoforms including estrogen receptor-a (Era) and estrogen receptor-b (ERb). After binding their ligand, these regulate genes expression participated in regulation of cell cycle through interaction with chromatin.[80] As aforementioned, hydroxytyrosol acts as antiestrogens that depict a significant antioxidant role through its prompt absorption by human intestine.[70] Similarly, it has been noticed that hydroxytyrosol modulates genes expression to participate in promyelocytes tumor cell proliferation.[51] Furthermore, hydroxytyrosol suppresses the progression of MCF-7 breast cancer cell lines due to presence of hydroxyl group showing antiestrogenic effects.[60] Oleuropein is another active phenolic compound from olive that has exhibited anticancer potential on human colon adenocarcinoma cell line (HT-29) and human leukemia cell line (HL-60) cancer cells.[51]

Hamdi and Castellon.[67] demonstrated that oleuropein suppressed the cell growth of human renal cancer cell line (786-O), human brain glioblastoma cell line (LN-18), renal cell adenocarcinoma, malignant melanoma of the human skin (RPMI 7951), skin–lymph node metastasis, colorectal adenocarcinoma cells, factor-independent variant isolated from factor-dependent TF-1 cell line (TF-1a), human breast cancer cell line (T-47D), and erythroleukemia.

The signaling pathways activate (NF)-κB and enhance inflammatory mediators, such as: IL-6, IL-1β, COX-2, TNF-α, inducible NOS (iNOS), and metalloproteinase (MMP) expression.[71,118] The oleuropein and hydroxytyrosol polyphenolic extracts inhibited the COX-2 and p38/CREB expression that caused colorectal cancer in large intestine.[34] Oleuropein showed anti-apoptotic effects on human colorectal cancer cells SW620 and also inhibited MCF-7 breast cancer cells proliferation.[125]

Likewise, hydroxytyrosol protects from MCF-7 breast cancer cells through: (1) induction of cell cycle arrest in G0/G1 phase; (2) reduction in expressions of peptidyl-prolyl cis–trans isomerase Pin1; and (3) enhancement in AP1 transcription factor member, c-Jun.

Oleuropein exhibits chemopreventive mechanisms against basal-like MDA-MB-231 cells through following strategies, such as: (1) showed specific cytotoxicity against breast cancer cells; (2) induced apoptosis; (3) suppressed cell proliferation; (4) up-regulated cyclin-dependent inhibitor; and (5) suppressed antiapoptosis and pro-proliferation protein NF-κB.[48]

In human colon adenocarcinoma cells (HT)-29, oleuropein and hydroxytyrosol suppressed cell growth, induced cell cycle arrest, enhanced apoptotic population, decreased HIF-1α protein, and upregulated the p53 protein expression, respectively.[25] Yao et al.[151] determined that administration of oleuropein (150–200 mM) for 24 h induced apoptotic cell death in HeLa cells by enhancing c-Jun NH2-terminal kinase protein, cytochrome c protein phosphorylated activating transcription factor 2 (ATF-2), cyclin-dependent kinase inhibitor 1 (p21), tumor protein (p53), and Bax in the cytoplasm. Oleuropein has chemopreventive effect on BPH-1 nonmalignant cells, human prostate cancer cell line (DU145), and androgen-sensitive human prostate adenocarcinoma cells (LNCaP) prostate cancer cell lines. It lowered the viability of cell and induced thiol group modifications, ROS, pAkt, heme oxygenase-1, and γ-glutamylcysteine synthetase.[2]

### *5.3.2 ANTIDIABETIC BENEFITS*

Diabetes mellitus (DM) is amongst serious metabolic health concerns around the globe.[109] The administration of hydroxytyrosol to diabetic animals employed a neuroprotective consequence on brain damage in hypoxia-reoxygenation model. The treatment of hydroxytyrosol significantly reduced oxidative and nitrosative stress and brain inflammatory mediators (prostaglandin E2 and interleukin 1ß concentration). It was recorded that cell death was reduced by 25.9%, 37.5%, and 41.0% after the administration of 1, 5, or 10 mg/kg/day.[113]

Hydroxytyrosol has shown reduction in plasma thiobarbituric acid-reactive substances (TBARS) and enhancement in nerve conduction velocity and thermal nociception impairment in diabetic rats. It also abolished reduction in sciatic nerve Na(+) and K(+)-ATPase activity.[115] Similarly, administration of oleuropein- and hydroxytyrosol-rich extracts (8 and 16 mg/kg body weight) decreased serum glucose, cholesterol, hepatic glycogen, TBARS, lipid peroxidation, and prevented from the depletion of antioxidant enzymes activities in diabetic rats induced by intraperitoneal injections of alloxan.[73]

Likewise, hydroxytyrosol also markedly increased insulin secretion, antioxidant enzymes, decreased glucose, alkaline phosphatases, lactate transaminases and aspartate activities, TBARS, creatinine, bilirubin, and urea levels.

Furthermore, it also lowered impaired lipid profile biomarkers and enhanced high-density lipoprotein (HDL) in alloxan-induced diabetic rats.

Conclusively, hydroxytyrosol has been proven effective to attenuate hyperglycemia and oxidative stress induced by diabetes.[66] Hydroxytyrosol (600 μM) exhibited strongest inhibitory activity against α-glucosidase and α-amylase by decreasing viability of Caco-2 cells. In immortalized mouse myoblast cell line (C2C12) cells, oleuropein has protective role against $H_2O_2$-induced stress in cells by increasing glucose consumption and phosphorylation of mitogen-activated protein kinases and AMP-activated protein kinase/acetyl-Co-enzyme A (CoA) carboxylase but not phosphatidylinositol 3-kinase. Moreover, it also improved insulin sensitivity via insulin-independent and -dependent pathways.[64]

It was reported that supplementation of hydroxytyrosol decreased concentration of plasma glucose level of treated rats up to 55% as compared to other diabetic rats. The hypoglycemic role of hydroxytyrosol in diabetic rats is delivered by[66]: (1) protection of pancreatic β-cells; (2) enhancement of insulin secretion by suppressing ATP-sensitive potassium K (ATP) channels; (3) increasing voltage-dependent calcium channel that plays a major role in insulin secretion; and (4) enhancing peripheral uptake of glucose. It was further observed that hydroxytyrosol acts as effective scavenger of 1,1-diphenyl-2-picrylhydrazyl (DPPH) radical that demonstrates its vital antioxidant ability.[66]

The antioxidant potential of hydroxytyrosol protects pancreatic β-cells from damage and death *via* enhanced insulin secretion and lowered glucose level in plasma. It also increases levels of various enzymes that produce pyruvate kinase and hexokinase as a result of phosphorylation of glucose. Likewise, it lowers the concentration of enzymes that catalyze the dephosphorylation of glucose-6-phosphate to glucose-6-phosphatase and fructose-1,6-bisphosphatase.[99]

Various research studies support hypoglycemic action of oleuropein in diabetic animals by two mechanisms[59]: (1) potentiation of glucose-induced insulin release and (2) enhanced peripheral glucose uptake. Al-Azzawie and Alhamdani[3] evaluated the hypoglycemic role of oleuropein active ingredient in alloxan-diabetic experimental rabbits. The researchers administrated 20 mg/kg body weight (BW) of oleuropein during 16 weeks to the rabbits and observed that oleuropein considerably lowered blood glucose levels after 8 weeks as compared to control rabbits.[3,85]

Similarly, hydroxytyrosol protects from high-fat-diet-induced obesity inbred mouse strain (C57BL/6J) mice through: suppression of the sterol regulatory element-binding protein-1c/Fatty acid synthase (SREBP-1c/FAS) pathway; increasing antioxidant enzyme activities; inhibiting apoptosis activation; ameliorating high-fat diet-induced oxidative stress by normalized expression of mitochondrial fission marker Drp1 and mitochondrial complex subunits. It also improves the concentration of lowered mitochondrial carbonyl protein and mitochondrial complex activities in muscle tissues. In addition, hydroxytyrosol reduced the fasting glucose and serum lipids in mice.[24] Both olive oil compounds inhibited lipid deposition and glycerol-3-phosphate dehydrogenase (GPDH) enzyme activity in a dose-dependent manner. The concentration of oleuropein (200 and 300 μM) and hydroxytyrosol (100 and 150 μM) suppressed the triglyceride (TG) accumulation by 40% and 70% and 55% and 70%, respectively.[114]

**TABLE 5.2** Health Benefits of Hydroxytyrosol and Oleuropein in Olive Oil.

| Health perspectives | Function/mechanism | References |
|---|---|---|
| **Cancer prevention** | Promoted apoptosis Reduced cyclin D1 expression Up-regulated cell cycle key modulator p21 | [132] |
| | Down-regulated cell proliferation and CB1 receptor gene expression Enhanced expression of FAS and LPL genes | [135] |
| | Inhibited proliferation of TFK-1, KMBC, and GBS-SD cell lines | [24] |
| | Down-regulated Wnt/β-catenin, nuclear factor-κB (NF-κB), signal transducer and activators of transcription (STAT)3 Inhibited Th17 response Lowered IL-17A and IFN-γ expression | [56,149] |
| | Induced apoptotic cell death in luminal MCF-7 cell lines Up-regulated p53 and Bax gene expression levels Down-regulated Bcl2 | [69] |
| | Reduced expressions of peptidyl-prolyl cis–trans isomerase Pin1 Increased AP1 transcription factor member, c-Jun | [22] |
| **Antidiabetic** | Reduced oxidative, nitrosative stress, and brain inflammatory mediators (prostaglandin E2 and interleukin 1ß concentration) | [113] |
| | Reduced plasma TBARS<br>Enhanced nerve conduction velocity and thermal nociception impairment Abolished reduction in sciatic nerve Na (+) and K (+)-ATPase activity | [115] |
| | Decreased serum glucose, cholesterol, hepatic glycogen, TBARS, lipid peroxidation Prevented from depletion of antioxidant enzymes activities | [73] |
| | Exhibited strongest inhibitory activity against α-glucosidase and α-amylase activities | [64] |

**TABLE 5.2** *(Continued)*

| Health perspectives | Function/mechanism | References |
|---|---|---|
| | Increased insulin secretion and antioxidant enzymes Decreased glucose, alkaline phosphatases, lactate transaminases, and aspartate activities Lowered TBARS, creatinine, bilirubin and urea levels | [66] |
| | Suppressed SREBP-1c/FAS pathway Normalized expression of mitochondrial fission marker Drp1 and mitochondrial complex subunits Improved mitochondrial complex activities | [24] |
| | Increased phosphorylation of MAPKs and AMPK Improved insulin sensitivity by insulin-independent (AMPK/ACC) and insulin-dependent (PI3 kinase/Akt) pathways | [64] |
| | Lowered blood glucose levels | [85] |
| | Inhibited lipid deposition and glycerol-3-phosphate dehydrogenase (GPDH) enzyme activity Suppressed triglyceride accumulation | [114] |
| | Enhanced insulin secretion through suppressing ATP-sensitive potassium K(ATP) channels Increased voltage-dependent calcium channel Enhanced the peripheral uptake of glucose | [66] |
| **Cardiovascular protection** | Lowered size of atherosclerotic lesions | [58,145] |
| | Inhibited the cyclooxygenase (COX)-2 via lowering thromboxane A2 blood levels Enhanced vascular nitric oxide production | [1] |
| | Lowered the level of creatinine phosphokinase (CPK) and lactate dehydrogenase (LDH) Decreased creatinine phosphokinase-MB (CPK-MB) and aspartate aminotransferase (AST) Reduced the content of protein carbonyls (PCs), nitrotyrosine (NT), iNOS | [8] |
| | Improved endothelial function with reduced systolic blood pressure Impaired glucose and insulin tolerance Reduced left ventricular fibrosis and diastolic stiffness | [105] |
| | Lowered the level of NO and iNOS Down-regulated the levels of membrane type-1 MMP (MT-1-MMP), MMP-2, MMP-9 Enhanced expression of tissue MMP inhibitor-1 (TIMP-1) | [78,98] |
| | Lowered systolic and diastolic blood pressure due to $Ca^{2+}$ channel agonist property Improved endothelial function | [16] |

**TABLE 5.2** *(Continued)*

| Health perspectives | Function/mechanism | References |
|---|---|---|
| | Modulated high-density lipoprotein (HDL) and apolipoprotein (Apo) A-I Inhibited vascular smooth muscle cell (SMC) proliferation and extracellular signal-regulated kinase 1/2 (ERK1/2) | [1] |
| **Oxidative stress** | Decreased protein carbonyl content, inducible NOS (iNOS) and endothelial NOS (eNOS) in liver | [5] |
| | Prevented cell damage, glutathione enzyme depletion, and IL-1β induced IL-8 synthesis | [110] |
| | Prevented DNA damage induced by Sudan I and acrylamide | [153,154] |
| | Induced phase II detoxifying enzymes Stimulated the mitochondrial biogenesis | [155] |
| | Modulated Nrf2-dependent gene expression Lowered concentrations of TBARS and protein carbonyls Enhanced paraoxonase-2 PON activity | [13] |
| **Antimicrobial** | Exhibited antimicrobial activity against *Salmonella enteritidis*, *Bacillus cereus* T spores, and *Staphylococcus aureus* Delayed growth and sporulation of *Aspergillus parasiticus* | [104] |
| | Antimicrobial activities against *Klebsiella pneumoniae*, *Staphylococcus aureus*, *Salmonella typhi*, *Bacillus cereus*, *Vibrio parahaemolyticus*, and *Escherichia coli* | [84] |
| | Antifungal activity against *Fusarium sambucinum*, *Verticillium dahliae*, and *Alternaria solani* | [150] |
| | Reduction of *Escherichia coli* 533 and 679 | [88] |
| | Antibacterial activity against *Staphylococcus epidermidis* Reduced bacterial adhesion | [38] |
| **Anti-aging** | Inhibited LDL efflux in rat brain slices | [14] |
| | Suppressed AP-1 and LPS-triggered NF-kB activation | [27,45] |
| | Inhibited the release of pro-inflammatory cytokines and chemokineses Reduced leukocyte infiltration | [8] |
| | Attenuated the $Fe^{2+}$- and nitric oxide (NO)-induced cytotoxicity | [123] |
| | Increased glutathione contents and suppressed NF-κB subunits | [128] |
| | Lowered the Aß42 deposits Enhanced the histone 3 and 4 acetylation Decreased histone deacetylase 2 expression | [81] |

### 5.3.3 PREVENTION FROM CARDIOVASCULAR DISEASES

Olive phenolic compounds have shown promising perspectives in alleviating cardiac health implications.[145] Gonzalez-Santiago et al.[58] reported that hydroxytyrosol lowered the size of atherosclerotic lesions and improved their antioxidant status in hyperlipidemic rabbits. Similarly, oleuropein promoted healthy dilatory action on coronary arteries and smooth muscles of rats.[104]

Adriamycin is a generally used antineoplastic agent but its utilization has been reported to cause cardiotoxicity.[74] The earlier findings of Andreadou et al.[8] manifested that oleuropein reduces toxic effects of doxorubicin (DXR) such as hepatotoxic and neurotoxic signs by lowering oxidative stress, and decreasing iNOS in cardiomyocytes and inhibiting lipid peroxidation products. Atherosclerosis damages vascular epithelium cells forming lesions that consequently increases platelet aggregation and endothelial adhesion molecule expression. Circulating monocytes are interacted with these molecules and differentiate into macrophages that scavenge oxidized LDL and TG-rich lipoproteins producing foam cells.[103]

Chang et al.[30] showed that oleuropein played a significant role in $Ca^{2+}$ mobilization of platelet aggregation and agonist-evoked $Ca^{2+}$ signaling. In type-2 diabetic patients, high concentration of free radicals induces dysfunction in $Ca^{2+}$ homeostasis of platelet aggregation.[111] Further, hydroxytyrosol also attenuated the release and entry of thrombin-activated $Ca^{2+}$. It has been anticipated that protein tyrosine phosphorylation is responsible for platelet aggregation.[20]

Visioli et al.[142] determined that both compounds suppress the copper sulfate-induced oxidation of LDL and total cholesterol alongside enhancing HDL and antioxidant enzymes concentration in hypocholesterolemic rats. Hydroxytyrosol and oleuropein decreased the lipopolysaccharide—stimulated expression of vascular adhesion molecule-1, monocyte cell adhesion to endothelial cells, and prevented from the vascular damages and platelet aggregation.[27]

In another study, Abe et al.[1] determined that oleuropein and hydroxytyrosol-inhibited COX-2 by lowering thromboxane A2 blood levels and enhancing vascular nitric oxide production. Both compounds inhibited the vascular smooth muscle cell (SMC) proliferation that associated with extracellular regulated kinase-1/2 activity.[1]

Manna et al.[83] determined that oleuropein significantly lowered the creatine kinase and oxidized glutathione, which are involved in the pathogenesis of atherosclerotic rats treated with DXR.[6] Similarly, Andreadou et al.[7] determined that utilization of oleuropein considerably lowered the infarct size, TG accumulation, and expression of proteins in hypercholesterolemic rabbits. Moreover, oleuropein lowers the level of creatine phosphokinase (CPK), lactate dehydrogenase (LDH), creatine phosphokinase-MB (CPK-MB), aspartate aminotransferase (AST), and alanine aminotransferase (ALT) in DXR-induced acute cardiotoxic rats. It also reduced the content of protein carbonyls (PCs), nitrotyrosine (NT), iNOS, and lipid peroxidation in myocardial tissue.[8]

Furthermore, Ebaid et al.[46] evaluated that oleuropein and hydroxytyrosol exhibited cardioprotective effect by enhancing oxygen consumption, myocardial beta-hydroxyacyl coenzyme-A dehydrogenase activity, and fat oxidation, and increasing catalase, superoxide dismutase, and glutathione peroxidase in myocardial tissue of rats. It has been anticipated that oral intake of oleuropein (500 mg twice daily for 56 days) considerably lowered the diastolic and systolic blood pressures and the levels of TG and low-density lipoprotein (LDL) cholesterol in hypertensive rats.[36]

Bester et al.[16] determined that oleuropein lowered the systolic and diastolic blood pressures due to $Ca^{2+}$ channel agonist property. It was further recorded that it also improved endothelial function by suppressing the formation of ROS that leads to NO-mediated vasorelaxation.[16] Furthermore, oleuropein (100 M) modulated apolipoprotein (Apo) A-I and HDL, inhibited vascular SMC proliferation, blocked cells in the G1-S phase, and constrained extracellular signal-regulated kinase 1/2 (ERK1/2). It also reduced atherogenesis and arterial injury in rats.[1] In addition, various concentrations of oleuropein (10, 20, and 50 g/mL) lowered the level of NO and iNOS in anoxia stress rats. It protected the proliferation and migration associated with anoxia in AEhy926 cells, down-regulated the levels of membrane type-1 MMP (MT-1-MMP), MMP-2, and MMP-9 and enhanced the expression of tissue MMP inhibitor-1 (TIMP-1).[78]

In another study, Poudyal et al.[105] investigated that supplemented hydroxytyrosol (20 mg/kg/d for 8 weeks) markedly (1) lowered the adiposity, (2) improved endothelial function, (3) reduced systolic blood pressure, (4) modulated glucose impairment and insulin tolerance, and (5) reduced the left ventricular fibrosis. The subsequent diastolic stiffness and abridged liver damage markers in a diet-induced metabolic syndrome in rat model have also been noticed. Further, it also decreased the infiltration of monocytes in heart that reduces the biomarkers of oxidative stress.[105]

In hypertensive rats (SHR) fed with oleuropein @ 60 mg/kg/day, there was decrease in blood pressure, expression of components of the renin–angiotensin system, and pro-inflammatory cytokines. It also regulated superoxide and enhanced antioxidant defense system. Moreover, oleuropein caused enhancement in mitochondrial biogenesis and also enhanced the phase II enzyme levels of nuclear factor-E2-related factor 2 (Nrf2) and its down-stream proteins HO-1 and NQO-1.[129]

### 5.3.4 OXIDATIVE STRESS

Imbalance between oxidizing and antioxidants moieties results in onset of oxidative stress, which is a cardinal factor in etiology of numerous degenerative disorders. Hence, it must be managed to attain oxidative homeostasis in the body.[9,109] In this context, oxidative stress markers (such as hydroxy-fatty acids, cholesterol-conjugated dienes) and oxidative DNA damage were shown to decline linearly with the ingestion of higher dose of olive oil phenolic compounds in healthy male subjects.[36,146] Supplementation of hydroxytyrosol and oleuropein were found to suppress copper sulfate-induced oxidation of LDL alongside reduction of malondialdehyde (MDA) levels.[138] Furthermore, these compounds showed capability to prevent tyrosine nitration and peroxynitrite-dependent DNA damage.[155] In addition, these phenolics also neutralized free radicals, mediated cytotoxicity, and also suppressed passive smoking-induced oxidative damage in erythrocytes cells of human.[141] Also, the administration of hydroxytyrosol decreased ROS production in human hepatoma human liver cancer cell line (HepG2) cells.[62,97]

Presence of oxidative stress can be determined through different markers including: lipid peroxides (LPO), oxidized glutathione (GSSG), glutathione peroxidase (GSH-Px), F2-isoprostanes, and reduced glutathione (GSH) F2-isoprostanes.[18,120] Visioli et al.[139] observed that utilization of a phenolic-rich hydroxytyrosol was linked with a momentous decrease in urinary excretion of F2-isoprostanes. Reportedly, hydroxytyrosol induced impairment in NO-mediated relaxation associated with oxidative stress and showed to protect the aorta. The cell surface expression of vascular cell adhesion molecule (VCAM-1) and intercellular adhesion molecule-1 (ICAM-1) were inhibited after hydroxytyrosol treatment in human umbilical vascular endothelial cells.[114]

It has been reported that in vitro antioxidant effect of hydroxytyrosol efficiently neutralizes the cytotoxicity of free radicals in melanoma and epithelial Caco-2 intestinal cells. Further, it also showed an effective role against MDA formation and $H_2O_2$-induced oxidative hemolysis in intact human erythrocytes.[39] Hydroxytyrosol also prevented DNA damage induced by Sudan-I and acrylamide in HepG2 cells.[153,154] Furthermore, it also prevented from oxidative damage in ARPE-19 human retinal pigment epithelial cells by inducing phase II detoxifying enzymes, stimulating the mitochondrial biogenesis, and inhibiting peroxisome proliferator-activated receptor coactivator-1 alpha and Nrf2 in ARPE-19 cells.[155]

Similarly, oleuropein (15 mg/kg) in ethanol-induced oxidative stressed male Sprague–Dawley rats showed positive effects on motility and plasma membrane integrity of the spermatozoa. It also enhanced the concentration of glutathione peroxidase and superoxide dismutase and lowered the TBARS.[4] Moreover, it has been anticipated that hydroxytyrosol prevents from oxidative stress by modulating Nrf2-dependent gene expression in the heart of senescence-accelerated-prone mouse (SAMP8). It also lowered the concentrations of PCs and TBARS in cardiac tissues. Additionally, it also induced Nrf2-dependent gene expression and enhances paraoxonase-2 paraoxonases (PON) activity.[13]

A group of researchers determined that administration of oleuropein (30 mg/kg/day) exhibited preventive effect against sodium arsenite-induced toxicity in rats by decreasing blood, liver, kidney, and brain MDA, PC content, iNOS, and endothelial NOS (eNOS) in liver.[5,94] Similarly, hydroxytyrosol also prevented from cell damage, glutathione enzyme depletion, and IL-1β-induced IL-8 synthesis in Caco-2 cells grown in hydroxytyrosol-enriched medium.[110]

### 5.3.5 ANTIMICROBIAL BENEFITS

Olive fruit bioactives (including protocatechuic, *p*-coumaric acids, hydroxytyrosol, *p*-hydroxybenzoic acid, oleuropein, tyrosol, elenolic acid, vanillic acid, caffeic, and quercetin) have shown antimicrobial activity against Gram-negative and Gram-positive bacteria.[89]

Hydroxytyrosol has been shown to possess anti-infective effect against yeasts, fungi, bacteria, and viruses.[152] Mechanistically, hydroxytyrosol disrupted cell wall synthesis, suppressed protein synthesis, blocked metabolic pathways, and subsequently, inhibited growth of microbes or destroyed them.[53] Similar results were reported by another group of researchers showing bioactive moieties in olive fruit such as oleuropein and hydroxytyrosol to be inhibiting the growth rate of bacteria and microfungi.[12,23] Additionally, hydroxytyrosol showed antimicrobial activities against numerous microorganisms including: *Klebsiella pneumoniae*, *S.*, *aureus*, *Salmonella typhi*,

*Bacillus cereus*, *V. parahaemolyticus*, and *Escherichia coli*.[84] Additionally, hydroxytyrosol also represented antifungal activity against *Fusarium sambucinum*, *Verticillium dahliae*, and *Alternaria solani*.[150]

It has been seen that hydroxytyrosol and oleuropein prevented HIV from entering into the host cell and bind the catalytic site of HIV-1 integrase. Hydroxytyrosol and its derivatives are also useful for preventing infections caused by viruses, bacteria, and fungi. Moreover, hydroxytyrosol inactivated influenza A virus.[148] Oleuropein, as an antimicrobial agent, enhanced the nitric oxide production in mouse macrophages against endotoxins by *Salmonella enteritidis*, *Bacillus cereus* T spores, and *S. aureus*. Additionally, it delayed the onset of growth and sporulation of *Aspergillus parasiticus*.[104] Recently, Medina-Martínez et al.[88] investigated that hydroxytyrosol (1000 μg/mL) depicted 15% reduction in *E. coli* 533 and 679. Hydroxytyrosol showed good scavenging and antibacterial activity versus a strain of *Staphylococcus epidermidis* alongside reducing bacterial adhesion significantly.[38]

### 5.3.6 ANTIAGING BENEFITS

Aging is associated with enhancement of mitochondrial enzymes and protein carbonylation and increased nuclear and mitochondrial DNA oxidation with reduced levels of endogenous scavengers.[57] Diet therapies especially consumption of fruits and vegetables play a protective role against toxic effects of aging on brain.[49] It has been shown in anesthetized mice that hydroxytyrosol from olive oil could pass through the blood–brain barrier and inhibited the effect of free radicals in blood and brain acting as an antioxidant.[147] Hence, hydroxytyrosol is the primary degradative product of oleuropein exhibiting neuroprotective effect in hypoxia-reoxygenation model and significantly inhibited LDL efflux both in vitro and in vivo in rat brain slices.[14]

RA is an autoimmune disease and produces prolonged inflammation and damage owing to an increase in concentration of neutrophils and macrophages due to high load of free radicals in the free-radical producing and synovial fluid enzymes.[41] Polyphenolic compounds showed a diverse range of indirect actions that may produce positive impact on health such as inhibition of enzymes and metabolic activation of procarcinogens and inhibition of platelet aggregation.[76] The oleuropein has been found to inhibit respiratory burst of neutrophils by scavenging hydroxyl radicals and superoxide anions.[77] Additionally, it also ameliorated arthritis induced by injection of collagen type II in experimental mice.[37]

Olive oil polyphenols exhibited a potent antioxidant activity and prevented ROS-mediated cell injury.[122] Supplementation of oleuropein also inhibited activator protein 1 (AP-1) and lipopolysaccharide (LPS) triggered NF-κB activation in human umbilical vein endothelial cells.[45] Interestingly, oleuropein also constrained the release of pro-inflammatory cytokines and chemokinases and reduction of leukocyte infiltration determined through myeloperoxidase activity.[8] Both polyphenols considerably attenuated inflammatory mediators expressions owing to their free radical scavenging and antioxidant properties in mouse model of carrageenan-induced pleurisy.[86]

Peroxynitrite (cytotoxic molecule) produces trauma in the spinal cord tissue through tyrosine nitration and lipid peroxidation and also causes DNA damage by activating poly(ADP-ribose) polymerase (PARP).[19] Oleuropein significantly attenuated expression of PARP and NT with further reduction in peroxynitrite formation.[44] Schaffer et al.[123] determined that administration of hydroxytyrosol (100 mg/kg BW) for 12 days attenuated $Fe^{2+}$- and nitric oxide-induced cytotoxicity via increasing oxidative stress resistance of dissociated brain cells in subjects.

The different concentrations of oleuropein (50, 75, and 100 mg/kg/day) enhanced the cholesterol, brain cholesterol ester, cerebroside, and phosphatidylcholine levels. It also mediated neuroprotection owing to enhancement of brain phosphatidylcholine levels. Oleuropein prevented formation of alkyl and peroxyl radicals and declined the expression of TNF-α activation to inhibit LPO in damaging brain.[145] Oleuropein may protect dopaminergic neuron loss in mid-brain of old people.[112] Similarly, Luccarini et al.[81] determined the effect of oleuropein against amyloid-ß fragments, pyroglutamylated-Aß peptides, and oligomeric Aß and autophagy dysfunction and epigenetic mechanisms. The oleuropein lowered the Aß42 deposits in the brain of young and middle-aged transgenic hemizygous CRND8 mouse (TgCRND8). It also enhanced the histone 3 and 4 acetylation, lowered the histone deacetylase 2 expression, and significantly improved synaptic function. On the other hand, hydroxytyrosol prevented from cell death by enhancing glutathione contents and suppressing NF-κB subunits.[128]

## 5.4 SUMMARY

Olive (*O. europaea*) is a promising source of hydroxytyrosol and oleuropein that exhibit cytoprotective effects against oxidative stress. These bioactive-based products have been gaining popularity and are being used to prevent and cure health-related disorders such as: cancer, cardiovascular diseases

(CVDs), aging, and viral and microbial infections. Additionally, these compounds provide protection to the hematological biomarkers against oxidative damage to inhibit tumor-cell proliferation, leukocyte leukotriene, peroxynitrite-dependent damage, and also reduce the production of superoxide anion in the human promonocyte cells.

Evidently, oleuropein exhibits health-promoting properties including, anti-arrhythmic, anti-inflammatory (responsible for inhibition of 5-lipoxygenase enzyme), immune-stimulant, hypoglycemic, antimicrobial (even against HIV), cytostatic (against McCoy cells), and enzyme-modulating effects due to its antioxidative properties. Oleuropein and hydroxytyrosol polyphenols lower lipid peroxidation and exhibit hypoglycemic and hypolipidemic effects through enhancement of antioxidant defense mechanism in diabetic animals. Conclusively in future, these bioactive compounds have broad spectrum of usage to effectively ameliorate various maladies. Further research is needed to explore therapeutic role of these phenolic compounds in humans with precise clinical models.

## KEYWORDS


- **olive (*Olea europaea*) oil**
- **oleuropein**
- **hydroxytyrosol**
- **anticancer**
- **antioxidant**
- **metabolic disorders**


## REFERENCES


1. Abe, R.; Beckett, J.; Abe, R.; Nixon, A.; Rochier, A.; Yamashita, N.; Sumpio, B. Olive Oil Polyphenol Oleuropein Inhibits Smooth Muscle Cell Proliferation. *Eur. J. Vasc. Endovasc. Surg.* **2011,** *41* (6), 814–820.
2. Acquaviva, R.; Di Giacomo, C.; Sorrenti, V.; Galvano, F.; Santangelo, R.; Cardile, V.; Gangia, S.; D'orazio, N.; Abraham, N. G.; Vanella, L. Antiproliferative Effect of Oleuropein in Prostate Cell Lines. *Int. J. Oncol.* **2012,** *41* (1), 31–38.
3. Al-Azzawie, H. F.; Alhamdani, M.-S. S. Hypoglycemic and Antioxidant Effect of Oleuropein in Alloxan-Diabetic Rabbits. *Life Sci.* **2006,** *78* (12), 1371–1377.

4. Alirezaei, M.; Dezfoulian, O.; Kheradmand, A.; Neamati, S.; Khonsari, A.; Pirzadeh, A. Hepatoprotective Effects of Purified Oleuropein from Olive Leaf Extract Against Ethanol-Induced Damages in the Rat. *Iranian J. Veter. Res.* **2012,** *13* (3), 218–226.
5. Alirezaei, M.; Rezaei, M.; Hajighahramani, S.; Sookhtehzari, A.; Kiani, K. Oleuropein Attenuates Cognitive Dysfunction and Oxidative Stress Induced by Some Anesthetic Drugs in the Hippocampal Area of Rats. *J. Physiol. Sci.* **2016,** *67* (1), 131–139.
6. Andreadou, I.; Benaki, D.; Efentakis, P.; Bibli, S.-I.; Milioni, A.-I.; Papachristodoulou, A.; Zoga, A.; Skaltsounis, A.-L.; Mikros, E.; Iliodromitis, E. The Natural Olive Constituent Oleuropein Induces Nutritional Cardioprotection in Normal and Cholesterol-fed Rabbits: Comparison with Preconditioning. *Planta Medica* **2014,** *81* (08), 655–663.
7. Andreadou, I.; Papaefthimiou, M.; Zira, A.; Constantinou, M.; Sigala, F.; Skaltsounis, A.-L.; Tsantili-Kakoulidou, A.; Iliodromitis, E. K.; Kremastinos, D. T.; Mikros, E. Metabonomic Identification of Novel Biomarkers in Doxorubicin Cardiotoxicity and Protective Effect of the Natural Antioxidant Oleuropein. *NMR Biomed.* **2009,** *22* (6), 585–592.
8. Andreadou, I.; Sigala, F.; Iliodromitis, E. K.; Papaefthimiou, M.; Sigalas, C.; Aligiannis, N.; Savvari, P.; Gorgoulis, V.; Papalabros, E.; Kremastinos, D. T. Acute doxorubicin Cardiotoxicity Is Successfully Treated with the Phytochemical Oleuropein Through Suppression of Oxidative and Nitrosative Stress. *J. Mol. Cellular Cardiol.* **2007,** *42* (3), 549–558.
9. Awan, K.; Butt, M.; Iahtisham-Ul-Haq; Ashfaq, F.; Suleria, H. Storage Stability of Garlic Fortified Chicken Bites. *J. Food Chem. Nanotechnol.* **2017,** *3* (2), 80–85.
10. Bai, C.; Yan, X.; Takenaka, M.; Sekiya, K.; Nagata, T. Determination of Synthetic Hydroxytyrosol in Rat Plasma by GC-MS. *J. Agric. Food Chem.* **1998,** *46* (10), 3998–4001.
11. Balsano, C.; Alisi, A. Antioxidant Effects of Natural Bioactive Compounds. *Curr. Pharm. Des.* **2009,** *15* (26), 3063–3073.
12. Battinelli, L.; Daniele, C.; Cristiani, M.; Bisignano, G.; Saija, A.; Mazzanti, G. In Vitro Antifungal and Anti-elastase Activity of Some Aliphatic Aldehydes from *Olea europaea* L. Fruit. *Phytomedicine* **2006,** *13* (8), 558–563.
13. Bayram, B.; Ozcelik, B.; Grimm, S.; Roeder, T.; Schrader, C.; Ernst, I. M. A.; Wagner, A. E.; Grune, T.; Frank, J.; Rimbach, G. A Diet rich in Olive Oil Phenolics Reduces Oxidative Stress in the Heart of SAMP8 Mice by Induction of Nrf2-Dependent Gene Expression. *Rejuven. Res.* **2012,** *15* (1), 71–81.
14. Bazoti, F. N.; Bergquist, J.; Markides, K. E.; Tsarbopoulos, A. Noncovalent Interaction between Amyloid-β-Peptide (1-40) and Oleuropein Studied by Electrospray Ionization Mass Spectrometry. *J. Am. Soc. Mass Spectr.* **2006,** *17* (4), 568–575.
15. Benavente-GarcíA, O.; Castillo, J.; Lorente, J.; Ortuño, A.; Del Rio, J. A. Antioxidant Activity of Phenolics Extracted from *Olea europaea* L. Leaves. *Food Chem.* **2000,** *68* (4), 457–462.
16. Bester, D.; Esterhuyse, A. J.; Truter, E. J.; Van Rooyen, J. Cardiovascular Effects of Edible Oils: A Comparison Between Four Popular Edible Oils. *Nutr. Res. Rev.* **2010,** *23* (2), 334–348.
17. Bianco, A.; Scalzo, R. L.; Scarpati, M. L. Isolation of Cornoside from *Olea europaea* and Its Transformation into Halleridone. *Phytochemistry* **1993,** *32* (2), 455–457.

18. Biswas, S. K.; Newby, D. E.; Rahman, I.; Megson, I. L. Depressed Glutathione Synthesis Precedes Oxidative Stress and Atherogenesis in Apo-E−/− Mice. *Biochem. Biophys. Res. Commun.* **2005,** *338* (3), 1368–1373.
19. Blanchard-Fillion, B.; Souza, J. M.; Friel, T.; Jiang, G. C. T.; Vrana, K.; Sharov, V.; Barrón, L.; Schöneich, C.; Quijano, C.; Alvarez, B.; Radi, R.; Przedborski, S.; Fernando, G. S.; Horwitz, J.; Ischiropoulos, H. Nitration and Inactivation of Tyrosine Hydroxylase by Peroxynitrite. *J. Biol. Chem.* **2001,** *276* (49), 46017–46023.
20. Bobe, R.; Bredoux, R.; Corvazier, E.; Andersen, J. P.; Clausen, J. D.; Dode, L.; Kovács, T.; Enouf, J. Identification, Expression, Function, and Localization of a Novel (Sixth) Isoform of the Human Sarco/Endoplasmic Reticulum Ca2+ATPase 3 gene. *J. Biol. Chem.* **2004,** *279* (23), 24297–24306.
21. Bonoli, M.; Bendini, A.; Cerretani, L.; Lercker, G.; Gallina Toschi, T. Qualitative and Semiquantitative Analysis of Phenolic Compounds in Extra Virgin Olive Oils as a Function of the Ripening Degree of Olive Fruits by Different Analytical Techniques. *J. Agric. Food Chem.* **2004,** *52* (23), 7026–7032.
22. Bouallagui, Z.; Han, J.; Isoda, H.; Sayadi, S. Hydroxytyrosol Rich Extract from Olive Leaves Modulates Cell Cycle Progression in MCF-7 Human Breast Cancer Cells. *Food Chem. Toxicol.* **2011**, *49* (1), 179–184.
23. Boudet, A.-M. Evolution and Current Status of Research in Phenolic Compounds. *Phytochemistry* **2007**, *68* (22–24), 2722–2735.
24. Cao, K.; Xu, J.; Zou, X.; Li, Y.; Chen, C.; Zheng, A.; Li, H.; Li, H.; Szeto, I. M.-Y.; Shi, Y.; Long, J.; Liu, J.; Feng, Z. Hydroxytyrosol Prevents Diet Induced Metabolic Syndrome and Attenuates Mitochondrial Abnormalities in Obese Mice. *Free Rad. Biol. Med.* **2014**, *67*, 396–407.
25. Cárdeno, A.; Sánchez-Hidalgo, M.; Rosillo, M. A.; De La Lastra, C. A. Oleuropein, a Secoiridoid Derived from Olive Tree, Inhibits the Proliferation of Human Colorectal Cancer Cell Through Down-regulation of HIF-1α. *Nutr. Cancer* **2013**, *65* (1), 147–156.
26. Cardoso, S. M.; Falcão, S. I.; Peres, A. M.; Domingues, M. R. Oleuropein/Ligstroside Isomers and Their dDerivatives in Portuguese Olive Mill Wastewaters. *Food Chem.* **2011**, *129* (2), 291–296.
27. Carluccio, M. A. Olive oil and Red Wine Antioxidant Polyphenols Inhibit Endothelial Activation: Antiatherogenic Properties of Mediterranean Diet Phytochemicals. *Arterioscler. Thromb. Vasc. Biol.* **2003**, *23* (4), 622–629.
28. Caruso, D.; Colombo, R.; Patelli, R.; Giavarini, F.; Galli, G. Rapid Evaluation of Phenolic Component Profile and Analysis of Oleuropein Aglycon in Olive Oil by Atmospheric Pressure Chemical Ionization-Mass Spectrometry (APCI− MS). *J. Agric. Food Chem.* **2000,** *48* (4), 1182–1185.
29. Casas, E. M.; Albadalejo, M. F.; Planells, M. I. C.; Colomer, M. F.; Raventós, R. M. L.; De La Torre Fornell, R. Tyrosol Bioavailability in Humans After Ingestion of Virgin Olive Oil. *Clin. Chem.* **2001,** *47* (2), 341–343.
30. Chang, M. C.; Uang, B. J.; Tsai, C. Y.; Wu, H. L.; Lin, B. R.; Lee, C. S.; Chen, Y. J.; Chang, C. H.; Tsai, Y. L.; Kao, C. J.; Jeng, J. H. Hydroxychavicol, a Novel Betel Leaf Component, Inhibits Platelet Aggregation by Suppression of Cyclooxygenase, Thromboxane Production and Calcium Mobilization. *Br. J. Pharmacol.* **2007**, *152* (1), 73–82.

31. Christian, M. S.; Sharper, V. A.; Hoberman, A. M.; Seng, J. E.; Fu, L.; Covell, D.; Diener, R. M.; Bitler, C. M.; Crea, R. The Toxicity Profile of Hydrolyzed Aqueous Olive Pulp Extract. *Drug Chem. Toxicol.* **2004,** *27* (4), 309–330.
32. Cicerale, S.; Conlan, X. A.; Sinclair, A. J.; Keast, R. S. J. Chemistry and Health of Olive Oil Phenolics. *Crit. Rev. Food Sci.Nutr.* **2008,** *49* (3), 218–236.
33. Cornwell, D. G.; Ma, J. Nutritional benefit of Olive Oil: the Biological Effects of Hydroxytyrosol and Its Arylating Quinone Adducts. *J. Agric. Food Chem.*, **2008,** *56* (19), 8774–8786.
34. Corona, G.; Deiana, M.; Incani, A.; Vauzour, D.; Assunta Dessì, M.; Spencer, J. P. E. Inhibition of p38/CREB Phosphorylation and COX-2 Expression by Olive Oil Polyphenols Underlies Their Anti-proliferative Effects. *Biochem. Biophys. Res. Commun.* **2007,** *362* (3), 606–611.
35. Corona, G.; Tzounis, X.; Assunta Dessì, M.; Deiana, M.; Debnam, E. S.; Visioli, F.; Spencer, J. P. E. The Fate of Olive Oil Polyphenols in the Gastrointestinal Tract: Implications of Gastric and Colonic Microflora-dependent Biotransformation. *Free Radical Res.* **2006**, *40* (6), 647–658.
36. Covas, M.-I.; Nyyssönen, K.; Poulsen, H. E.; Kaikkonen, J.; Zunft, H.-J. F.; Kiesewetter, H.; Gaddi, A.; De La Torre, R.; Mursu, J.; Bäumler, H.; Nascetti, S.; Salonen, J. T.; Fitó, M.; Virtanen, J.; Marrugat, J. The effect of Polyphenols in Olive Oil on Heart Disease Risk Factors. *Ann. Internal Med.* **2006,** *145* (5), 333–341.
37. Covas, M. I. Bioactive Effects of Olive Oil Phenolic Compounds in Humans: Reduction of Heart Disease Factors and Oxidative Damage. *Inflammopharmacology* **2008,** *16* (5), 216–218.
38. Crisante, F.; Taresco, V.; Donelli, G.; Vuotto, C.; Martinelli, A.; D'ilario, L.; Pietrelli, L.; Francolini, I.; Piozzi, A. Antioxidant Hydroxytyrosol-Based Polyacrylate with Antimicrobial and Antiadhesive Activity Versus *Staphylococcus Epidermidis. Advances in Experimental Medicine and Biology;* Springer International Publishing, Cham; 2015.
39. D'angelo, S.; Ingrosso, D.; Migliardi, V.; Sorrentino, A.; Donnarumma, G.; Baroni, A.; Masella, L.; Tufano, M. A.; Zappia, M.; Galletti, P. Hydroxytyrosol, a Natural antioxidant from Olive Oil, Prevents Protein Damage Induced by Long-wave ultraviolet radiation in Melanoma Cells. *Free Rad. Biol. Med.* **2005,** *38* (7), 908–919.
40. D'angelo, S.; Manna, C.; Migliardi, V.; Mazzoni, O.; Morrica, P.; Capasso, G.; Pontoni, G.; Galletti, P.; Zappia, V. Pharmacokinetics and Metabolism of Hydroxytyrosol, a Natural Antioxidant from Olive Oil. *Drug Metab. Dispos.* **2001,** *29* (11), 1492–1498.
41. Darlington, L. G.; Stone, T. W. Antioxidants and Fatty Acids in the Amelioration of Rheumatoid Arthritis and Related Disorders. *Br. J. Nutr.* **2001,** *85* (3), 251–269.
42. De Pinedo, A. T.; Penalver, P.; Morales, J. Synthesis and Evaluation of New Phenolic-Based Antioxidants: Structure-Activity Relationship. *Food Chem.* **2007,** *103* (1), 55–61.
43. Decker, E. A.; Warner, K.; Richards, M. P.; Shahidi, F. Measuring Antioxidant Effectiveness in Food. *J. Agric. Food Chem.* **2005,** *53* (10), 4303–4310.
44. Deiana, M.; Aruoma, O. I.; Maria De Lourdes, P. B.; Spencer, J. P.; Kaur, H.; Halliwell, B.; Aeschbach, R.; Banni, S.; Dessi, M. A.; Corongiu, F. P. Inhibition of Peroxynitrite Dependent DNA Base Modification and Tyrosine Nitration by the Extra Virgin Olive Oil-Derived Antioxidant Hydroxytyrosol. *Free Rad. Biol. Med.* **1999,** *26* (5), 762–769.

45. Dong, J.; Gao, Y.; Liu, Y.; Shi, J.; Feng, J.; Li, Z.; Pan, H.; Xue, Y.; Liu, C.; Shen, B. The Protective Antibodies Induced by a Novel Epitope of Human TNF-α Could Suppress the Development of Collagen-induced Arthritis. *PloS One* **2010,** *5* (1), e8920.
46. Ebaid, G. M.; Seiva, F. R.; Rocha, K. K.; Souza, G. A.; Novelli, E. L. Effects of Olive Oil and Its Minor Phenolic Constituents on Obesity-induced Cardiac Metabolic Changes. *Nutrition J.* **2010,** *9* (1), 46.
47. Edgecombe, S. C.; Stretch, G. L.; Hayball, P. J. Oleuropein, an Antioxidant Polyphenol from Olive Oil, Is Poorly Absorbed from Isolated Perfused Rat Intestine. *The J. Nutr.* **2000,** *130* (12), 2996–3002.
48. Elamin, M. H.; Hassan, Z. K.; Omer, S. A.; Daghestani, M. H.; Al-Olayan, E. M.; Virk, P.; Elobeid, M. A.; Mohammed, O. B. Apoptotic and Antiproliferative Activity of Olive Oil Hydroxytyrosol on Breast Cancer Cells. *J. Med. Plants Res.* **2013,** *7* (32), 2420–2428.
49. Everitt, A. V.; Hilmer, S. N.; Brand-Miller, J. C.; Jamieson, H. A.; Truswell, A. S.; Sharma, A. P.; Mason, R. S.; Morris, B. J.; Le Couteur, D. G. Dietary Approaches That Delay Age-Related Diseases. *Clin. Intervent. Aging* **2006**, *1* (1), 11–31.
50. Fabiani, R.; De Bartolomeo, A.; Rosignoli, P.; Servili, M.; Montedoro, G.; Morozzi, G. Cancer Chemoprevention by Hydroxytyrosol Isolated from Virgin Olive Oil Through G1 Cell Cycle Arrest and Apoptosis. *Eur. J. Cancer Prevent.* **2002,** *11* (4), 351–358.
51. Fabiani, R.; De Bartolomeo, A.; Rosignoli, P.; Servili, M.; Selvaggini, R.; Montedoro, G. F.; Di Saverio, C.; Morozzi, G. Virgin Olive Oil Phenols Inhibit proliferation of human Promyelocytic Leukemia Cells (HL60) by Inducing Apoptosis and Differentiation. *J. Nutr.* **2006,** *136* (3), 614–619.
52. Finn, R. S.; Zhu, A. X. Targeting Angiogenesis in Hepatocellular Carcinoma: Focus on VEGF and Bevacizumab. *Expert Rev. Anticancer Ther.* **2009**, *9* (4), 503–509.
53. Friedman, M.; Henika, P. R.; Mandrell, R. E. Bactericidal Activities of Plant Essential Oils and Some of Their Isolated Constituents Against *Campylobacter jejuni*, *Escherichia coli*, *Listeria Monocytogenes*, and *Salmonella Enterica*. *J. Food Protect.* **2002,** *65* (10), 1545–1560.
54. Fu, S.; Arráez-Román, D.; Menendez, J. A.; Segura-Carretero, A.; Fernández-Gutiérrez, A. Characterization of Isomers of Oleuropein Aglycon in Olive Oils by Rapid-resolution Liquid Chromatography Coupled to Electrospray Time-of-flight and Ion trap Tandem Mass Spectrometry. *Rapid Commun. Mass Spectr.* **2009,** *23* (1), 51–59.
55. Ghanbari, R.; Anwar, F.; Alkharfy, K. M.; Gilani, A.-H.; Saari, N. Valuable Nutrients and Functional Bioactives in Different Parts of Olive (*Olea europaea* L.): A Review. *Int. J. Mol. Sci.* **2012,** *13* (3), 3291–3340.
56. Giner, E.; Recio, M. C.; Ríos, J. L.; Cerdá-Nicolás, J. M.; Giner, R. M. Chemopreventive Effect of Oleuropein in Colitis-associated Colorectal Cancer in c57bl/6 Mice. *Mol. Nutr. Food Res.* **2016,** *60* (2), 242–255.
57. Giovannelli, L.; Decorosi, F.; Dolara, P.; Pulvirenti, L. Vulnerability to DNA Damage in the Aging Rat Substantia Nigra: A Study with the Comet Assay. *Brain Res.* **2003,** *969* (1), 244–247.
58. Gonzalez-Santiago, M.; Martín-Bautista, E.; Carrero, J.; Fonolla, J.; Baro, L.; Bartolomé, M.; Gil-Loyzaga, P.; Lopez-Huertas, E. One-month Administration of Hydroxytyrosol, a Phenolic Antioxidant Present in Olive Oil, to Hyperlipemic Rabbits Improves Blood Lipid Profile, Antioxidant Status and Reduces Atherosclerosis Development. *Atherosclerosis* **2006,** *188* (1), 35–42.

59. Gonzalez, M.; Zarzuelo, A.; Gamez, M.; Utrilla, M.; Jimenez, J.; Osuna, I. Hypoglycemic Activity of Olive Leaf. *Planta Medica* **1992,** *58* (6), 513–515.
60. Goulas, V.; Exarchou, V.; Troganis, A. N.; Psomiadou, E.; Fotsis, T.; Briasoulis, E.; Gerothanassis, I. P. Phytochemicals in Olive-leaf Extracts and Their Antiproliferative Activity Against Cancer and Endothelial Cells. *Mol. Nutr. Food Res.* **2009,** *53* (5), 600–608.
61. Goupy, P.; Fleuriet, A.; Amiot, M. J.; Macheix, J. J. Enzymic Browning, Oleuropein Content, and Diphenol Oxidase Activity in Olive Cultivars (*Olea europaea* L.). *J. Agric. Food Chem.* **1991,** *39* (1), 92–95.
62. Goyal, A.; Yusuf, S. The Burden of Cardiovascular Disease in the Indian Subcontinent. *Indian J. Med.Res.* **2006,** *124* (3), 235–244.
63. Gutierrez-Rosales, F.; Romero, M. P.; Casanovas, M. A.; Motilva, M. J.; MíNguez-Mosquera, M. I. Metabolites Involved in Oleuropein Accumulation and Degradation in Fruits of *Olea europaea* L.: Hojiblanca and Arbequina Varieties. *J. Agric. Food Chem.* **2010,** *58* (24), 12924–12933.
64. Hadrich, F.; Garcia, M.; Maalej, A.; Moldes, M.; Isoda, H.; Feve, B.; Sayadi, S. Oleuropein Activated AMPK and Induced Insulin Sensitivity in C2C12 Muscle Cells. *Life Sci.* **2016,** *151*, 167–173.
65. Hagiwara, K.; Goto, T.; Araki, M.; Miyazaki, H.; Hagiwara, H. Olive Polyphenol Hydroxytyrosol Prevents Bone Loss. *European J. Pharmacol.* **2011,** *662* (1), 78–84.
66. Hamden, K.; Allouche, N.; Damak, M.; Elfeki, A. Hypoglycemic and Antioxidant Effects of Phenolic Extracts and Purified Hydroxytyrosol from Olive Mill Waste *In Vitro* and in Rats. *Chemico-Biol. Interact.* **2009,** *180* (3), 421–432.
67. Hamdi, H. K.; Castellon, R. Oleuropein, a Non-toxic Olive Iridoid, Is an Anti-tumor Agent and Cytoskeleton Disruptor. *Biochem. Biophys. Res. Commun.* **2005,** *334* (3), 769–778.
68. Hasebe, Y.; Yokobori, K.; Fukasawa, K.; Kogure, T.; Uchiyama, S. Highly Sensitive Electrochemical Determination of *Escherichia Coli* Density Using Tyrosinase-based Chemically Amplified Biosensor. *Anal. Chimica Acta* **1997**, *357* (1), 51–54.
69. Hassan, Z. K.; Elamin, M. H.; Omer, S. A.; Daghestani, M. H.; Al-Olayan, E. S.; Elobeid, M. A.; Virk, P. Oleuropein Induces Apoptosis via the p53 Pathway in Breast Cancer Cells. *Asian Pacific J. Cancer Prevent.* **2013,** *14* (11), 6739–6742.
70. Hong, W. K. General Keynote: the Impact of Cancer Chemoprevention. *Gynecol. Oncol.* **2003,** *88* (1), S56–S58.
71. Impellizzeri, D.; Esposito, E.; Mazzon, E.; Paterniti, I.; Di Paola, R.; Bramanti, P.; Morittu, V. M.; Procopio, A.; Britti, D.; Cuzzocrea, S. The Effects of Oleuropein Aglycone, an Olive Oil Compound, in a Mouse Model of Carrageenan-induced Pleurisy. *Clin. Nutr.* **2011,** *30* (4), 533–540.
72. Japón-Luján, R.; Luque-Rodríguez, J.; De Castro, M. L. Dynamic Ultrasound-asisted Extraction of Oleuropein and Related Biophenols from Olive Leaves. *J. Chromatogr. A* **2006,** *1108* (1), 76–82.
73. Jemai, H.; El Feki, A.; Sayadi, S. Antidiabetic and Antioxidant Effects of Hydroxytyrosol and Oleuropein from Olive Leaves in Alloxan-Diabetic Rats. *J. Agric. Food Chem.* **2009,** *57* (19), 8798–8804.
74. Joshi, G.; Hardas, S.; Sultana, R.; St Clair, D. K.; Vore, M.; Butterfield, D. A. Glutathione Elevation by γ-glutamyl Cysteine Ethyl Ester as a Potential Therapeutic Strategy for Preventing Oxidative Stress in Brain Mediated by *In Vivo* Administration of Adriamycin: Implication for Chemobrain. *J. Neurosci. Res.* **2007,** *85* (3), 497–503.

75. Kane, S.; Seatter, M. J.; Gould, G. W. Functional Studies of Human GLUT5: Effect of PH on Substrate Selection and an Analysis of Substrate Interactions. *Biochem. Biophys. Res. Commun* **1997,** *238* (2), 503–505.
76. Kohyama, N.; Nagata, T.; Fujimoto, S.-I.; Sekiya, K. Inhibition of Arachidonate Lipoxygenase Activities by 2-(3, 4-dihydroxyphenyl) Ethanol, a Phenolic Compound from Olives. *Biosci. Biotechnol.Biochem.* **1997,** *61* (2), 347–350.
77. Kremastinos, D. T. Olive and Oleuropein. *Hellenic J. Cardiol.* **2008**, *49* (4), 295–296.
78. Lamy, S.; Ouanouki, A.; Béliveau, R.; Desrosiers, R. R. Olive oil Compounds Inhibit Vascular Endothelial Growth Factor Receptor-2 Phosphorylation. *Exp. Cell Res.* **2014,** *322* (1), 89–98.
79. Leger, C.; Carbonneau, M.; Michel, F.; Mas, E. A Thromboxane Effect of a Hydroxytyrosol-rich Olive Oil Wastewater Extract in Patients with Uncomplicated Type I Diabetes. *European J. Clin. Nutr.* **2005,** *59* (5), 727–730.
80. Levin, E. R. Integration of the Extranuclear and Nuclear Actions of Estrogen. *Mol. Endocrinol.* **2005,** *19* (8), 1951–1959.
81. Luccarini, I.; Dami, T. E.; Grossi, C.; Rigacci, S.; Stefani, M.; Casamenti, F. Oleuropein Aglycone Counteracts Aβ42 Toxicity in the Rat Brain. *Neuroscience Lett.* **2014,** *558*, 67–72.
82. Manna, C.; Galletti, P.; Maisto, G.; Cucciolla, V.; D'angelo, S.; Zappia, V. Transport Mechanism and Metabolism of Olive Oil Hydroxytyrosol in Caco-2 cells. *FEBS Lett.* **2000**, *470* (3), 341–344.
83. Manna, C.; Migliardi, V.; Golino, P.; Scognamiglio, A.; Galletti, P.; Chiariello, M.; Zappia, V. Oleuropein Prevents Oxidative Myocardial Injury Induced by Ischemia and Reperfusion. *The J. Nutr. Biochem.* **2004,** *15* (8), 461–466.
84. Markin, D.; Duek, L.; Berdicevsky, I. In Vitro Antimicrobial Activity of Olive Leaves. *Mycoses* **2003,** *46* (3–4), 132–136.
85. Martín-García, A.; Molina-Alcaide, E. Effect of Different Drying Procedures on the Nutritive Value of Olive (*Olea europaea* var. *europaea*) Leaves for Ruminants. *Animal Feed Sci. Technol.* **2008,** *142* (3), 317–329.
86. Martinínez-Domínguez, E.; De La Puerta, R.; Ruiz-Gutierrez, V. Protective Effects Upon Experimental Inflammation Models of a Polyphenol-Supplemented Virgin Olive Oil Diet. *Inflam. Res.* **2001,** *50* (2), 102–106.
87. Mateos, R.; Domínguez, M. M.; Espartero, J. L.; Cert, A. Antioxidant Effect of Phenolic Compounds, α-Tocopherol, and Other Minor Components in Virgin Olive Oil. *J. Agric. Food Chem.* **2003,** *51* (24), 7170–7175.
88. Medina-Martínez, M. S.; Truchado, P.; Castro-Ibáñez, I.; Allende, A. Antimicrobial Activity of Hydroxytyrosol: A Current Controversy. *Biosci. Biotechnol. Biochem.* **2016,** *80* (4), 801–810.
89. Moore, K. L.; Patel, J.; Jaroni, D.; Friedman, M.; Ravishankar, S. Antimicrobial Activity of Apple, Hibiscus, Olive, and Hydrogen Peroxide Formulations Against *Salmonella Enterica* on Organic Leafy Greens. *J. Food Protect.* **2011,** *74* (10), 1676–1683.
90. Moreno, J. J. Effect of Olive Oil Minor Components on Oxidative Stress and Arachidonic Acid Mobilization and Metabolism by Macrophages RAW 264.7. *Free Rad. Biol. Med.* **2003,** *35* (9), 1073–1081.
91. Mosca, L.; De Marco, C.; Visioli, F.; Cannella, C. Enzymatic Assay for the Determination of Olive Oil Polyphenol Content: Assay Conditions and Validation of the Method. *J. Agric. Food Chem.* **2000,** *48* (2), 297–301.

92. Murkovic, M.; Lechner, S.; Pietzka, A.; Bratacos, M.; Katzogiannos, E. Analysis of Minor Components in Olive Oil. *J. Biochem. Biophys. Methods* **2004,** *61* (1), 155–160.
93. Nomoto, M.; Yamada, K.; Haga, M.; Hayashi, M. Improvement of Intestinal Absorption of Peptide Drugs by Glycosylation: Transport of Tetrapeptide by the Sodium Ion-dependent D-glucose Transporter. *J. Pharm. Scie.* **1998,** *87* (3), 326–332.
94. Ogun, M.; Ozcan, A.; Karaman, M.; Merhan, O.; Ozen, H.; Kukurt, A.; Karapehlivan, M. Oleuropein Ameliorates Arsenic Induced Oxidative Stress in Mice. *J. Trace Elements Med. Biol.* **2016**, *36*, 1–6.
95. Ortega-García, F.; Blanco, S.; Peinado, M. Á.; Peragón, J. Polyphenol Oxidase and Its Relationship with Oleuropein Concentration in Fruits and Leaves of Olive (*Olea europaea*) cv.'Picual' Trees During Fruit Ripening. *Tree Physiol.* **2008,** *28* (1), 45–54.
96. Owen, R.; Mier, W.; Giacosa, A.; Hull, W.; Spiegelhalder, B.; Bartsch, H. Phenolic Compounds and Squalene in Olive Oils: The Concentration and Antioxidant Potential of Total Phenols, Simple Phenols, Secoiridoids, Lignansand Squalene. *Food Chem. Toxicol.* **2000,** *38* (8), 647–659.
97. Paiva-Martins, F.; Fernandes, J.; Rocha, S.; Nascimento, H.; Vitorino, R.; Amado, F.; Borges, F.; Belo, L.; Santos-Silva, A. Effects of Olive Oil Polyphenols on Erythrocyte Oxidative Damage. *Mol. Nutr. Food Res.* **2009,** *53* (5), 609–616.
98. Palmieri, D.; Aliakbarian, B.; Casazza, A. A.; Ferrari, N.; Spinella, G.; Pane, B.; Cafueri, G.; Perego, P.; Palombo, D. Effects of Polyphenol Extract from Olive Pomace on Anoxia-Induced Endothelial Dysfunction. *Microvasc. Res.* **2012,** *83* (3), 281–289.
99. Palsamy, P.; Subramanian, S. Modulatory Effects of Resveratrol on Attenuating the Key Enzymes Activities of Carbohydrate Metabolism in Streptozotocin–Nicotinamide-induced Diabetic Rats. *Chemico-Biol. Interact.* **2009,** *179* (2), 356–362.
100. Pang, R.; Poon, R. T. Angiogenesis and Antiangiogenic Therapy in Hepatocellular Carcinoma. *Cancer Lett.* **2006,** *242* (2), 151–167.
101. Panizzi, L.; Scarpati, M. Oriente, EG Structure of Oleuropein, Bitter Glycoside with Hypotensive Action of Olive Oil. Note II. *Gazzetta Chimica Italiana* **1960,** *90,* 1449–1485.
102. Pereira, A. P.; Ferreira, I. C.; Marcelino, F.; Valentão, P.; Andrade, P. B.; Seabra, R.; Estevinho, L.; Bento, A.; Pereira, J. A. Phenolic Compounds and Antimicrobial Activity of Olive (*Olea europaea* L. Cv. *Cobrançosa)* leaves. *Molecules* **2007,** *12* (5), 1153–1162.
103. Perona, J. S.; Cabello-Moruno, R.; Ruiz-Gutierrez, V. The Role of Virgin Olive Oil Components in the Modulation of Endothelial Function. *The J. Nutr. Biochem.* **2006,** *17* (7), 429–445.
104. Perrinjaquet-Moccetti, T.; Busjahn, A.; Schmidlin, C.; Schmidt, A.; Bradl, B.; Aydogan, C. Food Supplementation with an Olive (*Olea europaea* L.) Leaf Extract Reduces Blood Pressure in Borderline Hypertensive Monozygotic Twins. *Phytotherapy Res.* **2008,** *22* (9), 1239–1242.
105. Poudyal, H.; Lemonakis, N.; Efentakis, P.; Gikas, E.; Halabalaki, M.; Andreadou, I.; Skaltsounis, L.; Brown, L. Hydroxytyrosol Ameliorates Metabolic, Cardiovascular and Liver Changes in a Rat Model of Diet-induced Metabolic Syndrome:Pharmacological and Metabolism-based Investigation. *Pharmacol. Res.* **2017,** *117*, 32–45.
106. Procopio, A.; Alcaro, S.; Nardi, M.; Oliverio, M.; Ortuso, F.; Sacchetta, P.; Pieragostino, D.; Sindona, G. Synthesis, Biological Evaluation, and Molecular Modeling of Oleuropein and Its Semisynthetic Derivatives as Cyclooxygenase Inhibitors. *J. Agric. Food Chem.* **2009,** *57* (23), 11161–11167.

107. Pugh, C. W.; Ratcliffe, P. J. Regulation of Angiogenesis by Hypoxia: Role of the HIF System. *Nat. Med.* **2003,** *9* (6), 677–684.
108. Ranalli, A.; Contento, S.; Lucera, L.; Di Febo, M.; Marchegiani, D.; Di Fonzo, V. Factors Affecting the Contents of Iridoid Oleuropein in Olive Leaves (*Olea europaea* L.). *J. Agric. Food Chem.* **2006,** *54* (2), 434–440.
109. Raza, A.; Butt, M. S.; Iahtisham-Ul-Haq; Suleria, H. A. R. Jamun (*Syzygium cumini*) Seed and Fruit Extract Attenuate Hyperglycemia in Diabetic Rats. *Asian Pacific J. Trop. Biomed.* **2017**.
110. Reddy, K. V. K.; Naidu, K. A. Oleic Acid, Hydroxytyrosol and n-3 fatty Acids Collectively Modulate Colitis Through Reduction of Oxidative Stress and IL-8Synthesis; In Vitro and In Vivo Studies. *Int. Immunopharmacol.* **2016,** *35*, 29–42.
111. Redondo, P. C.; Salido, G. M.; Pariente, J. A.; Rosado, J. A. Dual Effect of Hydrogen Peroxide on Store-mediated Calcium Entry in Human Platelets. *Biochem. Pharmacol.* **2004,** *67* (6), 1065–1076.
112. Rehman, H. U.; Masson, E. A. Neuroendocrinology of Ageing. *Age Ageing* **2001,** *30* (4), 279–287.
113. Reyes, J. J.; Villanueva, B.; López-Villodres, J. A.; De La Cruz, J. P.; Romero, L.; Rodríguez-Pérez, M. D.; Rodriguez-Gutierrez, G.; FernáNdez-BolañOs, J.; González-Correa, J. A. Neuroprotective Effect of Hydroxytyrosol in Experimental Diabetes Mellitus. *J. Agric. Food Chem.* **2017,** *65* (22), 4378–4383.
114. Rietjens, S.; Bast, A.; De Vente, J.; Haenen, G. The Olive Oil Antioxidant Hydroxytyrosol Efficiently Protects Against the Oxidative Stress-induced Impairment of the NO' Response of Isolated Rat Aorta. *Am. J. Physiol.-Heart Circul. Physiol.* **2007,** *292* (4), H1931–H1936.
115. Ristagno, G.; Fumagalli, F.; Porretta-Serapiglia, C.; Orrù, A.; Cassina, C.; Pesaresi, M.; Masson, S.; Villanova, L.; Merendino, A.; Villanova, A. Hydroxytyrosol attenuates Peripheral Neuropathy in Streptozotocin-induced Diabetes in Rats. *J. Agric. Food Chem.* **2012,** *60* (23), 5859–5865.
116. Roginsky, V.; Lissi, E. A. Review of Methods to Determine Chain-breaking Antioxidant Activity in Food. *Food Chem.* **2005,** *92* (2), 235–254.
117. Romero-Segura, C.; Sanz, C.; Perez, A. G. Purification and Characterization of an Olive Fruit β-glucosidase Involved in the Biosynthesis of Virgin Olive Oil Phenolics. *J. Agric. Food Chem.* **2009,** *57* (17), 7983–7988.
118. Romier, B.; Schneider, Y.-J.; Larondelle, Y.; During, A. DietaryPolyphenols Can Modulate the Intestinal Inflammatory Response. *Nutr. Rev.* **2009,** *67* (7), 363–378.
119. Rosmorduc, O.; Housset, C. *Hypoxia: A Link Between Fibrogenesis, Angiogenesis, and Carcinogenesis in Liver Disease.* Thieme Medical Publishers, Paris, France; 2010, pp 258–270.
120. Ruano, J.; Lopez-Miranda, J.; Fuentes, F.; Moreno, J. A.; Bellido, C.; Perez-Martinez, P.; Lozano, A.; Gómez, P.; Jiménez, Y.; Jiménez, F. P. Phenolic Content of Virgin Olive Oil Improves Ischemic Reactive Hyperemia in Hypercholesterolemic Patients. *J. Am. Coll. Cardiol.* **2005,** *46* (10), 1864–1868.
121. Sabatini, N. Recent Patents in Olive Oil Industry: New Technologies for the Recovery of Phenols Compounds from Olive Oil, Olive Oil Industrial by-products and Waste Waters. *Recent Patents Food Nutr. Agric.* **2010,** *2* (2), 154–159.
122. Salvini, S.; Sera, F.; Caruso, D.; Giovannelli, L.; Visioli, F.; Saieva, C.; Masala, G.; Ceroti, M.; Giovacchini, V.; Pitozzi, V. Daily Consumption of a High-phenol

Extra-virgin Olive Oil Reduces Oxidative DNA Damage in Postmenopausal Women. *Br. J. Nutr.* **2006,** *95* (4), 742–751.

123. Schaffer, S.; Podstawa, M.; Visioli, F.; Bogani, P.; Müller, W. E.; Eckert, G. P. Hydroxytyrosol-rich Olive Mill Wastewater Extract Protects Brain Cells *In Vitro* and *Ex Vivo*. *J. Agric. Food Chem.* **2007,** *55* (13), 5043–5049.
124. Shah, N. A.; Ahmad, A.; Saleem, M.; Khair, S. M. Constraints and Opportunities in the Production and Marketing of Wild Olive in Highland Balochistan: Farmers Perception. *Asian J. Plant Sci.* **2002,** *1* (4), 370–372.
125. Sirianni, R.; Chimento, A.; De Luca, A.; Casaburi, I.; Rizza, P.; Onofrio, A.; Iacopetta, D.; Puoci, F.; Andò, S.; Maggiolini, M. Oleuropein and Hydroxytyrosol Inhibit MCF-7 Breast Cancer Cell Proliferation Interfering with ERK1/2 activation. *Mol. Nutr. Food Res.* **2010,** *54* (6), 833–840.
126. Soler-Rivas, C.; Espín, J. C.; Wichers, H. J. Oleuropein and Related Compounds. *J. Sci. Food Agric.* **2000**, *80* (7), 1013–1023.
127. Soni, M.; Burdock, G.; Christian, M.; Bitler, C.; Crea, R. Safety Assessment of Aqueous Olive Pulp Extract as an Antioxidant or Antimicrobial Agent in Foods. *Food Chem. Toxicol.* **2006,** *44* (7), 903–915.
128. St-Laurent-Thibault, C.; Arseneault, M.; Longpre, F.; Ramassamy, C. Tyrosol and Hydroxytyrosol Two Main Components of Olive Oil, Protect N2a Cells Against Amyloid-β-induced Toxicity. Involvement of the NF-κB Signaling. *Current Alzheimer Res.* **2011,** *8* (5), 543–551.
129. Sun, W.; Wang, X.; Hou, C.; Yang, L.; Li, H.; Guo, J.; Huo, C.; Wang, M.; Miao, Y.; Liu, J. Oleuropein Improves Mitochondrial Function to Attenuate Oxidative Stress by Activating the Nrf2 Pathway in the Hypothalamic Paraventricular Nucleus of Spontaneously Hypertensive Rats. *Neuropharmacology* **2017,** *113*, 556–566.
130. Tan, H.-W.; Tuck, K. L.; Stupans, I.; Hayball, P. J. Simultaneous Determination of Oleuropein and Hydroxytyrosol in Rat Plasma Using Liquid Chromatography with Fluorescence Detection. *J. Chromatogr. B* **2003,** *785* (1), 187–191.
131. Terzuoli, E.; Giachetti, A.; Ziche, M.; Donnini, S. Hydroxytyrosol, a Product from Olive Oil, Reduces Colon Cancer Growth by Enhancing Epidermal Growth Factor Receptor Degradation. *Mol. Nutr. Food Res.* **2016,** *60* (3), 519–529.
132. Toteda, G.; Lupinacci, S.; Vizza, D.; Bonofiglio, R.; Perri, E.; Bonofiglio, M.; Lofaro, D.; La Russa, A.; Leone, F.; Gigliotti, P. High doses of Hydroxytyrosol Induce Apoptosis in Papillary and Follicular Thyroid Cancer Cells. *J. Endocrinol. Invest.* **2017,** *40* (2), 153–162.
133. Tovar, M. J.; Motilva, M. J.; Romero, M. P. Changes in the Phenolic Composition of Virgin Olive Oil from Young Trees (*Olea europaea* L. cv. *Arbequina*) Grown Under Linear Irrigation Strategies. *J. Agric. Food Chem.* **2001,** *49* (11), 5502–5508.
134. Tuck, K. L.; Freeman, M. P.; Hayball, P. J.; Stretch, G. L.; Stupans, I. The In Vivo Fate of Hydroxytyrosol and Tyrosol, Antioxidant Phenolic Constituents of Olive Oil, After Intravenous and Oral Dosing of Labeled Compounds to Rats. *The J. Nutr.* **2001,** *131* (7), 1993–1996.
135. Tutino, V.; Orlando, A.; Russo, F.; Notarnicola, M. Hydroxytyrosol Inhibits Cannabinoid CB1 Receptor Gene Expression in 3T3-L1 Preadipocyte Cell Line. *J. Cell. Physiol.* **2016,** *231* (2), 483–489.

136. Vagelas, I.; Kalorizou, H.; Papachatzis, A.; Botu, M. Bioactivity of Olive Oil Mill Wastewater Against Plant Pathogens and Post-harvest Diseases. *Biotechnol. Biotechnol. Equip.* **2009**, *23* (2), 1217–1219.
137. Visioli, F.; Bellomo, G.; Galli, C. Free Radical-scavenging Properties of Olive Oil Polyphenols. *Biochem. Biophys. Res. Commun.* **1998**, *247* (1), 60–64.
138. Visioli, F.; Bellomo, G.; Montedoro, G.; Galli, C. Low Density Lipoprotein Oxidation Is Inhibited in Vitro by Olive Oil Constituents. *Atherosclerosis* **1995,** *117* (1), 25–32.
139. Visioli, F.; Caruso, D.; Grande, S.; Bosisio, R.; Villa, M.; Galli, G.; Sirtori, C.; Galli, C. Virgin Olive Oil Study (VOLOS): Vasoprotective Potential of Extra Virgin Olive Oil in Mildly Dyslipidemic Patients. *European J. Nutr.* **2005,** *44* (2), 121–127.
140. Visioli, F.; Galli, C.; Bornet, F.; Mattei, A.; Patelli, R.; Galli, G.; Caruso, D. Olive Oil Phenolics Are Dose-Dependently Absorbed in Humans. *FEBS Lett.* **2000,** *468* (2–3), 159–160.
141. Visioli, F.; Galli, C.; Plasmati, E.; Viappiani, S.; Hernandez, A.; Colombo, C.; Sala, A. Olive Phenol Hydroxytyrosol Prevents Passive Smoking-induced Oxidative Stress. *Circulation* **2000**, *102* (18), 2169–2171.
142. Visioli, F.; Poli, A.; Gall, C. Antioxidant and Other Biological Activities of Phenols from Olives and Olive Oil. *Med. Res. Rev.* **2002,** *22* (1), 65–75.
143. Visioli, F.; Wolfram, R.; Richard, D.; Abdullah, M. I. C. B.; Crea, R. Olive Phenolics Increase Glutathione Levels in Healthy Volunteers. *J. Agric. Food Chem.*, **2009,** *57* (5), 1793–1796.
144. Vissers, M. N.; Zock, P. L.; Roodenburg, A. J.; Leenen, R.; Katan, M. B. Olive Oil Phenols Are Absorbed in Humans. *J. Nutr.* **2002,** *132* (3), 409–417.
145. Wang, L.; Geng, C.; Jiang, L.; Gong, D.; Liu, D.; Yoshimura, H.; Zhong, L. The Anti-Atherosclerotic Effect of Olive Leaf Extract Is Related to Suppressed Inflammatory Response in Rabbits with Experimental Atherosclerosis. *European J. Nutr.* **2008,** *47* (5), 235–243.
146. Weinbrenner, T.; Fito, M.; De La Torre, R.; Saez, G. T.; Rijken, P.; Tormos, C.; Coolen, S.; Albaladejo, M. F.; Abanades, S.; Schroder, H. Olive Oils High in Phenolic Compounds Modulate Oxidative/Antioxidative Status in Men. *The J. Nutr.* **2004,** *134* (9), 2314–2321.
147. Wu, W.-S. The Signaling Mechanism of ROS in Tumor Progression. *Cancer Metastasis Rev.* **2006,** *25* (4), 695–705.
148. Yamada, K.; Ogawa, H.; Hara, A.; Yoshida, Y.; Yonezawa, Y.; Karibe, K.; Nghia, V. B.; Yoshimura, H.; Yamamoto, Y.; Yamada, M. Mechanism of the Antiviral Effect of Hydroxytyrosol on Influenza Virus Appears to Involve Morphological Change of the Virus. *Antiviral Res.* **2009,** *83* (1), 35–44.
149. Yan, C. M.; Chai, E. Q.; Cai, H. Y.; Miao, G. Y.; Ma, W. Oleuropein Induces Apoptosis Via Activation of Caspases and Suppression of Phosphatidylinositol 3-Kinase/Protein Kinase B Pathway in HepG2 Human Hepatoma Cell Line. *Mol. Med. Rep.* **2015,** *11* (6), 4617–4624.
150. Yangui, T.; Dhouib, A.; Rhouma, A.; Sayadi, S. Potential of Hydroxytyrosol-rich Composition from Olive Mill Wastewater as a Natural Disinfectant and Its Effect on Seeds Vigour Response. *Food Chem.* **2009,** *117* (1), 1–8.
151. Yao, J.; Wu, J.; Yang, X.; Yang, J.; Zhang, Y.; Du, L. Oleuropein Induced Apoptosis in HeLa Cells Via a Mitochondrial Apoptotic Cascade Associated with Activation of the C-Jun NH2-terminal Kinase. *J. Pharmacol. Sci.* **2014,** *125* (3), 300–311.

152. Yigit, A.; Sahan, Y.; Korukluoglu, M. In *Antimicrobial Substances Found in Olive Leaves and Olive*. 2nd International AltInoluk "Antandros" Olive Busines Symposium; Turkey; 2001; pp 12–20.
153. Zhang, X.; Jiang, L.; Geng, C.; Hu, C.; Yoshimura, H.; Zhong, L. Inhibition of Sudan I Genotoxicity in Human Liver-derived HepG2 Cells by the Antioxidant Hydroxytyrosol. *Free Rad. Res.* **2008,** *42* (2), 189–195.
154. Zhang, X.; Jiang, L.; Geng, C.; Yoshimura, H.; Zhong, L. Inhibition of Acrylamide Genotoxicity in Human Liver-derived HepG2 Cells by the Antioxidant Hydroxytyrosol. *Chemico-Biol. Interact.* **2008,** *176* (2), 173–178.
155. Zhu, L.; Liu, Z.; Feng, Z.; Hao, J.; Shen, W.; Li, X.; Sun, L.; Sharman, E.; Wang, Y.; Wertz, K. Hydroxytyrosol Protects Against Oxidative Damage by Simultaneous Activation of Mitochondrial Biogenesis and Phase II Detoxifying Enzyme Systems in Retinal Pigment Epithelial Cells. *The J. Nutr. Biochem.* **2010,** *21* (11), 1089–1098.

## CHAPTER 6

# PHYTOCHEMICALS FROM CITRUS PEEL: PERSPECTIVES AND ALLIED HEALTH CLAIMS

HUMAIRA ASHRAF, IAHTISHAM-UL-HAQ,
MASOOD SADIQ BUTT, MIR MUHAMMAD NASIR QAYYUM,
and HAFIZ ANSAR RASUL SULERIA

### ABSTARCT

Agricultural by-products are one of the major waste materials around the globe that may pose serious environmental issues if remain unhandled or poorly managed. On the other hand, this waste may also be exploited to isolate various bioactive compounds for their utilization in other products such as food additives. In this respect, citrus peel, a major by-product of citrus industry, can potentially be used to obtain biologically active agents, that is, flavonoids which are of primitive significance in the production of functional food items and nutraceutics to avoid various life-threatening disorders. Therefore, the use of cost effective dietary therapies as an adjunctive measure to restrain the onset of illnesses is of significant importance as diet plays basic role in overall health maintenance. Purposely, citrus peel flavonoids with promising antioxidant activities protect against degenerative disorders by neutralizing reactive oxygen species, which are involved in etiology of numerous physiological dysfunctions. In this context, this chapter summarizes the phytochemical, antioxidant, and health-promoting potentials of bioactive compounds from citrus peel.

### 6.1 INTRODUCTION

The survival of human being and sustenance of healthy life is dependent on quality of food and better nutritional practices. Recent era has witnessed an array of lifestyle-related disorders due to changing dietary habits and unhealthy living practices in individuals of different communities around the globe. Resultantly, rate of mortality and morbidity is increasing at alarming rate.[97] Especially, heart diseases are leading cause of mortality and morbidity among various age groups followed by arthritis, cancer, diabetes, and osteoporosis, worldwide. Allied to this, medical costs are escalating due to sophisticated technologies for the treatment of these disorders. Moreover, working capacities of the affected individuals are severely decreasing posing high economic burdens on the society.[115]

Recent scientific efforts have unveiled food and health linkage that is crucial approach to shield lifestyle-related disparities. Apart from latest medical technologies, dietary interventions are gaining more popularity due to their long-term administration consistency and safer nature. The diet of an individual is considered more important than medicines as it contains plethora of disease-fighting compounds, that is, nutraceutics which are of primitive reputation to avoid various health disparities.[96] Phytochemicals are plant-based bioactive components that exhibit free radical scavenging and antioxidant properties.[98,99]

In this milieu, the use of functional foods and nutraceutics is more cautious and sustainable approach for health management.[11,28] Moreover, dietary therapies are being recommended by the health practitioners and nutritionists to tackle the dilemma of dietary disorders in a safe and consistent manner. In diet-based therapies, functional, and nutraceutical foods share a major role to provide our body with required phytochemicals that are necessary to maintain antioxidant balance of the body.[14,34] In this respect, functional and nutraceutical foods not only fulfill nutritional requirements but also provide health benefits. Nowadays, these foods have become an integral component of the diet due to their positive therapeutic effects against hypercholesterolemia, cardiovascular diseases, hyperglycemia, and cancer.[85] Plant-based foods rich in polyphenols are more alluring to be effectively used against these disorders.[98]

Agricultural waste can pose serious threat to environment if left unhandled or poorly managed. On the other hand, this waste can be explored for obtaining valuable ingredients, which can further be utilized as additives in various products.[90] Furthermore, pervasiveness of diseases is mounting day by day among people of different age groups. In this scenario, food bioactive components are being exploited since they are stumpy cost, less toxic, and of natural origin.[13] In developing countries, a huge amount of agricultural produce goes as waste due to improper handling and management practices.[11,49] Presently, there is curiosity to use agricultural by-products for extraction and utilization of health-promoting ingredients due to diversity of bioactive components in them. In this context, citrus peel, which is a major by-product of citrus juice industry, is considered important especially due to presence of high flavonoids density.[78] Alongside, phenolic acids and flavonoids are one of the significant antioxidants in food systems that can promote health by scavenging free radicals.[25] Also, commercially available functional foods are mainly marketed for their antioxidant, cholesterol lowering, hypoglycemic, and anticarcinogenic perspectives as these innovative products are supplemented with bioactive ingredients to tackle these physiological threats.[11,49,56,84,88]

This chapter reviews phytochemistry and allied health benefits of citrus peel for better understanding the fate of its use in designing the dietary therapies.

## 6.2 CITRUS PEEL: AN OVERVIEW

Globally, nondebatable facts have promoted the use of functional and nutraceutical ingredients with increased life expectancy of individuals. There is a need to explore such constituents that are not only curative against disorders but also cost effective.[52] Nowadays, societies are demanding for safer and innovative products that have increased stress upon agroindustries to fulfill demands of growing population. This has derived agro-based industries to reduce wastage by recycling and adding value to the products.[93] The peels of numerous fruits contain higher amounts of bioactive compounds than their edible fleshy parts. For instance, peels from citrus, apples, peaches, and pears have been found to contain twofold the amount of total phenolics as compared to peeled fruits.[46]

Citrus cultivars are extensively grown in warm climate worldwide but tropical and subtropical regions are the ideal geographical locations for its production. Initially, it was originated in tropical areas around Southern Himalayas, Southeastern Asia, and Indonesia archipelago. Although a number of varieties are available in market yet sweet orange, mandarin, grapefruit, lemon, and lime are commercially adored for their flavor and taste.[23,59]

Globally, citrus production is estimated at >120 million metric tons out of which 50% is used for juice production.[36] Annually, citrus juice-processing industries produce large quantities of peels and seed residues, which may account for >60% of the total fruit weight.[18,36] Optimal utilization of waste material originating from citrus industry could be a major source of phenolic compounds. Due to rich phytochemistry, citrus peel has been used in various traditional Chinese medicines in the form of powder and oil. Research work has proven the health benefits of citrus peel against various maladies. Studies have also shown that citrus peel extract improves digestion, controls nausea, and vomiting and impart diuretic effects to detoxify body fluids. Citrus peel supplements help to strengthen heart function by reducing low-density lipoproteins (LDL) and very low-density lipoproteins (VLDL) in serum and regulate insulin level, hence controlling blood glucose level.[43,79] In Western culture, citrus powder has been used in various food formulations for its therapeutic effects.[82] The health-promoting perspectives of peel is accredited to its phenolic contents that are higher in flavedo portion as compared to edible parts. Gorinstein et al.[45] reported that the total phenolic content in peels of lemons, oranges, and grapefruit were 15% higher than in the peeled fruits. Among these, orange peel is a rich source of phenolic compounds that constitute about 179 mg/100 g of fresh orange fruit.[51]

Citrus peel polyphenols include two important classes of compounds, that is, flavonoids and phenolic acids. About a half century ago, information on exact mechanism of citrus flavonoids was not sufficient to prove its therapeutic effects in our body but now epidemiological studies have confirmed its health-promoting aspects.[18] An important property of citrus flavonoids [especially, glycosylated flavanones and polymethoxylated flavones (PMFs)] is to scavenge free reactive oxygen species (ROS) to mitigate various degenerative disorders. Pharmacological properties of flavonoids are dependent on total intake, bioavailability, absorption, and their conjugation in different body organs. Among different countries, average intake of flavonoids varies depending on their cultural habits and lifestyle patterns. Scientific evidences have confirmed that citrus peel-derived flavonoids have potential to alleviate mortality rate due to cardiovascular disorders in Japan, which signifies the positive role of citrus peel bioactives in improving health of the individuals.[8,9,48]

## 6.3 PHYTOCHEMISTRY OF CITRUS PEEL

The citrus peel has group of bioactive components that play a momentous role against various physiological threats.[12,27] Chemically, it has been estimated that citrus peel contains: total sugar content (16.5%), pectin (12.8%), crude fiber (8.6%), crude protein (4.2%), lignin (2.2%), total ash (2.1%), and crude fat (1.5%). Furthermore, phenolic compounds (179 mg/100 g), vitamin C (65 mg/100 g) and β-carotene (2.10 μg/g) on dry weight (DW) basis are also present in the peel portion of citrus.[5,45]

Flavonoids encompass an assembly of natural substances with variable phenolic structures. Depending upon molecular structures, flavonoids are divided into six classes[81]: flavones, flavanones, flavonols, isoflavones, anthocyanidins, and flavanols or catechins.

Total flavonoids content in the peel portions may fluctuate due to agronomic practices and climatic conditions. In orange peel, flavonoids are represented by two very important classes: glycosylated flavanones and PMFs.[18]

Flavanones are most abundant group of flavonoids present in citrus peel that rarely exists in other fruits in such high concentrations. The predominant flavanones present in orange peel (*Citrus sinensis*) are: hesperidin, naringin, and rutinose that make up to 6.5 g/L of essential oil.[43] Generally, they are glycosylated by a disaccharide at position 7, either a neohesperidose, which imparts a bitter taste such as naringin in grapefruit, hesperidin in orange peel, or a rutinose that is flavorless.[100] Hesperidin is a predominant glycosylated flavanone in orange that is mostly localized in flavedo portion that varies from 35% of the dry matter in immature fruits to 4% in the mature fruit, whereas naringin is present in core and segment membranes in addition to flavedo and albedo parts of grapefruit.[100,102]

PMFs from orange peel hold a mixture of nonhydroxylated PMFs (75.1%) and hydroxylated PMFs (5.44%) that explicit a broad spectrum of biological activity including antiproliferative and proapoptotic effects in cancer cells.[91] PMFs (such as nobiletin, sinenstin, and tangeretin) are mostly confined in essential oil of flavedo portion but they are lesser than flavonones.[83] Biological and pharmacological properties (such as antioxidant, anti-inflammatory, antiallergic, antiviral, antibacterial, antimutagenic, and anticarcinogenic activities) of citrus peel are associated with these two major classes of flavonoids. In the United States, previous studies have indicated that total intake of flavonoids should be 1 g/day for glycosides and 650 mg/day for aglycones.[44]

Keeping in view the health aspects associated with citrus peel flavonoids, food processors are engaged in preparation of peel extracts of variable concentrations to be utilized as food supplements for humans. In this milieu, hesperidin that is extracted from deoiled orange peel contains both classes of flavonoids (flavonones and flavones) that are being used in Western countries for various ailments.[53] Furthermore, citrus peel polyphenols are secondary metabolites with strong shield against oxidation in addition to vitamins C and E and carotenoids.[74]

Mostly natural antioxidants have multifunctional attributes in intricate heterogeneous food items and their potential cannot be assessed.[39,87] Gursoy et al.[47] determined antioxidant properties of orange peel oil. They analyzed that citrus oil has 96.8 ± 0.2% ability to hinder linoleic acid oxidation due to its polyphenolic constituents. de Moraes Barros et al.[29] have reported in vitro antioxidant power of various citrus fruit parts using ferric-ion reducing antioxidant capacity assay. Among five citrus cultivars, mandarin peels exhibited highest ferric-reducing potential (3897.90 μMTE/100 g) in contrast to its pulp portion (744.0 ± 12.7 μM TE/100 g of fruit weight). Likewise, Oboh and Ademosun[73] differentiated $ABTS^+$ scavenging capacity of orange, grapefruit, and shaddock. They narrated that orange peel-bound phenolics possess maximum antioxidant potential (6.03 mM TEAC/g of peel) than the shaddock peels that have least antioxidant power (1.23 mM TEAC/g). Similarly, free phenolics from grapefruit and orange peels have higher free radical scavenging activity, that is, 5.9 and 5.7 mM TEAC/g, respectively, as compared to shaddock peel exhibiting activity of 4.9 mM TEAC/g.

The HPLC characterization and spectral analysis of citrus peel bioactive have revealed presence of C-glycosylated flavones, O-glycosylated flavones, PMFs, flavonol, O-glycosylated flavonones, and phenolic acids.[6] Moreover, phenolic acids including hydroxycinnamics (caffeic, *p*-coumaric, sinapic, and ferulic acids) and hydroxybenzoics (protocatechuic, *p*-hydroxybenzoic, and vanillic acids) have also been found in citrus peel. It has further been seen that phenolics may disintegrate with extended time intervals at higher temperatures, whereas, hydrolysis of bound phenolic acids from citrus peel increases at ambient temperature under alkaline condition.[110] Furthermore, their antioxidant assays have also revealed their high radical scavenging activities.[7]

The ethyl acetate fraction of orange peel represented two glycosylated flavanones (narirutin and hesperidin) and two PMFs (PMF: nobiletin and tangeretin). It has been noticed that hesperidin caused difficulty to be extracted by hot water, while the contents of nobiletin were noticed to be substantially high in aqueous fractions. Increased extraction temperature may increase the yield of hesperidin with little impact on the yield of narirutin, nobiletin, and tangeretin. Hence, considerable amount of PMFs can be extracted in water at lower temperature except hesperidin that demonstrate low yields even at higher temperatures.[35]

In mandarin peel, Xu et al.[111] identified glycosylated flavanone and PMFs by employing high-performance liquid chromatography–photodiode array (HPLC-PDA) technique. They noticed that mandarin peels contain hesperidin (62.01 ± 0.24 mg/g DW) as major bioactive compound followed by narirutin, nobiletin, and tangeretinas (7.66, 0.31, and 0.16 mg/g DW, respectively). Likewise, methanolic extracts of citrus leaves revealed the presence of flavonols such as apigenin (88.7 ± 2.8 mg/100 g), kaempferol (64.2 ± 1.7 mg/100 g), and rutin (134.6 ± 3.6 mg/100 g). However, citrus peels demonstrated higher concentrations of quercetin, which is a potent antioxidant. Additionally, nobiletin and hesperidin contents were 0.2 ± 0.01 mg/100 g and 0.05 ± 0.002 mg/100 g, respectively, in citrus fruit peels.[63]

Essential oils from citrus peels are one of the important by-products that encompass 85–99% volatile and 1–15% nonvolatile components. The volatile constituents are a combination of monoterpene (limonene), sesquiterpene hydrocarbons, and their oxygenated derivatives including aldehydes (citral), ketones, acids, alcohols (linalool), and esters,[37] whereas, nonvolatile portion of citrus oil contains flavanoids that are deprived off lavonone glycosides but rich in hydrophobic flavone aglycons particularly nobiletin. Hexane-extracted oil showed extremely high concentrations of nobiletin (32%) and other flavone aglycons.[66]

## 6.4 HEALTH CLAIMS

The lifestyle of human populations around the globe has changed significantly in 21st century. Sedentariness and escalated consumption of junk food items have triggered the prevalence of various health problems such as cardiovascular diseases, hypertension, diabetes, obesity, osteoporosis, and cancer causative of mortalities.[31,114] Furthermore, oxidative stress has been identified as a cardinal factor in etiology of numerous health disorders.[96,108] In this context, dietary and lifestyle modifications can compensate oxidative stress-mediated malfunctions.[17]

Presently, research outcomes have unveiled natural therapeutic sources that are cost effective, easily accessible, and less toxic in nature. Fruits and vegetables and their peels are reliable source of bioactive components. Citrus peel has also been proven to possess certain therapeutic agents, which are of significant consideration for biomedical applications. Consequently, its bioactives have been demonstrated to serve against hepatotoxicity, mutagenicity,[72] hypertension, hyperlipidemia,[43,62] hyperglycemia,[79] inflammation, microbial infection, and oxidation.[113] These therapeutic effects are attributed to the flavonoids in citrus fruits that have ability to scavenge free radicals and shield cells and tissues from degeneration. In this respect, some of the potential health benefits of citrus peel bioactives are summarized in this section.

### 6.4.1 HYPERCHOLESTEROLEMIA

Cholesterol is essential to maintain normal integrity of cellular membranes. Its main function in body is not only to provide cell integrity but to facilitate in transportation of phospholipids, synthesis of aldosterone, cortisol, and secretion of hormones.[104] However, its level in the body must be maintained in normal limits so that to avoid any discomfort in routine physiological processes.[76]

In atherosclerosis, oxidation of LDL causes their accumulation in interior lining of blood vessels and deposits in arteries. Further oxidation of LDL by free radicals modifies deposit structures that cause foaming and leads to the formation of hard plaque, hence retarding normal blood flow through the vessels. Under this situation, the supply of oxygen to heart, brain, and other organs of the body is limited, which can further increase the risk of heart attack and stroke. In this way, cardiovascular diseases are causative of numerous mortalities around the globe especially among the people with high LDL levels.[21] Paradoxically, high-density lipoproteins (HDL) help to remove excess cholesterol from atherosclerotic deposits and prevent plaque formation thus minimizing chances for cardiovascular disorders.[22]

In the human body, 80% of the required cholesterol is synthesized by liver cells while 20% comes from dietary sources. For synthesis and esterification of the cholesterol in the body, two key enzymes are of prime importance: (1) 3-hydroxy-3-methylglutaryl coenzyme A (HMG-CoA) reductase that is the rate-limiting enzyme engaged in cholesterol biosynthesis and (2) acyl CoA: cholesterol *O*-acyltransferase (ACAT), which participates in esterification of cholesterol in body tissues. Under normal circumstances, cholesterol biosynthesis by HMG-CoA reductase is switched off to maintain its normal body levels when enough cholesterol has been produced to meet body's requirements, whereas these enzyme systems are compromised due to changing lifestyle and dietary trends as unnecessary cholesterol tends to come from the diet that has increased risk of heart diseases and atherosclerosis.[20]

Recent advances in medical research and health expertise has suggested various drugs to normalize body cholesterol levels. Alongside, HMG-CoA reductase inhibitors are being utilized in hypocholesterolemic drugs such as statins that facilitates in removing circulatory LDL by inhibiting activities of HMG-CoA reductase thus lowering cholesterol synthesis by liver. ACAT aids in esterification of cholesterol within cells that assist in its assimilation, VLDL production from liver, and development of plaque in blood arteries.

For the purpose, ACAT inhibitors are being widely used as cholesterol-lowering and anti-atherosclerotic agents in pharmaceutical products.[40]

It has been seen that long-term administration of statins (lovastatin) for LDL cholesterol lowering exhibits various side effects. Among these, inhibition of HMG-CoA reductase activity by statins reduces CoQ10 (an integral compound of respiratory chain) levels that cause cardiac stress thereby increasing chances of heart attack. Moreover, chronic use of statins also causes liver dysfunctions.[16,24] In this context, there is a dire necessity to identify cost-effective natural remedies with therapeutic effects against hypercholesterolemia with safer, nontoxic, and readily available nature. It has been reported that bioflavonoids from plant sources are effective against life-threatening diseases such as cancer, diabetes, and hyperlipidemia. Also, citrus peel powder and extract are of crucial importance due to its wide biological spectrum.[27,28,30–32]

Abdelbaky et al.[1] have revealed the hypocholesterolemic effects of orange peel powder and its methanolic extract using rat modeling. Supplementation of diet with 5% orange peel powder and 1% orange peel extract revealed that peel extract was more effective in lowering serum total cholesterol and LDL than orange peel powder, which showed negative impact on triglycerides and VLDL but improved HDL levels.[1] Citrus peel-glycosylated flavonones and PMFs are of vital importance to lower the risk of hypercholesterolemia and cardiovascular diseases. Among citrus flavanones, hesperidin from oranges, and naringen from grapefruits are well known for maintaining normal blood lipid profile. These are naturally found in their glycosylated forms, that is, hesperetin and naringenin and become available in colon after deglycosylation by glycosidase or intestinal microflora.[33]

Hepatoma HepG2 cells are involved in synthesis and catabolism of apolipoprotein B (ApoB) having lipoproteins such as LDL and VLDL. In vitro study relating to cholesterol-lowering potential of both hesperidin and naringen on hepatoma HepG2 cells of human body has confirmed that citrus peel bioflavonoids can reduce secretion of LDL-linked ApoB and decrease its $IC_{50}$ value as well. The reported $IC_{50}$ concentrations were 142 and 178 μM for hesperidin and naringen, respectively.[57,58] In a rat model trial, cholesterol-lowering effects of tangerine peel extract and mixture of bioflavonoids were investigated by Kurowska and Manthey.[58] They suggested that biosynthesis of hepatic cholesterol and its esterification was reduced by 28.3% and 23.7%, respectively, in rats group on diet supplemented with mixture of hesperidin and naringenin (0.05g + 0.05 g/100 g diet). While, rats fed with tangerine peel extract supplemented diet exhibited 37% and 32% decrease in hepatic cholesterol and esterification of liver cholesterol, respectively. It was concluded that hypolipidemic potential of citrus peel was due to its bioactive components that have an inhibitory effect for HMG-CoA reductase. Although high cholesterol diet reduces activity of HMG-CoA reductase, ACAT still remains active.[58]

In another report, Lee et al.[60] reported cholesterol-lowering effect of hesperetin that did not affect blood lipid profile. However, it decreased plasma cholesterol concentration in hypercholesterolemic rats. According to the authors, citrus bioflavonoids supplemented diet inactivates ACAT that is involved in formation of cholesterol esters needed for VLDL packing in liver for its secretion in blood. Likewise, Wang et al.[103] have reported that hesperidin (0.08%) can reduce weight of fatty tissues and liver, hepatic steatosis, retinol-binding protein (involved in lipid metabolism), and total plasma cholesterol in hypercholesterolemic rats.

Nobiletin is PMF that is mostly found in orange peel and is effective against inflammation, cancer, and hyperlipidemia, and provides neuroprotective effects in Alzheimer diseases.[41] In a model feed trial, 0.1% nobiletin in their diet reduced white adipose tissue with a significant increase in HDL and apolipoprotein A-1 without modifying triglyceride levels.

These findings justify the hypolipidemic effect of nobiletin under hypercholesterolemic conditions.[71] Similarly, Kurowska and Manthey[58] studied in vivo effect of citrus flavonones and PMFs in hamster for cholesterol-lowering aptitude. Dietary treatment with 1% nobiletin resulted in considerable reduction in serum LDL and VLDL (32%) and total cholesterol (19%) as compared to 0.25% nobiletin treated animals. On the other hand, treatment with 1% PMF reduced total triglycerols LDL, and VLDL by 44%, 40% and 27%, respectively. In another experiment, supplementation of hesperidin (3%) significantly decreased VLDL and LDL by 28% and 38%, respectively, along with 57% reduction in serum triglycerides. These outcomes revealed that glycosylated flavonones are more effective to manage serum lipid profile as compared to PMF.[58]

Functional product development is a complicated process that depends on customer satisfaction, their eating habits, and lifestyle patterns. Supplementation of food product with citrus peel is a novel strategy for new-generation therapies such as citrus peel-enriched biscuits. A study has revealed that supplementation of biscuits with 10% navel peel powder may significantly reduce total cholesterol, total hepatic lipids, glutamic pyruvic, and glutamic oxaloacetic transaminase by 32.35%, 33.58%, 23.81%, and 26.66%, respectively.[64] In a nutshell, citrus peel bioactives are quite promising in alleviating hypercholesterolemic conditions, and these compounds can be exploited in dietary therapies.

### 6.4.2 HYPERGLYCEMIA

Hyperglycemia generally refers to increased blood glucose levels and leads to diabetes mellitus if it persists for longer durations.[50] World Health Organization (WHO) has estimated that 336 million persons in the world will be victimized by diabetes in 2030, which demands immediate control over this situation.[106]

Pancreatic α-amylase and α-glucosidase enzymes are responsible for catabolizing complex sugars to their monosaccharide units to be readily absorbed by the small intestine. Among these, glucose is the main need of each cell in the body for energy production to support normal body functions. Furthermore, its ingress in cells is one of the limiting factors for its utilization that is driven by insulin produced in β-cells of pancreases.[69] Impaired functioning of β-cells reduces the insulin levels that ultimately hinder glucose uptake by the cells that leads to elevated blood glucose levels, that is, hyperglycemic condition. Additionally, various other complications may also develop in body of a diabetic patient such as cardiovascular disorders, immune dysfunction, and oxidative stress.[26]

Diabetes is subdivided into two major types: insulin dependent (type I) and noninsulin dependent (type II). For type I diabetes, insulin is being used for its management, while type II diabetes cannot be treated with insulin administration.[107] Various drugs have been recommended to control blood glucose level for hyperglycemic people, but they are effective only to a certain extent having side effects. One of the cost-effective and readily available approaches to treat diabetic patient is to take flavonoids-rich food due to their antioxidant properties and hypoglycemic potential. Citrus peel exhibits a heterogeneous group of flavonoids having broad spectrum of biological activity against cardiovascular disorders, cancer, diabetes, hypercholesterolemia, and oxidative stress. Main bioactive components of citrus peel, that is., hesperidin and naringin deliver antioxidant effects at early stage of diabetes mellitus and hinder the severity of the condition.[78]

Bioactive ingredients of orange peel provide antidiabetic role by decreasing activity of glucose-6-phosphatase and phosphoenol pyruvate, and increasing glucose kinase activity, hepatic glycogen, and blood insulin levels in response to elevation in glucose.[53] Furthermore, the diabetic potential of citrus peel extract was confirmed in alloxan-induced rats that was mediated by repairing damage in β-cells that secrete insulin and by inhibition of α-amylase activity.[79,80] Similarly, Tundis et al.[101] reported in vitro hypoglycemic activity of citrus peel extract by α-amylase and α-glucosidase inhibition method. Orange peel extract restrained both α-amylase and α-glucosidase having $IC_{50}$ values of 258.7 and 263.2 μg/mL, respectively. These results are five to seven folds less than acarbose-treated patients for which $IC_{50}$ value was 50 μg/mL for α-amylase and 35.5 μg/mL for α-glucosidase.[69]

In vivo study for effect of citrus peel extract (300 and 600 mg/kg) on diabetic rats has shown a significant reduction in blood glucose level by controlling glucose regulatory enzyme of body.[69] Likewise, oral administration of peel extract (100–600 mg/kg/day) to rats for 30 days maintained blood glucose level and mitigated progression of liver dysfunction caused by diabetes even after taking a high-glucose diet that revealed improvement in glucose tolerance in hyperglycemic rats.[75] In previous studies, hesperidin, naringin, and rutin each at dose level of 0.05% of diet reduced blood glucose level by 18%, 16%, and 21%, respectively, in streptozotocin-induced diabetic rats. At the same time, daily oral administration of these bioflavonoids to human @ 5, 10, and 15 mg/kg of body weight has shown significantly decreased blood glucose level by 17%, 23%, and 33%, correspondingly.[19] Antidiabetic potential of hesperidin was further investigated by Ibrahim,[50] who employed rodent feed trial using rats. The hesperidin supplemented diet administration for 35 days restored normal blood glucose level and reduced complications due to diabetes in diabetic rats. The protective effects were attributed to ability of hesperidin to increase liver glycolysis, glycogen synthesis, and decrease in gluconeogenesis.

According to Parmar and Kar,[79] administration of citrus *sinensis* extract (25 mg/kg) normalized adverse health effects caused by diabetes in male Sprague–Dawley rats. It was further confirmed by Adeneye[3] in alloxan-induced diabetic rats. Moreover, Magda et al.[64] reported that biscuits containing 10% navel peel powder reduced blood glucose level of diabetic rats by 20.53%. They further stated that hypoglycemic effects may partially be attributed to fiber contents of citrus peel too as it decreases absorption of carbohydrates through small intestines by lowering activity of α-amylase involved in sugars metabolism.

Both in vitro as well as in vivo studies have depicted that hyperglycemic condition increases oxidative stress and weakens body intracellular antioxidant defense mechanism that increase chances for type II diabetes. Inclusion of flavonoids-rich food items in daily dietary plan is leading step toward health maintenance and disease prevention.[92] In this respect, taking citrus peel nutraceutics-supplemented food products and its various preparation endow body with natural antioxidant that has strong free radical scavenging activities.

### 6.4.3 OXIDATIVE STRESS

Naturally, body has been provided with physiological processes to keep itself in a state of homeostasis. However, imbalance between free radical production and their neutralization by antioxidants in body causes serious damages to biological molecules, impairs mitochondrial function and weakens the immune system rendering the body more vulnerable for diseases. This whole phenomenon is generally referred as oxidative stress that leads to cardiovascular disorders, diabetic nephropathy, cancer, and deoxyribose nucleic acid (DNA) damage.[17,38,96,108]

The ROS are formed by glycoxidation, irradiation, inflammation, lipid oxidation, pro-oxidative enzyme systems, detoxification of pollutants, and smoking.[94] ROS interact with DNA, lipids, protein, carbohydrates, and other cell entities and cause mutation in genes that leads to aging, cancer, cardiovascular diseases, diabetes mellitus, hyperthyroidism, and other cell degenerative disorders,[77] Hence, the control of oxidative stress at cellular levels is necessary to avoid onset of these degenerative disorders. In this context, intake of polyphenols-rich diet is one of the promising practical approaches to minimize injurious effects allied to reactive free radicals. Polyphenols are secondary metabolites present in fruits and vegetables and exhibit antioxidant potential. They are recognized for scavenging free radicals, chelating metal ions, inhibiting oxidase, and absorbing free radical.[73,74]

Among agricultural by-products, phenols are abundantly present in citrus plants especially in their peel portion. Flavonoids and phenolic acids constitute a major portion of citrus phenols that link to provide defense mechanisms against free radicals. It has been unveiled that citrus polyphenols represent health-promoting attributes such as inhibiting unwanted cell proliferation, inducing apoptosis, modulating enzymes activities, and antioxidant capacity.[95] Moreover, studies have revealed that citrus peel glycosylated flavonones and PMFs possess a broad spectrum of biological activity that assist in reducing oxidative stress by antioxidant mechanisms. In vivo studies have demonstrated redox potential of flavonoids. In liver, metabolism of bioactive components changes their function, whereas flavonoids modify into phenolic acids in colon that exert beneficial antioxidant effects.[70,86,105]

Researchers have also confirmed that citrus flavonoids are more potent to scavenge free radicals as compared to vitamins C and E.[41,54] Moreover, Sawalha et al.[89] studied main phenolic compounds in sweet and bitter orange peels. The identified compounds were: hesperidin, naringin, neohesperidin, and narirutin; these compounds demonstrated significant in vitro antioxidant potentials. Furthermore, antioxidant effects of citrus bioactives have been endorsed to protect against hyperlipidemia.[1] Besides, hypoglycemic and hypolipidemic effects of citrus peel methanolic extracts have also been found as a result of antioxidant potential of citrus phytoceutics.[3] Bocco et al.[18] reported high antioxidant activities of phenolics present in citrus peels and seeds. Asikin et al.[10] have also depicted promising antioxidant abilities of aroma components in flavedo peel extract of citrus fruits. Similarly, Ghasemi et al.[42] illustrated the phenol and flavonoid contents of various citrus species and elaborated their high antioxidant activities that strengthen their role in defense mechanism against oxidation.

Scientific investigations have revealed that defect in body defense mechanism and formation of ROS among diabetic patients leads to other complications such as: nephropathy, hypertriglyceridemia, and hypercholesterolemia[3,4,38] Oxidative stress due to diabetes causes oxidation of LDL that is facilitated by 15-lipoxygenase (15-LO) present in liver. As a result, hard plaques in blood vessel appear which increases risk for atherosclerosis and stroke. Also, HDL levels were decreased due to which removal of excess cholesterol from atherosclerosis plaque is disturbed, thus maximizing chances for cardiovascular disorders.[2] However, citrus peel flavonoids especially flavonones, flavones, and flavonol have strong inhibitory effects on 15-LO enzyme which is of primitive importance in controlling the oxidation of LDL.[65]

Inflammatory conditions and cancer occur as a result of imbalance between ROS and body's antioxidant potential. Both in vitro and in vivo studies have elucidated the antioxidant activity of citrus peel against different types of cancer such as breast, colon, liver, and lungs.[41] It has been identified that antitumor effect of polymethoxylated flavones was stronger against colon Seoul National University-Cellosaurus cell line (SNU-C4) cells as compared to glycosylated flavonones due to the ability to activate proapoptotic genes *Bax* and *Caspase* by scavenging free radicals that induce mutation in gene.[78]

Furthermore, Parmar and Kar[80] have reported thyroregulatory aspects of citrus bioactive components. In this regard, orange peel flavones have shown to regulate thyroid glands function via inactivating thyroid peroxidase.

Nevertheless, citrus flavonoids provide a number of beneficial effects especially due to their high antioxidant and modulatory attributes (Table 6.1).

### 6.4.4 CANCER RISKS

Epidemiological studies have depicted that incidence of cancer is less prominent among people who rely on fruits and vegetables, due to presence of bioactive compounds from plant foods such as: flavonoids. Various evidences have proved chemopreventive potential of flavonoids.[112] Flavonoids protect DNA from oxidative damage by scavenging free radicals produced in the vicinity of DNA and interact with carcinoma produced from detoxification process.[15]

**TABLE 6.1** Health Claims for Phytochemicals in Citrus Peel.

| Phytochemicals | Activity | References |
|---|---|---|
| 5-hydroxy-3,6,7,8,3,4`-hexamethoxyflavone | Anticancer | [109] |
| Aromatic components | Antioxidant | [10] |
| Citrus flavonoids | Hepatoprotective, antimutagenic, hypotensive, hypolipidemic, hypoglycemic, anticancer, anti-inflammatory, antioxidant, antimicrobial, free radical scavengers, inhibitor of 15-lipoxygenase | [41, 43, 54, 62, 65, 67, 72, 79, 113] |
| Citrus peel oil | Antioxidant | [47] |
| Glycosylated flavones | Hypocholesterolemic, liver detoxification, anticancer | [1, 33, 55, 58, 64] |
| Hesperidin, naringen, quercetin | Normalize blood lipid profile, cholesterol lowering, hypoglycemic, antioxidant | [1, 33, 57, 58, 69, 75, 78, 89, 101, 103] |
| Hesperidin, hesperetin | Cholesterol lowering, anti-diabetic | [50, 60] |
| Nobiletin | Anti-cancer, hypolipidemic, neuroprotective, anti-inflammatory | [41, 71, 109] |
| Nobiletin, sinenstin, tangeretin | Antioxidant, anti-inflammatory, antiallergic, antiviral, antibacterial, antimutagenic, anticarcinogenic | [41, 83] |
| Phenolic acids | Antioxidant | [18, 70, 86, 105] |
| Polymethoxylated flavones | Antioxidant, antithrombotic, chemopreventive, antiproliferative, proapoptotic, *Bax* and *Caspase* activator, cholesterol lowering | [55, 61, 68, 78, 91] |

PMF are distinctive group of flavonoids that are solely present in citrus fruits especially in orange and mandarin peels in higher quantities. PMF acquire a wide spectrum of therapeutic effects such as antioxidant attributes, antithrombotic property, chemopreventive actions, and cholesterol-lowering aptitude.[61,68] The chemopreventive attributes against biosynthesis of adhesion molecules, appearance of tumor necrosis factor-R (TNFR), spreading of tumor to surrounding tissues by enhancing apoptosis, minimizing lymphocytes propagation, and platelet aggregation have been demonstrated in various investigations.

PMF are more permeable and readily absorbed via small intestine into blood circulation due to their lipophilic nature.[61,70] Manthey and Guthrie[67] identified cancer-preventive effect of citrus flavonoids on six human cancer cell lines including colon, lungs, melanoma, prostate, and estrogen receptor negative and estrogen receptor positive breast cancers. Momentous differences were illuminated in mode of action of each flavonoid constituent against these cell lines. The outcomes exposed that PMFs are stronger to prevent propagation of cancer whereas glycosylated flavonones only have antiproliferative effects at initiation stage. Glycosylated flavonones such as hesperidin and diosmin activate liver detoxification enzymes and scavenge free radicals. Citrus PMFs have strong antiproliferative activity against a broad spectrum of other cancer lines such as: gastric cancer, leukemia HL-60, T-cell leukemia, and lymph nodes metastases.[55]

Sergeev et al.[91] reported antiproliferative and proapoptotic properties of orange peel PMFs in human breast cancer cells. Orange peel extract containing a mixture of nonhydroxylated PMFs (75.1%) and hydroxylated PMFs (5.44%) stimulated $Ca^{2+}$-mediated apoptosis in cancer cells. Hydroxylated PMFs at concentration of 10 μg/mL were more potent for treatment of Michigan Cancer Foundation-7 cell lines (MCF-7) breast cancer cells as evident from reduced cancer cells, increased apoptosis, and $Ca^{2+}$ intracellular concentration as compared to nonhydroxylated PMFs. More than 20 types of PMFs have been characterized for their therapeutic effects. Among these, citrus peel PMFs (nobiletin: a chemopreventive bioactive compound) occupy major portion of aged orange peel extract taken from cold-pressed citrus peel oil. Nobiletin inhibits nitrite production and decreases level of both protein and ribonucleic acid in cycloxygenase-2. On the other hand, 5-hydroxy-3,6,7,8,3`,4`-hexamethoxyflavone has stronger inhibitory effect against breast cancer and leukemia cells as compared to nobiliten.[109]

Various drugs have been recommended as antiproliferative and proapoptotic agents for various cancer treatments; however, their high dosage and long-term administration may cause toxicity which is a limiting factor for their proper usage. Therefore, dietary components such as citrus peel bioactive ingredients are being exploited as therapeutic agents because of their more economical, effective, and practical aptitudes to ameliorate life-threatening disorders.

## 6.5 SUMMARY

Citrus peel flavonoids consisting of polymethoxylated and glycosylated flavones have potential for utilization in designing dietary control of numerous degenerative disorders. Their main function is to reduce oxidative stress that is involved in the etiology of other diseases such as cardiovascular disorders, diabetes, cancer, and immune-related malfunctions. The exploitation of citrus peel for isolating bioactive ingredients not only provides plenty of compounds for value addition in functional food market but also reduces the environmental pollution. Nevertheless, potential use of agricultural by-products for value addition will open new horizons of dietary products capable of assisting in health maintenance and disease management.

## KEYWORDS

- **citrus peel**
- **phytochemicals**
- **phenolic acids**
- **polyphenols**
- **antioxidants**

## REFERENCES

1. Ashraf H.; Butt M. S.; Iqbal M. J.; Suleria H. A. R. Citrus Peel Extract and Powder Attenuate Hypercholesterolemia and Hyperglycemia Using Rodent Experimental Modeling. *Asian Pac. J. Trop. Biom.* **2017,** *7* (10), 870–80.
2. Abou-Seif, M. A.; Youssef, A. A. Evaluation of Some Biochemical Changes in Diabetic Patients. *Clin. Chim. Acta.* **2004,** *346* (2), 161–170.

3. Adeneye, A. Hypoglycemic and Hypolipidemic Effects of Methanol Seed Extract of *Citrus paradisi* Macfad (*Rutaceae*) in Alloxan-induced Diabetic Wistar Rats. *Niger. Q. J. Hosp. Med.* **2007,** *18* (4), 211–215.
4. Ahn, J.; Choi, W.; Kim, S.; Ha, T. Anti-diabetic Effect of Watermelon (*Citrullus vulgaris* Schrad) on Streptozotocin-induced Diabetic Mice. *Food Sci. Biotechnol.* **2011,** *20* (1), 251–254.
5. Ali, J.; Abid, H.; Hussain, A. Study of Some Macronutrients Composition in Peels of Different Citrus Fruits Grown in NWFP. *J. Chem. Soc. Pak.* **2010,** *32* (1), 83–86.
6. Anagnostopoulou, M. A.; Kefalas, P.; Kokkalou, E.; Assimopoulou, A. N.; Papageorgiou, V. P. Analysis of Antioxidant Compounds in Sweet Orange Peel by HPLC-diode Array Detection-electrospray Ionization Mass Spectrometry. *Biomed. Chromatogr.* **2005,** *19* (2), 138–148.
7. Anagnostopoulou, M. A.; Kefalas, P.; Papageorgiou, V. P.; Assimopoulou, A. N.; Boskou, D. Radical Scavenging Activity of Various Extracts and Fractions of Sweet Orange Peel (*Citrus sinensis*). *Food Chem.* **2006,** *94* (1), 19–25.
8. Arai, Y.; Watanabe, S.; Kimira, M.; Shimoi, K.; Mochizuki, R.; Kinae, N. Dietary Intakes of Flavonols, Flavones and Isoflavones by Japanese Women and the Inverse Correlation Between Quercetin Intake and Plasma LDL Cholesterol Concentration. *J. Nutr.* **2000,** *130* (9), 2243–2250.
9. Asami, D. K.; Hong, Y.-J.; Barrett, D. M.; Mitchell, A. E. Comparison of the Total Phenolic and Ascorbic Acid Content of Freeze-dried and Air-dried Marionberry, Strawberry, and Corn Grown Using Conventional, Organic, and Sustainable Agricultural Practices. *J. Agric. Food Chem.* **2003,** *51* (5), 1237–1241.
10. Asikin, Y.; Taira, I.; Inafuku, S.; Sumi, H.; Sawamura, M.; Takara, K.; Wada, K. Volatile Aroma Components and Antioxidant Activities of the Flavedo Peel Extract of Unripe Shiikuwasha (*Citrus depressa* Hayata). *J. Food Sci.* **2012,** *77* (4), C469–75.
11. Awan, K.; Butt, M.; Haq, I.; Ashfaq, F.; Suleria, H. Storage Stability of Garlic Fortified Chicken Bites. *J. Food Chem. Nanotechnol.* **2017,** *3* (2), 80–85.
12. Balasundram, N.; Sundram, K.; Samman, S. Phenolic Compounds in Plants and Agri-industrial By-products: Antioxidant Activity, Occurrence, and Potential Uses. *Food Chem.* **2006,** *99* (1), 191–203.
13. Barta, Z.; Tóth, L.; Zeher, M. Pulse Cyclophosphamide Therapy for Inflammatory Bowel Disease. *World J. Gastroenterol.* **2006,** *12* (8), 1278.
14. Bech-Larsen, T.; Scholderer, J. Functional Foods in Europe: Consumer Research, Market Experiences and Regulatory Aspects. *Trends Food Sci. Technol.* **2007,** *18* (4), 231–234.
15. Benavente-García, O.; Castillo, J.; Marin, F. R.; Ortuño, A.; Del Río, J. A. Uses and Properties of Citrus Flavonoids. *J. Agric. Food Chem.* **1997,** *45* (12), 4505–4515.
16. Bermudez, E. A.; Ridker, P. M. C-reactive Protein, Statins, and the Primary Prevention of Atherosclerotic Cardiovascular Disease. *Preventive Cardiol.* **2002,** *5* (1), 42–46.
17. Betteridge, D. J. What Is Oxidative Stress? *Metabolism* **2000,** *49* (2), 3–8.
18. Bocco, A.; Cuvelier, M.-E.; Richard, H.; Berset, C. Antioxidant Activity and Phenolic Composition of Citrus Peel and Seed Extracts. *J. Agric. Food Chem.* **1998,** *46* (6), 2123–2129.
19. Bok, S. H.; Jeong, T. S.; Bae, K. H.; Park, Y. B.; Choi, M. S.; Moon, S. S.; Kwon, Y. K.; Lee, E. S.; Hyun, B. H.; Choi, Y. K. Method for Lowering Blood Glucose Level by the Administration of Bioflavonoid. Google Patent, 2000.

20. Bok, S. H.; Lee, S. H.; Park, Y. B.; Bae, K. H.; Son, K. H.; Jeong, T. S.; Choi, M. S. Plasma and Hepatic Cholesterol and Hepatic Activities of 3-Hydroxy-3-Methyl-Glutaryl-CoA Reductase and Acyl CoA: Cholesterol Transferase Are Lower in Rats Fed Citrus Peel Extract or a Mixture of Citrus Bioflavonoids. *J. Nutr.* **1999,** *129* (6), 1182–1185.
21. Brescianini, S.; Maggi, S.; Farchi, G.; Mariotti, S.; Di Carlo, A.; Baldereschi, M.; Inzitari, D. Low Total Cholesterol and Increased Risk of Dying: Are Low Levels Clinical Warning Signs in the Elderly? Results from the Italian Longitudinal Study on Aging. *J. Am. Geriatr. Soc.* **2003,** *51* (7), 991–996.
22. Brown, M. D.; Jin, L.; Jien, M. L.; Matsumoto, A. H.; Helm, G. A.; Lusis, A. J.; Frank, J. S.; Shi, W. Lipid Retention in the Arterial Wall of Two Mouse Strains with Different Atherosclerosis Susceptibility. *J. Lipid Res.* **2004,** *45* (6), 1155–1161.
23. Butt, M. S.; Siddiq, M.; Ahmed, W. Tangerine, Mandarin and Clementine. *Tropical and Subtropical Fruits: Postharvest Physiology, Processing and Packaging*; Wiley-Blackwell, 2012; pp 419. https://www.wiley.com/en-au/Tropical+and+Subtropical+Fruits%3A+Postharvest+Physiology%2C+Processing+and+Packaging-p-9780813811420
24. Chang, J. T.; Staffa, J. A.; Parks, M.; Green, L. Rhabdomyolysis with HMGCoA Reductase Inhibitors and Gemfibrozil Combination Therapy. *Pharmacoepidemiol. Drug Saf.* **2004,** *13* (7), 417–426.
25. Chen, M. L.; Yang, D. J.; Liu, S. C. Effects of Drying Temperature on the Flavonoid, Phenolic Acid and Antioxidative Capacities of the Methanol Extract of Citrus Fruit (*Citrus sinensis* L. Osbeck) Peels. *Int. J. Food Sci. Technol.* **2011,** *46* (6), 1179–1185.
26. Conforti, F.; Statti, G. A.; Tundis, R.; Loizzo, M. R.; Menichini, F. In Vitro Activities of *Citrus medica* L. cv. Diamante (Diamante citron) Relevant to Treatment of Diabetes and Alzheimer's Disease. *Phytother. Res.* **2007,** *21* (5), 427–433.
27. Corradini, C.; Borromei, C.; Cavazza, A.; Merusi, C.; De Rossi, A.; Nicoletti, I. Determination of Flavanones in Citrus Byproducts and Nutraceutical Products by a Validated RP-HPLC Method. *J. Liq. Chromatogr. Relat. Technol.* **2009,** *32* (10), 1448–1462.
28. Das, L.; Bhaumik, E.; Raychaudhuri, U.; Chakraborty, R. Role of Nutraceuticals in Human Health. *J. Food Sci. Technol.* **2012,** *49* (2), 173–183.
29. de Moraes Barros, H. R.; De Castro Ferreira, T. A. P.; Genovese, M. I. Antioxidant Capacity and Mineral Content of Pulp and Peel from Commercial Cultivars of Citrus from Brazil. *Food Chem.* **2012,** *134* (4), 1892–1898.
30. Demonty, I.; Lin, Y.; Zebregs, Y. E.; Vermeer, M. A.; Van Der Knaap, H. C.; Jäkel, M.; Trautwein, E. A. The Citrus Flavonoids Hesperidin and Naringin Do not Affect Serum Cholesterol in Moderately Hypercholesterolemic Men and Women. *J. Nutr.* **2010,** *140* (9), 1615–1620.
31. Donaldson, M. S. Nutrition and Cancer: A Review of the Evidence for an Anti-cancer Diet. *Nutr. J.* **2004,** *3* (1), 19.
32. Ebrahimzadeh, M. A.; Pourmorad, F.; Bekhradnia, A. R. Iron Chelating Activity, Phenol and Flavonoid Content of Some Medicinal Plants from Iran. *Afr. J. Biotechnol.* **2008,** *7* (18), 3188–3192.
33. Erlund, I.; Silaste, M.; Alfthan, G.; Rantala, M. Plasma concentrations of the Flavonoids Hesperetin, Naringenin and Quercetin in Human Subjects Following Their Habitual Diets, and Diets High or Low in Fruit and Vegetables. *Eur. J. Clin. Nutr.* **2002,** *56* (9), 891.
34. Espín, J. C.; García-Conesa, M. T.; Tomás-Barberán, F. A. Nutraceuticals: Facts and Fiction. *Phytochemistry* **2007,** *68* (22), 2986–3008.

35. Fabre, N.; Rustan, I.; De Hoffmann, E.; Quetin-Leclercq, J. Determination of Flavone, Flavonol, and Flavanone Aglycones by Negative Ion Liquid Chromatography Electrospray Ion Trap Mass Spectrometry. *J. Am. Soc. Mass Spectr.* **2001,** *12* (6), 707–715.
36. Faiz, F.; Khan, M. I.; Butt, M. S.; Nawaz, H. Effects of Dietary Natural Antioxidants from Citrus Waste on Growth and Blood Antioxidants Status of the Broilers. *Sarhad J. Agric.* **2017,** *33* (3), 371–376.
37. Fisher, K.; Phillips, C. Potential Antimicrobial Uses of Essential Oils in Food: Is Citrus the Answer? *Trends Food Sci. Technol.* **2008,** *19* (3), 156–164.
38. Forbes, J. M.; Coughlan, M. T.; Cooper, M. E. Oxidative Stress as a Major Culprit in Kidney Disease in Diabetes. *Diabetes* **2008,** *57* (6), 1446–1454.
39. Frankel, E. N.; Meyer, A. S. The Problems of Using One-dimensional Methods to Evaluate Multifunctional Food and Biological Antioxidants. *J. Sci. Food Agric.* **2000,** *80* (13), 1925–1941.
40. Garg, A.; Garg, S.; Zaneveld, L.; Singla, A. Chemistry and Pharmacology of the Citrus Bioflavonoid Hesperidin. *Phytother. Res.* **2001,** *15* (8), 655–669.
41. Ggyawali, R.; Kim, K. S. Anticancer Phytochemicals of Citrus Fruits: A Review. *J. Anim. Res.* **2014,** *4* (1), 85–95.
42. Ghasemi, K.; Ghasemi, Y.; Ebrahimzadeh, M. A. Antioxidant Activity, Phenol and Flavonoid Contents of 13 Citrus Species Peels and Tissues. *Pak. J. Pharmacol. Sci.* **2009,** *22* (3), 277–281.
43. Gil-Izquierdo, A.; Gil, M. I.; Ferreres, F.; Tomás-Barberán, F. A. In Vitro Availability of Flavonoids and Other Phenolics in Orange Juice. *J. Agric. Food Chem.* **2001,** *49* (2), 1035–1041.
44. Gonzalez-Gallego, J.; Sánchez-Campos, S.; Tunon, M. Anti-inflammatory Properties of Dietary Flavonoids. *Nutric. Hospital.* **2007,** *22* (3), 287–93.
45. Gorinstein, S.; Martın-Belloso, O.; Park, Y.-S.; Haruenkit, R.; Lojek, A.; ĈíŽ, M.; Caspi, A.; Libman, I.; Trakhtenberg, S. Comparison of Some Biochemical Characteristics of Different Citrus Fruits. *Food Chem.* **2001,** *74* (3), 309–315.
46. Gorinstein, S.; Martin-Belloso, O.; Lojek, A.; Číž, M.; Soliva-Fortuny, R.; Park, Y. S.; Caspi, A.; Libman, I.; Trakhtenberg, S. Comparative Content of Some Phytochemicals in Spanish Apples, Peaches and Pears. *J. Sci. Food Agric.* **2002,** *82* (10), 1166–1170.
47. Gursoy, N.; Tepe, B.; Sokmen, M. Evaluation of the Chemical Composition and Antioxidant Activity of the Peel Oil of *Citrus Nobilis*. *Int. J. Food Prop.* **2010,** *13* (5), 983–991.
48. Hassimotto, N. M. A.; Genovese, M. I.; Lajolo, F. M. Antioxidant Activity of Dietary Fruits, Vegetables, and Commercial Frozen Fruit Pulps. *J. Agric. Food Chem.* **2005,** *53* (8), 2928–2935.
49. Iahtisham-Ul-Haq; Butt, M. S.; Suleria, H. A. R.; Ahmed, W. Physicochemical Behavior of Zinc-fortified, Sodium Caseinate-based, Edible-coated Apricots During Storage in Controlled Atmosphere. *J. Food Process. Preserv.* **2015,** *39* (6), 2431–2441.
50. Ibrahim, S. S. Protective Effect of Hesperidin, a Citrus Bioflavonoid, on Diabetes-induced Brain Damage in Rats. *J. Appl. Sci. Res.* **2008,** *4* (1), 84–95.
51. Jeong, S. M.; Kim, S. Y.; Kim, D. R.; Jo, S. C.; Nam, K.; Ahn, D.; Lee, S. C. Effect of Heat Treatment on the Antioxidant Activity of Extracts from Citrus Peels. *J. Agric. Food Chem.* **2004,** *52* (11), 3389–3393.

52. Jones, P. J.; Jew, S. Functional Food Development: Concept to Reality. *Trends Food Sci. Technol.* **2007,** *18* (7), 387–390.
53. Jung, U. J.; Lee, M. K.; Jeong, K. S.; Choi, M. S. The Hypoglycemic Effects of Hesperidin and Naringin are Partly Mediated by Hepatic Glucose-regulating Enzymes in C57BL/KsJ-db/db Mice. *J. Nutr.* **2004,** *134* (10), 2499–2503.
54. Kamaraj, S.; Anandakumar, P.; Jagan, S.; Ramakrishnan, G.; Devaki, T. Hesperidin Attenuates Mitochondrial Dysfunction During Benzo (a) Pyrene-induced Lung Carcinogenesis in Mice. *Fund. Clin. Pharmacol.* **2011,** *25* (1), 91–98.
55. Kawaii, S.; Tomono, Y.; Katase, E.; Ogawa, K.; Yano, M. Effect of Citrus Flavonoids on HL-60 Cell Differentiation. *Anticancer Res.* **1999,** *19* (2A), 1261–1269.
56. Khossousi, A.; Binns, C.; Dhaliwal, S.; Pal, S. The Acute Effects of Psyllium on Postprandial Lipaemia and Thermogenesis in Overweight and Obese Men. *Br. J. Nutr.* **2008,** *99* (5), 1068–1075.
57. Chen, Z. Y.; Jiao, R.; Ma, K. Y. Cholesterol-lowering Nutraceuticals and Functional Foods. *J. Agric. Food Chem.* **2008,** *56* (19), 8761–8773.
58. Kurowska, E. M.; Manthey, J. A. Hypolipidemic Effects and Absorption of Citrus Polymethoxylated Flavones in Hamsters with Diet-induced Hypercholesterolemia. *J. Agric. Food Chem.* **2004,** *52* (10), 2879–2886.
59. Laszlo, P. *Citrus: A History*; University of Chicago Press; 2008.
60. Lee, S. H.; Jeong, T. S.; Park, Y. B.; Kwon, Y. K.; Choi, M. S.; Bok, S. H. Hypocholesterolemic Effect of Hesperetin Mediated by Inhibition of 3-Hydroxy-3-Methylgultaryl Coenzyme-A Reductase and Acyl Coenzyme-A: Cholesterol Acyltransferase in Rats Fed High-cholesterol Diet. *Nutr. Res.* **1999,** *19* (8), 1245–1258.
61. Li, S.; Lo, C. Y.; Ho, C. T. Hydroxylated Polymethoxyflavones and Methylated Flavonoids in Sweet Orange (*Citrus sinensis*) Peel. *J. Agric. Food Chem.* **2006,** *54* (12), 4176–4185.
62. Li, S.; Wang, Z.; Sang, S.; Huang, M. T.; Ho, C. T. Identification of Nobiletin Metabolites in Mouse Urine. *Mol. Nutr. Food Res.* **2006,** *50* (3), 291–299.
63. Loizzo, M. R.; Tundis, R.; Bonesi, M.; Menichini, F.; De Luca, D.; Colica, C.; Menichini, F. Evaluation of *Citrus Aurantifolia* Peel and Leaves Extracts for Their Chemical Composition, Antioxidant and Anti-cholinesterase Activities. *J. Sci. Food Agric.* **2012,** *92* (15), 2960–2967.
64. Magda, R.; Awad, A.; Selim, K. Evaluation of Mandarin and Navel Orange Peels as Natural Sources of Antioxidant in Biscuits. *Alexandria J. Food Sci. Technol.* **2008,** 75–82.
65. Malterud, K. E.; Rydland, K. M. Inhibitors of 15-lipoxygenase from Orange Peel. *J. Agric. Food Chem.* **2000,** *48* (11), 5576–5580.
66. Manthey, J. A. Fractionation of Orange Peel Phenols in Ultrafiltered Molasses and Mass Balance Studies of Their Antioxidant Levels. *J. Agric. Food Chem.* **2004,** *52* (25), 7586–7592.
67. Manthey, J. A.; Guthrie, N. Antiproliferative Activities of Citrus Flavonoids Against Six Human Cancer Cell Lines. *J. Agric. Food Chem.* **2002,** *50* (21), 5837–5843.
68. Manthey, J. A.; Guthrie, N.; Grohmann, K. Biological Properties of Citrus Flavonoids Pertaining to Cancer and Inflammation. *Curr. Med. Chem.* **2001,** *8* (2), 135–153.
69. Menichini, F.; Loizzo, M. R.; Bonesi, M.; Conforti, F.; De Luca, D.; Statti, G. A.; De Cindio, B.; Menichini, F.; Tundis, R. Phytochemical Profile, Antioxidant, Anti-Inflammatory and Hypoglycemic Potential of Hydroalcoholic Extracts from *Citrus*

*Medica* L. cv Diamante Flowers, Leaves and Fruits at Two Maturity Stages. *Food Chem. Toxicol.* **2011,** *49* (7), 1549–1555.
70. Murakami, A.; Ohigashi, H. Polymethylated Flavonoids: Cancer Preventive and Therapeutic Potentials Derived from Anti-inflammatory and Drug Metabolism-modifying Properties. *Oxid. Stress Dis.* **2004,** *12*, 187–212.
71. Nagata, E.; Ichi, I.; Kataoka, R.; Matsushima, M.; Adachi, N.; Kitamura, Y.; Sasaki, T.; Kojo, S. Effect of Nobiletin on Lipid Metabolism in Rats. *J. Health Sci.* **2010,** *56* (6), 705–711.
72. Nichols, L. A.; Jackson, D. E.; Manthey, J. A.; Shukla, S. D.; Holland, L. J. Citrus Flavonoids Repress the mRNA for Stearoyl-CoA Desaturase, a Key Enzyme in Lipid Synthesis and Obesity Control, in Rat Primary Hepatocytes. *Lipids Health Dis.* **2011,** *10* (1), 36.
73. Oboh, G.; Ademosun, A. Characterization of the Antioxidant Properties of Phenolic Extracts from Some Citrus Peels. *J. Food Sci. Technol.* **2012,** *49* (6), 729–736.
74. Oboh, G.; Rocha, J. Antioxidant in Foods: A New Challenge for Food Processors. *Lead. Edge Antioxid. Res.* **2007,** 35–64.
75. Olufunmilayo, A. O.; Adewale, A. A. The Role of Phytotherapy in the Management of Diabetes Mellitus. *Phytother. Manage. Diab. Hypertens.* **2012,** *1*, 131.
76. Onder, G.; Landi, F.; Volpato, S.; Fellin, R.; Carbonin, P.; Gambassi, G.; Bernabei, R. Serum Cholesterol Levels and In-hospital Mortality in the Elderly. *Am. J. Med.* **2003,** *115* (4), 265–271.
77. Ott, M.; Gogvadze, V.; Orrenius, S.; Zhivotovsky, B. Mitochondria, Oxidative Stress and Cell Death. *Apoptosis*. **2007,** *12* (5), 913–922.
78. Parmar, H. S.; Dixit, Y.; Kar, A. Fruit and Vegetable Peels: Paving the Way Towards the Development of New Generation Therapeutics. *Drug Discov. Therap.* **2010,** *4* (5), 314–325.
79. Parmar, H. S.; Kar, A. Antidiabetic Potential of *Citrus sinensis* and *Punica granatum* Peel Extracts in Alloxan Treated Male Mice. *Biofactors* **2007,** *31* (1), 17–24.
80. Parmar, H. S.; Kar, A. Protective Role of *Citrus sinensis*, *Musa paradisiaca* and *Punica granatum* Peels Against Diet-induced Atherosclerosis and Thyroid Dysfunctions in Rats. *Nutr. Res.* **2007,** *27* (11), 710–718.
81. Peterson, J.; Dwyer, J. Flavonoids: Dietary Occurrence and Biochemical Activity. *Nutr. Res.* **1998,** *18* (12), 1995–2018.
82. Proteggente, A. R.; Saija, A.; De Pasquale, A.; Rice-Evans, C. A. The Compositional Characterisation and Antioxidant Activity of Fresh Juices from Sicilian Sweet Orange (*Citrus sinensis* L. Osbeck) Varieties. *Free Radic. Res.* **2003,** *37* (6), 681–687.
83. Ram, L.; Singh, S. Medicinal Importance of Citrus Products and By-products: A Review. *Agric. Rev.* **2006,** *27* (3), 170–180.
84. Ramkumar, D.; Rao, S. S. Efficacy and Safety of Traditional Medical Therapies for Chronic Constipation: Systematic Review. *Am. J. Gastroenterol.* **2005,** *100* (4), 936–971.
85. Raza, A.; Butt, M. S.; Iahtisham-Ul-Haq; Suleria, H. A. R. Jamun (*Syzygium cumini*) Seed and Fruit Extract Attenuate Hyperglycemia in Diabetic Rats. *Asian Pac. J. Trop. Biomed.* **2017,** *7* (8), 750–754.
86. Rizza, S.; Muniyappa, R.; Iantorno, M.; Kim, J. A.; Chen, H.; Pullikotil, P.; Senese, N.; Tesauro, M.; Lauro, D.; Cardillo, C. Citrus Polyphenol Hesperidin Stimulates Production of Nitric Oxide in Endothelial Cells While Improving Endothelial Function

and Reducing Inflammatory Markers in Patients with Metabolic Syndrome. *J. Clin. Endocrinol. Metab.* **2011,** *96* (5), 782–792.

87. Sanchez-Moreno, C. Methods Used to Evaluate the Free Radical Scavenging Activity in Foods and Biological Systems. *Revista de Agaroquimica y Tecnologia de Alimentos* **2002,** *8* (3), 121–137.
88. Sánchez-Moreno, C.; Plaza, L.; De Ancos, B.; Cano, M. P. Quantitative Bioactive Compounds Assessment and Their Relative Contribution to the Antioxidant Capacity of Commercial Orange Juices. *J. Sci. Food Agric.* **2003,** *83* (5), 430–439.
89. Sawalha, S. M.; Arráez-Román, D.; Segura-Carretero, A.; Fernández-Gutiérrez, A. Quantification of Main Phenolic Compounds in Sweet and Bitter Orange Peel using CE-MS/MS. *Food Chem.* **2009,** *116* (2), 567–574.
90. Schieber, A.; Stintzing, F.; Carle, R. By Products of Plant Food Processing as a Source of Functional Compounds: Recent Developments. *Trend. Food Sci. Technol.* **2001,** *12* (11), 401–413.
91. Sergeev, I. N.; Ho, C. T.; Li, S.; Colby, J.; Dushenkov, S. Apoptosis-inducing Activity of Hydroxylated Polymethoxyflavones and Polymethoxyflavones from Orange Peel in Human Breast Cancer Cells. *Mol. Nutr. Food Res.* **2007,** *51* (12), 1478–1484.
92. Raza, A., Butt, M. S., Iahtisham Ul, H., & Suleria, H. A. R. Jamun (Syzygium cumini) Seed and Fruit Extract Attenuate Hyperglycemia in Diabetic Rats. *Asian Pac. J. Trop. Biomed.* **2017,** 7 (8), 750–754.
93. Spatafora, C.; Tringali, C. Valorization of Vegetable Waste: Identification of Bioactive Compounds and Their Chemo-enzymatic Optimization. *Open Agric. J.* **2012,** *6*, 9–16.
94. Stief, T. W. The physiology and Pharmacology of Singlet Oxygen. *Med. Hypoth.* **2003,** *60* (4), 567–572.
95. Su, M. S.; Shyu, Y. T.; Chien, P. J. Antioxidant Activities of Citrus Herbal Product Extracts. *Food Chem.* **2008,** *111* (4), 892–896.
96. Suleria, H. A. R.; Butt, M. S.; Anjum, F. M.; Saeed, F.; Khalid, N. Onion: Nature Protection Against Physiological Threats. *Critic. Rev. Food Sci. Nutr.* **2015,** *55* (1), 50–66.
97. Suleria, H. A. R.; Osborne, S.; Masci, P.; Gobe, G. Marine-based Nutraceuticals: An Innovative Trend in the Food and Supplement Industries. *Marine Drugs* **2015,** *13* (10), 6336–6351.
98. Sultan, M. T.; Buttxs, M. S.; Qayyum, M. M. N.; Suleria, H. A. R. Immunity: Plants as Effective Mediators. *Critic. Rev. Food Sci. Nutr.* **2014,** *54* (10), 1298–1308.
99. Tiwari, P.; Kumar, B.; Kaur, M.; Kaur, G.; Kaur, H. Phytochemical Screening and Extraction: A Review. *Intern. Pharm. Sci.* **2011,** *1* (1), 98–106.
100. Tomás-Barberán, F. A.; Clifford, M. N. Flavanones, Chalcones and Dihydrochalcones-Nature, Occurrence and Dietary Burden. *J. Sci. Food Agric.* **2000,** *80* (7), 1073–1080.
101. Tundis, R.; Loizzo, M. R.; Bonesi, M.; Menichini, F. *Citrus medica* L. cv Diamante: An Overview on the Phytochemistry and Potential Health Benefits. *Molecular Phylogeny, Antioxidant Properties and Medicinal Uses*, 2014, pp 125.
102. Vega-Villa, K. R.; Remsberg, C. M.; Takemoto, J. K.; Ohgami, Y.; Yáñez, J. A.; Andrews, P. K.; Davies, N. M. 2011. Stereospecific Pharmacokinetics of Racemic Homoeriodictyol, Isosakuranetin, and Taxifolin in Rats and Their Disposition in Fruit. *Chirality* **2011,** *23* (4), 339–348.
103. Wang, X.; Hasegawa, J.; Kitamura, Y.; Wang, Z.; Matsuda, A.; Shinoda, W.; Miura, N.; Kimura, K. Effects of Hesperidin on the Progression of Hypercholesterolemia and Fatty Liver Induced by High-cholesterol Diet in Rats. *J. Pharmacol. Sci.* **2011,** *117* (3), 129–138.

104. Weverling-Rijnsburger, A. W.; Jonkers, I. J.; Van Exel, E.; Gussekloo, J.; Westendorp, R. G. High-density vs Low-density Lipoprotein Cholesterol as the Risk Factor for Coronary Artery Disease and Stroke in Old Age. *Arch. Internal Med.* **2003,** *163* (13), 1549–1554.
105. Whitman, S. C.; Kurowska, E. M.; Manthey, J. A.; Daugherty, A. Nobiletin, a Citrus Flavonoid Isolated from Tangerines, Selectively Inhibits Class A Scavenger Receptor-mediated Metabolism of Acetylated LDL by Mouse Macrophages. *Atherosclerosis* **2005,** *178* (1), 25–32.
106. WHO. Prevention of Blindness from Diabetes Mellitus. Report of a WHO Consultation; World Health Organization, Geneva, Switzerland; November 9–11, 2005; pp 23–28.
107. Wild, S. H.; Roglic, G.; Green, A.; Sicree, R.; King, H. Global Prevalence of Diabetes: Estimates for the Year 2000 and Projections for 2030: Response to Rathman and Giani. *Diabetes Care* **2004,** *27* (10), 2569–2569.
108. Wright, G.; Terada, K.; Yano, M.; Sergeev, I.; Mori, M. Oxidative Stress Inhibits the Mitochondrial Import of Preproteins and Leads to Their Degradation. *Exp. Cell Res.* **2001,** *263* (1), 107–117.
109. Xiao, H.; Yang, C. S.; Li, S.; Jin, H.; Ho, C. T.; Patel, T. Monodemethylated Polymethoxyflavones from Sweet Orange (*Citrus sinensis*) Peel Inhibit Growth of Human Lung Cancer Cells by Apoptosis. *Mol. Nutr. Food Res.* **2009,** *53* (3), 398–406.
110. Imran, M.; Butt, M. S.; Iqbal, M. J.; Gilani, S. A.;Basharat, S.; Saeed, F.; Suleria, H. A. R. Antioxidant Potential: Physicochemical and Sensory Attributes of Cookies Supplemented with Mosambi Peel Extract. *Int. J. Fruit Sci.* **2016,** 16 (3), 341–349.
111. Xu, G.; Ye, X.; Chen, J.; Liu, D. Effect of Heat Treatment on the Phenolic Compounds and Antioxidant Capacity of Citrus Peel Extract. *J. Agric. Food Chem.* **2007,** *55* (2), 330–335.
112. Yang, C. S.; Landau, J. M.; Huang, M. T.; Newmark, H. L. Inhibition of Carcinogenesis by Dietary Polyphenolic Compounds. *Annu. Rev. Nutr.* **2001,** *21* (1), 381–406.
113. Yang, E. J.; Kim, S.; Moon, J. Y.; Oh, T. H.; Baik, J.; Lee, N.; Hyun, C. G. Inhibitory Effects of *Fortunella japonica* var. Margarita and *Citrus sunki* Essential Oils on Nitric Oxide Production and Skin Pathogens. *Acta Microb. Immunol. Hungar.* **2010,** *57* (1), 15–27.
114. Youssef, S. A. Evaluation of Composted Chicken Manure in Biocontrolling *Fusarium* Wilt on Tomato. *Egypt. J. Phytopathol.* **2007,** *35* (1), 61–72.
115. Zhang, F.; Aravanis, A. M.; Adamantidis, A.; De Lecea, L.; Deisseroth, K. Circuit-Breakers: Optical Technologies for Probing Neural Signals and Systems. *Nat. Rev. Neurosci.* **2007,** *8* (8), 577–581.

# CHAPTER 7

# PHYTOCHEMISTRY OF GRAPES (*Vitis vinifera* L.): FUNCTIONAL AND NUTRACEUTICAL ATTRIBUTES

HAFIZ ANSAR RASUL SULERIA, FARHAN SAEED, TABUSSAM TUFAIL, SAIRA SULTAN, and HAFSA DAUD

## ABSTRACT

Modern research has revitalized conventional fruits to retrieve from various disorders by using contemporary tools such as diet-based therapy and other regimens. Grapes (*Vitis vinifera* L.) are the archetypal paradigms of fruits used for not only nutritional purpose but also for exclusive therapeutic use owing to their antimicrobial, antioxidant, and anti-inflammatory perspective. The most imperative phytochemicals in grapes are polyphenols to provide various health benefits. Resveratrol, a phytoalexin antioxidant found in red grapes, has both chemopreventive and therapeutic effects against various ailments, including those of skin. This chapter elaborates the health claims of grapes and their functional roles, with special reference to antioxidant potential, immunonutrition, anticancer perspectives, and cardiovascular cure. Although present review chapter has summarized the literature pertaining to grapes and their bioactive compounds, yet further research should be instantly carried out for diligence and exploration.

## 7.1 INTRODUCTION

In the modern era, diet-based therapy is gaining popularity worldwide to provide basic requirements of nutrients for our body metabolism. Extraction of bioactive moieties and their impacts on human metabolism need systematic research to attain meticulous and beneficial perspectives for consumers. Grape (*Vitis vinifera* L.) belongs to the genus *Vitis*, and is the world's leading fruit with a global production of around 69 million tons, out of which about 80% is used for wine making.[22,35]

Grapes and their products have several trade names such as grape seed, extract, and activin.[61] Grapevine seeds and foliage are utilized in formulation of medication. Grapes are also used as dietary supplement; fresh and dried forms are also utilized in confectionary industries. Composition profiles for grapes show that grapes are rich source of carbohydrates, vitamins, and minerals (Table 7.1).

**TABLE 7.1** Nutritional Profiles of *Vitis vinifera L.*

| Chemical analysis | Value per 100 g |
|---|---|
| **Vitamins (μg)** | |
| Vitamin B1, thiamine | 69 (6%) |
| Vitamin B2, riboflavin | 70 (6%) |
| Vitamin B3, niacin | 188 (1%) |
| Vitamin B5, pantothenic acid | 50 (1%) |
| Vitamin B6, pyridoxamine | 86 (7%) |
| Vitamin B9, folate | 2 (1%) |
| Vitamin C, ascorbic acid | 10,800 (13%) |
| Vitamin K, phylloquinone | 22 (21%) |
| **Macronutrients (mg)** | |
| Carbohydrates ($CH_2O$) | 18,100 |
| Dietary fiber | 900 |
| Fat | 160 |
| Protein | 720 |
| Sugars | 15,480 |
| Energy | 69 Kcal |
| **Minerals (μg)** | |
| Calcium | 10,000 |
| Manganese | 71 |
| Phosphorus | 20,000 |
| Potassium | 191,000 |
| Sodium | 3020 |
| Zinc | 70 |

Grape seeds have numerous phytochemical compounds, such as: proanthocyanidins (typically hexamers) or procyanidins.[45] The acylated procyanidin from extract of grape seeds is an ester of gallic acid. The deliberate investigations are being carried on 1 tetrameric, 11 trimeric, and 14 dimeric procyanidins.[54,55]

The most imperative phytochemicals in grapes are polyphenols that provide numerous biological activities and health-promoting potentials.[9] Primarily, the phenolic compounds comprise of phenolic acids, flavonols, anthocyanins, stilbenes (resveratrol), and flavanols (Fig. 7.1). Grape skin is primarily composed of anthocyanins pigments.[13] Flavonoids are extensively dispersed in the seeds and stems of grapes, and predominantly comprise (+)-catechins, (−)-epicatechin, and procyanidin polymers.

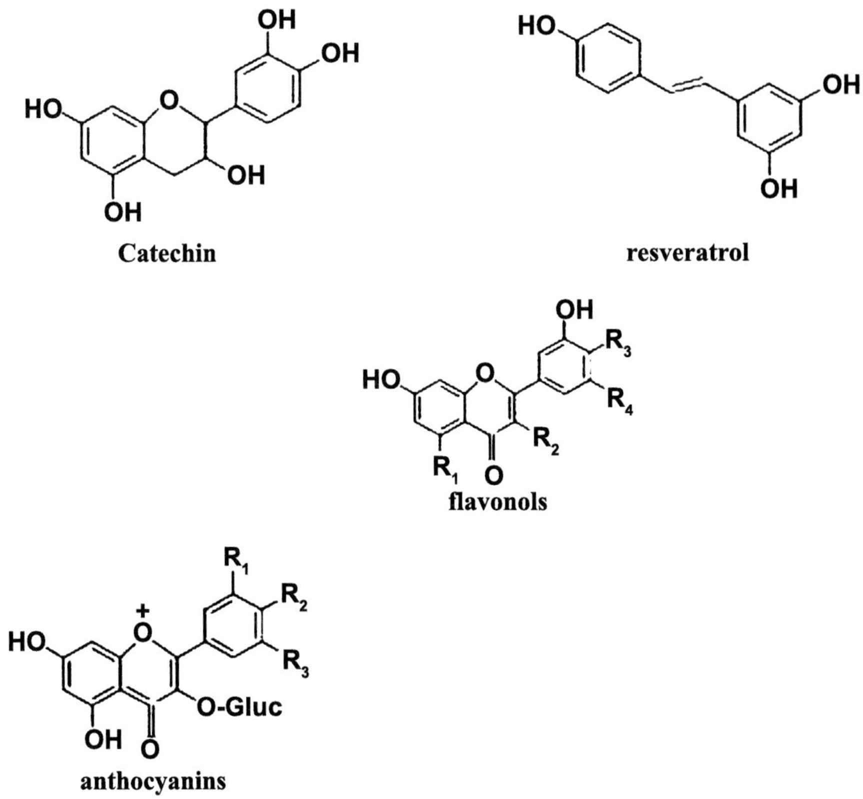


**FIGURE 7.1** Chemical structure of catechin, resveratrol, flavonols, and anthocyanins.

The polyphenolic compounds in red grapes are anthocyanins, while white grapes are rich in flavan-3-ols. Starting from "French paradox," numerous investigations have been carried out to explore composition and properties of polyphenolics from grapes and red wines. A number of degenerative diseases (i.e., cardiovascular and various types of cancers insurgence) can be cured by using polyphenols from different fruits in diet. Polyphenols reduce plasma oxidation stress and slow down the process of aging.[7,19]

Preservative effects of phenolic compounds are also observed in microorganisms and oxidative changes in foods.[54] Phenolic compounds are bioavailable[39] and these may also show negative effects on health when consumed at higher concentrations.[29] Various structures in particular higher molecular weight phenolic compounds may promote the negative effects, especially improper absorption in body.[44]

## 7.2 PHENOLIC COMPOUNDS IN GRAPES

Grape plant and its components are rich in numerous phenolic compounds.[124] The total phenolic compounds gallic acid equivalent is about 2178.8 mg/g of seeds, 374.6 mg/g of skin, 23.8 mg/g of flesh, and 351.6 mg/g of leaves.[85] Total phenolic content is dependent on the cultivar, soil composition and texture, climatic conditions, crop growing practices, geographic location, and infections due to fungi. Various phenolics of grapes are: anthocyanins, flavanols, flavonols, proanthocyanidins, resveratrols, and phenolic acids while proanthocyanidins are the abundant in skin and seed of grape.[65] Research studies on polyphenols' anticancer activities have revealed that the grape extracts and other grape products have positive effects on cancer patients as mentioned in Table 7.2.

**TABLE 7.2** Resources of Phenolic Derivatives from *Vitis vinifera* L.

| Resource | Phenolic derivatives | Reference |
|---|---|---|
| Leaves | Gallic acid, myricetin, ellagic acid, kaempferol, quercetin | [101] |
| Raisins | Hydroxymethylfurfural, hydroxycinnamic acid | [72] |
| Seeds | Catechin, gallic acid, epicatechin, dimeric procyanidin, proanthocyanins | [101] |
| Skin | Quercetin, proanthocyanins, ellagic acid, myricetin, kaempferol, trans-resveratrol | [65] |
| Stem | Trans-astilbin, resveratrol, quercetin 3-O-glucuronide, rutin, | [85] |
| Wine (red) | Three glucosides of petunidin, cyanidin, peonidin, malvidin, quercetin, hydroxycinnamic acid, resveratrol, catechin | [110] |

It has been explored that the grape skin extract is capable of inducing prostate tumor cell lines' apoptosis.[21] Grape juice phenolics can also regulate carcinogen-induced DNA and repress DNA amalgamation in breast cancer cells. Phenolic compounds have twin effects on cells and cell proliferation depending upon dosages. Antiaging effects of polyphenolics are advantageous in avoiding neuronal and behavioral aging processes.[47] Owing to polyphenolics' antioxidant properties, free radical scavenging activity, organs and tissues oxidative damages can be prevented while some body mechanism of redox status can be modified. Striatal slices discharge dopamine, and cognitive actions have been observed in rats after consumption of 10% of grape juice while action ability was increased by consuming 50% grape juice. The central nervous system of developed rats showed less lipid peroxidation induced by free radicals.

### 7.2.1 POLYPHENOLS

One of the prime issues regarding the advantage of polyphenols is based on their biological availability and destiny after metabolism. Bioavailability of dietary compounds depends upon their digestive immovability, liberation from the food, and the effectiveness of their transepithelial channel.[87] Bioavailability of polyphenols can be determined by the gastro-intestinal situation and the release from food matrix. For example, biological availability of anthocyanins can be recognized due to slow movement of these molecules in the small intestine.[27] The bioavailability of polyphenols is elaborated in Table 7.3.

**TABLE 7.3** Functions of Phenolic Compounds from *Vitis vinifera* L.

| Resource | Phenolic compounds | Reference |
|---|---|---|
| Dietary fiber concentrate of white grape | Antioxidant activity of polyunsaturated fatty acid in oil | [113] |
| Food prepared with grape seed specifically its extract | For preservation of fish flesh and oil | [82] |
| Fruit beverage | Induction of *Saccharomyces cerevisiae* from $H_2O_2$ | [133] |
| Grape + orange + apricot | Hypercholesterolemic protection of hamster against aortic fatty streak accumulation from grape wine | [114] |
| Grape seed | Reduction in the level of oxidized LDL in plasma | [114] |
| Juice | Reduction in the level of oxidized LDL in plasma | [56] |
| Milled and defatted grape seed | Deals with Adriamycin; an anticancer chemical work in reduction of GSH, ATP, and TBAS. Normally produced due to oxidant stress | [129] |
| Red wine | Membrane oxidation protection | [36] |

The absorption of phenolic compounds is usually slow and small, not more than the concentration of plasma that is 10. This low absorption is due to structural variation of polyphenols that decide their absorption in gut.[87] The absorption of polyphenols takes place by the process of passive diffusion transversely through the gut epithelial cells membranes.

During this process, a large number of polyphenols may get diffused into the gut wall due to their hydrophilic character.[86] Moreover, the majority of the polyphenols present in foodstuff are not absorbed in their indigenous form, that is, esters, glycosides, or polymers. Small intestine can merely absorb aglycones and some glucosides.[87]

### *7.2.2 FLAVONOIDS*

Among the naturally occurring phenolic secondary metabolites, flavonoids are ordinary group of aromatic compounds.[117] Flavonoids are phenylpropanoid derivatives that exist in the form of "$C_6$–$C_3$–$C_6$": a tri-circle configuration. In flavonoid structure, level of oxidation and substitution are noted on the basis of third central ring. The most significant classes of flavonoids are flavonols and flavan-3-ols while proanthocyanidins, which are condensated form of tannins and anthocyanins, also fall in this category.

The structural modification of flavonoid is possible, which may result in diversity of various products present in overall plant kingdom.[64,76] Structural composition of flavonoids shows that the flavonoids and their conjugates can exist in various families of plants and these compounds can be accumulated in plants based on location of tissue, type of cell, and phenological period of the plant.[117] Several important flavonoids classes represent natural molecules such as anthocyanins, main colorant, which provide attractive coloration (red to purple) to plant. The proanthocyanidins based on the polymerization of flavan-3-ols are colorless, but on oxidation they give dark-brown color to most of the seed coats.

All biocompounds belonging to flavonoid group play imperative physiological role in different regions of plants, that is, roots and leaves. In plants, vegetative part flavonoids are used to gather in reaction to numerous abiotic and biotic stress situations, or nodule formation in legume.[59] Flavonoids perform the role of straight ultraviolet screens, and provide antioxidants in opposition to reactive $O_2$ and forecast molecules. The prime function of colorant flavonoids is to give color to reproductive parts (flowers, fruits, and seeds) for appealing pollinators and seed-dispersal agents but on the other hand, tannins having no or brownish color grant opposition to infections caused by microbes and fungi.[76] To exhibit such functions, flavonoids are gathered predominantly in the plant organs margins like cells in the hypodermis.

Grapevine (*V. vinifera* L.) contains sufficient levels of flavonoids and it is characterized by dual reproductive organs, berries and seeds of fruit, restraining a huge quantity of soluble flavonoids.[32] In reality, soluble phenolics are present in fruit berries of grapes. The leading flavonoids categories are flavonols, flavan-3-ols, anthocyanins and tannins.[1] These chemical compounds along with benzoic and hydroxycinnamic acids (phenolic acids) are involved in the organoleptic features of wine and other by-products.[104] Grape fruits have immense economical significance and nutraceutical value due to presence of certain phenolics, therefore it is necessary to carry out research for the presence of flavonoids in different tissues during the ripening process.[32]

It is clear from the anatomical aspect that the grape flavonoids are concentrated exclusively in both the peripheral layers of pericarp of berries and in few layers of seed coat.[27] The berries mesocarp (pulp) has phenolic hydroxycinnamates and insufficient quantity of flavonoids. The grape skin acts as hydrophobic barricade of the pericarp and is composed of two discernible tissues.[1] The epidermal layers of grape fruit are cutinized, while the interior hypodermal layers are thickened and restrain the majority of the skin flavonoids.[104] This major class of flavonoids can be represented by anthocyanins, tannins, and negligible amount of flavonols and flavan-3-ols.

Grape skin is rich in tannins, which is formed by condensated monomeric units of flavan-3-ols, primarily composed of epicatechin and epigallocatechin and catechin.[41] The grape skin tannins differ from grape seed tannins having mean degree of polymerization of 28 and are bigger in size and less in amounts of galloylated units.[123] Organoleptic properties of grape skin are based on these characteristics that is why astringency and bitterness of phenolic acids (PAs) are inversely associated to the degree of polymerization. Various varieties of red grape contain anthocyanins in the order of 11.5–29.8 mg/g in the layers of hypodermis along with tannins. Skin of grape fruit also contains minor quantities of monomeric units of flavan-3-ol monomers, such as epicatechin and catechin, acting as precursors in the polymeric tannins condensation.

It has been observed that the third flavonoid constituent of skin is flavonol, which is highly representative having glucuronides of quercetin, galactosides, glucosides, myricetin, and kaempferol.[67] Flavonols, particularly quercetin, are able to act as UV-protectants and can take part in copigmentation with anthocyanins. Skin of both red and white grapes has same flavonoid composition in addition to the presence of anthocyanins in red cultivars.[134] It has been investigated that the flavonoid composition varies during various maturation stages of berries.[89] In reality, PA accretion takes place during setting of fruits up to 1–2 weeks subsequent to veraison and after that their amounts lower between veraison and harvest.[42] The total content of flavonoids has been noticed to be superior at the preliminary stages of berry formation.[41]

In grape seeds, mainly flavan-3-ols and minute quantities of flavonols and flavonoids are present.[104] The grape seeds in red wine make a payment primarily to condensed tannins and catechins, but maturation of grape and maceration during vinification step of processing is an indication of flavanol concentration in the red wine.[23] Flavonoids other than pericarp are found both in seed coat's facade tissues and in the inner layers. There is also dissimilarity in the constitution of skin and minor tannins of seed.[104] Flavonoids composition varies during the maturation of seed and tissues, for example, hardness and the color. At the time of veraison, levels of flavanol-3-ols can reach maximum, and after that these are decreased gradually forthcoming maturation; 90% decline in monomers and 60% decline in PAs have been observed.

### *7.2.3 FLAVONOLS*

The antioxidative properties of flavonols have been evaluated and the mode of action of flavonols with special reference to oxidation inhibition is not yet comprehensive. It is assumed that it may inhibit oxidation by: (1) acting as chain-breaking antioxidants as they can scavenge lipid alkoxyl and peroxyl radicals, for example, donors of hydrogen; (2) chelation of metallic ions; and (3) producing $\alpha$ tocopherol by reducing $\alpha$ tocopheroxyl radical. The flavones effectiveness as antioxidative is significantly dependent on their chemical configuration; being three structural confirmations, phenols are most effective against various maladies.[18]

Concord and De Chaunac grape samples contain an average of 34.95 mg/kg of quercetin 3-O-galactoside.[100] Similarly, grapes from Napoleon cultivar have 21.6 mg/kg of quercetin 3-O-glucuronide and quercetin 3-O-glucoside.[25] In a research study, two grape samples from Cabernet Sauvignon and two samples of Merlot grapes from Chile were examined for presence of flavonols, which may be free and conjugated (quercetin, myricetin, kaempferol, and isorhamnetin); and it was assumed that their content ranged from 84.6 to 327.9 nmol/g.[22] Extracts of white grapes (var. Sauvignon

Blanc, Thompson Seedless, and Chardonnay) were examined and it was observed that the total amount of flavonols compounds ranged from 4.8 to 10.4 mg/L. The flavonols examination of five white Muscadine cultivars (*Vitis rotundifolia* sp.), after glycosides hydrolysis, exhibited that total quercetin, myricetin, and weight mean were 6.56 mg/100 g, however, it was poorer and ranged from 1.6 to 3.5 mg/100 g in red Muscadine cultivars. The research study of Chardonnay pomace exposed some remarkable properties of flavonols. It was reported that excess of polyphenolic compounds were in Charisma including flavonol glycosides, quercetin 3-O-glucoside, quercetin 3-O-glucuronide, kaempferol 3-O-glucoside, and kaempferol 3-O-galactoside. Merlot grape stems contained variety of diversified flavonol glycosides,[124] however, the myricetin 3-O-glucuronide, myricetin 3-O-glucoside, quercetin 3-O-glucoside, quercetin 3-O-glucuronide, and kaempferol 3-O-glucoside usually were in trace amounts, that is, 218 mg/kg. In a study of three Sultana grapes (raisin samples), quercetin glycosides ranged from 82.1 to 121.8 mg/kg.[72] It was also reported that quercetin is present in molasses @ 1.69 mg/L in Sultana grape but kaempferol was absent.[73] It has been reported that quercetin glycoside in minute quantity of 7.2–9 mg/L was also present in grape juice.[125] Moreover, it was also analyzed that the juice samples have flavonol glycoside compounds in a range of 5.7–8.6 mg/L, even though the data regarding individual flavonols in grapes are not available.[49] Muscadine grape juice had quercetin, myricetin, and kaempferol in range of approx. 9.9 to 67 mg/L while in some cases it was reported to vary from 13.4 to 100.9 mg/L.[126] Diagnostic examination on 92 samples of vinegar has indicated that quercetin and isorhamnetin were present in minute levels of 1.53 mg/L.[57]

### *7.2.4 RESVERATROL (3, 5, 4′-TRIHYDROXY-TRANS-STILBENE)*

Resveratrol is a natural phenol and stilbenoid. It is phytoalexin produced in various plants by pathogenic attack of bacteria or fungi. The red grapes skin is significantly rich in trans-3, 5, 4′-trihydroxy-trans-stilbene that is essential to cure three basic stages of carcinogenesis. The resveratrol was discovered from the poisonous grandiflorum medicinal *Veratrum album* in 1939. The name resveratrol actually originated from the derivative of resorcinol of *Veratrum* species.

Plants produce resveratrol in response to pathogenic attack of *Botrytis cinerea*.[38] Different conditions of stress such as climatic fluctuations, exposure to $O_3$, UV light, and certain heavy metals also produce it.[16] Almost 70 species of plants are found to contain it, especially grapes, peanuts, berries, and pines. Skin of fresh grapes has resveratrol in range of 50–100 μg/g of wet fruit weight.[14] Trans-resveratrol is also termed as 3,5,4′-trihydroxystilbene. It is a member of class stilbene from polyphenolic compounds and present in cis- and transisomeric forms as well. Glycosylated piceid (3-O-B-d-glucosides) is dominant form of resveratrol in different varieties of plants. Some other negligible forms are conjugated, having a fatty acid, sulfate group (trans-resveratrol-3-sulfate) or 1–2 methyl groups (pterostilbene). Oxidative deprivation of resveratrol can be avoided by a process of glycosylation. That is why the glycosylated resveratrol is highly resistant, and it can be easily solubilized and absorption can occur in gastrointestinal tract (GIT).[107]

In the gut of human, glycosylated resveratrol is absorbed and metabolized in liver by the activity of enzymes of phase-2 associated in drug metabolism; water-soluble trans-resveratrol-3-O-glucuronide and trans-resveratrol-3-O-sulfate are formed and even excreted through urine.[131] Plasma half life of resveratrol exhibits is 8–14 min and that of its metabolic by-products is 9.2 h.[131] On the other hand, bioavailability and efficacy of resveratrol metabolites are anonymous.[15] Comparison of resveratrol with quercetin and catechin showed that absorption of its trans-resveratrol compound is more efficient when taken orally.[122]

In frequent experimentations, a wide range of resveratrol concentrations have been employed, and it was revealed that the biological performance of resveratrol is mostly dependent on different types of cells and tissues as well. Resveratrol concentration of 32 nM and 100 μM was used to investigate numerous resveratrol effects in vitro and 100–1500 mg/kg in in vivo studies.[11] Rodents' preclinical examination followed by HPLC techniques has recommended incorporation of 20 mg/kg trans-resveratrol into gastric system to produce 1.2 μM resveratrol in plasma.[7] In another experiment, it has been explained that the treatment of mice with 300, 1000, and 3000 mg/kg body weight/day is able to get resveratrol plasma concentrations of 576, 991, and 2728 ng/mL, correspondingly, and it was 333, 704, and 1137 ng/mL in female rats, respectively.[34]

The 14C trans-resveratrol given orally to male Balb/c mice can exhibit better radiolabeled resveratrol binding in various visceral tissues.[130] In these tissues mutually, the native chemicals and the metabolic by-products of phase-2 have also been examined.[130] Resveratrol dose @ 24.6% applied orally to human objects enabled to be present in the urine, together with other metabolites[122]; plasma screened-off area can be achieved by merely introducing 1.5% of resveratrol into gastric system.[7]

## 7.3 HEALTH ENDORSING POTENTIALS

### 7.3.1 ANTIOXIDANT EFFECTS OF EXTRACT OF GRAPE SEEDS

Grape seed extract has numerous antioxidants that may exhibit scavenging activity for free radicals.[24,69] Procyanidin is one of the components of *V. vinifera* seed not only scavenging free radicals activity but it also inhibits xanthine oxidase activity.[46] Peroxidation of polyunsaturated fatty acid can be repressed by Squat grape seed proanthocyanidins concentrations @ 2 mg/L.[19] Grape seed proanthocyanidin extract (GSPE) can be used to obtain in vitro free radical scavenging assay @ 50 mg/L and it was concluded that GSPE is superior than other natural antioxidants, that is, Vitamin C and Vitamin E.[11] Furthermore, low concentrations of catechin, procyanidin B4, and gallic acid (10 mol/L, 25 mol/L) are excellent anti-cellular factors in opposition to damage caused to DNA by oxidation. But, such chemicals may cause damage to cell DNA at elevated levels of 150 mol/L.[47]

Medical and biochemical researches have revealed the considerable antioxidant characteristics of grape seed oligomeric proanthocyanidins. Tannins, polyphenols, and polyunsaturated fatty acids have inhibitory activities against various diseases such as tumors, cardiopathy, and other physiological disorders. Oil is obtained after grinding grape seeds and it is used in cosmetics and other skincare products. Many individuals used these extracts for health benefits. Oil from grape seeds is distinguished for abundance of phytosterols, tocopherols (vitamin E), and polyunsaturated fatty acids such as oleic acid, linoleic acid, and alpha-linolenic acid.

Moreover, GSPE demonstrated remarkable shielding property in leukocytes of rats against deterioration caused by oxidation.[93] It has been mentioned that combined application of extract of grape seed (75 mg/kg) and marjoram volatile oil (0.16 mL/kg) can stop damage due to oxidation. In a research study on rats, they were provided with a dose of 25% v/v ethyl alcohol (10 mL/kg body weight) on regular basis for 10 weeks. Ethanol-induced interruption of embryonic enlargement can be prevented by resveratrol (10 mol) pretreatment in ESC-B5 stem cells of embryo and blastocysts. Antioxidants characteristics have been reported to reperfuse ischemia in rats in skeletal muscles.[44]

### 7.3.2 CARDIOPROTECTIVE PERSPECTIVES

Cardioprotection can be obtained by consuming 100 and 200 mg/kg extract of grapes as it enhances the postischemic ventricular recuperation and myocardial rat infarction.[35]

During in vivo trials on aortic rings of rats, Ex-Grape seeds (7 g/mL) influenced 77% endothelium-reliant leisure. Tannins from seed of grape (2% monomers or 2% polymers for 3 or 9 weeks) have antihypercholesterolemic potential, ensuing better cholesterol transportation and absorption of intestinal cholesterol and augmentation in excretion of bile acid in mice.[127] In animal modeling such as rat and rabbit with low reperfusion, damages of heart were observed on procyanidin supplementation allied with increased antioxidant activity of blood plasma.[17] It was also observed that a peroxynitrite attack on vascular cells can be prevented by placing layers on the endothelial cells of coronary, and increasing the endothelial nitric oxide (NO) synthase-mediated in aortic rings of inner mammary in humans.[6]

It has also been exhibited that the catechin and epicatechin did not affect the vascular endothelium; and procyanidin from grape seeds and anthocyanins were together were involved in integrity of endothelium and emancipation of NO formation.[91] Extracts obtained from seeds of grape having polyphenolic complexes help in relaxing of blood vessel endothelium. It was assumed that the AKT/PI3 kinase is involved in endothelium relaxation induced by the grape seed extract. Redox-sensitive method ensures the signaling pathway based on AKT/PI3 kinase in phosphorylating endothelial nitric oxide synthase (eNOS) rabbit aortic rings by ingestion of grape seed extract.[43] In the same case, grape seed extracts rich in proanthocyanidins had cardioprotective properties against reperfusion-induced damages in hearts of rats.[102] Collagen and platelet combination is synergistically regulated by quercetin (50–100 mol/L) and catechin (10–20 mol/L)[103] as well as in vitro and in vivo inhibition of platelet combination at (10–1000 mol/L) and (4 mg/kg/d), respectively[132]

Purple grape juice administration lowered the low-density lipoprotein (LDL) oxidative chances and flow-mediated vasodilatation (FMD) was improved in the case of coronary artery disease. Similarly, enhancement of FMD was also observed in brachial artery of heart patients by consuming purple grape juice @ 4–8 mL/kg for 4 weeks. Application of juice extracted from purple grapes at 7 mL/kg/day for about 14–20 days to healthy subjects can decrease superoxide release, can augment platelet-derived NO production, and inhibit platelet aggregation.[50]

Additionally, improved FMD was observed when purple grape juice (500 mL/day) was given to patients with hypercholesterolemia for 14–16 days without other risk factors.[31] Recently, it was concluded that by consumption of juice of red grapes at 50 mL/two times a day for 2 weeks enhanced the antioxidant capability of plasma, reduced the oxidized LDL concentration, and regulated the cholesterol-standardized tocopherol concentration equally in healthy individuals and patients suffering from hemodialysis. The reduction of plasma monocyte chemoattractant protein 1 associated with cardiovascular disease menace was reported by consumption of red grape juice.[28] No significant results were obtained on the aspirin-consuming patients for antithrombotic. Decrease in antioxidant plasma level and stress of postprandial oxidation was observed in vigilants consuming 300 mg of a proanthocyanidin-rich grape seed extract. Thus, the oxidation variability of LDL was amplified.[96] The 600 mg dose of extract from red grape containing polyphenol to patients suffering from coronary heart disease can help in improvement of endothelial activities.

Dilatation based on flow mediation was inspected in post fast conditions at 30, 60, and 120 min after consumption of grape extract. Red grape polyphenol extract can improve flow-mediated dilatation, and was maximum at 60 min, which was appreciably superior to the placebo ($P = 0.001$). No change was observed in data of FMD after the ingestion of placebo during the entire period. In another experiment, the dose of flavanol at the levels of 400 mg obtained from extracts of grape seed for 8 weeks resulted in positive values for platelet counts in females with post-menopause.[120] On the other hand, the grape juice (from Lakewood, FL) containing substantial levels of fructose and glucose enhanced adenine nucleotide dis-possession and lactic acid fabrication, which enhance the concentration of urate plasma.[99] Antitumor induced by doxorubicin was enhanced by using the grape seeds @ 12.5 and 25 mg/L in vitro and 10 mg/kg in vivo for developing drug from proanthocyanidin and increasing intracellular doxorubicin. This research was designed to investigate the transplantation of Sarcoma 180 (S180) and Hepatoma 22 (H22) in mice. It was revealed that elicit from seeds of grapes can aggreviate the action of amphotericin-B for antifungal infection in rats.[63]

### 7.3.3 ANTIDIABETIC EFFECTS

It has been investigated that GSPE is efficient in curing diabetic nephropathy by changing body functional proteins. Treatment of diabetic rats with GSPE caused the nine proteins in kidney to their usual levels and the linkage was reported between them with stress caused by oxidation, damage associated with glycosylation, and metabolism of amino acid.[80] Glycation-based heart problems in diabetic rats were improved by using GPSE @ 250 mg/kg of body weight per day.[30]

### 7.3.4 CANCER PERSPECTIVES OF GRAPES AND THEIR BY-PRODUCTS

#### 7.3.4.1 NONMELANOMA SKIN CANCER

The main category of sarcoma is skin cancer, that is, nonmelanoma and about one million cases are being identified in the United States every year. It has been reported that resveratrol is efficient in reducing various cutaneous abnormalities such as skin cancer.[3,13] Phototoxicity induced by ultraviolet-B (180 mJ/cm$^2$) in mice can be reduced by using resveratrol (25 μmol) to concentrated silicon and potassium with humic acid (SKH-1) and augmented bifold thickness and edema of skin.[4] When the skin is exposed to ultraviolet-B (180 mJ/cm$^2$) for 7 alternative days, the skin hyperplasia is stimulated. The skin hyperplasia can be cured by provision of resveratrol @ 10 μmol/animal for 30 min to every UVB contact. Resveratrol has anti-proliferative properties due to regulatory proteins in cell cycles.[106]

The improvement of p21WAF1/CIP1 was observed by use of resveratrol while multiplication cell nuclear antigen and cyclins D1, D2, Cdk 2, 4, 6 were decreased. Moreover, anti-apoptotic proteins also have inhibiting expressions, for example, survivin, tumor promotion markers, ornithine decarboxylase (ODC), and cyclooxygenase COX-2.[9] Chemoprevention of resveratrol was examined and it was concluded that it can provide research potential for study on UVB-mediated skin tumor formation in the SKH-1 bald mouse model. Tumor formation can be postponed or repressed by applying topical resveratrol either before or after UVB. Up to 98% decrease in skin cancer in mouse was observed due to tissue plasminogen activator (TPA)-promoted murine skin cancer exhibiting and 7,12-Dimethylbenz[a]anthracene (DMBA)-initiated.[67,71] Oral administration of resveratrol may cause inhibition of DMBA/croton in oil instigated skin papillomas of mouse, associated with protracting the embryonic cycle of tumors incidence and hindering croton oil-induced augmentation of epidermal ODC behaviors.[51]

It has been clearly implicit in vitro and in vivo study on squamous cell and carcinoma cells that resveratrol enables to induce G1-phase cell cycle capture, together with p21WAF1/CDK-interacting protein-1 (CIP1) stimulation, decrease the cyclins D1, D2, E, and C (dks, hyperphosphorylated pRb proteins, MEK1 > ERK 1/2, and AP-1 signaling) and cell cycle regulators.[2,5,75] A great deal of study concerning skin tumor formation has paid attention on UVB radiation effects, whereas fewer studies have been reported on UVA persuaded signaling of mechanisms and their performance in tumor endorsement. Cutaneous squamous cell carcinomas formation is observed to be caused by UVA, which is above 90% of striking solar radiation from earth.[10]

#### 7.3.4.2 MAMMARY TUMOR

Resveratrol may enhance production of 8-dihydro-2′-deoxyguanosine and 8-oxo-7 by HaCat human keratinocyte cells in UVA-irradiated genomic DNA.[119] In the presence of resveratrol, UVA enabled to cure the stimulation of DNA strand splintering and cell death.[119] Resveratrol's properties and phenomenon of differentiation in UV spectrum have been revealed in vitro. Development of mammary tumor can be regulated by resveratrol supplementation (1 mg/L; μg/mouse).

It is assumed that the reduced metastasizing capability in HER-2/neu can express transgenic mice production of multiple mammary tumors at premature age. This anticancer property is supposed to be linked with regulation of HER-2/neu.[105] Inferior tumor growth reduces angiogenesis while enhances apoptotic index that have been observed in nude mice having treatment with resveratrol @ 25 mg/kg/day for MDA-MB-231 tumors and human breast cancer xenografts.[58] Growth inhibition in metastatic 4T1 murine mammary tumor cells has been experienced by giving treatment of resveratrol @ 1–5 mg/kg day for 23 days.[20] The dissimilarities in these xenograft experiments were mysterious but attributed to varying doses.

The existing verification based on experiments in breast cancer jeopardy and utilization of estrogenic chemicals for short period showed that the timing and echelon of estrogenic chemicals were imperative hazards. Tumorigenesis induced by methylnitrosourea (MNU) in mammary of female rats of Sprague Dawley during their pre-puberty was augmented by giving elevated levels of resveratrol at 100 mg/kg without estrogen.[115] The available data revealed that the application of resveratrol before puberty can influence endocrine functioning. Resveratrol's method that speeds up the incidence of mammary tumor in rats required exploration with special reference to their segregation into alveolar buds.[111,115]

There is uncertainty about the dose of resveratrol for anticancer effects but a dose of 100 mg/kg/day of resveratrol symbolizes 5000 times than taking red wine by a person at one glass per day.[70] Resveratrol has been effective in restraining replication of *Helicobacter pylori*,[83] and it offers to fight against gastric cancer.[8] A variety of cells can react against resveratrol by capturing cell cycle and mediation of apoptosis during NO development.[66] Resveratrol treatment influences the signaling of the protein kinase C (PKC) alpha and delta.[8,109]

### 7.3.4.3 COLON CANCER

Resveratrol can trigger a variety of apoptosis divisions in cancer cells of colon, which may occupy the growth of the proapoptotic proteins Bax and Bak that reorganize the rafts membranes' FAS receptor.[39] Telomerase activity can be significantly normalized by relatively high concentrations of resveratrol.[52] The resveratrol effectiveness reported in two animals having cancer of colorectal (transformed Min mice and dimethylhydrazine) persuaded azoxymethane (AOM). AOM-influenced cancer contributed to lot of histopathologic resemblances with tumors formed in human bodies, and they can alter in K-ras and β-catenin genes contrasting cancer in humans.[33]

Oral administration of 200 μg/kg/day resveratrol along $H_2O$ can reduce a lot of AOM-induced anomalous crypt foci connected with alterations in p21 and Bax appearances.[128] Min mice getting resveratrol (0.01% in water for 7 weeks) can exhibit a 70% decrease in tumor development in small intestine and prevention of colon cancer growth. Treatment with resveratrol can regulate genes linked with succession of cell cycle or propagation of cells (binding protein of Y-box, D1 and D2 cyclins and DP1 transcription factor) and the enhancement of genes regulations. These genes are supposed to be implicated in the mobilization and beginning of cells of immune system and in the reticence of the cancer progression and enlargement of tumor as well as signifying the surges involved in signals and multiplication of the molecular targets.[116]

Resveratrol @ 500–1500 mg/kg via direct injections on tumor body for 2 days can produce inhibitory effects on xenograft gastric tumor models.[137] The analog of artificial resveratrol 3,4,5,4′-tetramethoxystilbene (DMU-212) also introverted the progress of adenomas in the Apc(Min+) mouse.[112]

Another findings revealed that resveratrol dose of 8 mg/kg body weight, given daily for 30 weeks to Wistar rats can noticeably reduce cancer prevalences and the incidents of histological lacerations and tumor sizes in 1, 2-dimethylhydrazine-based colon tumorogenesis.[118] Administration of resveratrol in the drinking water can powerfully suppress the development of tumors in colons and small intestines.[116]

Esophageal tumorigenesis commonly has insufficient prediction due to delayed diagnosis. Smoking along with polyaromatic hydrocarbons such as benzo[a]pyrene (BaP) exposure is considered as a major risk factor. Resveratrol can restrain an esophageal cancer cell line EC-9706 augmentation.[138]

Donryu rats entrenched with ascites hepatoma cell line of AH109A exhibited antitumor and antimetastatic properties by administration of 10–50 ppm of resveratrol through diet.[92] In an additional experiment, resveratrol administration @ 1 mg per kg for 7 days has proved to be valuable in controlling ascites hepatoma in rats.[26]

#### 7.3.4.4 PANCREATIC CANCER

Many in vitro studies have explored that trans-resveratrol can attenuate the process of apoptosis in cancer cells of pancreas, and depolarize the mitochondria and cytochrome-c discharge pursued by caspase-3 creation.[94] AsPC-1, PANC-1, and cancer of human pancreas are inhibited by the resveratrol (100 μM) and also amplify the division of sub-G0/G1 cells.[40] Additionally, MiaPaCa2 sensitization for tumor necrosis factor related to apoptosis-inducing ligand occurred due to resveratrol treatment. Src tyrosine kinase commotion is hampered by resveratrol and thus can obstruct the modulator of transcription 3 (Stat3) protein commencements constitutive signal transducer and.[53,77]

Dietic administration of 10 ppm of resveratrol concentration may illustrate considerable properties on N-nitrosobis(2-oxopropyl) amine instigation of cancer formation in pancreatic hamster.[78] Additional research work is requisite for novel preclinical assessment of resveratrol effectiveness on pancreatic cancer.

#### 7.3.4.5 NEUROBLASTOMA

Resveratrol can successfully hinder caspase-7 activation of neuronal cells, that is, human neuroblastoma SH-SY5Y, in addition to poly-(ADP-ribose)-polymerase dilapidation that is present in paclitaxel (anticancer drug).[98] The activity of resveratrol is also neuroprotective in nature, and this property of resveratrol takes place during signal pathways modulation that consign apoptosis of these neuronal-like cells. S-phase capture can be persuaded by utilizing resveratrol that can avert mitosis SH-SY5Y.[108] Additionally, Bcl-2 and Jun N-terminal kinase/stress-activated protein kinase (JNK/SAPK) phosphorylation can also be reversed by consuming resveratrol. This phosphorylation exclusively takes place subsequent to paclitaxel exposure.[97] The neuroblastoma cells treated with resveratrol for larger periods augmented the delineation situation of the cells.[90]

Resveratrol can decrease viability of cells during stage 4 v-myc myelocytomatosis viral related oncogene, neuroblastoma derived (MYCN)-amplified neuroblastoma cell lines. In the same way, it can stimulate cell cycle seize and apoptosis along with transient upregulation of p53 appearance and nuclear translocation of p53 and p21 (WAF 1/CIP 1) initiation and expression of Bax.[81] The growth rate can be suppressed in mice by daily administration of resveratrol injections at 40 mg/kg of body weight for a period of 28 days to cure subcutaneous neuroblastomas and may result in 70% endurance for a long period of time.[29]

#### *7.3.4.6 FIBROSARCOMA*

In chick chorioallantoic membrane tumor model, resveratrol can slow down fibroblast growth factors (FGF) 2-induced angiogenesis and considerably it can hinder the human colon platelet/fibrin clot and enlargement of fibrosarcoma tumor.[95] Orally administrated resveratrol dose of 1 mg/kg/day can reduce the intensification of a murine T241 fibrosarcoma implanted in C57Bl6/J mice.[21]

### *7.3.5 MISCELLANEOUS BENEFICIAL EFFECTS*

The GSPE @ 5000 ppm of resveratrol has potential to speed up wound in mice. Topical GSPE assists in oxidant-induced vascular endothelial growth factor appearance in keratinocytes by transformation of mechanism, which is persistent equally to hydrogen peroxide and a signaling tumor necrosis factor (TNF).[74]

Development of cataract due to antioxidative accomplishment can be prevented by elutes of grape seed administration having 38.5% procyanidins to hereditary cataractous rats (ICR/f rats).

Diet supplementation with GSPE may enhance bone mandible health.[62] Pre-eclampsia in rats can be cured by extract from *Vinifera* grape skin (200 mg/kg/day) dose during experimentation. It is assumed that two factors play vital role in controlling the effects of grapefruit seed extract (GSE) in experimental pre-eclampsia, that is, vasodilator effects, which are endothelium-dependent and antioxidant actions.[37] Concord grape juice have been proved against the inception cancer stage, aggregation of platelets and atherosclerosis, failure to perform physically, and mental perspicacity throughout aging and hypertension in humans. Diet supplementation with GSE for 8 weeks helps in tumbling cellular DNA damage of lymphocytes and in scavenging free radicals of plasma.

## 7.4 SUMMARY

At present, dependence on intrinsic foodstuffs is accomplishing popularity to struggle for assorted physiological threats including cardiovascular disorders, cancer insurgence, and immune dysfunction. The utilization of conventional therapies may come across more persistently owing to an assortment of systematic and technological evidences in their support. Grapes hold an inimitable site in history and were renowned for their remedial perspectives. Current progressions in the field of immuno-nutrition, pharmacology, and physiology further explored their significance as a nutraceutical food against various ailments. Much research work has been carried out on the health-promoting potential of grapes, often referred to their bioactive component, that is, resveratrol. Grapes and their components can scavenge free radicals and protect membranes from damage and maintain cell integrity. Similarly, antiplatelet aggregation, antithrombus, decreasing homocysteine level, and cholesterol lowering are also their possible targets resulting in reduced risks of atherosclerosis and related cardiovascular disorders. Certainly, an assortment of evidences has been presented in its favor but some vague reports demand further scientific researches prior to claim its enthusiasm.

## KEYWORDS

- **antioxidant potential**
- **bioactive compounds**
- **functional foods**
- **grapes**
- **nutraceuticals**
- **phytochemicals**

## REFERENCES

1. Adams, D. O. Phenolics and Ripening in Grape Berries. *Am. J. Clin. Nutr.* **2006,** *57* (3), 249–256.
2. Adhami, V. M.; Afaq, F.; Ahmad, N. Involvement of the Retinoblastoma (pRb)-E2F/DP Pathway During Antiproliferative Effects of Resveratrol in Human Epidermoid Carcinoma (A431) Cells. *Biochem. Biophysiol. Res. Commun.* **2001,** *288* (3), 579–585.
3. Afaq, F.; Adhami, V. M.; Ahmad, N.; Mukhtar, H. Botanical Antioxidants for Chemoprevention of Photocarcinogenesis. *Front. BioSci.* **2002,** *7* (1), 784–792.
4. Afaq, F.; Adhami, V. M.; Ahmad, N. Prevention of Short-term Ultraviolet B Radiation-mediated Damages by Resveratrol in SKH-1 Hairless Mice. *Toxicol. Appl. Pharmacol.* **2003,** *186* (1), 28–37.
5. Ahmad, N.; Adhami, V. M.; Afaq, F.; Feyes, D. K.; Mukhtar, H. Resveratrol Causes WAF-1/p21-mediated G1-phase Arrest of Cell Cycle and Induction of Apoptosis in Human Epidermoid Carcinoma A431. *Cells* **2001,** *7* (5), 1466–1473.
6. Aldini, G.; Carini, M.; Piccoli, A.; Rossoni, G.; Facino, R. M. Procyanidins from Grape Seeds Protect Endothelial Cells from Peroxynitrite Damage and Enhance Endothelium-dependent Relaxation in Human Artery: New Evidences for Cardio-protection. *Life Sci.* **2003,** *73* (22), 2883–2898.
7. Asensi, M.; Medina, I.; Ortega, A.; Carretero, J.; Bano, M. C.; Obrador, E.; Estrela, J. M. Inhibition of Cancer Growth by Resveratrol is Related to its Low Bioavailability. *Free Radic. Biol. Med.* **2002,** *33* (3), 387–398.
8. Atten, M. J.; Godoy-Romero, E.; Attar, B. M.; Milson, T.; Zopel, M.; Holian, O. Resveratrol Regulates Cellular PKC Alpha and Delta to Inhibit Growth and Induce Apoptosis in Gastric Cancer Cells. *Invest. New Drugs* **2005,** *23* (2), 111–119.
9. Aziz, M. H.; Afaq, F.; Ahmad, N. Prevention of Ultraviolet-B Radiation Damage by Resveratrol in Mouse Skin is Mediated Via Modulation in Surviving. *Photochemistry Photobiol.* **2005,** *81* (1), 25–31.
10. Bachelor, M. A.; Bowden, G. T. UVA-mediated Activation of Signaling Pathways Involved in Skin Tumor Promotion and Progression. *Semin. Cancer Biol.* **2004,** *14* (2), 131–138.
11. Bagchi, D.; Bagchi, M.; Stohs, S. J.; Das, D. K.; Ray, S. D.; Kuszynski, C. A.; Joshi, S. S.; Pruess, H. G. Free Radicals and Grape Seed Proanthocyanidin Extract: Importance in Human Health and Disease Prevention. *Toxicol.* **2000,** *148* (2–3), 187–197.
12. Bagchi, D.; Sen, C. K, Bagchi M, Atalay M. Anti-angiogenic, Antioxidant, and Anticarcinogenic Properties of a Novel Anthocyanin-rich Berry Extract Formula. *Biochemistry* **2004,** *69* (1), 75–80.
13. Baliga, M. S.; Katiyar, S. K. Chemoprevention of Photocarcinogenesis by Selected Dietary Botanicals. *Photochem. Photobiol. Sci.* **2006,** *5* (2), 243–253.
14. Baliga, M. S.; Meleth, S.; Katiyar, S. K. Growth Inhibitory and Antimetastatic Effect of Green Tea Polyphenols on Metastasis-specific Mouse Mammary Carcinoma 4T1 Cells In Vitro and In Vivo Systems. *Clin. Cancer Res.* **2005,** *11* (5), 1918–1927.
15. Baur, J. A.; Sinclair, D. A. Therapeutic Potential of Resveratrol: The In Vivo Evidence. *Nat. Rev. Drug Discov.* **2006,** *5* (6), 493–506.
16. Bavaresco, L. Role of Viticultural Factors on Stilbene Concentrations of Grapes and Wine. *Drugs Exp. Clin. Res.* **2003,** *299* (5–6), 181–187.

17. Berti, F.; Manfredi, B.; Mantegazza, P.; Rossoni, G. Procyanidins from *Vitis vinifera* Seeds Display Cardioprotection in an Experimental Model of Ischemia-reperfusion Damage. *Drugs Exp. Clin. Res.* **2003,** *29* (5–6), 207–216.
18. Bors, W.; Heller, W.; Michel, C.; Saran, M. Flavonoids as Antioxidants: Determination of Radical Scavenging Efficiencies. In *Methods in Enzymology*; Academic Press: New York, 1990; Vol. 186, pp 343–355.
19. Bouhamidi, R.; Prevost, V.; Nouvelot, A. High Protection by Grape Seed Proanthocyanidins (GSPC) of Polyunsaturated Fatty Acids Against UV-C Induced Peroxidation. *CR Acad. Sci.* III **1998,** *321* (1), 31–38.
20. Bove, K.; Lincoln, D, W.; Tsan, M. F. Effect of Resveratrol on Growth of 4T1 Breast Cancer Cells In Vitro and In Vivo. *Biochem. Biophysiol. Res. Commun.* **2002,** *291* (4), 1001–1005.
21. Brakenhielm, E.; Cao, R.; Cao, Y. Suppression of Angiogenesis, Tumor Growth, and Wound Healing by Resveratrol, a Natural Compound in Red Wine and Grapes. *FASEB J.* **2001,** *15* (10), 1798–1800.
22. Burns, J.; Gardner, P. T.; O'Neil, J.; Crawford, S.; Morecroft, I.; McPhail, D. B.; Lister, C.; Matthews, D.; MacLean, M. R.; Lean, M. E.; Duthie, G. G.; Crozier, A. Relationship Among Antioxidant Activity, Vasodilation Capacity, and Phenolic Content of Red Wines. *J. Agric. Food Chem.* **2000,** *48* (2), 220–230.
23. Cadot, Y.; Miñana-Castelló, M. T.; Chevalier, M. Anatomical, Histological, and Histochemical Changes in Grape Seeds from *Vitis vinifera* L. cv Cabernet Franc During Fruit Development. *J. Agric. Food Chem.* **2006,** *54* (24), 9206–9215.
24. Caillet, S.; Salmieri, S.; Lacroix, M. Evaluation of Free Radical-scavenging Properties of Commercial Grape Phenol Extracts by a Fast Colorimetric Method. *Food Chem.* **2006,** *95* (1), 1–8.
25. Cantos, E.; Garcia-Viguera, C.; de Pascual-Teresa, S.; Tomas-Barberan, F. A. Effect of Postharvest Ultraviolet Irradiation on Resveratrol and Other Phenolics of cv. Napoleon Table Grapes. *J. Agric. Food Chem.* **2000,** *48* (10), 4606–4612.
26. Carbo, N.; Costelli, P.; Baccino, F. M.; Lopez-Soriano, F. J.; Argiles, J. M. Resveratrol, a Natural Product Present in Wine, Decreases Tumour Growth in a Rat Tumour Model. *Biochem. Biophysiol. Res. Commun.* **1999,** *254* (3), 739–743.
27. Castellarin, S. D.; Di Gaspero, G.; Marconi, R.; Nonis, A.; Peterlunger, E.; Paillard, S.; Adam-Blondon, A. F.; Testolin, R. Color Variation in Red Grapevines (*Vitis vinifera* L.): Genomic Organisation, Expression of Flavonoid 3′-Hydroxylase, Flavonoid 3′,5′-hydroxylase Genes and Related Metabolite Profiling of Red Cyanidin-/Blue Delphinidin-based Anthocyanins in Berry Skin. *BMC Genom.* **2006,** *7* (12), 17–29
28. Castilla, P.; Echarri, R.; Davalos, R.; Cerrato, F.; Ortega, H.; Teruel, J. L.; Lucas, M. F.; Gomez-Coronado, D.; Ortuno, J.; Lasuncion, M. A. Concentrated Red Grape Juice Exerts Antioxidant, Hypolipidemic, and Antiinflammatory Effects in both Hemodialysis Patients and Healthy Subjects. *Am. J. Clin. Nutr.* **2006,** *84* (1), 252–262.
29. Chen, Y.; Tseng, S. H.; Lai, H. S.; Chen, W. J. Resveratrol-induced Cellular Apoptosis and Cell Cycle Arrest in Neuroblastoma Cells and Antitumor Effects on Neuroblastoma in Mice. *Surgery* **2004,** *136* (2), 57–66.
30. Cheng, M.; Gao, H. Q.; Xu, L.; Li, B. Y.; Zhang, H.; Li, X. H. Cardioprotective Effects of Grape Seed Proanthocyanidins Extracts in Streptozocin Induced Diabetic Rats. *J. Cardiovasc. Pharmacol.* **2007,** *50* (12), 503–509.

31. Coimbra, S. R.; Lage, S. H.; Brandizzi, L.; Yoshida, V.; da Luz, P. L. The Action of Red Wine and Purple Grape Juice on Vascular Reactivity is Independent of Plasma Lipids in Hypercholesterolemic Patients. *Braz. J. Med. Biol. Res.* **2005,** *38* (1), 1339–1347.
32. Conde, C.; Silva, P.; Fontes, N, Dias, A. C. P.; Tavares, R. M.; Sousa, M. J.; Agasse, A.; Delrot, S.; Geros, H. Biochemical Changes Throughout Grape Berry Development and Fruit and Wine Quality. *Food* **2007,** *1* (1), 1–22.
33. Corpet, D. E.; Pierre, F. Point: From Animal Models to Prevention of Colon Cancer. Systematic Review of Chemoprevention in Min Mice and Choice of the Model System. *Cancer Epidemiol. Biomarkers Prev.* **2003,** *12* (5), 391–400.
34. Crowell, J. A.; Korytko, P. J.; Morrissey, R. L.; Booth, T. D.; Levine, B. S. Resveratrol-associated Renal Toxicity. *Toxicology Sci.* **2004,** *82* (2), 614–619.
35. Cui, J.; Cordis, G. A.; Tosaki, A.; Maulik, N.; Das, D. K. Reduction of Myocardial Ischemia Reperfusion Injury with Regular Consumption of Grapes. *Ann. NY Acad. Sci.* **2002,** *957* (1), 302–307.
36. Dani, C.; Bonatto, D.; Salvador, M.; Pereira, M. D.; Henriques, J. P.; Eleutherio, E. Antioxidant Protection of Resveratrol and Catechin in *Saccharomyces cerevisiae*. *J. Agric. Food Chem.* **2008,** *56* (11), 4268–4272.
37. De Moura, R. S.; Resende, A. C.; Moura, A.; Maradei, M. F. Protective Action of a Hydroalcoholic Extract of a Vinifera Grape Skin on Experimental Preeclampsia in Rats. *Hypertens. Pregn.* **2007,** *26* (1), 89–100.
38. Delmas, D.; Lancon, A.; Colin, D.; Jannin, B.; Latruffe, N. Resveratrol as a Chemopreventive Agent: A Promising Molecule for Fighting Cancer. *Curr. Drug Targets* **2006,** *7* (4), 423–442.
39. Delmas, D.; Rebe, C.; Lacour, S. Resveratrol-induced Apoptosis is Associated with Fas Redistribution in the Rafts and the Formation of a Death-inducing Signaling Complex in Colon Cancer Cells. *J. Biol. Chem.* **2003,** *278* (1), 41482–41490.
40. Ding, X. Z.; Adrian, T. E. Resveratrol Inhibits Proliferation and Induces Apoptosis in Human Pancreatic Cancer Cells. *Pancreas* **2002,** *25* (4), 71–76.
41. Doshi, P.; Adsule, P.; Banerjee, K. Phenolic Composition and Antioxidant Activity in Grapevine Parts and Berries (*Vitis vinifera* L.) cv Kishmish Chornyi (Sharad Seedless) During Maturation. *Int. J. Food Sci. Technol.* **2006,** *41* (1), 1–9.
42. Downey, M. O.; Harvey, J. S.; Robinson, S. P. Synthesis of Flavonols and Expression of Flavonol Synthase Genes in the Developing Grape Berries of Shiraz and Chardonnay (*Vitis vinifera* L.). *Austr. J. Grape Wine Res.* **2003,** *9* (2), 110–121.
43. Edirisinghe, I.; Burton-Freeman, B.; Kappagoda, T. The Mechanism of the Endothelium Dependent Relaxation Evoked by a Grape Seed Extract. *Clin. Sci.* **2008,** *14* (4), 331–337.
44. Elmali, N.; Esenkaya, I.; Karadag, N.; Tas, F.; Elmali, N. Effects of Resveratrol on Skeletal Muscle in Ischemia-reperfusion Injury. *Ulus Travma Acil Cerrahi Derg* **2007,** *13* (11), 247–280.
45. Escribano-Baiton, T.; Gutierrez-Fernandez, Y.; Rivas-Gonzalo, J. C.; Santos-Buelga, C. Characterization of Procyanidins of *Vitis vinifera* Variety Tintal Del Pais Grape Seeds. *J. Agric. Food Chem.* **1992,** *40* (1), 1794–1799.
46. Facino, M. R.; Carini, M.; Aldini, G.; Calloni, M. T.; Bombardelli, E.; Morazzoni, P. Sparing Effect of Procyanidins from *Vitis vinifera* on Vitamin E: In Vitro Studies. *Planta Med.* **1998,** *64* (4), 343–347.

47. Fan, P.; Lou, H. X. Effects of Polyphenols from Grape Seeds on Oxidative Damage to Cellular DNA. *Mol. Cell Biochem.* **2004,** *267* (1–2), 67–74.
48. Faria, A.; Calhau, C.; De Freitas, V.; Mateus, N. Procyanidins as Antioxidants and Tumor Cell Growth Modulators. *J. Agric. Food Chem.* **2006,** *54* (19), 2392–2397.
49. Frankel, E. N.; Bosanek, C. A.; Meyer, A. S.; Silliman, K.; Kirk, L. L. Commercial Grape Juices Inhibit the In Vitro Oxidation of Human Low Density Lipoproteins. *J. Agric. Food Chem.* **1998,** *46* (11), 834–838.
50. Freedman, J. E.; Parker, C.; Li, L.; Perlman, J. A.; Frei, B.; Ivanov, V.; Deak, L. R.; Iafrati, M. D.; Folts, J. D. Select Flavonoids and Whole Juice from Purple Grapes Inhibit Platelet Function and Enhance Nitric Oxide Release. *Circulation* **2001,** *103* (23), 2792–279.
51. Fu, Z. D.; Cao, Y.; Wang, K. F.; Xu, S. F.; Han, R. Chemopreventive Effect of Resveratrol to Cancer. *Chin. J. Cancer* **2004,** *23* (8), 869–873.
52. Fuggetta, M. P.; Lanzilli, G.; Tricarico, M.; Cottarelli, A.; Falchetti, R.; Ravagnan, G.; Bonmassar, E. Effect of Resveratrol on Proliferation and Telomerase Activity of Human Colon Cancer Cells In Vitro. *J. Exp. Clin. Cancer Res.* **2006,** *25* (1), 189–193.
53. Fulda, S.; Debatin, K. M. Sensitization for Tumor Necrosis Factor-related Apoptosis-inducing Ligand-induced Apoptosis by the Chemopreventive Agent Resveratrol. *Cancer Res.* **2004,** *64* (1), 337–346.
54. Fuleki, T, Ricardo-da-Silva, J. M. Catechin and Procyanidin Composition of Seeds from Grape Cultivars Grown in Ontario. *J. Agric. Food Chem.* **1997,** *45* (1), 1156–116.
55. Gabetta, B.; Fuzzati, N.; Griffini, A.; Lolla, E.; Pace, R.; Ruffilli, T.; Peterlongo, F. Charcterisation of Proanthocyanidins from Grape Seeds. *Fitoterapia* **2000,** *71* (2), 162–175.
56. Garcia-Alonso, J.; Ros, G.; Vidal-Guevara, M. L.; Periago, M. J. Acute Intake of Phenolic-rich Juice Improves Antioxidant Status in Healthy Subjects. *Nutr. Res.* **2006,** *26* (11), 330–339.
57. Garcia-Parilla, M. C.; Gonzalez, G. A.; Heredia, F. J.; Troncoso, A. M. Differentiation of Wine Vinegars based on Phenolic Composition. *J. Agric. Food Chem.* **1997,** *45* (1), 3487–3492.
58. Garvin, S.; Ollinger, K.; Dabrosin, C. Resveratrol Induces Apoptosis and Inhibits Angiogenesis in Human Breast Cancer Xenografts In Vivo. *Cancer Lett.* **2006,** *231* (1), 113–122.
59. Gould, K. S.; Lister, C. Flavonoid Functions in Plants. In *Flavonoids: Chemistry, Biochemistry and Applications*; Andersen, Ø. M., Markham, K. R., Eds.; Boca Raton, FL: CRC Press, Taylor and Friends Group; 2006, pp 397–442.
60. Greenblatt, D. J.; Von-Moltke, L. L.; Perloff, E. S.; Luo, Y.; Harmatz, J. S.; Zinny, M. A. Interaction of Flurbiprofen with Cranberry Juice, Grape Juice, Tea, and Fluconazole: In Vitro and Clinical Studies. *Clin. Pharmacol. Ther.* **2006,** *79* (19), 125–133.
61. Gruenwald, J.; Brendler, B. A.; Jaenicke, C. *PDR for Herbal Medicines*, 3rd ed.; Thomson PDR: Montvale, NJ, 2004.
62. Gunjima, M.; Tofani, I.; Kojima-Maki, K.; Kimura, M. Mechanical Evaluation of Effect of Grape Seed Proanthocyanidins Extract on Debilitated Mandibles in Rats. *Dent. Matern. J.* **2004,** *23* (2), 67–74.
63. Han, Y. Synergic Effect of Grape Seed Extract with Amphotericin B Against Disseminated Candidiasis due to Candida albicans. *Phytomedicine* **2007,** *14* (9), 733–738.
64. Harborne, J. B.; Williams, C. A. Advances in Flavonoid Research Since 1992. *Phytochemistry* **2000,** *55* (6), 481–504.

65. Hernandez-Jimenez, A.; Gomez-Plaza, E.; Martinez-Cutillas, A.; Kennedy, J. A. Grape Skin and Seed Proanthocyanidins from Monastrell x Syrah Grapes. *J. Agric. Food Chem.* **2009,** *57* (1), 10798–10803.
66. Holian, O.; Wahid, S.; Atten, M.; Attar, B. M. Inhibition of Gastric Cancer Cell Proliferation by Resveratrol: Role of Nitric Oxide. *Am. J. Physiol.* **2002,** *282* (5), G809–G816.
67. Hollman, P. C. H.; Arts, I. C. W. Flavonols, Flavones and Flavanols—Nature, Occurrence and Dietary Burden. *J. Sci. Food Agric.* **2000,** *80* (4), 1081–1093.
68. Jang, M.; Cai, L.; Udeani, G. O. Cancer Chemopreventive Activity of Resveratrol, A Natural Product Derived from Grapes. *Science*. **1997,** *275* (5297), 218–220.
69. Jayaprakasha, G. K.; Selvi, T.; Sakariah, K. K. Antibacterial and Antioxidant Activities of Grape (*Vitis vinifera*) Seed Extracts. *Food Res. Int.* **2003,** *36* (2), 117–122.
70. Juan, M. E.; Vinardell, M. P.; Planas J. M. The Daily Oral Administration of High Doses of Trans-resveratrol to Rats for 28 Days is not Harmful. *J. Nutr.* **2002,** *132* (1), 257–260.
71. Kapadia, G. J.; Azuine, M. A.; Tokuda, H.; Takasaki, M.; Mukainaka, T.; Konoshima, T.; Nishino, H. Chemopreventive Effect of Resveratrol, Sesamol, Sesame Oil and Sunflower Oil in the Epstein–Barr Virus Early Antigen Activation Assay and the Mouse Skin Two-stage Carcinogenesis. *Pharmacol. Res.* **2002,** *45* (21), 499–505.
72. Karadeniz, F.; Durst, R. W.; Wrolstad, R. E. Polyphenolic Composition of Raisins. *J Agric. Food Chem.* **2000,** *48* (11), 5343–5350.
73. Karakaya, S.; Nehir, E. L. S. Quercetin, Luteolin, Apigenin and Kaempferol Contents of Some Foods. *Food Chem.* **1999,** *66* (1), 289–292.
74. Khanna S, Venojarvi M, Roy S, Sharma N, Trikha P, Bagchi D, Bagchi M, Sen CK. Dermal Wound Healing Properties of Redox-active Grape Seed Proanthocyanidins. *Free Rad. Biol. Med.* **2002,** *33* (8), 1089–1096.
75. Kim AL, Zhu Y, Zhu H, Han L, Kopelovich L, Bickers DR, Athar M. Resveratrol Inhibits Proliferation of Human Epidermoid Carcinoma A431 Cells by Modulating MEK1 and AP-1 Signalling Pathways. *Exp. Dermatol.* **2006,** *15* (1), 538–546.
76. Kitamura S. Transport of Flavonoids: From Cytosolic Synthesis to Vacuolar Accumulation. In *Science of Flavonoids*; Grotewold, E., Ed.; Berlin D: Springer, 2006; 123–146.
77. Kotha, A.; Sekharam, M.; Cilenti, L. Resveratrol Inhibits Src and Stat3 Signaling and Induces the Apoptosis of Malignant Cells Containing Activated Stat3 Protein. *Mol. Cancer Ther*. **2006,** *5* (1), 621–629.
78. Kuroiwa, Y.; Nishikawa, A.; Kitamura, Y.; Kanki, K.; Ishii, Y.; Umemura, T.; Hirose, M. Protective Effects of Benzyl Isothiocyanate and Sulforaphane but not Resveratrol Against Initiation of Pancreatic Carcinogenesis in Hamsters. *Cancer Lett.* **2006,** *241* (2), 275–280.
79. Kuwajerwala, N.; Cifuentes, E.; Gautam, S.; Menon, M.; Barrack, E. R.; Reddy, G. P. V. Resveratrol Induces Prostate Cancer Cell Entry into S Phase and Inhibits DNA Synthesis. *Cancer Res.* **2002,** *62* (11), 2488–2492.
80. Li, B. Y.; Cheng, M.; Gao, H. Q.; Ma, Y. B.; Xu, L.; Li, X. H.; Li, X. L.; You, B. A. Back-regulation of Six Oxidative Stress Proteins with Grape Seed Proanthocyanidin Extracts in Rat Diabetic Nephropathy. *J. Cell Biochem.* **2008,** *104* (1), 668–679.
81. Liontas, A.; Yeger, H. Curcumin and Resveratrol Induce Apoptosis and Nuclear Translocation and Activation of p53 in Human Neuroblastoma. *Anticancer Res.* **2004,** *24* (3), 987–998.
82. Luther, M.; Parry, J.; Moore, J.; Meng, J. H.; Zhang, Y. F.; Cheng, Z. H.; Yu, L. Inhibitory Effect of Chardonnay and Black Raspberry Seed Extracts on Lipid Oxidation in Fish

Oil and Their Radical Scavenging and Antimicrobial Properties. *Food Chem.* **2007,** *104* (8), 1065–1073.

83. Mahady, G. B.; Pendland, S. L. Resveratrol Inhibits the Growth of *Helicobacter pylori* In Vitro. *Am. J. Gastroenterol.* **2000,** *95* (7), 1849.
84. Makris, D. P.; Kallithrakab, S.; Kefalas, P. Flavonols in Grapes, Grape Products and Wines: Burden, Profile and Influential Parameters. *J. Sci. Food Compos. Anal.* **2006,** *19* (9), 396–404.
85. Makris, D. P.; Boskou, G.; Andrikopoulos, N. K.; Kefalas, P. Characterisation of Certain Major Polyphenolic Antioxidants in Grape (*Vitis vinifera*) Stems by Liquid Chromatography-mass Spectrometry. *Eur. Food Res. Technol.* **2008,** *226* (1), 1075–1079.
86. Manach, C.; Scalbert, A.; Morand, C.; Rémésy, C.; Jiménez, L. Polyphenols: Food Sources and Bioavailability. *Am. J. Clin. Nutr.* **2004,** *79* (11), 727–747.
87. Manach, C.; Williamson, G.; Morand, C.; Scalbert, A.; Remesy, C. Bioavailability and Bioefficacy of Polyphenols in Humans. I. Review of 97 Bioavailability Studies. *Am. J. Clin. Nutr.* **2005,** *81* (2), 2305–2425.
88. Mantena, S. K.; Baliga, M. S.; Katiyar, S. K. Grape Seed Proanthocyanidins Induce Apoptosis and Inhibit Metastasis of Highly Metastatic Breast Carcinoma Cells. *Carcinogenesis* **2006,** *27* (8), 1682–1691.
89. Masa, A.; Vilanova, M. Flavonoid and Aromatic Characterisation of cv. Albarín blanco (*Vitis vinifera* L.). *Food Chem.* **2008,** *107* (11), 273–281.
90. Melzig, M. F.; Escher, F. Induction of Neutral Endopeptidase and Angiotensin-converting Enzyme Activity of SK-N-SH Cells In Vitro by Quercetin and Resveratrol. *Diet. Pharmaceut.* **2002,** *57* (21), 556–558.
91. Mendesa, A.; Desgranges, C.; Catherine, C.; Vercauteren, J.; Freslon, J. L. Vasorelaxant Effects of Grape Polyphenols in Rat Isolated Aorta. Possible Involvement of a Purinergic Pathway. *Fund. Clin. Pharmacol.* **2003,** *17* (1), 673–68.
92. Miura, D.; Miura, Y.; Yagasaki K. Hypolipidemic Action of Dietary Resveratrol, A Phytoalexin in Grapes and Red Wine, in Hepatoma-bearing Rats. *Life Sci.* **2003,** *73* (11), 1393–1400.
93. Morin, B.; Narbonn, J. F.; Ribera, D.; Badouard, C.; Ravanat, J. L. Effect of Dietary Fat-soluble Vitamins A and E and Proanthocyanidin-rich Extract from Grape Seeds on Oxidative DNA Damage in Rats. *Food Chem. Toxicol.* **2008,** *46* (21), 787–796.
94. Mouria, M.; Gukovskaya, A. S.; Jung, Y.; Buechler, P.; Hines, O. J.; Reber, H. A.; Pandol, S. J. Food-derived Polyphenols Inhibit Pancreatic Cancer Growth Through Mitochondrial Cytochrome C Release and Apoptosis. *Int. J. Cancer* **2002,** *98* (9), 761–769.
95. Mousa, S. S.; Mousa, S. S.; Mousa, S. A. Effect of Resveratrol on Angiogenesis and Platelet/Fibrin-accelerated Tumor Growth in the Chick Chorioallantoic Membrane Model. *Nutr. Cancer* **2005,** *52* (1), 59–65.
96. Natella, F.; Belelli, F.; Gentili, V.; Ursini, F.; Scaccini, C. Grape Seed Proanthocyanidins Prevent Plasma Postprandial Oxidative Stress in Humans. *J. Agric. Food Chem.* **2002,** *18* (3), 7720–7725.
97. Nicolini, G.; Rigolio, R.; Scuteri, A.; Miloso, M.; Saccomanno, D.; Cavaletti, G.; Tredici, G. Effect of Trans-resveratrol on Signal Transduction Pathways Involved in Paclitaxel-induced Apoptosis in Human Neuroblastoma SH-SY5Y Cells. *Neurochem. Int.* **2003,** *42* (5), 419–429.
98. Nicolini, G.; Rigolio, R.; Miloso, M.; Bertelli, A. A.; Tredici, G. Anti-apoptotic Effect of Trans-resveratrol on Paclitaxel-induced Apoptosis in the Human Neuroblastoma SH-SY5Y Cell Line. *Neurosci. Lett.* **2001,** *302* (1), 41–44.

99. Ohno, M.; Ka, T.; Inokuchi, T.; Moriwaki, Y.; Yamamoto, A.; Takahashi, S.; Tsutsumi, Z.; Tsuzita, J.; Yamamoto, T.; Nishiguchi, S. Effects of Exercise and Grape Juice Ingestion in Combination on Plasma Concentrations of Purine Bases and Uridine. *Clin. Chem. Acta.* **2008,** *388* (12), 167–172.
100. Oszmianski, J.; Lee, C. Y. Isolation and HPLC Determination of Phenolic Compounds in Red Grapes. *Am. J. Enol. Viticulture.* **1990,** *41* (7), 204–206.
101. Pastrana-Bonilla, E.; Akoh, C. C.; Sellappan, S.; Krewer, G. Phenolic Content and Antioxidant Capacity of Muscadine Grapes. *J. Agric. Food Chem.* **2003,** *51* (21), 5497–5503.
102. Pataki, T.; Bak, I.; Kovacs, P.; Bagchi, D.; Das, D. K.; Tosaki, A. Grape Seed Proanthocyanidins Improved Cardiac Recovery During Reperfusion After Ischemia in Isolated Rat Hearts. *Am. J. Clin. Nutr.* **2002,** *75* (5), 894–899.
103. Pignatelli, P.; Pulcinelli, F. M.; Celestini, A.; Lenti, L.; Ghiselli, A.; Gazzaniga, P. P.; Violi, F. The Flavonoids Quercetin and Catechin Synergistically Inhibit Platelet Function by Antagonizing the Intracellular Production of Hydrogen Peroxide. *Am. J. Clin. Nutr.* **2000,** *72* (5), 1150–1155.
104. Pinelo, M.; Arnous, A.; Meyer, A. S. Upgrading of Grape Skins: Significance of Plant Cell-wall Structural Components and Extraction Techniques for Phenol Release. *Trends Food Sci. Technol.* **2006,** *17* (9), 579–590.
105. Provinciali, M.; Re, F.; Donnini, A.; Orlando, F.; Bartozzi, B.; Di Stasio, D.; Smorlesi, A. Effect of Resveratrol on the Development of Spontaneous Mammary Tumors in HER-2/neu Transgenic Mice. *Int. J. Cancer* **2005,** *115* (2), 36–45.
106. Reagan-Shaw, S.; Afaq, F.; Aziz, M. H.; Ahmad, N. Modulations of Critical Cell Cycle Regulatory Events During Chemoprevention of Ultraviolet B-mediated Responses by Resveratrol in SKH-1 Hairless Mouse Skin. *Oncogene* **2004,** *23* (30), 5151–5160.
107. Regev-Shoshani, G.; Shoseyov, O.; Bilkis, I.; Kerem, Z. Glycosylation of Resveratrol Protects it from Enzymic Oxidation. *Biochem. J.* **2003,** *374* (21), 157–163.
108. Rigolio, R.; Miloso, M.; Nicolini, G.; Villa, D.; Scuteri, A.; Simone, M.; Tredici, G. Resveratrol Interference with the Cell Cycle Protects Human Neuroblastoma SH-SY5Y Cell from Paclitaxel-induced Apoptosis. *Neurochem. Int.* **2005,** *46* (1), 205–211.
109. Riles, W. L.; Erickson, J.; Nayyar, S.; Atten, M. J.; Attar, B. M.; Holian, O. Resveratrol Engages Selective Apoptotic Signals in Gastric Adenocarcinoma Cells. *World J. Gastroenterol.* **2006,** *12* (35), 5628–5634.
110. Rivero-Perez, M. D.; Muniz, P.; Gonzalez-Sanjose, M. L. Contribution of Anthocyanin Fraction to the Antioxidant Properties of Wine. *Food Chem Toxicol.* **2008,** *46* (8), 2815–2822.
111. Russo, J.; Russo, I. H. Biological and Molecular Bases of Mammary Carcinogenesis. *Lab Invest. J. Technol. Methods Pathol.* **1987,** *57* (1), 112–137.
112. Sale, S.; Tunstall, R. G.; Ruparelia, K. C.; Potter, G. A.; Steward, W. P.; Gescher, A. J. Comparison of the Effects of the Chemopreventive Agent Resveratrol and its Synthetic Analog Trans 3,4,5,4′-tetramethoxystilbene (DMU-212) on Adenoma Development in the Apc(Min+) Mouse and Cyclooxygenase-2 in Human-derived Colon Cancer Cells. *Int. J. Cancer* **2005,** *115* (12), 194–201.
113. Sanchez-Alonso, I.; Borderias, J.; Larsson, K.; Undeland, I. Inhibition of Hemoglobin-mediated Oxidation of Regular and Lipid-fortified Washed Cod Mince by a White Grape Dietary Fiber. *J. Agric. Food Chem.* **2007,** *55* (13), 5299–5305.
114. Sano, A.; Uchida, R.; Saito, M.; Shioya, N.; Komori, Y.; Tho, Y.; Hashizume, N. Beneficial Effects of Grape Seed Extract on Malondialdehyde-modified LDL. *J. Nutr. Sci. Vitaminol.* **2007,** *53* (2), 174–182.

115. Sato, M.; Pei, R. J.; Yuri, T.; Danbara, N.; Nakane, Y.; Tsubura, A. Prepubertal Resveratrol Exposure Accelerates N-methyl-N-nitrosourea-induced Mammary Carcinoma in Female Sprague–Dawley Rats. *Cancer Lett.* **2003,** *202* (2), 137–145.
116. Schneider, Y.; Duranton, B.; Gosse, F.; Schleiffer, R.; Seiler, N.; Raul, F. Resveratrol Inhibits Intestinal Tumorigenesis and Modulates Host-defense-related Gene Expression in an Animal Model of Human Familial Adenomatous Polyposis. *Nutr. Cancer* **2001,** *39* (1), 102–107.
117. Schwinn, K. E.; Davies, K. M. Flavonoids. In *Plant Pigments and Their Manipulation. Annual Plant Reviews*; Davies, K. M., Ed.; Boca Raton, FL: CRC Press, Blackwell Publishing, 2004; Vol. 14, pp 92–149.
118. Sengottuvelan, M.; Viswanathan, P.; Nalini, N. Chemopreventive Effect of Trans-resveratrol—A Phytoalexin Against Colonic Aberrant Crypt Foci and Cell Proliferation in 1,2-dimethylhydrazine Induced Colon Carcinogenesis. *Carcinogenesis* **2006,** *27* (1), 1038–1046.
119. Seve, M.; Chimienti, F.; Devergnas, S. Resveratrol Enhances UVA-induced DNA Damage in HaCaT Human Keratinocytes. *Med. Chem.* **2005,** *1* (6), 629–633.
120. Shenoy, S. F.; Keen, C. L.; Kalgaonkar, S.; Polagruto, J. A. Effects of Grape Seed Extract Consumption on Platelet Function in Postmenopausal Women. *Thromb. Res.* **2007,** *121* (3), 431–432.
121. Shih, P. H.; Yeh, C. T.; Yen, G. C. Anthocyanins Induce the Activation of Phase II Enzymes Through Antioxidant Response Element Pathway Against Oxidative Stress-induced Apoptosis. *J. Agric. Food Chem.* **2007,** *55* (23), 9427–9435.
122. Soleas, G. J.; Yan, J.; Goldberg, D. M. Ultrasensitive Assay for Three Polyphenols (Catechin, Quercetin and Resveratrol) and their Conjugates in Biological Fluids Utilizing Gas Chromatography with Mass Selective Detection. *J. Chromatogr.* **2001,** *757* (1), 161–172.
123. Souquet, J. M.; Cheynier, V.; Brossaud, F.; Moutounet, M. Polymeric Proanthocyanidins from Grape Skins. *Phytochemistry.* **1996,** *43* (2), 509–512.
124. Souquet, J. M.; Labarbe, B.; Le Guerneve, C.; Cheynier, V.; Moutounet, M. Phenolic Composition of Grape Stems. *J. Agric. Food Chem.* **2000,** *48* (4), 1076–1080.
125. Spanos, G. A.; Wrolstad, R. E. Phenolics of Apple, Pear, and White Grape Juices and their Changes with Processing and Storage—A Review. *J. Agric. Food Chem.* **1992,** *40* (9), 1478–1487.
126. Talcott, S. T.; Lee, J. H. Ellagic Acid and Flavonoid Antioxidant Content of Muscadine Wine and Juice. *J. Agric. Food Chem.* **2002,** *50* (11), 3186–3192.
127. Tebib, K.; Besana, P.; Besanacon, P.; Rouanet, J. M. Dietary Grape Seed Tannins Affect Lipoproteins, Lipoprotein Lipases and Tissue Lipids in Rats Fed Hypercholesterolemic Diets. *J. Nutr.* **1994,** *124* (12), 2452–2457.
128. Tessitore, L.; Davit, A.; Sarotto, I.; Caderni, G. Resveratrol Depresses the Growth of Colorectal Aberrant Crypt Foci by Affecting Bax and 21(CIP) Expression. *Carcinogenesis* **2000,** *21* (8), 1619–1622.
129. Valls-Belles, V.; Torres, M. C.; Muniz, P.; Beltran, S.; Martinez-Alvarez, J. R.; Codoner-Franch, P. Defatted Milled Grape Seed Protects Adriamycin-treated Hepatocytes Against Oxidative Damage. *Eur. J. Clin. Nutr.* **2006,** *45* (5), 251–258.
130. Vitrac, X.; Desmouliere, A.; Brouillaud, B. Distribution of [14C]-Trans-resveratrol, a Cancer Chemopreventive Polyphenol, in Mouse Tissues After Oral Administration. *Life Sci.* **2003,** *72* (20), 2219–2233.

131. Walle, T.; Hsieh, F.; DeLegge, M. H.; Oatis, Jr J. E.; Walle, U. K. High Absorption but Very Low Bioavailability of Oral Resveratrol in Humans. *Drug Metab. Dispos.* **2004,** *32* (12), 1377–1382.
132. Wang, Z.; Zou, Z.; Haung, Y.; Cao, K.; Xu, Y.; Wu, J. M. Effect of Resveratrol on Platelet Aggregation In Vivo and In Vitro. *Chin. Med. J.* **2002,** *115* (3), 378–380.
133. Wenzel, U.; Kuntz, S.; Brendel, M. D.; Daniel, H. Dietary Flavone is a Potent Apoptosis Inducer in Human Colon Carcinoma Cells. *Cancer Res.* **2000,** *60* (14), 3823–3831.
134. Winkel-Shirley, B. Biosynthesis of Flavonoids and Effects of Stress. *Curr. Opin. Plant Biol.* **2002,** *5* (3), 218–223
135. Yi, W. G.; Fischer, J.; Akoh, C. C. Study of Anticancer Activities of Muscadine Grape Phenolics In Vitro. *J. Agric. Food Chem.* **2005,** *53* (22), 8804–8812.
136. Yu, L.; Sun, Z. J.; Wu, S. L.; Pan, C. E. Effect of Resveratrol on Cell Cycle Proteins in Murine Transplantable Liver Cancer. *World J. Gastroenterol.* **2003,** *9* (10), 2341–2343.
137. Zhou, H. B.; Chen, J. J.; Wang, W. X.; Cai, J. T.; Du, Q. Anticancer Activity of Resveratrol on Implanted Human Primary Gastric Carcinoma Cells in Nude Mice. *World J. Gastroenterol.* **2005,** *11* (2), 280–284.
138. Zhou, H. B.; Yan, Y.; Sun, Y. N.; Zhu, J. R. Resveratrol Induces Apoptosis in Human Esophageal Carcinoma Cells. *World J. Gastroenterol.* **2003,** *9* (3), 408–441.

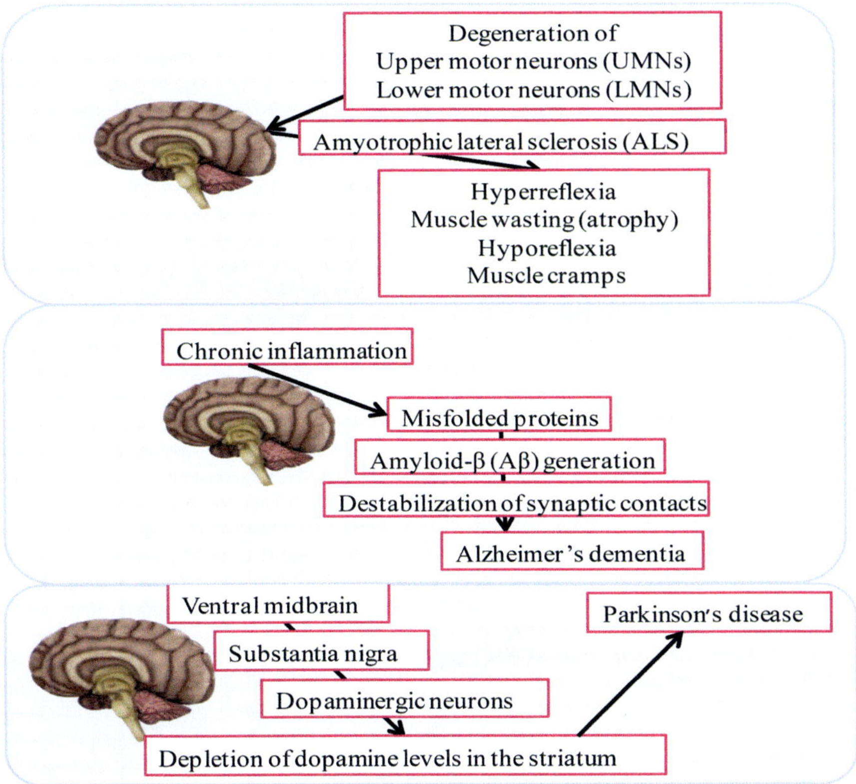


**FIGURE 9.1** Mechanism of Parkinson's disease, Alzheimer's disease, and amyotrophic lateral sclerosis.

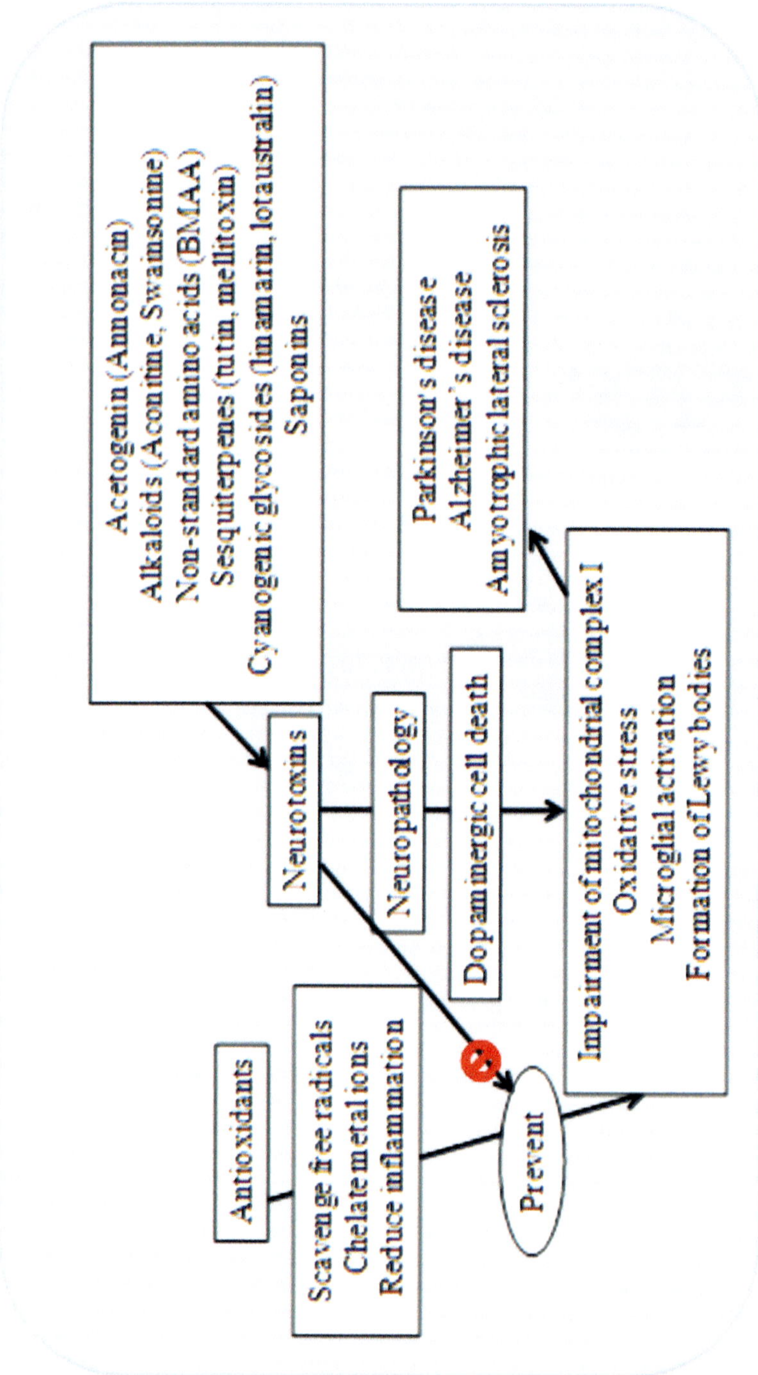


**FIGURE 9.2** Neuroprotectant as well as neurotoxic property of phytochemicals.

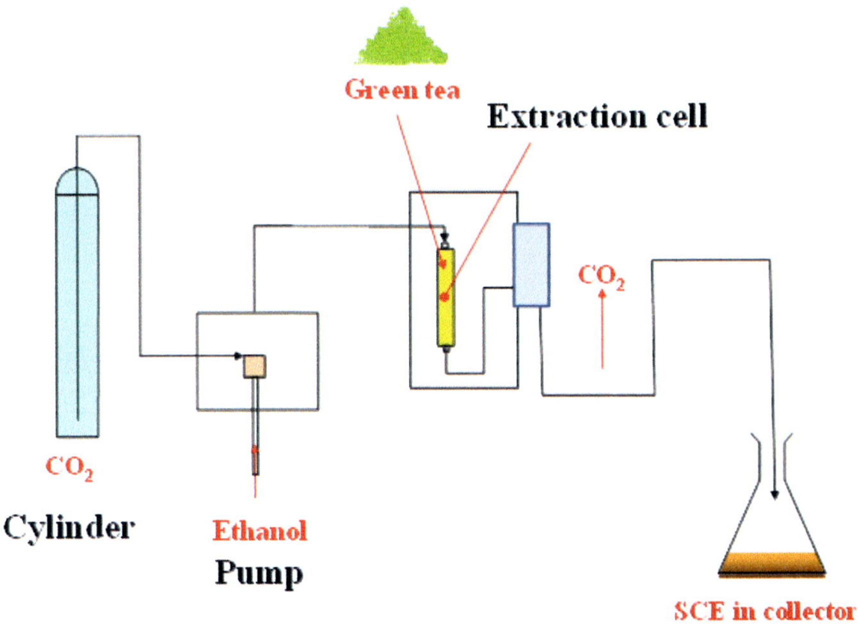


**FIGURE 13.2** Extraction of green tea polycatechins using supercritical carbon dioxide.

# PART III

# Pharmacological Aspects of Natural Products

## CHAPTER 8

# PICEATANNOL: A REVIEW ON NATURAL SOURCES, EXTRACTION METHODS, AND BIOLOGICAL ACTIVITIES

JULIANE VIGANÓ, ANDRESSA MARA BASEGGIO, LARISSA AKEMI KIDO, and JULIAN MARTÍNEZ

## ABSTRACT

Piceatannol or 3,4′,3′,5-trans-trihydroxystilbene is a metabolite and analog of resveratrol. The difference between both molecules is the piceatannol hydroxyl in position of 3′. Therefore, in terms of antioxidant activity, piceatannol is more potent than resveratrol. Although less-known than the resveratrol, piceatannol is a correspondingly attractive natural compound that possesses similar biological activity. This chapter reviews natural sources of piceatannol, its recovery techniques, and its health benefits on obesity and cancer.

Passion fruit and species of genus *Rheum* are most important natural sources of piceatannol, which can also be found in red wine, grapes, *Rhodomyrtus tomentosa*, and among others. Methods to obtain piceatannol from the raw materials and to purify it are future perspectives, especially when fulfilling the goals of green chemistry. Recent research data reports that piceatannol is largely metabolized after oral administration, especially in the liver, demonstrating strong antioxidant activity, and modulation of inflammatory and proliferative pathways in biological models. Several studies highlight the role of this phenolic compound in the prevention and inhibition of the progression of diseases, such as: cancer, inflammation, and metabolic syndrome. In this sense, the use of piceatannol in therapeutic approach emerges as a possible alternative method among chemopreventive treatments.

## 8.1 INTRODUCTION

The worldwide market for phytonutrients may reach $4.63 billion with a compound annual growth rate of 7.2% between 2015 and 2020.[92] Indeed, it is remarkable to note the increasing number of research studies reporting the use of these compounds. The main influencing factors of the global interest are health benefits against cardiovascular diseases, cancer, obesity, and type-2 diabetes. Moreover, phytonutrients market growth is also due to demand by aging population and increased awareness about health and wellness.[92]

Piceatannol is a polyphenol belonging to the class of stilbene, and is a resveratrol analogue.[86] Since 1956, several vegetable species have been identified as sources of piceatannol; however, it is most abundant in species of the genus *Rheum* and *Passiflora.* Although there is scarcity of research studies on extraction techniques to recover piceatannol from vegetable matrices, still some researchers point out emergent techniques such as pressurized fluid extraction to obtain piceatannol on large scale.[117–119]

The consumption of foods rich in phenolic compounds has been receiving great attention these days, because the action of piceatannol is able to decrease the risk of development and mitigate the effects of cancer, diabetes, obesity, and cardiovascular and inflammatory diseases.[58,63,129] Strong biological activities have been reported on the use of piceatannol, such as anticancer[53,60,109] and anti-inflammatory[7,111]; and increased interest of this compound for pharmaceutical or nutraceutical applications. However, a gap in the knowledge of piceatannol health effects is their bioavailability, since many studies about the compound's benefits have been performed only in vitro. Few studies identified a large amount of piceatannol metabolites after oral or intravenous administration, which includes products of phase I and II reactions in liver.[97,98,108] The formed metabolites are higher than those yet identified for resveratrol, the most studied stilbene.[108,122]

This review chapter focuses on piceatannol, which has called attention of researchers due to its high biological activity. The natural sources, piceatannol extraction techniques, its bioavailability, and action on obesity and cancer are also discussed.

## 8.2 PICEATANNOL: THE STRUCTURE AND PHYSICAL PROPERTIES

Piceatannol ($C_{14}H_{12}O_4$) belongs to stilbenoid class, which includes nonflavanoid polyphenolic compounds.[78] Piceatannol is also known as "3,5,3′,4′-tetrahydroxystilbene; astringinin; 3′-hydroxyresveratol; 5-[2-(3,4-dihydroxyphenyl) ethenyl]benzene-1,3-diol; 3,5,3′,4′-tetrahydroxy-*trans*-stilbene; 3,3′,4,5′-tetrahydroxystilbene; 3,5,3′,4′-tetrahydroxystilbene; 3,3′,4′,5-tetrahydroxystilbene; dimethyl isorhapontigenin; and (E)-4-[2-(3,5-dihydroxyphenyl)ethenyl]-1,2-benzenediol."[86] Figure 8.1 shows that the structure of piceatannol is similar to resveratrol, aside from the single hydroxyl group at the 3′-carbon (3,5,4′-trihydroxystilbene). Piceatannol refers to the glycoside of piceatannol-3′-O-β-D-glucopyranoside, and is thus sometimes known as astringin.[86,109]

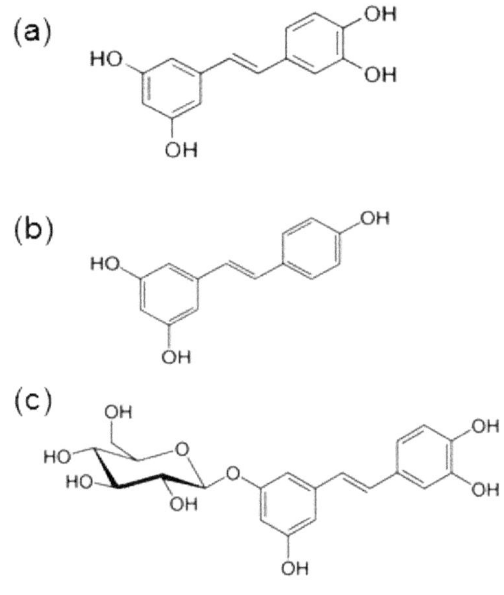


**FIGURE 8.1** Molecular structure of: (a) piceatannol, (b) resveratrol, and (c) astringin.

Physically, piceatannol at room temperature is an almost white powder, its melting point is around 226–223°C, and it has slightly higher molecular weight (244.24 kDa) than that of resveratrol (253–255°C and 228.24 kDa, respectively).[86,109] Piceatannol exists either in the *cis* or *trans* form; and most research studies refer to the chemically stable trans form of piceatannol.[58] Both forms are soluble in ethanol and dimethyl sulfoxide (DMSO), but not in water.[86,109]

## 8.3 NATURAL SOURCES OF PICEATANNOL

Piceatannol has biological activity similar to resveratrol, though it is still less known and studied. A search was done in the Web of Science Core

Collection database using the topics "piceatannol" and "resveratrol." The number of results evidenced the above statement; the term piceatannol showed 734 results, while the topic "resveratrol" showed 14,435 results. Piceatannol was first isolated from natural sources by King et al.,[50] Graßmann et al.,[35] and Grassmann et al.,[36] who identified this stilbene in *Vouacapoua macropetala* and spruce bark, respectively. After some years, Cunningham et al.[26] synthesized and described the structure of this compound; and Rao and Rajadurai[90] identified and isolated piceatannol from *Cassia marginata* heartwood. Table 8.1 indicates vegetable sources of piceatannol.

The most common source of stilbenes is grapes. Therefore, grapes represent until now the largest studied piceatannol source. In Table 8.1, grapes represent 32% of the piceatannol sources, followed by species of the genus *Rheum* (12%), passion fruit (8%), wines (7%), species of the genus *Vaccinium* (5%), *Polygonum* (5%), *Arachis* (5%), and *Euphorbia* (3%), while other sources achieved 23% only.

The research data in Table 8.1 deserves attention. The identification of piceatannol has been the major aim of research studies, which is not surprising, since piceatannol is less known than other stilbenes such as resveratrol. Therefore, the identification of plant sources that synthesize piceatannol is still a field of research. Other two common goals were: (1) the study of some effects of plant extracts containing piceatannol and (2) the development of analytic methods to detect and quantify such compounds. Moreover, some studies indicated the enrichment of piceatannol concentration in grapes and wine using ultraviolet-C (UV-C) light irradiation.[20,21,38–40]

Table 8.1 also shows the extraction method and the amount of piceatannol quantified in each research study. Based on data in Table 8.1, it is hard to determine, which one is the best source of piceatannol, since the extraction methods were different in each research study. However, the magnitude of the values allows the conclusion that species of genus *Rheum* and *Passiflora* are potential natural sources of piceatannol. Regarding the extraction of piceatannol, Table 8.1 presents the most commonly used solvents, that is, methanol, ethanol, acetone, and water, while other extraction parameters have not been explored enough to establish the most appropriate extraction method to obtain piceatannol. Therefore, strategies to recover piceatannol from plant sources become suitable and important research objects.

**TABLE 8.1** Piceatannol: Natural Sources, Amount, and Extraction Method.

| Piceatannol | | | References |
|---|---|---|---|
| **Source** | **Amount** | **Extraction information** | |
| *Aiphanes aculeata* Willd. (Arecaceae) Seeds | 43 μg/g dry sample | Using methanol by maceration | [66] |
| *Arachis hypogaea* Calluses of peanuts | 2.17–5.31 μg/g sample | 1 g sample + 1 mL methanol | [57] |
| *Arachis hypogaea* Roots | 0.55 μg/g dry sample | Medium sample (40 mL) was extracted twice with 50 mL of EtOA c in a separatory funnel | [127] |
| *Cassia marginata* Roxb. Heartwood | na | 2 kg sample + 4 L acetone for 8 h at room temperature | [90] |
| *Cercis chinensis Bunge* Roots | 150 μg/g dry sample | 2.0 kg were extracted with 95% ethanol under reflux | [70] |
| *Cissus quadrangularis* Climbing stems | 2 μg/g dry sample | Sample + ethanol–water (4:1) | [2] |
| *Euphorbia lagascae* Seeds | 89 μg/g dry sample | 1.511 kg sample was extracted in sequence with 30–60°C petroleum ether (Soxhlet), $C_6H_6$ (Soxhlet), CHCI (percolation, 5 L), and ethanol (percolation, 10 L) | [33] |
| *Euphorbia lagascae* Spreng. Seeds | 417 μg/dry sample | 1400 g sample + 7 L *n*-hexane at room temperature | [30] |
| Grape—Three varieties Skin | 3.51 μg/g fresh sample | 5 mL of diethyl ether + 3 g of grape skin, stirring at 1200 rpm for 20 min | [39] |
| Grape: Uvalino Skin | 10 μg/g fresh sample | Extraction with tartaric buffer at 25°C for 4 h | [83] |
| Grape juice Concord and Isabel | na | na | [34] |
| Grape submitted to UV-C irradiation Skin | na | 1 g samples + 4 mL of a solution of methanol/formic acid (97:3) and homogenization at 24,000 rpm for 1 min | [21] |
| Grape submitted to UV-C light Skin and cane | 17 μg/g fresh sample | Cane: 200 mg sample + 10 mL of an acetone/water mixture (6:4) overnight at room temperature; skin: 5 mL of diethyl ether + 3 g of grape skin, stirring at 1200 rpm for 20 min | [38] |
| Grape var. Monastrell Berries | 655 μg/L sample | Traditional maceration wine-making | [20] |
| Grapes Cane | na | 200 mg sample + 10 mL of acetone/water mixture (60:40 v/v) overnight at room temperature | [37] |

**TABLE 8.1** *(Continued)*

| Piceatannol | | | References |
|---|---|---|---|
| **Source** | **Amount** | **Extraction information** | |
| Grapes var. Raboso Piave and Corvina Berries | 4.35 μg/g dry sample | Methanol + sample in ratio 2:1 (v/w) under stirring for 20 min | [96] |
| Grapes var. Campbell early and Kyoho Skin | 8.53 μg/g fresh sample | 1 g sample + 2 mL of methanol, and mixed vigorously by vortex for 3 min | [46] |
| Passion fruit Pulp | na | 6 g of sample + 30 mL of aqueous acetone (70%, v/v) containing HCl 1.2 M, after incubation at 4 °C for 90 min | [106] |
| Passion fruit Seeds | 37.06 mg/g dry extract | 30% 1,3-butylene glycol | [74] |
| Passion fruit Seeds | 4.8 mg/g dry sample | Samples extracted twice using 10-fold amounts of 80%(v/v) ethanol, by shaking at room temperature | [75] |
| Passion fruit (*Passiflora edulis* sp.) Seeds | 18.59 mg/g dry sample | Dynamic PLE at 10 MPa, 70°C, and ethanol 50%, solvent/sample ratio 120 | [117] |
| Passion fruit (*Passiflora edulis*) Seeds | 5.7 mg/g dry sample | 100 g sample were refluxed at 90 °C with 90% ethanol for 90 min | [101] |
| Peanut Skin | na | Bath shaker set at 150 rpm and 45°C for 30 min | [73] |
| *Picea jezoensis* Bark, needles, and roots | 0.01–6.07 mg/g dry sample | na | [51] |
| *Polygonum cuspidatum* Roots | 2.76 mg/g dry sample | Using methanol in a Soxhlet extraction apparatus | [116] |
| *Polygonum cuspidatum* Sieb. and Zucc. and *Polygonum sachalinensis* F. Schmidt Roots and stems | na | $CH_2Cl_2$ (3 × 24 h) followed by methanol (3 × 24 h) at room temperature | [31] |
| *Polygonum multiflorum* Roots | na | Methanol at room temperature | [100] |
| Propolis Brazilian and Australian propolis | na | Solid–liquid extraction Solvent: ethanol Contact: mixture was stirred for 24 h | [1] |

**TABLE 8.1** *(Continued)*

| | Piceatannol | | References |
|---|---|---|---|
| **Source** | **Amount** | **Extraction information** | |
| *Prunus dulcis* Almonds | na | 10 g of sample + 20 mL ethanol and water (80:20 v/v) at 4°C for 5 min using homogenization | [126] |
| *Rheum acuminatum* Roots | 8.66 mg/g fresh sample | 0.5 mg sample + 15 mL of 80% (v/v) methanol/water in an ultrasonic bath at 25°C for 5 min under argon, followed by maceration overnight | [95] |
| *Rheum australe* Roots | 13.99 mg/g dry sample | 0.5 mg sample + 15 mL of 80% (v/v) methanol/water in an ultrasonic bath at 25 °C for 5 min under argon, followed by maceration overnight | [95] |
| *Rheum emodi* Wall Rhizomes | 99.2 mg/g dry sample | 1 kg sample extracted with 95% ethanol under reflux (3 × 5 L, 2 h each time) | [23] |
| *Rheum rhabarbarum L. and Rheum rhaponticum L.* Leaves, petioles, and rhizomes | na | Acetone/water (7:3 v/v) at 5°C for 24 h under argon atmosphere | [56] |
| *Rheum rhaponticum* L. Roots | na | 1 g were macerated with 10-fold excess (v/w) of methanol for 72 h at room temperature with periodic shaking | [89] |
| *Rheum undulatum* Rhizomes | na | Sample + hot water for 4 h | [54] |
| *Rheum undulatum* | na | Hot methanol 70% | [55] |
| *Rheum undulatum* Rhizomes | 2 μg/g dry sample | 6 kg sample was cut into small pieces and soaked in methanol at room temperature for 5 days | [128] |
| *Rhodomyrtus tomentosa* Sim fruit | 2.3 mg/g dry sample | 0.4 g of sample + 8 mL of acetone: water: acetic acid (50:49:1; v/v/v) and shaken for 1 h at 37°C | [62] |
| *Rhodomyrtus tomentosa* Sim fruit | na | 0.4 g of sample + 8 mL of acetone:water:acetic acid (50:49:1, v/v/v) and shaken for an hour at 37 °C | [61] |
| *Rhodomyrtus tomentosa* (rose myrtle) Fruit | 11.4 and 24.2 μg/g sample | 100 g sample + 1 L 80% ethanol | [110] |

**TABLE 8.1** *(Continued)*

| | **Piceatannol** | | **References** |
|---|---|---|---|
| **Source** | **Amount** | **Extraction information** | |
| *Robinia pseudoacacia* Heartwood | 263 and 650 μg/g sample | 80 g sample + 500 mL of ethanol–water (1:1), maceration at 50°C for 4 h under moderate agitation | [107] |
| *Scirpus californicus* Rhizomes | na | Sample + methanol | [104] |
| *Sophora yunnanensis* Roots | 3.29 μg/g dry sample | Extraction using 80% methanol at room temperature | [27] |
| *Stuhlmannia moavi* Roots | 7.46 μg/g dry sample | Ethanol at room temperature | [72] |
| Sugarcane (*Saccharum* sp.) infected with *Colletotrichum falcatum* Inner tissues of the internodes | na | Washing with methanol, $CHCl_3$ and EtOAc | [18] |
| *Vaccinium corymbosum* and *V. stamineum* Berries | 195–422 ng/g dry sample | 51 g sample + three volumes of 40:40:20:0.1 (v/v/v/v) methanol–acetone–water–formic acid, kept for 30 min, then ground in a VirTis homogenizer | [94] |
| *Vaccinium corymbosum* L. (highbush blueberry) Berries | 186–422 ng/g dry sample | 51 g sample + three volumes of 40:40:20:0.1 (v/v/v/v) methanol–acetone–water–formic acid, kept for 30 min, then ground in a VirTis homogenizer | [93] |
| *Vaccinium stamineum* L. (deer berry) Berries | 138–195 ng/g dry sample | 50 g sample + three volumes of 40:40:20:0.1 (v/v/v/v) methanol–acetone–water–formic acid, kept for 30 min, then ground in a VirTis homogenizer | [93] |
| *Vateria indica* Leaves | na | 250 g sample successively extracted with acetone (3 × 1 L × 24 h), methanol (3 × 1 L × 24 h), and 70% methanol (2 × 1 L × 24 h) at room temperature | [47] |
| *Vitis sylvestris* Leaves | ~50 μg/g fresh sample | 300 mg sample + 1 mL of methanol and homogenized for 10 min on a platform vortexer | [29] |
| *Vitis vinifera* Seeds and stalks | na | Maceration and lixiviation in open column | [28] |
| *V. vinifera* Grapevine shoots | na | na | [22] |

**TABLE 8.1** *(Continued)*

| | Piceatannol | | References |
|---|---|---|---|
| **Source** | **Amount** | **Extraction information** | |
| *V. vinifera* Canes from eight cultivars | ~600 μg/g dry sample | 50 mg sample + 1 mL ethanol/water (60:40; v/v), shaken for 30 min at 1400 rpm at 83°C | [42] |
| *V. vinifera cv.* Semilignified shoots and canes | 4.2– 25.8 μg/g fresh sample | 1 g sample + 7 mL methanol in cooled ultrasonic bath for 1 h with 7 | [99] |
| *V. vinifera* L. Grape skin from 21 red cultivars | 72.1 μg/g fresh sample | 1 g sample + 40 mL methanol + 50 μL of hydrochloric acid. Homogenization for 1 min with Ultra-Turrax, the samples were maintained for 48 h in closed containers under stirring at room temperature in the dark | [121] |
| *V. vinifera* L. Canes from 16 cultivars | 0.19– 1.71 mg/g dry sample | 200 mg sample + 10 mL of acetone/water mixture (60:40 v/v) overnight at room temperature | [64] |
| *V. vinifera* L. cv. Cabernet Sauvignon Roots, stems, leaves, berry skins and seeds | ~0.1 to ~350 μg/g dry sample | 300 g sample + 6 mL methanol by ultrasonication at 40 kHz for 12 min and then incubated in the dark for 1 h at 4 °C without agitation | [123] |
| *V. vinifera* L. cv. Malbec Pomace | 38.8 μg/g extract | Solid–liquid extraction. Solvent: ethanol:water, 50:50 v/v. Solvent-to-sample (dried weight, DW) ratio: 25:1. Contact: during 120 min under continuous stirring at 60°C | [6] |
| *V. vinifera L. cv.* Cabernet Sauvignon Berries | 0.052 μg/g fresh wt | 20 g sample + 30 mL methanol, shaking for 20 min at room temperature | [12] |
| Wine—Enriched Grape skin and pomace | 0.4–2.95 μg/g fresh sample | 5 mL of diethyl ether was added to 3 g of sample and stirred at 1200 rpm for 20 min | [40] |
| Wine—Italian | 0.54–5.22 mg/L | na | [19] |
| Wines Four red Burgundy | na | na | [16] |
| Wines and grapes Wine (red, rosé, white, and sweet) and whole grapes (red and white) | 6.1–14 ng/mL wine and 1.2– 24 ng/g grape | Solid-phase microextraction | [120] |

na, not available; EtOAc, ethyl acetate; PLE, pressurized liquid extraction; UV-C, ultraviolet-C.

## 8.4 EXTRACTION METHODS

Table 8.1 focuses only on extraction methods of piceatannol. Viganó et al.[117] established a sequential extraction process that gave higher piceatannol yields than the single extraction process and conventional extraction methods (maceration and Soxhlet). The used raw material was passion fruit bagasse, which is a by-product from passion fruit processing composed of seeds and little amount of remaining pulp. The sequential extraction process was performed by applying three steps of supercritical fluid extraction (SFE), and one more step of pressurized liquid extraction (PLE). The authors[117] used carbon dioxide as solvent in SFE to obtain nonpolar extracts composed of tocotrienols, fatty acids, and carotenoids. The defatted passion fruit bagasse resulting from SFE was used to study the piceatannol extraction by PLE. The authors evaluated two extraction parameters, temperature and solvent composition, and achieved the highest piceatannol yield when PLE was performed at 70°C using 50% (w/w) ethanol as solvent.

Viganó et al.[117] compared PLE to maceration and Soxhlet and concluded that PLE was able to recover about 1.6 and 2.5 times more piceatannol than maceration and Soxhlet, respectively. The maceration experiment was performed keeping all the conditions of the best PLE experiment constant to recover total phenolics, except atmospheric pressure. The PLE was also performed on non-defatted raw material in order to verify the influence of SFE process, and the piceatannol yield was 1.6 times higher when the defatted passion fruit was used. Therefore, the previous SFE process allowed increase in piceatannol yield, because according to the authors, it was responsible for removing the non-polar phase that could have hampered the phenolic compounds desorption from the raw material.[117]

### *8.4.1 SFE AND PLE*

SFE and PLE have been considered as emerging techniques to recover phytochemicals from plant sources due to their technical and environmental advantages.[119] SFE is commonly used to obtain low polarity compounds because the most employed solvent is $CO_2$. Changes in solvent polarity can be achieved using a cosolvent such as water and/or ethanol, and consequently the extraction target compounds will be more polar. However, high extraction rates of polar compounds can be obtained using PLE when the liquid solvent has polar characteristic. Indeed, ethanol and water have been extensively employed in PLE because high yields of the target compound are produced, and such solvents are generally recognized as safe (GRAS). Since $CO_2$ is used in SFE, the extraction is considered secure and nontoxic. Moreover, $CO_2$ is universally available at high purity, and generally at low cost. The critical point of $CO_2$ (7.34 MPa and 31.06°C) avoid damage of thermosensible compounds during extraction.[103] Supercritical $CO_2$ extraction process (SC-$CO_2$) has another important advantage due the tunable $CO_2$ property. In other words, SC-$CO_2$ can solubilize different classes of compounds by changing the pressure and temperature of the extraction process, which allows obtaining enriched fractions of target compounds.[8,71,88] The most important variables in SFE are pressure and temperature. However, other variables have also been studied such as: solvent flow rate, sample particle size, extraction time, and extraction bed properties such as height/diameter ratio. Temperatures between 40°C and 70°C are usually used because these are higher than the critical $CO_2$ temperature and target properties such as density and viscosity can be reached. High temperatures decrease the $CO_2$ density; consequently, lower extraction yields are reached. The range of explored pressure is from 10 to 50 MPa. The $CO_2$ density increases as the pressure is increased, resulting in better solvation capacity. Pressures above 50 MPa can lead to high energy costs and special design of the extractor may be required.[119]

PLE is performed by using solvents at high pressure and temperature, but below the critical point. It is known for promoting the compounds extraction from solid or semisolid matrices in short time and using less amount of solvent. The main PLE characteristic is that the solvent is kept in the liquid state even above the boiling point because of the high pressure applied, which allows the improvement of the solvent power. Moreover, the extraction is performed in an environment without oxygen and light.[43,102,119] The most important variables in PLE are temperature and solvent composition. High temperatures improve the desorption of target compounds from the raw material by: (1) decreasing the surface tension of the solvent, solute, and matrix and (2) reducing the viscosity of liquid solvent, reaching high mass transfer rates. However, high temperatures decrease the extraction selectivity because the amount of coextracted analytes might increase. In addition, high temperatures might be negative for thermosensible compounds.[119] Regarding the solvent composition, ethanol, water, and their mixture are the most used solvents in PLE. According to Mustafa and Turner,[81] applying two solvent mixture can enhance the solubility and increase the interaction with the target compound, for example, one solvent acts in the solubility and the other in the solute desorption from the raw material. Regarding pressure, high pressure enhances mass transfer of the target compounds from the raw material to the solvent by the disruption of cells of raw material.

## 8.5 BIOLOGICAL EFFECTS OF PICEATANNOL

### *8.5.1 BIOAVAILABILITY OF PHENOLIC COMPOUNDS: FOCUS ON STILBENES*

According to health-promoting effects, polyphenols can be classified as flavonoids (such as isoflavones, flavones, and anthocyanidins) or nonflavonoids (phenolic acids and stilbenes).[14,17] Several research studies show that those phenolic compounds reduce the risk of diseases due to possible interactions in enzyme activities and cellular signaling pathways. Bioactive polyphenols are found in diets rich in fruits and vegetables. However, researchers want to know about the compounds or conjugate compounds that will have the highest activity in one food matrix. Vegetable extracts are subject of intense research both in pharmaceutical and food science.[14,44]

Thus, the metabolism and bioavailability of specific polyphenols are keys to the success in their biological functions. Bioavailability refers to the fraction of a nutrient or nonnutrient that enters the bloodstream and reaches specific tissues, being available for cellular functions. For polyphenol oral consumption, this process involves gastrointestinal digestion, where the compound can undergo changes in native structure or even be conjugated with others substances, leading to the formation of metabolites.[44] Studies have revealed a wide range of biological applications for stilbenes (a class of nonflavonoid polyphenol). Stilbenes are naturally found in different dietary sources, such as: grape skins, berries, red wine, peanuts, and medicinal plants, and the most known compound of this class is resveratrol.[122] In vitro and in vivo assays indicate that resveratrol has many systemic effects, such as: antioxidant, anti-inflammatory, antimicrobial, and anticarcinogenic.[11] As the most studied stilbene, the mechanisms of bioavailability of resveratrol and formed metabolites are most described in the literature. It is known that resveratrol is hydroxylated, glucuronidated, or sulfated in liver,[84] and the oral bioavailability seems to be low in humans (less than 1% of resveratrol or their metabolites in plasma, even after extensive metabolism in the liver and intestine), despite a considerable absorption of approximately 75%.[122] In this context, experiments for the observation of resveratrol pharmacokinetics after oral ingestion of a single dose or short-time administration (daily) have been performed to justify its possible chemoprotective effect.

Using an ample total dose range (150 mg at 900 mg/day), Almeida et al.[4] evaluated the pharmacokinetic and safety profile for oral trans-resveratrol consumption in healthy volunteers for 13 days. The results indicate that individuals at highest dose (900 mg/day) showed larger quantity of trans-resveratrol in blood, although the amount of recovered compound in plasma still was little, confirming the hypothesis of low bioavailability of this substance. Another important described result was that for all doses of the trans-resveratrol, the half-life in plasma was low: approximately 1–3 h, with the maximum plasma concentration after 0.8–1.5 h postdose. On the other hand, Boocock et al.[15] used higher doses of resveratrol in their study (at 0.5–5 g/day) in a single dose, with an interval time of 14 days between each oral ingestion in healthy individuals to verify from which dose it would be possible to recover 5 μmol/L in plasma. This level appears to be required for chemopreventive effects in cancer cells at in vitro models. Despite the high compound amount administered in the last dose, the results showed that it was not possible to detect the desired concentration in the blood, even in short time after the administration. However, this study also identified resveratrol metabolites, specially generated after sulfation and glucuronidation in the liver (products from biotransformation by phase II reactions), in considerable amounts. The pharmacological properties of these compounds are still little known, but have speculated some activity in cellular mechanisms,[11] as previously observed for other polyphenols.[32,130]

An important factor for interindividual variability for stilbenes metabolism could be the gut microbiota. A recent study conducted by Bode, et al.[13] demonstrated that the bioconversion of trans-resveratrol in human health individuals is closely related to intestinal diversity, showing three different microbial trans-resveratrol metabolites in feces. It is interesting to mention that in samples with high abundance of bacteria from *Bacteroidetes* filo, large amounts from 3,4′-dihydroxy-trans-stilbene and lunularin were detected but not dihydro resveratrol, whereas in samples with more amount of bacteria from *Firmicutes* filo the opposite was verified: the most abundant metabolite found was dyhydro resveratrol. Moreover, differences in chemical structure of stilbenes can induce significant changes in the metabolism and bioavailability parameters of these compounds.[98] However, a few studies evaluated the metabolism of other stilbenes such as piceatannol.

Resveratrol and piceatannol differ due to a hydroxyl group, creating a structure called catechol. However, this difference can be the key to more biologically effects attributed for this compound: hydroxylated analogs of resveratrol seem to present a stronger in vitro antioxidant capacity than this stilbene[80,82] and effects pronounced in cell mechanisms, such as antilipolytic,[68] anti-inflammatory[25] and antitumor.[80] A study using microsomes of human lymphoblast showed that resveratrol may be hydroxylated in piceatannol through the cytochrome P450 CYP1B1, an enzyme found in wide variety of tumors, whose activation for prodrugs is considered a cancer treatment or a form for tumor prevention.[87] The presence of a catechol ring in the piceatannol structure could make the metabolism more complicated due to the presence of sites for methylation available, which increase the number of pathways for its metabolism.[108]

### 8.5.2 BIOAVAILABILITY OF PICEATANNOL

Roupe et al.[97] described the first study about the piceatannol pharmacokinetics and metabolism characterization in vivo, intending to recover the compound in Sprague–Dawley rat serum in different times after a single dosage of piceatannol (10 mg/kg) administered intravenously. Besides the serum, the animal liver was collected for microsomes extraction and study of hepatic metabolism. The results showed that the half-life of the parent compound in serum was 2 h and two unidentified glucuronidated metabolites were present. As for the hepatic metabolism by microsomes, the identification of extra peak after induction of phase I reactions (with NADPH) and piceatannol addition (10 μg/mL) suggests oxidative metabolism. For the evaluation of products by phase II metabolism, the uridine 5'-diphospho-glucuronosyltransferase (UGT) enzyme was used to induce the reactions. In this case, two peaks indicating glucuronidation were verified, but unidentified. According to the chromatograms, similar peaks were observed in the serum samples.

Roupe et al.[98] verified that after a single intravenous piceatannol dose (10 mg/kg) in rats, it was still possible to identify the presence of this native compound in the plasma at 4.23 ± 1.25 h. Also, an enterohepatic recycling of piceatannol was suggested by the authors. Therefore, the liver seems to have importance in piceatannol metabolism: a significant amount of stilbenes excretion is hepatic, via bile.

In general, stilbenes are poorly soluble in water and present a good volume of distribution in tissues. In this study, the authors found that the volume of distribution of piceatannol in body tissues was higher than other stilbenes. Once more, unidentified glucuronidated metabolites were detected in plasma and urine rapidly after piceatannol intravenous administration. In both studies, the authors identified an important step in the piceatannol metabolism in the liver: the glucuronidation.

All classes of drugs, recognized as xenobiotic, undergo this process that is able to increase the polarity of hydrophobic compounds.[124] The glucuronidation strongly affects the bioavailability of compounds, because it is crucial in substance elimination,[76] but paradoxically some pharmacologically active drug glucuronides have been recognized. Glucuronides may also indirectly contribute to compound action when there is a cleavage of the active parent substance in body tissues by β-glucuronidase.[112] Miksits et al.[76] evaluated the piceatannol glucuronidation in microsomes derivate by human liver cells and identified three monoglucuronides, all metabolized by *UGT1A1*.

Another important metabolic pathway for piceatannol in the liver is sulfation, which is a way to increase the polarity of hydrophobic compounds, leading to biological activity decrease or generating more active compounds.[5,77] Miksits et al.[77] identified piceatannol metabolites formed in cytosol hepatocytes in the presence of 30-phosphoadenoseine-50-phosphosulfate (PAPS). The results showed that piceatannol is extensively metabolized in three sulfated conjugates: a disulfate and two monosulfates. The formation of disulfate was correlated with lower piceatannol concentrations, being that higher levels generated a substrate inhibition. Enzymes of the sulfotransferase family are found in other organs of the gastrointestinal tract, such as gut. Therefore, the extrahepatic sulfation of piceatannol may contribute to compound metabolism after oral ingestion.

However, studies about piceatannol bioavailability after oral ingestion are scarce. Setoguchi et al.[108] observed the metabolism of piceatannol and resveratrol in rodents after oral ingestion. The animals received piceatannol or resveratrol diluted in polyethylene glycol 400 and saline solution (50% v/v) in different doses. The effect of stilbenes coadministration was also evaluated, through a mixture of both compounds in equal diluent solutions. The parent piceatannol maximum concentration in plasma was reached within 15 min of administration, in dose-dependent response. The authors detected nine peaks of piceatannol conjugates through mass spectrometry, being one of these a diglucuronide, two of monoglucuronides, three *O*-methyl piceatannol-monoglucuronide, and one of *O*-methyl piceatannol-monosulfate, isorhapontigenin, and rhapontigenin each. Although the half-life of conjugates glucuronidated was 2 h for lower doses and 4 h for the higher dose, *O*-methyl piceatannol conjugates were detected at 24 h after compound oral ingestion in high doses. The stilbene coadministration promoted a greater absorption of piceatannol than resveratrol in intact form.

An interesting observation about differences between piceatannol and resveratrol bioavailability is the higher amount of piceatannol methylated metabolites found in plasma, in contrast to any methylated resveratrol molecules. Furthermore, two stilbenes were detected as piceatannol metabolites (isorhapontigenin and rhapontigenin): those compounds present significant biological effects, such as apoptosis stimulation, inhibition of prostate cancer cells proliferation,[131] and improvement of serum lipid profile.[48] Together, these data predict a higher bioavailability of piceatannol compared to the most known stilbene, resveratrol. A lot of metabolites may be related with many piceatannol metabolic pathways.

Inagaki et al.[45] analyzed the piceatannol absorption when complexed with cyclodextrin. This substance is widely used by pharmaceutical industry for the improvement of biological activity of drugs over wide pH ranges (as those found in the gastrointestinal tract) due to the high stability.[79] The study evaluated the effect of gastric conditions through their simulation in vitro and the bioavailability assessment of the animal research. The inclusion of cyclodextrin increases the solubility of piceatannol under acidic conditions (mimetizing a gastric juice) and enhances the area under curve (AUC) in first 3 h after administration of parent piceatannol and derivates (isorhapontigenin, conjugates, and O-methyl-conjugates) in plasma of rats compared to control group (only piceatannol). The maximum concentration observed in piceatannol complexed with cyclodextrin was approximately two-high folds than that of the parent piceatannol as well as a higher level for piceatannol conjugates.

New studies about the piceatannol bioavailability after oral ingestion are required to indicate interindividual variability in metabolites formation, possible effects of gut microbiota in their metabolism, and synergism with other phenolic compounds that are also found in vegetable extracts which are considered source of piceatannol. However, few in vivo studies have already demonstrated that piceatannol administration is able to improve pathologic conditions.

## 8.6 OBESITY: A NEW TARGET FOR PICEATANNOL POTENTIAL

According to Adhami and Mukhtar,[3] chemoprevention by nontoxic agents, more advantageously from dietetic sources, is a good strategy in the fight against cancer progression. As stated before, piceatannol is a resveratrol analogue, which is widely known for its antioxidant, anti-obesity, anti-inflammatory, antimutagenic, antiproliferative, and cardioprotective properties.[91,109] This was confirmed in review articles by Piotrowska et al. [85] and Seyed et al.[109] However, few studies have focused on the effects of piceatannol on obesity.

A recent study verified piceatannol activity in high-fat diet (HFD)-induced animal model, comparing to the performance of resveratrol, and established that piceatannol can be used in obesity treatment and prevention.[114] This work showed a decrease in size and weight of perigonadal and retroperitoneal adipose tissue, and food efficiency ratio was decreased after treatments with resveratrol and piceatannol. Regarding intracellular energy balance and lipid metabolism, these authors also verified that piceatannol decreased adipogenic transcription factors levels related to adipocyte differentiation, especially the peroxisome proliferator-activated receptor gamma (PPARγ), indicating an increase of fatty acid oxidation-related protein. Piceatannol also promoted body weight reduction via adenosine 5′-monophosphate-activated protein kinase alpha (AMPKα) activation, and decreased lipogenesis in adipose tissue and liver when compared to resveratrol. Moreover, piceatannol activity seemed to have better results than the resveratrol supplementation, probably due to its bioavailability in the body,[114] as previously discussed in this chapter. In contrast to this study, Uchida-Maruki et al.[115] showed that piceatannol did not modify the body weight gain, subcutaneous and visceral fat gain, and food intake in HFD mice model. However, piceatannol was able to reduce blood glucose levels in HFD and diabetic mice model, indicating its potential application in the prevention of diabetes.[115]

Another study has shown that piceatannol had no apparent anti-obesity effect in obese Zucker rats.[41] Although serum metabolic parameters have not changed, yet nonesterified fatty acids and low-density lipoprotein cholesterol were reduced after piceatannol treatment at dose of 45 mg/kg. This result indicates that piceatannol may attenuate the toxicity from circulating lipids accompanying excessive fat accumulation in this mice model, but without mitigating hyperinsulinemia obesity and cardiac hypertrophy.[41] According to Les et al.,[69] piceatannol has the ability to change different human adipose tissue functions, such as decreasing hydrogen peroxide release and limiting free fatty acid release. In other words, this polyphenol has an antilipolytic feature, attenuating the lipotoxicity derivable of fat depots and contributing to alleviate obesity complications.

Despite the divergences found in the literature on piceatannol performance on obesity, this bionutrient appears to regulate key stages of lipid metabolism. Notwithstanding, it is important to consider differences between each study, such as: piceatannol purity, experimental design, dose, period of treatment, and the evaluated living system models. Literature is limited on this issue, and additional studies are required for a better understanding of piceatannol efficacy on obesity disorders.

## 8.7 CANCER

### *8.7.1 PICEATANNOL: ROLE IN DIFFERENT TYPES OF CANCER*

Literature has demonstrated that piceatannol and resveratrol have anticancer properties.[85,109] It is known that cancer-blocking mechanisms by medicines or natural substances/factors may extend the life expectancy.[52] Based on cancer research, piceatannol mechanism of action is summarized in some action fronts: cell cycle arrest, proapoptotic, anti-inflammatory, antiangiogenic, and antimetastasic activities.[113] As a consequence of these properties, piceatannol is able to modulate several pathways, depending on the cancer type.

Piceatannol was able to decrease the invasion and migration potentials of androgen-insensitive DU145 prostate cancer cells by reducing metalloproteinase-9 (MMP-9), vascular endothelial growth factor, and urokinase type plasminogen activator,[60] possibly by blocking signal transducer and activator of transcription 3 (STAT-3) activation through IL-6 secretion inhibition. It is known that IL-6/STAT-3 pathway is constitutively activated in tumor development, including prostate tumors, where it plays a crucial role in proliferation, metastasis, and angiogenesis processes.[9,65,105] According to Kwon et al.,[60] bioactive substances that react with STAT-3 can be employed in therapeutic approaches in prostate cancer metastasis, which can be a good strategy if considered the high mortality index associated to this disease. Furthermore, piceatannol treatment in vivo was able to inhibit lung metastasis from cancerous prostate cells. Similarly, other authors have found that piceatannol increased the cell death in tumors by means of the activation of the death receptor and mitochondrial-dependent pathways.[49]

In breast cancer, piceatannol reduced tumor growth mediated by MMP-9 activity inhibition without affecting metalloproteinase (TIMP-2), its natural inhibitor. These results were accompanied by phosphatase and tensin homolog (PTEN) increase, and phosphoinositide-3 kinase (PI3K), protein kinase B (AKT), and mammalian target of rapamycin protein kinase (mTOR) inhibition.[53] The PI3K/AKT pathway is commonly activated in breast cancer and promotes cancer cell invasion and metastasis through upregulation of MMP-9.[10,24] Moreover, piceatannol also interfered in the main pro-inflammatory mediator, nuclear kappa B transcription factor (NFκB), which in fact led to the blockade of MMP-9 activity because NFκB is a downstream target of AKT and one of the main MMP-9 promoter.[53]

Similar results were observed in breast nonmalignant cells treated with the prototypic tumor promote (TPA) and after piceatannol.[111]

The TPA-induced activation of NFκB and expression of cyclooxygenase-2 (COX-2) were suppressed by piceatannol, which was able to block migration and transformation of these cells due to its specific interaction with critical catalytic subunits of protein kinases responsible for IKKB phosphorylation, which plays a central role in coordinating the NFκB signaling pathway.[111] Furthermore, AH109A hepatoma cells maintained in oxidative medium (reactive oxygen species, ROS) were also affected by piceatannol treatment, which was able to suppress the ROS-potentiated invasive activity of malignant cells.[52] Therefore, piceatannol appeared to be a blocker of invasive ability cancer cells, as well as including invasion and migration, mainly due to its action on pathways linked to MMP-9.

Data regarding piceatannol effects on cell cycle arrest showed that cyclin D1, cyclin B1, cdk 4, and p27Kip1 were downregulated, whereas the increase in percentage of human colon carcinoma cell line (CaCo-2) in the G1 phase and accumulation in the S phase was accentuated after piceatannol administration.[125] Cyclin-dependent kinases (cdk) 2 and 6, as well as cdc2, have no changes after treatment.[125] Piceatannol also blocked cell cycle progression in the G0/G1 phase in human bladder cancer cells (T24 and HT1376), due to the increase of cyclin-dependent kinase inhibitor p21/WAF1, which is thought to be responsible for G0/G1 cell cycle arrest.[59]

In DU145 cells, piceatannol treatment increased the number of cells in G1 phase, at the same time decreased cyclin A, cyclin D1, and cyclin dependent kinase cdk2 and cdk4 expression, as well as decreased cdk2 and cdk4 activity.[67] However, in this study piceatannol had no effect on the levels of p21WAF1/CIP1 or p27KIP1, as observed in the CaCo-2 and cells.[125] Kita et al.[52] demonstrated the piceatannol role as an antiproliferative and proapoptotic compound, by means of retard of cell cycle in G2/M phase and cell death stimulation.

Based on these reports, it is possible to state that the biological role of piceatannol in different cancer cell lines achieved the desirable effects, considering the mandatory properties that a medicine must have. Despite the literature data indicating the application of this substance as an active biomolecule with nutritional/pharmacological effects,[52,85] more studies are needed to better understand the biological effects of this piceatannol, especially using in vivo models.

## 8.8 SUMMARY

The benefits of piceatannol presented in this chapter encourage its recovery on industrial scale process for future applications in foods and beverages, drugs, cosmetics, and nutraceuticals. Although analytical aspects have already been explored, yet few studies have been carried out to optimize the extraction parameters of this natural compound. Moreover, there are not many available studies reporting the scale-up of piceatannol extraction. Therefore, the investigation of extraction and purification processes from potential piceatannol sources, as well as the scale-up and economic evaluation of the involved processes, are new trends. In addition, the application of extraction techniques that use pressurized fluids is internationally acceptable, since they are considered environmentally friendly and enable the use of GRAS solvents, being therefore adequate choices to recover piceatannol from natural sources.

Beside efficient extraction forms, other important aspect about piceatannol potential use is an effective mode for oral administration, to make this compound more stable in a matrix and therefore more bioavailable and bioactive. Although not well elucidated, yet the stilbene oral bioavailability appears to be low. Few studies evaluated the compounds generated through piceatannol metabolism, and it was verified that phase I and II reactions in liver are responsible for different compounds formation, such as methylated and glucuronidated derivatives. However, many studies have identified piceatannol benefits using the pure compound in cells, which makes the evaluation of the substance effect in vivo a gap. Therefore, the search for new techniques for piceatannol oral administration could be interesting to enhance the recovery of the intact compound in plasma and its delivery to action site, possibly approaching the effects observed in cell studies. Thus, the piceatannol encapsulation or incorporation in polymeric or biopolymeric matrices, using biodegradable materials, is an interesting form to ensure its functional mechanisms.

The piceatannol effectiveness in cancer and metabolic diseases in vitro studies make it more efficient than resveratrol. In this sense, the detailed knowledge of piceatannol mechanisms of action through in vivo biological models is extremely important to reflect the pathological conditions affected in humans. Another challenge for piceatannol use as a chemopreventive is the determination of the optimal safe dose in humans. For this purpose, the compound chemical characteristics, the delivery into its target tissue, and the required amount for a biologically relevant action are aspects that should be explored together in future studies.

## KEYWORDS

- **antioxidant capacity**
- **bioavailability**
- **phytonutrients**
- **polyphenol**
- **pressurized liquid extraction**
- **stilbenes**
- **supercritical fluid extraction**

## REFERENCES

1. Abu-Mellal, A.; Koolaji, N.; Duke, R. K.; Tran, V. H.; Duke, C. C. Prenylated Cinnamate and Stilbenes from Kangaroo Island Propolis and their Antioxidant Activity. *Phytochemistry* **2012,** *77* (1), 251–259.
2. Adesanya, S. A.; Nia, R.; Martin, M.-T.; Boukamcha, N.; Montagnac, A.; Païs, M. Stilbene Derivatives from Cissus Quadrangularis. *J. Nat. Prod.* **1999,** *62* (12), 1694–1695.
3. Adhami, V. M.; Mukhtar, H. Human Cancer Chemoprevention: Hurdles and Challenges. *Top. Curr. Chem.* **2013,** *329*, 203–20.
4. Almeida, L.; Vaz-da-Silva, M.; Falcão, A.; Soares, E.; Costa, R.; Loureiro, A. I.; Fernandes-Lopes, C.; Rocha, J. F.; Nunes, T.; Wright, L.; Soares-da-Silva, P. Pharmacokinetic and Safety Profile of Trans-resveratrol in a Rising Multiple-dose Study in Healthy Volunteers. *Mol. Nutr. Food Res.* **2009,** *53* (1), S7–S15.
5. Anderson, R. J.; Kudlacek, P. E.; Clemens, D. L. Sulfation of Minoxidil by Multiple Human Cytosolic Sulfotransferases. *Chem. Biol. Interact.* **1998,** *109* (1–3), 53–67.
6. Antoniolli, A.; Fontana, A. R.; Piccoli, P.; Bottini, R. Characterization of Polyphenols and Evaluation of Antioxidant Capacity in Grape Pomace cv. Malbec. *Food Chem.* **2015**, *178*, 172–8.
7. Ashikawa, K.; Majumdar, S.; Banerjee, S.; Bharti, A. C.; Shishodia, S.; Aggarwal, B. B. Piceatannol Inhibits TNF-Induced NF-B Activation and NF-B-Mediated Gene Expression Through Suppression of I B Kinase and p65 Phosphorylation. *J. Immunol.* **2002,** *169* (11), 6490–6497.
8. Babovic, N.; Djilas, S.; Jadranin, M.; Vajs, V.; Ivanovic, J.; Petrovic, S.; Zizovic, I. Supercritical Carbon Dioxide Extraction of Antioxidant Fractions from Selected Lamiaceae Herbs and Their Antioxidant Capacity. *Innov. Food Sci. Emerg. Technol.* **2010,** *11* (1), 98–107.
9. Barton, B. E.; Karras, J. G.; Murphy, T. F.; Barton, A.; Huang, H. F.; Signal Transducer and Activator of Transcription 3 (STAT3) Activation in Prostate Cancer: Direct STAT3 Inhibition Induces Apoptosis in Prostate Cancer Lines. *Mol. Cancer Therapeut.* **2004,** *3* (1), 11–20.

10. Baselga, J. Targeting the Phosphoinositide-3 (PI3) Kinase Pathway in Breast Cancer. *Oncologist* **2011,** *16* (1), 12–9.
11. Baur, J. A.; Sinclair, D. A. Therapeutic Potential of Resveratrol: The in Vivo Evidence. *Nat. Rev. Drug Discov.* **2006,** *5* (6), 493–506.
12. Bavaresco, L.; Fregoni, M.; Trevisan, M.; Mattivi, F.; Vrhovsek, U.; Falchetti, R. The Occurrence of the Stilbene Piceatannol in Grapes. *Vitis* **2002,** *41* (3), 133–136.
13. Bode, L. M.; Bunzel, D.; Huch, M.; Cho, G.-S.; Ruhland, D.; Bunzel, M.; Bub, A.; Franz, C. M.; Kulling, S. E. In Vivo and in Vitro Metabolism of Trans-resveratrol by Human Gut Microbiota. *Am. J. Clin. Nutr.* **2013,** *97* (2), 295–309.
14. Fawad, S. A.; Khalid, N.; Asghar, W.; Suleria, H. A. R. In Vitro Comparative Study of Bougainvillea Spectabilis "Stand" Leaves and Bougainvillea Variegata Leaves in Terms of Phytochemicals and Antimicrobial Activity. *Chin. J. Nat. Med.* **2012,** *10* (6), 441–447.
15. Boocock, D. J.; Faust, G. E. S.; Patel, K. R.; Schinas, A. M.; Brown, V. A.; Ducharme, M. P.; Booth, T. D.; Crowell, J. A.; Perloff, M.; Gescher, A. J.; Steward, W. P.; Brenner, D. E. Phase I Dose Escalation Pharmacokinetic Study in Healthy Volunteers of Resveratrol, a Potential Cancer Chemopreventive Agent. *Cancer Epidemiol. Biomark. Prev.* **2007,** *16* (6), 1246–1252.
16. Boutegrabet, L.; Fekete, A.; Hertkorn, N.; Papastamoulis, Y.; Waffo-Teguo, P.; Merillon, J. M.; Jeandet, P.; Gougeon, R. D.; Schmitt-Kopplin, P. Determination of Stilbene Derivatives in Burgundy Red Wines by Ultra-high-pressure Liquid Chromatography. *Anal. Bioanal. Chem.* **2011,** *401* (5), 1513–1521.
17. Saklani, S.; Mishra, A.; Chandra, H.; Atanassova, M.; Stankovic, M.; Sati, B.; Suleria, H. Comparative Evaluation of Polyphenol Contents and Antioxidant Activities Between Ethanol Extracts of *Vitex negundo* and *Vitex trifolia* L. Leaves by Different Methods. *Plants* **2017,** *6* (4), 45.
18. Brinker, A. M.; Seigle, D. S. Isolation and Identification of Piceatannol as a Phytoalexin from Sugarcane. *PhytoChemistry* **1991,** *30* (10), 3229–3232.
19. Buiarelli, F.; Coccioli, F.; Jasionowska, R.; Merolle, M.; Terracciano, A. Analysis of Some Stilbenes in Italian Wines by Liquid Chromatography/Tandem Mass Spectrometry. *Rapid Commun. Mass Spectrom.* **2007,** *21* (18), 2955–2964.
20. Cantos, E.; Espín, J. C.; Fernández, M. J.; Oliva, J.; Tomás-Barberán, F. A. Postharvest UV-C-irradiated Grapes as a Potential Source for Producing Stilbene-enriched Red Wines. *J. Agric. Food Chem.* **2003,** *51* (5), 1208–1214.
21. Cantos, E.; Espín, J. C.; Tomás-Barberán, F. A. Postharvest Stilbene-enrichment of Red and White Table Grape Varieties Using UV-C Irradiation Pulses. *J. Agric. Food Chem.* **2002,** *50* (22), 6322–6329.
22. Chaher, N.; Arraki, K.; Dillinseger, E.; Temsamani, H.; Bernillon, S.; Pedrot, E.; Delaunay, J. C.; Merillon, J. M.; Monti, J. P.; Izard, J. C.; Atmani, D.; Richard, T. Bioactive Stilbenes from *Vitis vinifera* Grapevine Shoots Extracts. *J. Sci. Food Agric.* **2014,** *94* (5), 951–954.
23. Chai, Y. Y.; Wang, F.; Li, Y. L.; Liu, K.; Xu, H. Antioxidant Activities of Stilbenoids from *Rheum emodi* Wall. *Evid.-Based Complementary Altern. Med.* **2012,** *2012* (1), 603–678.
24. Chen, J. S.; Wang, Q.; Fu, X. H.; Huang, X. H.; Chen, X. L.; Cao, L. Q.; Chen, L. Z.; Tan, H. X.; Li, W.; Bi, J.; Zhang, L. J. Involvement of PI3K/PTEN/AKT/mTOR Pathway in Invasion and Metastasis in Hepatocellular Carcinoma: Association with MMP-9. *Hepatology Res.* **2009,** *39* (2), 177–186.

25. Choo, Q.-Y.; Yeo, S. C. M.; Ho, P. C.; Tanaka, Y.; Lin, H.-S. Pterostilbene Surpassed Resveratrol for Anti-inflammatory Application: Potency Consideration and Pharmacokinetics Perspective. *J. Funct. Foods* **2014,** *11*, 352–362.
26. Cunningham, J.; Haslam, E.; Haworth, R. 535. The Constitution of Piceatannol. *J. Chem. Soc.* **1963,** 2875–2883.
27. Daikonya, A.; Kitanaka, S. Polyphenols from Sophora Yunnanensis, and Their Inhibitory Effects on Nitric Oxide Production. *Chem. Pharm. Bull.* **2011,** *59* (12), 1567–1569.
28. Delaunay, J.-C.; Castagnino, C.; Chèze, C.; Vercauteren, J. Preparative Isolation of Polyphenolic Compounds from *Vitis vinifera* by Centrifugal Partition Chromatography. *J. Chromatogr. A* **2002,** *964* (1), 123–128.
29. Duan, D.; Halter, D.; Baltenweck, R.; Tisch, C.; Troster, V.; Kortekamp, A.; Hugueney, P.; Nick, P. Genetic Diversity of Stilbene Metabolism in Vitis Sylvestris. *J. Exp. Bot.* **2015,** *66* (11), 3243–3257.
30. Duarte, N.; Kayser, O.; Abreu, P.; Ferreira, M. J. Antileishmanial Activity of Piceatannol Isolated from Euphorbia Lagascae Seeds. *Phytother. Res.* **2008,** *22* (4), 455–457.
31. Fan, P.; Hay, A.-E.; Marston, A.; Lou, H.; Hostettmann, K. Chemical Variability of the Invasive Neophytes Polygonum Cuspidatum Sieb. and Zucc. and Polygonum Sachalinensis F. Schmidt ex Maxim. *Biochem. Syst. Ecol.* **2009,** *37* (1), 24–34.
32. Fernandes, I.; Faria, A.; Calhau, C.; de Freitas, V.; Mateus, N. Bioavailability of Anthocyanins and Derivatives. *J. Funct. Foods* **2014,** *7*, 54–66.
33. Ferrigni, N.; McLaughlin, J.; Powell, R.; Smith Jr., C. Use of Potato Disc and Brine Shrimp Bioassays to Detect Activity and Isolate Piceatannol as the Antileukemic Principle from the Seeds of Euphorbia Lagascae. *J. Nat. Products* **1984,** *47* (2), 347–352.
34. Gollücke, A. P. B.; Catharino, R. R.; de Souza, J. C.; Eberlin, M. N.; de Queiroz Tavares, D. Evolution of Major Phenolic Components and Radical Scavenging Activity of Grape Juices Through Concentration Process and Storage. *Food Chem.* **2009,** *112* (4), 868–873.
35. Graßmann, W.; Deffner, G.; Schuster, E.; Pauckner, W. Über den Gerbstoff der Fichtenrinde. *Chemische Berichte* **1956,** *89* (11), 2523–2535.
36. Grassmann, W.; Endres, H.; Pauckner, W.; Mathes, H. Über Die Gerbstoffe der Fichtenrinde, III: Chromatographische Trennung von Gerbstoffen an Einer Polyamidsäule: Isolierung Eines Zweiten Aglucons aus Dem Fichtenrindengerbstoff. *Chemische Berichte*, **1957,** *90* (6), 1125–1128.
37. Guerrero, R. F.; Biais, B.; Richard, T.; Puertas, B.; Waffo-Teguo, P.; Merillon, J. M.; Cantos-Villar, E. Grapevine Cane's Waste is a Source of Bioactive Stilbenes. *Ind. Crops. Products* **2016**, *94*, 884–892.
38. Guerrero, R. F.; Cantos-Villar, E.; Puertas, B.; Richard, T. Daily Preharvest UV-C Light Maintains the High Stilbenoid Concentration in Grapes. *J. Agric. Food Chem.* **2016,** *64* (25), 5139–5147.
39. Guerrero, R. F.; Puertas, B.; Fernández, M. I.; Palma, M.; Cantos-Villar, E. Induction of Stilbenes in Grapes by UV-C: Comparison of Different Subspecies of Vitis. *Innov. Food Sci. Emerg. Technol.* **2010,** *11* (1), 231–238.
40. Guerrero, R. F.; Puertas, B.; Jiménez, M. J.; Cacho, J.; Cantos-Villar, E. Monitoring the Process to Obtain Red Wine Enriched in Resveratrol and Piceatannol Without Quality Loss. *Food Chem.* **2010,** *122* (1), 195–202.
41. Hijona, E.; Aguirre, L.; Perez-Matute, P.; Villanueva-Millan, M. J.; Mosqueda-Solis, A.; Hasnaoui, M.; Nepveu, F.; Senard, J. M.; Bujanda, L.; Aldamiz-Echevarria, L.; Llarena, M.; Andrade, F.; Perio, P.; Leboulanger, F.; Hijona, L.; Arbones-Mainar, J. M.; Portillo,

M. P.; Carpene, C. Limited Beneficial Effects of Piceatannol Supplementation on Obesity Complications in the Obese Zucker Rat: Gut Microbiota, Metabolic, Endocrine, and Cardiac Aspects. *J. Physiol. BioChem.* **2016,** *72* (3), 567–582.

42. Houille, B.; Besseau, S.; Courdavault, V.; Oudin, A.; Glevarec, G.; Delanoue, G.; Guerin, L.; Simkin, A. J.; Papon, N.; Clastre, M.; Giglioli-Guivarc'h, N.; Lanoue, A. Biosynthetic Origin of E-Resveratrol Accumulation in Grape Canes During Postharvest Storage. *J. Agric. Food Chem.* **2015,** *63* (5), 1631–1638.
43. Howard, L.; Pandjaitan, N. Pressurized Liquid Extraction of Flavonoids from Spinach. *J. Food Sci.* **2008,** *73* (3), C151–C157.
44. Hu, B.; Liu, X.; Zhang, C.; Zeng, X. Food Macromolecule Based Nanodelivery Systems for Enhancing the Bioavailability of Polyphenols. *J. Food Drug Anal.* **2017,** *25* (1), 3–15.
45. Inagaki, H.; Ito, R.; Setoguchi, Y.; Oritani, Y.; Ito, T. Administration of Piceatannol Complexed with α-Cyclodextrin Improves Its Absorption in Rats. *J. Agric. Food Chem.* **2016**, *64* (18), 3557–3563.
46. Islam, M.; Ahn, S. Y.; Islam, M.; Kim, S.; Yun, H. Accumulation of Stilbene Compounds and Induction of Related Gene Expression by Hairy Vetch and ryegrass Extracts in Grape Berries. *Biologia* **2014,** *69* (12), 1693–1700.
47. Ito, T.; Abe, N.; Masuda, Y.; Nasu, M.; Oyama, M.; Sawa, R.; Takahashi, Y.; Iinuma, M. Two Novel Resveratrol Derivatives from the Leaves of Vateria Indica. *Helv. Chim. Acta* **2009,** *92* (1), 195–208.
48. Jo, S. P.; Kim, J. K.; Lim, Y. H. Antihyperlipidemic Effects of Rhapontin and Rhapontigenin from *Rheum undulatum* in Rats Fed a High-cholesterol Diet. *Planta Medica* **2014,** *80* (13), 1067–1071.
49. Kim, E. J.; Park, H.; Park, S. Y.; Jun, J. G.; Park, J. H. The Grape Component Piceatannol Induces Apoptosis in DU145 Human Prostate Cancer Cells Via the Activation of Extrinsic and Intrinsic Pathways. *J. Med. Food* **2009,** *12* (5), 943–951.
50. King, F.; King, T.; Godson, D.; Manning, L. The Chemistry of Extractives from Hardwoods. Part XXVIII. The Occurrence of 3: 4: 3′: 5′-Tetrahydroxy-and 3: 4: 5: 3′: 5′-Pentahydroxy-Stilbene in Vouacapoua Species. *J. Chem. Soc. (Summary)* **1956,** 4477–4480.
51. Kiselev, K. V.; Grigorchuk, V. P.; Ogneva, Z. V.; Suprun, A. R.; Dubrovina, A. S. Stilbene Biosynthesis in the Needles of Spruce Picea Jezoensis. *PhytoChemistry* **2016,** *131*, 57–67.
52. Kita, Y.; Miura, Y.; Yagasaki, K. Antiproliferative and Anti-invasive Effect of Piceatannol, a Polyphenol Present in Grapes and Wine, Against Hepatoma AH109A cells. *J. Biomed. Biotechnol.* **2012,** *2012*, 672–686.
53. Ko, H. S.; Lee, H. J.; Kim, S. H.; Lee, E. O. Piceatannol Suppresses Breast Cancer Cell Invasion Through the Inhibition of MMP-9: Involvement of PI3K/AKT and NF-KappaB Pathways. *J. Agric. Food Chem.* **2012,** *60* (16), 4083–4089.
54. Ko, S. K. A New Stilbene Diglycoside from *Rheum undulatum. Arch. Pharm. Res.* **2000,** *23* (2), 159–162.
55. Ko, S. K.; Whang, W. K.; Kim, I. H. Anthraquinone and Stilbene Derivatives from the Cultivated Korean Rhubarb Rhizomes. *Arch. Pharmacal. Res.* **1995,** *18* (4), 282–288.
56. Krafczyk, N.; Kötke, M.; Lehnert, N.; Glomb, M. A. Phenolic Composition of Rhubarb. *Eur. Food Res. Technol.* **2008,** *228* (2), 187–196.
57. Ku, K.-L.; Chang, P.-S.; Cheng, Y.-C.; Lien, C.-Y. Production of Stilbenoids from the Callus of Arachis Hypogaea: A Novel Source of the Anticancer Compound Piceatannol. *J. Agric. Food Chem.* **2005,** *53* (10), 3877–3881.

58. Kukreja, A.; Wadhwa, N. Therapeutic Role of Resveratrol and Piceatannol in Disease Prevention. *J. Blood Disord. Transf.* **2014,** *05* (09), 1–6.
59. Kuo, P. L.; Hsu, Y. L. The Grape and Wine Constituent Piceatannol Inhibits Proliferation of Human Bladder Cancer Cells Via Blocking Cell Cycle Progression and Inducing Fas/Membrane Bound Fas Ligand-mediated Apoptotic Pathway. *Mol. Nutr. Food Res.* **2008,** *52* (4), 408–418.
60. Kwon, G. T.; Jung, J. I.; Song, H. R.; Woo, E. Y.; Jun, J. G.; Kim, J. K.; Her, S.; Park, J. H. Piceatannol Inhibits Migration and Invasion of Prostate Cancer Cells: Possible Mediation by Decreased Interleukin-6 Signaling. *The J. Nutr. BioChem.* **2012,** *23* (3), 228–238.
61. Lai, T. N.; Andre, C.; Rogez, H.; Mignolet, E.; Nguyen, T. B.; Larondelle, Y. Nutritional Composition and Antioxidant Properties of the Sim Fruit (*Rhodomyrtus tomentosa*). *Food Chem.* **2015,** *168*, 410–6.
62. Lai, T. N.; Herent, M. F.; Quetin-Leclercq, J.; Nguyen, T. B.; Rogez, H.; Larondelle, Y.; Andre, C. M. Piceatannol, a Potent Bioactive Stilbene, as Major Phenolic Component in *Rhodomyrtus tomentosa. Food Chem.* **2013,** *138* (2–3), 1421–1430.
63. Imran, M.; Nadeem, M.; Khan, M. A.; Ahmed, S.; Imran, A.; Amir, R. M.; Arshad, M. U.; Rauf, A.; Saeed, F.; Khan, S.; Mehmood, Z.; Gilani, S. A.; Sulerial, H. A. R. Curcumin and Its Allied Analogues: Epigenetic and Health Perspectives: A Review. *Czech J. Food Sci.* **2017,** *35* (4), 285–310.
64. Lambert, C.; Richard, T.; Renouf, E.; Bisson, J.; Waffo-Teguo, P.; Bordenave, L.; Ollat, N.; Merillon, J. M.; Cluzet, S. Comparative Analyses of Stilbenoids in Canes of Major *Vitis vinifera* L. Cultivars. *J. Agric. Food Chem.* **2013,** *61* (47), 11392–11399.
65. Lavecchia, A.; Di Giovanni, C.; Novellino, E. STAT-3 Inhibitors: State of the Art and New Horizons for Cancer Treatment. *Curr. Med. Chem.* **2011,** *18* (16), 2359–2375.
66. Lee, D.; Cuendet, M.; Vigo, J. S.; Graham, J. G.; Cabieses, F.; Fong, H. H.; Pezzuto, J. M.; Kinghorn, A. D. A Novel Cyclooxygenase-inhibitory Stilbenolignan from the Seeds of *Aiphanes aculeata. Organic Lett.* **2001,** *3* (14), 2169–2171.
67. Lee, K. W.; Kim, H. J. Acteoside Inhibits Human Promyelocytic HL-60 Leukemia Cell Proliferation Via Inducing Cell Cycle Arrest at G0/G1 Phase and Differentiation into Monocyte. *Carcinogenesis* **2007,** *28* (9), 1928–1936.
68. Lee, Y. M.; Lim, D. Y.; Cho, H. J.; Seon, M. R.; Kim, J. K.; Lee, B. Y.; Park, J. H. Piceatannol, a Natural Stilbene from Grapes, Induces G1 Cell Cycle Arrest in Androgen-insensitive DU145 Human Prostate Cancer Cells Via the Inhibition of CDK Activity. *Cancer Lett.* **2009,** *285* (2), 166–173.
69. Les, F.; Deleruyelle, S.; Cassagnes, L. E.; Boutin, J. A.; Balogh, B.; Arbones-Mainar, J. M.; Biron, S.; Marceau, P.; Richard, D.; Nepveu, F.; Mauriège, P.; Carpéné, C. Piceatannol and Resveratrol Share Inhibitory Effects on Hydrogen Peroxide Release, Monoamine Oxidase and Lipogenic Activities in Adipose Tissue, but Differ in Their Antilipolytic Properties. *Chem.-Biol. Interact.* **2016,** *258*, 115–125.
70. Li, Y.; ZHANG, D. M.; YU, S. S. A New Stilbene from *Cercis chinensis* Bunge. *J. Integr. Plant Biol.* **2005,** *47* (8), 1021–1024.
71. Liu, G.; Xu, X.; Gong, Y.; He, L.; Gao, Y. Effects of Supercritical $CO_2$ Extraction Parameters on Chemical Composition and Free Radical-scavenging Activity of Pomegranate (*Punica granatum* L.) Seed Oil. *Food. Bioproducts Process.* **2012,** *90* (3), 573–578.
72. Liu, Y.; Harinantenaina, L.; Brodie, P. J.; Bowman, J. D.; Cassera, M. B.; Slebodnick, C.; Callmander, M. W.; Randrianaivo, R.; Rakotobe, E.; Rasamison, V. E.; Applequist,

W.; Birkinshaw, C.; Lewis, G. P.; Kingston, D. G. Bioactive Compounds from Stuhlmannia Moavi from the Madagascar Dry Forest. *Bioorganic Med. Chem.* **2013,** *21* (24), 7591–7594.

73. Ma, Y.; Kosinska-Cagnazzo, A.; Kerr, W. L.; Amarowicz, R.; Swanson, R. B.; Pegg, R. B. Separation and Characterization of Phenolic Compounds from Dry-blanched Peanut Skins by Liquid Chromatography-electrospray Ionization Mass Spectrometry. *J. Chromatogr. A* **2014,** *1356*, 64–81.
74. Maruki-Uchida, H.; Kurita, I.; Sugiyama, K.; Sai, M.; Maeda, K.; Ito, T. The Protective Effects of Piceatannol from Passion Fruit (*Passiflora edulis*) seeds in UVB-irradiated Keratinocytes. *Biol. Pharmaceut. Bull.* **2013,** *36* (5), 845–849.
75. Matsui, Y.; Sugiyama, K.; Kamei, M.; Takahashi, T.; Suzuki, T.; Katagata, Y.; Ito, T. Extract of Passion Fruit (*Passiflora edulis*) Seed Containing High Amounts of Piceatannol Inhibits Melanogenesis and Promotes Collagen Synthesis. *J. Agric. Food Chem.* **2010,** *58* (20), 11112–11118.
76. Miksits, M.; Maier-Salamon, A.; Vo, T. P.; Sulyok, M.; Schuhmacher, R.; Szekeres, T.; Jager, W. Glucuronidation of Piceatannol by Human Liver Microsomes: Major Role of UGT1A1, UGT1A8 and UGT1A10. *J. Pharm. Pharmacol.* **2010,** *62* (1), 47–54.
77. Miksits, M.; Sulyok, M.; Schuhmacher, R.; Szekeres, T.; Jager, W. In-vitro Sulfation of Piceatannol by Human Liver Cytosol and Recombinant Sulfotransferases. *J. Pharm. Pharmacol.* **2009,** *61* (2), 185–191.
78. Moss, R.; Mao, Q.; Taylor, D.; Saucier, C. Investigation of Monomeric and Oligomeric Wine Stilbenoids in Red Wines by Ultra-high-performance Liquid Chromatography/Electrospray Ionization Quadrupole Time-of-Flight Mass Spectrometry. *Rapid Commun. Mass Spectrom.* **2013,** *27* (16), 1815–1827.
79. Muankaew, C.; Jansook, P.; Stefánsson, E.; Loftsson, T. Effect of γ-Cyclodextrin on Solubilization and Complexation of Irbesartan: Influence of pH and Excipients. *Int. J. Pharm.* **2014,** *474* (1–2), 80–90.
80. Murias, M.; Jäger, W.; Handler, N.; Erker, T.; Horvath, Z.; Szekeres, T.; Nohl, H.; Gille, L. Antioxidant, Prooxidant and Cytotoxic Activity of Hydroxylated Resveratrol Analogues: Structure–Activity Relationship. *Biochem. Pharmacol.* **2005,** *69* (6), 903–912.
81. Mustafa, A.; Turner, C. Pressurized Liquid Extraction as a Green Approach in Food and Herbal Plants Extraction: A Review. *Anal. Chim. Acta* **2011,** *703* (1), 8–18.
82. Ovesna, Z.; Kozics, K.; Bader, Y.; Saiko, P.; Handler, N.; Erker, T.; Szekeres, T. Antioxidant Activity of Resveratrol, Piceatannol and 3,3',4,4',5,5'-Hexahydroxy-trans-stilbene in Three Leukemia Cell Lines. *Oncol. Rep.* **2006,** *16* (3), 617–624.
83. Pervaiz S. Resveratrol: From Grapevines to Mammalian Biology. *FASEB J.* **2003,** *17* (1), 1975–1985
84. Piano, F.; Bertolone, E.; Pes, D.; Asproudi, A.; Borsa, D. Focusing on Bioactive Compounds in Grapes: Stilbenes in *Uvalino* cv. *Eur. Food Res. Technol.* **2013,** *237* (6), 897–904.
85. Piotrowska, H.; Kucinska, M.; Murias, M. Biological Activity of Piceatannol: Leaving the Shadow of Resveratrol. *Mutat. Res.* **2012,** *750* (1), 60–82.
86. Piver, B.; Fer, M.; Vitrac, X.; Merillon, J. M.; Dreano, Y.; Berthou, F.; Lucas, D. Involvement of Cytochrome P450 1A2 in the Biotransformation of Trans-resveratrol in Human Liver Microsomes. *Biochem. Pharmacol.* **2004,** *68* (4), 773–782.
87. Potter, G. A.; Patterson, L. H.; Wanogho, E.; Perry, P. J.; Butler, P. C.; Ijaz, T.; Ruparelia, K. C.; Lamb, J. H.; Farmer, P. B.; Stanley, L. A.; Burke, M. D. The Cancer Preventative

Agent Resveratrol is Converted to the Anticancer Agent Piceatannol by the Cytochrome P450 Enzyme CYP1B1. *Br. J. Cancer* **2002,** *86* (5), 774–778.

88. Prado, I. M.; Prado, G. H. C.; Prado, J. M.; Meireles, M. A. A. Supercritical $CO_2$ and Low-pressure Solvent Extraction of Mango (*Mangifera indica*) Leaves: Global Yield, Extraction Kinetics, Chemical Composition and Cost of Manufacturing. *Food Bioprod. Process.* **2013,** *91 (4)*, 656–664.
89. Pussa, T.; Raudsepp, P.; Kuzina, K.; Raal, A. Polyphenolic Composition of Roots and Petioles of Rheum Rhaponticum L. *Phytochem. Anal.* **2009,** *20* (2), 98–103.
90. Rao, V. S. S.; Rajadurai, S. Isolation of 3,4,3',5'-Tetrahydroxystilbene (Piceatannol) from *Cassia marginata* Heartwood. *Aust. J. Chem.* **1968,** *21*, 1921–1922.
91. Renaud, S.; de Lorgeril, M. Wine, Alcohol, Platelets, and the French Paradox for Coronary Heart Disease. *Lancet* **1992,** *339* (8808), 1523–6.
92. Reuters (Research and Markets). Phytonutrients Market, 2015: Carotenoids, Phytosterols, Flavonoids, Phenolic Compounds, and Vitamins: Global Trends and Forecast to 2020. http://www.reuters.com/article/researchandmarketsidUSnBw146199a+100+BSW20150714 (accessed April 12, 2017).
93. Rimando, A. M.; Cody, R. Determination of Stilbenes in Blueberries. *LCGC North America* **2005,** *23* (11), 1192–1200.
94. Rimando, A. M.; Kalt, W.; Magee, J. B.; Dewey, J.; Ballington, J. R. Resveratrol, Pterostilbene, and Piceatannol in Vaccinium berries. *J. Agric. Food Chem.* **2004,** *52* (15), 4713–4719.
95. Rokaya, M. B.; Maršík, P.; Münzbergová, Z. Active Constituents in *Rheum acuminatum* and *Rheum australe* (Polygonaceae) Roots: A Variation Between Cultivated and Naturally Growing Plants. *Biochem. System. Ecol.* **2012,** *41*, 83–90.
96. Rosso, M. D.; Soligo, S.; Panighel, A.; Carraro, R.; Vedova, A. D.; Maoz, I.; Tomasi, D.; Flamini, R. Changes in Grape Polyphenols (*V. vinifera* L.) as a Consequence of Post-harvest Withering by High-resolution Mass Spectrometry: Raboso Piave Versus Corvina. *J. Mass Spectrom.* **2016,** *51* (9), 750–760.
97. Roupe, K.; Teng, X. W.; Fu, X.; Meadows, G. G.; Davies, N. M. Determination of Piceatannol in Rat Serum and Liver Microsomes: Pharmacokinetics and Phase I and II Biotransformation. *Biomed. Chromatogr. BMC* **2004,** *18* (8), 486–491.
98. Roupe, K. A.; Yáñez, J. A.; Teng, X. W.; Davies, N. M. Pharmacokinetics of Selected Stilbenes: Rhapontigenin, Piceatannol and Pinosylvin in Rats. *J. Pharm. Pharmacol.* **2006,** *58* (11), 1443–1450.
99. Rusjan, D.; Mikulic-Petkovsek, M. Phenolic Responses in 1-Year-Old Canes of *Vitis vinifera* cv. Chardonnay Induced by Grapevine Yellows (Bois noir). *Aust. J. Grape Wine Res.* **2015,** *21* (1), 123–134.
100. Ryu, G.; Ju, J. H.; Park, Y. J.; Ryu, S. Y.; Choi, B. W.; Lee, B. H. The Radical Scavenging Effects of Stilbene Glucosides from Polygonum Multiflorum. *Arch. Pharmacal Res.* **2002,** *25* (5), 636–639.
101. Sano, S.; Sugiyama, K.; Ito, T.; Katano, Y.; Ishihata, A. Identification of the Strong Vasorelaxing Substance Scirpusin B, a Dimer of Piceatannol, from Passion Fruit (*Passiflora edulis*) Seeds. *J. Agric Food Chem.* **2011,** *59* (11), 6209–13.
102. Santos, D. T.; Veggi, P. C.; Meireles, M. A. A. Optimization and Economic Evaluation of Pressurized Liquid Extraction of Phenolic Compounds from Jabuticaba Skins. *J. Food Eng.* **2012,** *108* (3), 444–452.

103. Santos, P.; Aguiar, A. C.; Barbero, G. F.; Rezende, C. A.; Martinez, J. Supercritical Carbon Dioxide Extraction of Capsaicinoids from Malagueta Pepper (*Capsicum frutescens* L.) Assisted by Ultrasound. *Ultrason. SonoChem.* **2015,** *22*, 78–88.
104. Schmeda-Hirschmann, G.; Gutikrrez, M. I.; Loyola, J. I.; Zúniga, J. Biological Activity and Xanthine Oxidase Inhibitors from *Scirpus californicus* (CA Mey.) Steud. *Phytother. Res.* **1996,** *10*, 683–685.
105. Imran, M.; Ullah, A.; Saeed, F.; Nadeem, M.; Arshad, M. U. Suleria, H. A. R. Cucurmin, Anticancer, & Antitumor Perspectives: A Comprehensive Review. *Crit. Rev. Food Sci. Nutr.* **2016,** *7* (1), 1–23.
106. Septembre-Malaterre, A.; Stanislas, G.; Douraguia, E.; Gonthier, M. P. Evaluation of Nutritional and Antioxidant Properties of the Tropical Fruits Banana, Litchi, Mango, Papaya, Passion Fruit and Pineapple Cultivated in Reunion French Island. *Food Chem.* **2016,** *212*, 225–233.
107. Sergent, T.; Kohnen, S.; Jourez, B.; Beauve, C.; Schneider, Y.-J.; Vincke, C. Characterization of Black Locust (Robinia Pseudoacacia L.) Heartwood Extractives: Identification of Resveratrol and Piceatannol. *Wood Sci. Technol.* **2014,** *48 (5)*, 1005–1017.
108. Setoguchi, Y.; Oritani, Y.; Ito, R.; Inagaki, H.; Maruki-Uchida, H.; Ichiyanagi, T.; Ito, T. Absorption and Metabolism of Piceatannol in Rats. *J. Agric. Food Chem.* **2014,** *62* (12), 2541–2548.
109. Seyed, M. A.; Jantan, I.; Bukhari, S. N.; Vijayaraghavan, K. A Comprehensive Review on the Chemotherapeutic Potential of Piceatannol for Cancer Treatment, with Mechanistic Insights. *J. Agric. Food Chem.* **2016,** *64* (4), 725–737.
110. Shiratake, S.; Nakahara, T.; Iwahashi, H.; Onodera, T.; Mizushina, Y. Rose Myrtle (*Rhodomyrtus tomentosa*) Extract and Its Component, Piceatannol, Enhance the Activity of DNA Polymerase and Suppress the Inflammatory Response Elicited by UVB Induced DNA Damage in Skin Cells. *Mol. Med. Rep.* **2015,** *12* (4), 5857–5864.
111. Son, P. S.; Park, S. A.; Na, H. K.; Jue, D. M.; Kim, S.; Surh, Y. J. Piceatannol, a Catechol-type Polyphenol, Inhibits Phorbol Ester-induced NF-{kappa}B Activation and Cyclooxygenase-2 Expression in Human Breast Epithelial Cells: Cysteine 179 of IKK{beta} as a Potential Target. *Carcinogenesis* **2010,** *31* (8), 1442–1449.
112. Sperker, B.; Backman, J. T.; Kroemer, H. K. The Role of β-Glucuronidase in Drug Disposition and Drug Targeting in Humans. *Clin. Pharm.* **1997,** *33* (1), 18–31.
113. Surh, Y. J.; Na, H. K. Therapeutic Potential and Molecular Targets of Piceatannol in Chronic Diseases. *Adv. Exp. Med. Biol.* **2016,** *928*, 185–211.
114. Tung, Y. C.; Lin, Y. H.; Chen, H. J.; Chou, S. C.; Cheng, A. C.; Kalyanam, N.; Ho, C. T.; Pan, M. H. Piceatannol Exerts Anti-obesity Effects in C57BL/6 Mice Through Modulating Adipogenic Proteins and Gut Microbiota. *Molecules* **2016,** *21* (11), E1419.
115. Uchida-Maruki, H.; Inagaki, H.; Ito, R.; Kurita, I.; Sai, M.; Ito, T. Piceatannol Lowers the Blood Glucose Level in Diabetic Mice. *Biol. Pharm. Bull.* **2015,** *38* (4), 629–633.
116. Vastano, B. C.; Chen, Y.; Zhu, N.; Ho, C.-T.; Zhou, Z.; Rosen, R. T. Isolation and Identification of Stilbenes in Two Varieties of *Polygonum Cuspidatum. J. Agric. Food Chem.* **2000,** *48* (1), 253–256.
117. Viganó, J.; Aguiar, A. C.; Moraes, D. R.; Jara, J. L. P.; Eberlin, M. N.; Cazarin, C. B. B.; Maróstica, M. R.; Martínez, J. Sequential High Pressure Extractions Applied to Recover Piceatannol and Scirpusin B from Passion Fruit Bagasse. *Food Res. Int.* **2016,** 85, 51–58.
118. Viganó, J.; Coutinho, J. P.; Souza, D. S.; Baroni, N. A. F.; Godoy, H. T.; Macedo, J. A.; Martínez, J. Exploring the Selectivity of Supercritical CO2 to Obtain Nonpolar Fractions of Passion Fruit Bagasse Extracts. *The J. Supercrit. Fluids* **2016,** *110*, 1–10.

119. Viganó, J.; Martinez, J. Trends for the Application of Passion Fruit Industrial by-Products: A Review on the Chemical Composition and Extraction Techniques of Phytochemicals. *Food Public Health* **2015,** *5* (5), 164–173.
120. Vinas, P.; Campillo, N.; Martinez-Castillo, N.; Hernandez-Cordoba, M. Solid-phase Microextraction On-fiber Derivatization for the Analysis of Some Polyphenols in Wine and Grapes Using Gas Chromatography–Mass Spectrometry. *J. Chromatogr. A* **2009,** *1216* (9), 1279–1284.
121. Vincenzi, S.; Tomasi, D.; Gaiotti, F.; Lovat, L.; Giacosa, S.; Torchio, F.; Segade, S. R.; Rolle, L. Comparative Study of the Resveratrol Content of Twenty-one Italian Red Grape Varieties. *South African J. Enol. Viticult.* **2016,** *34* (1), 30–35.
122. Rauf, A.; Imran, M.; Suleria, H. A. R.; Ahmad, B.; Peters, D. G.; Mubarak, M. S. A Comprehensive Review of the Health Perspectives of Resveratrol. *Food Funct.* **2017,** *8* (12), 4284–4305.
123. Wei, Y. J.; Zhao, S. R.; Li, J. M.; Xue, B. Stilbene Profiles in Different Tissues of *Vitis vinifera* L. cv. Cabernet Sauvignon and a Comparison of Their Antioxidant Activity. *Aust. J. Grape Wine Res.* **2016,** *22* (2), 226–231.
124. Wells, P. G.; Mackenzie, P. I.; Roy Chowdhury, J.; Guillemette, C.; Gregory, P. A.; Ishii, Y.; Hansen, A. J.; Kessler, F. K.; Kim, P. M.; Roy Chowdhury, N.; Ritter, J. K. Glucuronidation and the UDP-glucuronosyltransferases in Health and Disease. *Drug Metab. Disp.* **2004,** *32* (3), 281–290.
125. Wolter, F.; Clausnitzer, A.; Akoglu, B.; Stein, J. Piceatannol, a Natural Analog of Resveratrol, Inhibits Progression Through the S Phase of the Cell Cycle in Colorectal Cancer Cell Lines. *J. Nutr.* **2002,** *132* (2), 298–302.
126. Xie, L.; Bolling, B. W. Characterisation of Stilbenes in California Almonds (*Prunus dulcis*) by UHPLC-MS. *Food Chem.* **2014,** *148*, 300–306.
127. Yang, T.; Fang, L.; Nopo-Olazabal, C.; Condori, J.; Nopo-Olazabal, L.; Balmaceda, C.; Medina-Bolivar, F. Enhanced Production of Resveratrol, Piceatannol, Arachidin-1, and Arachidin-3 in Hairy Root Cultures of Peanut Co-Treated with Methyl Jasmonate and Cyclodextrin. *J. Agric. Food Chem.* **2015,** *63* (15), 3942–3950.
128. Yoo, M. Y.; Oh, K. S.; Lee, J. W.; Seo, H. W.; Yon, G. H.; Kwon, D. Y.; Kim, Y. S.; Ryu, S. Y.; Lee, B. H. Vasorelaxant Effect of Stilbenes from Rhizome Extract of Rhubarb (*Rheum undulatum*) on the Contractility of Rat Aorta. *Phytother. Res.* **2007,** *21* (2), 186–189.
129. Yoon, J. H.; Baek, S. J. Molecular Targets of Dietary Polyphenols with Anti-inflammatory Properties. *Yonsei Med. J.* **2005,** *46* (5), 585–596.
130. Yoshizumi, M.; Tsuchiya, K.; Suzaki, Y.; Kirima, K.; Kyaw, M.; Moon, J.-H.; Terao, J.; Tamaki, T. Quercetin Glucuronide Prevents VSMC Hypertrophy by Angiotensin II Via the Inhibition of JNK and AP-1 Signaling Pathway. *Biochem. Biophys. Res. Commun.* **2002,** *293* (5), 1458–1465.
131. Zhu, C.; Zhu, Q.; Wu, Z.; Yin, Y.; Kang, D.; Lu, S.; Liu, P. Isorhapontigenin Induced Cell Growth Inhibition and Apoptosis by Targeting EGFR-related Pathways in Prostate Cancer. *J. Cell. Physiol.* **2018,** *233* (2), 1104–1119.

## CHAPTER 9

# PLANT-BASED NEUROTOXINS: IMPACT ON NEURAL PATHOLOGIES

SEEMA PATEL, GIRISH KUMAR GUPTA, ABDUR RAUF, HAROON KHAN, and HAFIZ ANSAR RASUL SULERIA

### ABSTRACT

Several plant-based phytochemicals are neurotoxins which can cause chronic or acute degenerative locomotor diseases such as Parkinson's disease, Alzheimer's disease, and amyotrophic lateral sclerosis. Hallucinations, seizures, convulsions, coma, and death have been linked to the neurotoxins belonging to alkaloid, nonstandard amino acids, sesquiterpenes, acetogenins, saponins, cyanogenic glycosides, etc. Literature review indicates that there is only a blurred line between plant-derived neuroprotectants and neurotoxins; therefore, one ought to be highly aware of the risk while using them as foods or medications. Dosage of toxins is one cardinal factor that can inhibit or excite sodium–potassium pumps and impact neurons. Well-studied neurotoxins are β-N-oxalyl-L-α, β-diaminopropionic acid, annonacin, anisatin, tutin, aconitine, tropane alkaloids, pyrethrins, and grayanotoxin. However, most of them are yet to be characterized. This review chapter presents critical insights on plant-based neurotoxins and their role in neural health aggravation.

### 9.1 INTRODUCTION

Neural pathologies include dementia, mania, bipolar disorder, seizure, trigeminal neuralgia, psychosis, schizophrenia, depression, myalgic encephalomyelitis, Alzheimer's disease, Parkinson's disease, amyotrophic lateral sclerosis (ALS), autism, attention deficit hyperactivity disorder (ADHD), multiple sclerosis, epilepsy, ischemic brain stroke, subarachnoid hemorrhage. Parkinson's disease or paralysis agitansis is mostly a geriatric, neurodegenerative disease[58]; and it affects motor abilities like walking, balancing, swallowing, tremor, bradykinesia (slow spontaneous movement), and pain.[73] Comorbidities include cognitive decline, osteoporosis, etc.[53] Degeneration of dopaminergic neurons in midbrain *substantia nigra* initiates this disease. Consequently, cholinergic circuits and prefrontal dopaminergic systems malfunction, leading to low dopamine level.[49] Low level of the neurotransmitter dopamine is one hallmark of this disease. Mutations in profilin, *PINK-1,* and *Parkin* gene, abnormality of mitochondria functions, ubiquitin–proteasome, and autophagy system are causes of this disease.[22,38,50,78]

Administration of L-3,4-dihydroxyphenylalanine (L-DOPA) (a levodopa), dopaminergic agonists (dopamine receptor agonists) and monoamine oxidase-B (MAO-B) inhibitors are possible therapies for Parkinson's disease.[20] Ergot alkaloid derivatives are one source of therapeutics. However, the available therapeutic options have side effects like dyskinesia and dystonia. Alzheimer's disease is a cognitive-degeneration disease that is caused by amyloid beta production and uncontrolled posttranslational modification of *tau* protein.[29]

*Tau* protein is needed for the stabilization of microtubules. However, the hyperphosphorylation and aggregation of *tau* results in neural pathology (taupathy).[29] ALS is a late-onset progressive, fatal (death within 2–5 years of symptom onset) neurodegenerative disease that destroys motor neurons, both upper and lower.[23,72] Among multiple etiologies of this pathology, the variation in mutations in superoxide dismutase 1 (SOD1) has been identified as one cause.[81] The degeneration of cortical and spinal motor neurons is common hallmark of this disease.[81] The loss of anterior horn cells affects multiple functions. Weakness, cramps, muscle atrophy, hyperreflexia, and hyporeflexia are some of the symptoms. Pathologic mechanisms of these three neural diseases have been illustrated in Figure 9.1.

## 9.2 PLANTS AS NEUROPROTECTANTS

Medicinal plants have been identified based on pharmacopeia (ayurvedic, traditional Chinese medicine (TCM), and others), and ethnobotanical surveys followed by biological assays.[62] Velvet bean (*Mucuna pruriens* L.) and Indian ginseng (*Withania somnifera* L.) have been validated as protectants of neurons.[25,53] Neuroactive properties of eugenol, pulegone, and citronellal have been found through insect models. These essential oil components interfere with central nervous function of the tested flies by blocking sodium channels and facilitating gamma-amino butyric acid

(GABA) action.[75] Tetramethylpyrazine (TMP), an alkaloid, is isolated from *Ligusticum chuanxiong*, which inhibits the production of nitric oxide, tumor necrosis factor α (TNF-α), interleukin 1β (IL-1β), monocyte chemoattractant protein-1, and intracellular reactive oxygen species from primary microglial cells.[28] The in vitro study has shown that TMP significantly blocked reactive oxygen species (ROS) generation and phosphorylation of Akt, which might be therapeutic for Alzheimer's disease.[28]

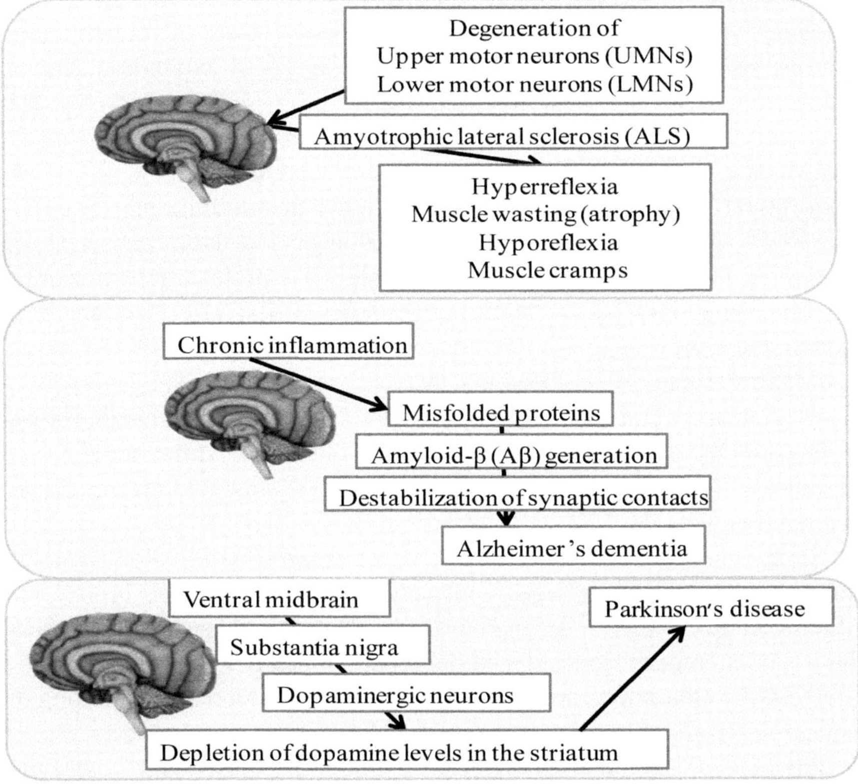


**FIGURE 9.1 (See color insert.)** Mechanism of Parkinson's disease, Alzheimer's disease, and amyotrophic lateral sclerosis.

Alkaloids from *Dendrobium* at a dose of 2.5 mg/L prevented caspase activity, lowered apoptosis, blocked the decrease in mitochondrial membrane potential, and increased Ca(2+) homeostasis.[77] Harmine alkaloids from the seeds of *Peganum harmala*, a Zygophyllaceae family member, at a dose of 5 mg/kg, prevented DNA fragmentation in the frontotemporal cortex of the brain in mice model. Also, the alkaloid inhibited enzyme MAO-A that metabolizes epinephrine, serotonin, and other monoamines, thus promoting brain activity. This effect might be promising for Alzheimer's disease management.[2]

*Combretum leprosum* ethanolic extract prevented dopamine deficit and consequent motor anomalies.[46] Tenuifolin, derived from the hydrolysate of polygalasaponins from *Polygala tenuifolia,* protected against amyloid beta25-35 (Aβ25-35) peptides in vivo[39,80] Ginseng (*Panax* sp.), *Eleutherococcus*, and *Rhodiola rosea* extracts have been identified as neuroprotective in Parkinson's disease.[3] Panaxatriol saponins from notoginseng (*Panax notoginseng*) conferred neuroprotection against loss of dopaminergic neurons which are involved the increment in Trx-1 expression and the suppression of cyclooxygenase-2 (COX-2).[40]

In TCM medication, plants from *Acanthopanax*, *Alpinia*, and *Astragalus* genus have been used for Parkinson therapy.[36] Curcumin, a polyphenol from turmeric (*Curcuma longa*), can cross the blood–brain barrier (BBB) and exerts neuroprotective effect on Parkinson's disease.[48] Piperine, an alkaloid in fruit of *Piper nigrum*, has been shown to protect rat hippocampus by decreasing lipid peroxidation and acetylcholinesterase enzyme.[7] A rat model study showed that curcumin along with piperine exhibited protection against olfactory bulbectomy-induced depression.[60] The intervention mechanism was found to be via modulation of oxidative-nitrosative stress-induced neuroinflammation and apoptosis.[60] Standardized flavonoid extract of safflower (*Carthamus tinctorius* L.) protected against Parkinson's disease in animal models.[59]

Senegenin from root extracts of *Polygala tenuifolia* (radix Polygalae) has shown neuroprotective effects. Senegenin-rich extract is used in TCM for cognitive improvement.[79] In another study, denegenin was able to attenuate the ischemia–reperfusion-induced neuronal apoptosis by upregulating Rho GDP dissociation inhibitor alpha (Rho-GDIα) expression and inhibiting the Jun amino-terminal kinases (JNK) pathway.[37]

Phytochemicals (such as resveratrol, curcumin, ginsenoside, catechins, etc.) have been validated to protect dopaminergic neurons.[16,70] The mechanisms involve prevention of anti-inflammatory cytokines and inhibition of microglial activation.[16] The neuroprotective activity of green tea polyphenol (-)-epigallocatechin-3-gallate (EGCG) has been widely reviewed.[41,52] A polyphenolic-rich extract derived from elderberry (*Sambucus nigra*) protected the neuronal cells against oxidizing agents.[70] The extract from the fruit and root bark of boxthorn (*Lycium chinense* Miller) protected neurons against rotenone-induced toxicity as studied through pheochromocytoma

(PC12) cell line.[21] The neuroprotective mechanism was via attenuation of caspase activation and mitochondrial membrane depolarization.[21]

In an in vivo study, polyethoxylated flavones, procyanidins, and isoflavones exposure protected the nigrostriatal dopaminergic neurons in Parkinson's brain.[11] Anthocyanin- and proanthocyanidin-rich botanical extracts were able to alleviate neurodegeneration in Parkinson's disease by optimizing mitochondrial functions.[71] Plant-derived natural products as therapy for Parkinson's disease has been reviewed.[65] Protective role of phytochemicals has been illustrated in Figure 9.2.

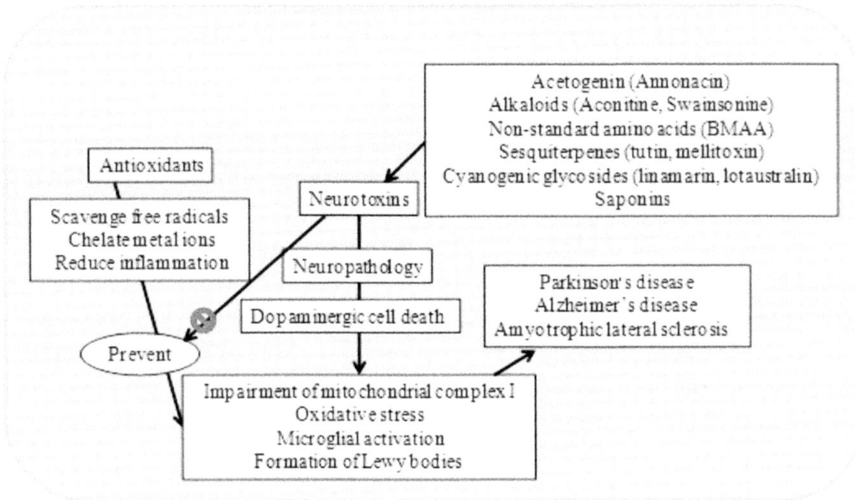


**FIGURE 9.2 (See color insert.)** Neuroprotectant as well as neurotoxic property of phytochemicals.

## 9.3 PHYTOCHEMICALS AS NEUROTOXINS

Plants express neurotoxins in response to herbivory.[9] Several plants have been identified as sources of neurotoxins in mammals. Unfortunately, people ignorant of the threats keep subsisting on them, or take as medications. Some of the key plant-derived neurotoxins are discussed in this section. Lathyrism is neural pathology caused by excessive consumption of *Lathyrus sativus* (grass pea or *khesari dal*). In India, Nepal, Bangladesh, and Ethiopia, low-income people consume it as staple food and they have been victim of the toxicity.[27] Now the role of β-N-oxalyl-L-α, β-diaminopropionic acid (L-ODAP; a non-standard amino acid) in the neurotoxicity has been proven. L-ODAP occurs in seeds of legume. Another excitatory amino acid, β-N-oxalyl amino-L-alanine (L-BOAA), has also been implicated in neurolathyrism. These toxins affect upper motor neurons and cause spastic paraparesis of the lower limbs, a flaccid paralysis.[68]

Annonacin from Annonaceae family has been validated as a neurotoxin. Tropical fruit plants such as *Annona muricata* (soursop), *Asimina triloba* (pawpaw), etc. have this toxin in leaves, raw fruit, seeds, roots, and bark.[56] This toxin, an acetogenin (polyketide), is lipophilic and it acts as mitochondrial complex I inhibitor. The toxin can traverse BBB and can damage cortical neurons and basal ganglia.[32] Dopaminergic neurons (in the *substantia nigra*) and cholinergic, GABAergic neurons (in the striatum) are affected.[5] Even nanomolar concentrations of annonacin is toxic to mitochondria.[32] Like most acetogenins, it hampers mitochondrial energy production. In the Caribbean island of Guadeloupe, a typical parkinsonism (levodopa-unresponsive, akinetic-rigid syndrome) has been linked to annonacins.[5,32] Seventy percent of Parkinsonian conditions in Guadeloupe is attributed to annonacin ingestion.[33]

Other acetogenin compounds are asimicin, bullatacin, and bullatalicin.[55] Asimicin and bullatacin have bis-tetrahydrofuran core and high cytotoxicity against cancer cell lines.[69] The Chamorro people of Guam (a US island territory in the Western Pacific) often develop the cluster of neurodegenerative diseases such as Alzheimer, Parkinson, and ALS.[4] The causes have been attributed to the consumption of cycad (*Cycas micronesica*) seeds, because the flour of cycad contains neurotoxins.

The source of the toxins has been found to be the symbiotic blue–green alga *Nostoc* inhabiting the specialized roots of *Cycas* plant. The alga generates a nonprotein amino acid β-N-methylamino-L-alanine (BMAA).[4] Also, it is mentioned that people consume flying foxes, who forage on these seeds.[10] An in vivo study showed that cycad flour-fed rats suffered the symptoms of parkinsonism.[66] Direct and biomagnification incorporation of BMAA is attributed to the adverse neural effects. Apart from BMAA, the sterols and steryl glucosides in the *Cycas* megagametophyte tissue have been proven to be neurotoxic.[42]

The fruits of Japanese star anise (*Illicium anisatum*) from family Schisandraceae possess neurotoxin anisatin. People get exposed to the sesquiterpene dilactone when it is used as an adulterant in Chinese star anise (*Illicium verum*). Anisatin and its derivatives act as GABA antagonists.[64] This toxin has the potential to cause epileptic convulsions, hallucinations, and nausea.[43,67] Palmyrah (*Borassus flabellifer* L.) shoot flour has been verified to possess neurotoxic effects. Water-soluble saponins have been identified as the neurotoxins.[26]

Honey is considered a wholesome food and therapeutic agent. But, honey contaminated with tutin, a sesquiterpenoid, is detrimental for health. Tutin, an oxygenated sesquiterpene picrotoxane, is a neurotoxin, which when ingested causes nausea, vomiting, dizziness, or seizures.[15] Some plants with tutin include: shrub tutu (*Coriaria ruscifolia* subsp. Ruscifolia) from the family Coriariaceae.[17,54] Role of bee in the transformation of tutin has been reported.[34] Essential oils of Myristicaceae family plants have hallucinogenic and delirium-inducing alkaloids via neural manipulations. Myristin in nutmeg (*Myristica frgarans*) has been reported to cause hallucinations.[12]

Aconitine, an arrhythmogenic site 2 neurotoxin in monkshood (*Aconitum napellus*) root triggers voltage-gated $Na^+$ channels (acts as a sodium channel agonist), leading to poisoning.[57] Some Aconitum species also contain an alkaloid called lappaconitine, structurally similar to aconitine. Cassava (*Manihot esculenta* Crantz) root is used as a food source.[18] However, it contains neurotoxic cyanogenic glycoside (linamarin and lotaustralin) which can cause cyanide poisoning.[18] In rat models, linamarin exerted damage to central nervous system (CNS) motor activity.[61]

Tropane alkaloids from *Brugmansia* spp. and several other Solanaceae family members meddle with neuroreceptors and ion channels of insects.[8] Tropane alkaloids like atropine, hyoscyamine, and scopolamine from *Datura* affect the CNS.[30] Swainsonine, an indolizidine triol alkaloid from locoweed (*Astragalus sp.*) from Fabaceae family, when chronically consumed by livestock leads to the inhibition of lysosomal α-mannosidase (AMA) and exerts toxic effect on dopaminergic cells.[35] Pyrethrins from Asteraceae plant family target sodium channels of honeybees and other insects, by inhibiting the voltage-gated sodium channels in nerve axons, thus hampering their mobility.[14] Therefore, pyrethrins are also potential neurotoxins for humans. In fact, synthetic version of pyrethrins, the pyrethroids used as insecticides (such as allethrin) are toxic to mammals.[76]

Lycopsamine and allied pyrrolizidine alkaloids in the Boraginaceae family and in the Eupatorieae family have been linked to encephalitis.[13] *Ericaceae* family *plants (such as Rhododendron, Pieris, Agarista,* and *Kalmia)* contain grayanotoxins. This diterpene has been associated with dizziness, mood alterations, and vascular manipulations via recurrent sodium ion channels excitations.[19,24] Phytochemicals as neurotoxins are presented in Figure 9.2.

The Table 9.1 indicates some key neurotoxins, their biochemical class, and plant families.

**TABLE 9.1** Plant Neurotoxins: Families and Their Biochemical Class.

| Neurotoxins | Plant family | Plant parts | References |
|---|---|---|---|
| Aconitine<br>Lappaconitine | Ranunculaceae | Root | [57] |
| Anisatin | Schisandraceae | Fruits | [64] |
| Annonacin | Annonaceae | Seeds, leaves raw, fruit, roots, bark | [32] |
| Beta-N-methylamino-L-alanine (BMAA)<br>Sterols Steryl glucosides | Cycadaceae<br>(blue–green algae Nostoc) | Seeds | [4]<br>[42] |
| Beta-N-oxalyl-L-α, β-diaminopropionic acid Beta-N-oxalyl amino-L-alanine | Fabaceae | Pulse | [68] |
| Cyanogenic glycosides (linamarin and lotaustralin) | Euphorbiaceae | | [61] |
| Essential oil | Myristicaceae | Seed | [12] |
| Grayanotoxins | *Ericaceae* | | [19,24] |
| Lycopsamine | Boraginaceae<br>Eupatorieae | | [13] |
| Pyrethrins | Asteraceae | – | [14] |
| Saponins | Arecaceae | Shoot | [26] |
| Swainsonine | Fabaceae | – | [35] |
| Tropane alkaloids (atropine, hyoscyamine, scopolamine) | Solanaceae | – | [30] |
| Tutin | Coriariaceae | – | [17] |

## 9.4 DUAL EFFECTS ON NEURAL SYSTEM

Wormwood (*Artemisia absinthium* L.) has demonstrated both neurotoxic and neuroprotective properties. Thujone, a monoterpene ketone, in its essential oil exhibit neurotoxic properties, manifested in tonic–clonic seizures.[31] Its neural mechanism has been attributed to GABA type A receptor manipulation.

### 9.4.1 NEUROACTIVE MECHANISMS

Neuroinflammation is central to many neural pathologies. Microglial cells are activated by oxidative stress-stimulated cytokines.[28] Amyloid β (Aβ) peptides, the features of Alzheimer's disease, are known to activate cultured microglial cells. The excitatory amino acid L-ODAP in *L. sativus* L. caused neurolathyrism has been found to deregulate $Ca^{++}$ homeostasis. Activation of protein kinase C and perturbation of the redox balance leads to neuronal cell death.[45] Glutaredoxin 1 (Grx1), a cytosolic thiol disulfide oxido-reductase, is required for the redox balance of proteins and mitochondrial integrity; and Grx1 prevents mitochondrial membrane potential loss caused by oxidative stressors.[63]

L-ODAP (L-BOAA) is one of the stressor disturbing the membrane potential. In vitro studies reveal that Grx1 expression is regulated by estrogen in the CNS.[63] Research study indicated that annonacin increases dopaminergic neuronal apoptosis by inhibiting energy production by mitochondrial respiratory chain complex-I.[33] Thus, annonacin is responsible for neuronal death by impairment of ATP generation.[33] BMAA in *Cycas* plant extract, when ingested, leads to progressive neurodegeneration in the hippocampus, causing intracellular fibrillar inclusions. The pathological mechanism is by the excitation of glutamate receptors such as N-methyl-D-aspartate (NMDA) and calcium-dependent α-amino-3-hydroxy-5-methyl-4-isoxazolepropionic acid (AMPA). Incorporation of BMAA in place of L-serine leads to protein misfolding, paving the path for number of neural diseases.[1] Tutin leads to augmentation of synaptic activity and phosphorylated cyclic adenosine monophosphate (cAMP) response element binding protein (CREB) levels. The uncontrollable neuronal excitability is attributed to the toxicity of tutin.[17]

Cycad extract causes a selective loss of dopamine neurons and alpha-syn aggregates in the substantia nigra.[66] The neurotoxicity of aconitine and related alkaloids (mesaconitine and hypaconitine) are due to sodium channel manipulations (longer sodium influx during the action potential) and resultant excitation of target tissues.[6] By activation of the ventromedial nucleus of the hypothalamus, these alkaloids block neuromuscular transmission by anticholinergic effect.[6,74]

## 9.5 DISCUSSION: NEURAL PATHOLOGIES

There are thousands of research articles reporting neuroprotective role of plant parts. However, they have hardly been exploited for the therapy of neural pathologies. It is because the results are based on in vitro or in vivo models. Both paradigms have severe limitations as they cannot simulate the complex human system and only a few variables are considered while deriving inferences. Even if epidemiological studies are considered, generalization of inferences is not possible because the impact of a compound depends on health status, age, gender, and comorbidities.

The consumption methods of the plant parts determine the exposure to the neuromodulators. Cooking process can deplete some of the neuroprotective attributes and can also destroy some neurotoxins. Most of the plant diet-based neurotoxicity cases have emerged from low-economic strata which indicated the deadly combination of ignorance, poverty, and poor nonvaried diet. Recreational experiments by youth have led to some of the fatalities.

Furthermore the available literature appears to be biased, as most of these report only the neuroprotective attributes, and any neurotoxicity aspects are not reported. Liquid chromatography (LC-MS), high-performance liquid chromatography (HPLC), capillary electrophoresis (CE), nuclear magnetic resonance (NMR) spectroscopy, and other metabolomic studies can help to identity the neurotoxins.

Neural manipulation is an extremely complex process. Number of systems interact to maintain neural homeostasis, most importantly, endocrine and immune components. Perturbation of one can initiate neurological morbidities. Hypertension can agitate renin–angiotensin–aldosterone system (RAAS), which can complicate the cerebral vasculature.[51]

It may not appear encouraging but as life span has increased therefore more instances of geriatric issues have emerged. To treat them, drugs are being used, which unfortunately are adding to the oxidative stress and inflammation, leading to exacerbation of the ailments, including Parkinson's and Alzheimer's disease.

Therefore, what makes plant alkaloids neurotoxins? Number of plant alkaloids (such as vinblastine, vincristine nicotine, berberine, codeine, thebaine, scopolamine, ibogaine, and harmaline) are validated as therapeutic compounds.[44,47,82] Annonacin has been reported to exert antitumor, pesticidal, antimalarial, anthelmintic, piscicidal, antiviral, and antimicrobial effects.[55] If a phytochemical is therapeutic for one illness but is causing another disorder, how rational is it to consider it safe? A phytochemical or plant extract might be protective in in vitro studies, but it might be detrimental in vivo or in long-term exposure. The results of biological assays apart from the limitations of models have other issues. One extract might be neuroprotective, but it might have estrogenic, cardio/renal-damaging effect. In that case, it will lead to other issues. Therefore, holistic assessment of an extract is required. Also, dosage is a major factor in deciding toxicity or protectant property of an extract. In fact, this is why herbal products show conflicting effects in biological studies. If a drug is taken in high dosage, then it is as bad as a poison. Illiteracy and poverty are key risk factors in vulnerability to the plant-based neurotoxins. Unaware of threats, they subsist on these plants for prolonged period.

Based on the past literature, it can be suggested that most plant products should not be detrimental for brain health while consumed in moderation and occasionally. However, educating self before venturing on foraging on unknown plants should be emphasized. Lots of potentially deadly plant products are sold online, which one should be aware of and stay away from. One should be alert enough to understand that not all herbal products and honey are health providers.

## 9.6 SUMMARY

The literature review suggests that alkaloids, acetogenins, sequiterpenes, saponins, cyanogenic glycosides, and nonstandard amino acids in plants may act as neurotoxins. They affect the neurons by various pathways, but mostly by manipulating the ion-exchange pumps and the neuroreceptors. It is the dosage of plant diet intake that decides the neuroprotective or neuropathogenic property. Awareness of the toxic parts of a plant can spare an individual from chronic or acute neural illnesses.

## KEYWORDS


- **alkaloids**
- **annonacin**
- **β-N-methylamino-L-alanine**
- **lathyrism**
- **neurototoxins**
- **neurodegeneration**


## REFERENCES


1. Banack, S. A.; Caller, T. A.; Stommel, E. W. The Cyanobacteria Derived Toxin Beta-N-Methylamino-L-Alanine and Amyotrophic Lateral Sclerosis. *Toxins* (Basel) **2010,** *2* (12), 2837–2850.

2. Biradar, S. M.; Joshi, H.; Tarak. K. C. Cerebroprotective Effect of Isolated Harmine Alkaloids Extracts of Seeds of Peganum harmala L. on Sodium Nitrite-Induced Hypoxia and Ethanol-induced Neurodegeneration in Young Mice. *Pakistan J. Biol. Sci.* **2013,** *16* (23), 1687–1697.
3. Bocharov, E. V.; Ivanova-Smolenskaya, I. A.; Poleshchuk, V. V.; Kucheryanu, V. G.; Il'enko, V. A.; Bocharova, O. A. Therapeutic Efficacy of the Neuroprotective Plant Adaptogen in Neurodegenerative Disease (Parkinson's Disease as an Example). *Bull. Exp. Biol. Med.* **2010,** *149* (6), 682–684.
4. Bradley, W. G.; Mash, D. C. Beyond Guam: Cyanobacteria/BMAA Hypothesis of the Cause of ALS and Other Neurodegenerative Diseases. *Amyotr. Lateral Scler.* **2009,** *10* (Suppl 2), 7–20.
5. Champy, P.; Höglinger, G. U.; Féger, J.; Gleye, C.; Hocquemiller, R.; Laurens, A.; Guérineau, V.; Laprévote, O.; Medja, F.; Lombès, A.; Michel. P. P.; Lannuzel, A.; Hirsch, E. C.; Ruberg, M. Annonacin, a Lipophilic Inhibitor of Mitochondrial Complex I, Induces Nigral And Striatal Neurodegeneration in Rats: Possible Relevance for Atypical Parkinsonism in Guadeloupe. *J. Neurochem.* **2004,** *88* (1), 63–69.
6. Chan, T. Y. K. Aconite Poisoning. *Clin. Toxicol.* (Phila) **2009,** *47* (4), 279–285.
7. Chonpathompikunlert, P.; Wattanathorn, J.; Muchimapura, S. Piperine, the Main Alkaloid of Thai Black Pepper, Protects Against Neurodegeneration and Cognitive Impairment in Animal Model of Cognitive Deficit Like Condition of Alzheimer's Disease. *Food Chem. Toxicol.* **2010,** *48* (3), 798–802.
8. Chowański, S.; Adamski, Z.; Marciniak, P.; Rosiński, G.; Büyükgüzel, E.; Büyükgüzel, K.; Falabella, P.; Scrano, L.; Ventrella, E.; Lelario, F.; Bufo, S. A. Review of Bioinsecticidal Activity of Solanaceae Alkaloids. *Toxins* (Basel) **2016,** *8* (3), E60.
9. Cox, P. A. Botanical Neurotoxins. *Reference Module in Biomedical Sciences*; Elsevier: New York; 2014, 465–475.
10. Cox, P. A.; Sacks, O. W. Cycad Neurotoxins, Consumption of Flying Foxes, and ALS-PDC Disease in Guam. *Neurology* **2002,** *58* (6), 956–959.
11. Datla, K. P.; Zbarsky, V.; Rai, D.; Parkar, S.; Osakabe, N.; Aruoma, O. I.; Dexter, D. T. Short-term Supplementation with Plant Extracts Rich in Flavonoids Protect Nigrostriatal Dopaminergic Neurons in a Rat Model of Parkinson's Disease. *J. Am. Coll. Nutr.* **2007,** *26* (4), 341–349.
12. Dawidowicz, A. L.; Dybowski, M. P. Simple and Rapid Determination of Myristicin in Human Serum. *Forensic Toxicol.* **2013,** *31* (1), 119–123.
13. Diaz, G. J. Toxicosis by Plant Alkaloids in Humans and Animals in Colombia. *Toxins* (Basel) **2015,** *7* (12), 5408–5416.
14. Duke, S. O.; Cantrell, C. L.; Meepagala, K. M.; Wedge, D. E.; Tabanca, N.; Schrader, K. K. Natural Toxins for Use in Pest Management. *Toxins* (Basel) **2010,** *2* (8), 1943–1962.
15. Fields, B. A.; Reeve, J.; Bartholomaeus, A.; Mueller, U. Human Pharmacokinetic Study of Tutin in Honey: Plant-derived Neurotoxin. *Food Chem. Toxicol.* **2014,** *72*, 234–241.
16. Fu, W.; Zhuang, W.; Zhou, S; Wang, X. Plant-derived Neuroprotective Agents in Parkinson's Disease. *American J. Trans. Res.* **2015,** *7* (7), 1189–1202.
17. Fuentealba, J.; Guzmán, L.; Manríquez-Navarro, P.; Pérez, C.; Silva, M.; Becerra, J.; Aguayo, L. G. Inhibitory Effects of Tutin on Glycine Receptors in Spinal Neurons. *Eur. J. Pharmacol.* **2007,** *559* (1), 61–64.

18. Gleadow, R,; Pegg, A.; Blomstedt, C. K. Resilience of Cassava (*Manihot esculenta Crantz*) to Salinity: Implications for Food Security in Low-Lying Regions. *J. Exp. Bot.* **2016,** *67* (18), 5403–5413.
19. Gunduz, A.; Turedi, S.; Russell, R. M.; Ayaz, F. A. Clinical Review of Grayanotoxin/Mad Honey Poisoning Past and Present. *Clin. Toxicol.* (Philadelphia) **2008,** *46* (5), 437–442.
20. Houghton, P. J.; Howes, M. J. Natural Products and Derivatives Affecting Neurotransmission Relevant to Alzheimer's and Parkinson's Disease. *Neurosignals* **2005,** *14* (1–2), 6–22.
21. Im, A. R,; Kim, Y. H.; Uddin, M. R.; Chae, S.; Lee, H. W.; Kim, Y. S.; Lee, M. Y. Neuroprotective Effects of *Lycium Chinense Miller* Against Rotenone-induced Neurotoxicity in PC12 Cells. *Am. J. Chin. Med.* **2013,** *41* (6), 1343–1359.
22. Imai, Y,; Lu, B. Mitochondrial Dynamics and Mitophagy in Parkinson's Disease: Disordered Cellular Power Plant Becomes a Big Deal in a Major Movement Disorder. *Curr. Opin. Neuro. Biol.* **2011,** *21* (6), 935–941.
23. Ingre, C.; Roos, P. M.; Piehl, F.; Kamel, F.; Fang, F. Risk Factors for Amyotrophic Lateral Sclerosis. *Clin. Epidemiol.* **2015,** *7*, 181–193.
24. Jansen, S. A.; Kleerekooper, I.; Hofman, Z. L. M.; Kappen, I. F. P. M.; Stary-Weinzinger, A.; van der Heyden, M. A. G. Grayanotoxin Poisoning: Mad Honey Disease and Beyond. *Cardiovasc. Toxicol.* **2012,** *12* (3), 208–215.
25. Katzenschlager, R.; Evans, A.; Manson, A.; Patsalos, P. N.; Ratnaraj, N.; Watt, H.; Timmermann, L.; Van der Giessen, R.; Lees, A. J. *Mucuna pruriens* in Parkinson's Disease: A Double Blind Clinical and Pharmacological Study. *J. Neurol. Neurosurg. Psychiatry* **2004,** *75* (12), 1672–1677.
26. Keerthi, A. A. P.; Ekanayake, S.; Jansz, E. R. Review of the Neurotoxic Effect of Palmyrah Flour. *Int. J. Food Sci. Nutr.* **2009,** *60* (Suppl 4), 306–316.
27. Khandare, A. L.; Babu, J. J.; Ankulu, M.; Aparna, N.; Shirfule, A.; Rao, G. S. Grass Pea Consumption & Present Scenario of Neurolathyrism in Maharashtra State of India. *Indian J. Med. Res.* **2014,** *140* (1), 96–101.
28. Kim, M.; Kim, S. O.; Lee, M.; Lee, J. H.; Jung, W. S.; Moon, S. K.; Kim, Y. S.; Cho, K. H.; Ko, C. N.; Lee, E. H. Tetramethylpyrazine, a Natural Alkaloid, Attenuates Pro-Inflammatory Mediators Induced By Amyloid β and Interferon-γ in Rat Brain Microglia. *Eur. J. Pharmacol.* **2014,** *740*, 504–511.
29. Kolarova, M.; García-Sierra, F.; Bartos, A.; Ricny, J.; Ripova, D. Structure and Pathology of Tau Protein in Alzheimer Disease. *Int. J. Alzheimer's Dis.* **2012,** 2012, 1–13.
30. Kumar, G. P.; Khanum, F. Neuroprotective Potential of Phytochemicals. *Pharmacog. Rev.* **2012,** *6* (12), 81–90.
31. Lachenmeier, D. W. Wormwood (*Artemisia absinthium* L.): Curious Plant with Both Neurotoxic and Neuroprotective Properties. *J. Ethnopharmacol.* **2010,** *131* (1), 224–227.
32. Lannuzel, A.; Höglinger, G. U.; Champy, P.; Michel, P. P.; Hirsch, E. C.; Ruberg, M. Is Atypical Parkinsonism in the Caribbean Caused by the Consumption of Annonacae? *J. Neural Trans. Suppl.* **2006,** *70*, 153–157.
33. Lannuzel, A.; Michel, P. P.; Höglinger, G. U.; Champy, P.; Jousset, A.; Medja, F.; Lombès, A.; Darios, F.; Gleye, C.; Laurens, A.; Hocquemiller, R.; Hirsch, E. C.; Ruberg, M. The Mitochondrial Complex-I Inhibitor Annonacin Is Toxic to Mesencephalic Dopaminergic Neurons by Impairment of Energy Metabolism. *Neuroscience* **2003,** *121* (2), 287–296.

34. Larsen, L.; Joyce, N. I.; Sansom, C. E.; Cooney, J. M.; Jensen, D. J.; Perry, N. B. Sweet Poisons: Honey Contaminated with Glycosides of the Neurotoxin Tutin. *J. Nat. Products* **2015,** *78*, 1363–1369.
35. Li, Q.; Wang, Y.; Moldzio, R.; Lin, W.; Rausch, W. D. Swainsonine as a Lysosomal Toxin Affects Dopaminergic Neurons. *J. Neural Trans.* **2012,** *119* (12), 1483–1490.
36. Li, X. Z.; Zhang, S. N.; Liu, S. M.; Lu, F. Recent Advances in Herbal Medicines Treating Parkinson's Disease. *Fitoterapia* **2013,** *84*, 273–285.
37. Li, X.; Zhao, Y.; Liu, P.; Zhu, X.; Chen, M.; Wang, H.; Lu, D.; Qi, R. Senegenin Inhibits Hypoxia/Reoxygenation-Induced Neuronal Apoptosis By Upregulating RhoGDIα. *Molecular Neurobiol.* **2015,** *52* (3), 1561–1571.
38. Lin, F.; Qin Z. H. Degradation of Misfolded Proteins by Autophagy: Is It a Strategy for Huntington's Disease Treatment? *J. Huntington's Dis.* **2013,** *2* (2), 149–157.
39. Liu, Y.; Li, Z.; Hu, H.; Xu, S.; Chang, Q.; Liao, Y.; Pan, R.; Liu, X. Tenuifolin, A Secondary Saponin from Hydrolysates of Polygalasaponins, Counteracts the Neurotoxicity Induced by Aβ25–35 Peptides In Vitro and In Vivo. *Pharmacol. Biochem. Behav.* **2015,** *128*, 14–22.
40. Luo, F. C.; Wang, S. D.; Qi, L.; Song, J. Y.; Lv, T.; Bai, J. Protective Effect of Panaxatriol Saponins Extracted from Panax notoginseng Against MPTP-Induced Neurotoxicity In Vivo. *J. Ethnopharmacol.* **2011,** *133* (2), 448–453.
41. Mandel, S.; Weinreb, O.; Amit, T.; Youdim, M. B. H. Cell Signaling Pathways in the Neuroprotective Actions of the Green Tea Polyphenol (-)-epigallocatechin-3-gallate: Implications for Neurodegenerative Diseases. *J. Neurochem.* **2004,** *88* (6), 1555–1569.
42. Marler, T. E.; Shaw, C. A. Phenotypic Characteristics as Predictors of Phytosterols in Mature Cycas micronesica Seeds. *HortScience* **2009,** *44* (3), 725–729.
43. Mathon, C.; Bongard, B.; Duret, M.; Ortelli, D.; Christen, P.; Bieri, S. Analysis of the Neurotoxin Anisatin in Star Anise by LC-MS/MS. *Food Add. Contaminants: Part A: Chem. Anal. Control Exp. Risk Assess.* **2013,** *30* (9), 1598–1605.
44. Moloudizargari, M.; Mikaili, P.; Aghajanshakeri, S.; Asghari, M. H.; Shayegh, J. Pharmacological and Therapeutic Effects of *Peganum harmala* and Its Main Alkaloids. *Pharmacog. Rev.* **2013,** *7* (14), 199–212.
45. Moorhem, M.; Decrock, E.; De Vuyst, E.; De Bock, M.; Wang, N.; Lambein, F.; Van Den Bosch, L.; Leybaert, L. L-β-N-oxalyl-α,β-Diaminopropionic Acid Toxicity in Motor Neurons. *NeuroReport* **2011,** *22* (3), 131–135.
46. Moraes, L. S.; Rohor, B. Z.; Areal, L. B.; Pereira, E. V.; Santos, A. M. C.; Facundo, V. A.; Santos, A. R. S.; Pires, R. G. W., Martins-Silva, C. Medicinal Plant Combretum Leprosum Mart Ameliorates Motor, Biochemical and Molecular Alterations in a Parkinson's Disease Model Induced by MPTP. *J. Ethnopharmacol.* **2016,** *185*, 68–76.
47. Moudi, M.; Go, R.; Yien, C. Y. S.; Nazre, M. Vinca Alkaloids. *Int. J. Prevent. Med.* **2013,** *4* (11), 1231–1235.
48. Mythri, R. B.; Bharath, M. M. S. Curcumin: A Potential Neuroprotective Agent in Parkinson's Disease. *Current Pharm. Design* **2012,** *18* (1), 91–99.
49. Narayanan, N. S.; Rodnitzky, R. L.; Uc, E. Y. Prefrontal Dopamine Signaling and Cognitive Symptoms of Parkinson's Disease. *Rev. Neurosci.* **2013,** *24* (3), 267–278.
50. Otomo, A.; Pan, L.; Hadano, S. Dysregulation of the Autophagy-Endolysosomal System in Amyotrophic Lateral Sclerosis and Related Motor Neuron Diseases. *Neurol. Res. Int.* **2012,** *2012*, 1–12.

51. Pacurari, M.; Kafoury, R.; Tchounwou, P. B.; Ndebele, K. The Renin-Angiotensin-Aldosterone System in Vascular Inflammation and Remodeling. *Int. J. Inflam.* **2014,** *2014*, 1–13.
52. Patel, S. Green Tea as a Nutraceutical: The Latest Developments. *Food Sci. Technol. Res.* **2013,** *19* (6), 923–932.
53. Pathak-Gandhi, N.; Vaidya, A. D. B. Management of Parkinson's disease in Ayurveda: Medicinal Plants and Adjuvant Measures. *J. Ethnopharmacol.* **2017,** *197*, 46–51.
54. Pérez, C.; Becerra, J.; Manríquez-Navarro, P.; Aguayo, L. G.; Fuentealba, J.; Guzmán, J. L.; Joseph-Nathan, P.; Jiménez, V.; Muñoz, M. A.; Silva, M. Inhibitory Activities on Mammalian Central Nervous System Receptors and Computational Studies of Three Sesquiterpene Lactones from *Coriaria ruscifolia* subsp. *Ruscifolia. Chem. Pharm. Bull.* (Tokyo) **2011,** *59*, 161–165.
55. Pomper, K. W.; Lowe, J. D.; Crabtree, S. B.; Keller, W. Identification of Annonaceous Acetogenins in the Ripe Fruit of the North American Pawpaw (Asimina triloba). *J. Agric. Food Chem.* **2009,** *57* (18), 8339–8343.
56. Potts, L. F.; Luzzio, F. A.; Smith, S. C.; Hetman, M.; Champy, P.; Litvan, I. Annonacin in *Asimina triloba* Fruit: Implication for Neurotoxicity. *Neurotoxicology* **2012,** *33* (1), 53–58.
57. Pullela, R.; Young, L.; Gallagher, B.; Avis, S. P.; Randell, E. W. A Case of Fatal Aconitine Poisoning by Monkshood Ingestion. *J. Forensic Sci.*, **2008,** *53* (2), 491–494.
58. Reeve, A.; Simcox, E.; Turnbull, D. Ageing and Parkinson's Disease: Why Is Advancing Age the Biggest Risk Factor? *Ageing Res. Rev.* **2014,** *14*, 19–30.
59. Ren, R.; Shi, C.; Cao, J.; Sun, Y.; Zhao, X.; Guo, Y.; Wang, C. Neuroprotective Effects of A Standardized Flavonoid Extract of Safflower Against Neurotoxin-induced Cellular and Animal Models of Parkinson's Disease. *Scient. Rep.* **2016,** *6*, 22135.
60. Rinwa, P.; Kumar, A.; Garg, S. Suppression of Neuroinflammatory and Apoptotic Signaling Cascade by Curcumin Alone and in Combination with Piperine in Rat Model of Olfactory Bulbectomy Induced Depression. *PLoS One* **2013,** *8* (4), e61052.
61. Rivadeneyra-Domínguez, E.; Vázquez-Luna, A.; Rodríguez-Landa, J. F.; Díaz-Sobac, R. Neurotoxic Effect of Linamarin in Rats Associated with Cassava (Manihot esculenta Crantz) Consumption. *Food Chem. Toxicol.* **2013,** *59*, 230–235.
62. Rus Jacquet, A.; Subedi, R.; Ghimire, S. K.; Rochet, J-C. Nepalese Traditional Medicine and Symptoms Related to Parkinson's Disease and other Disorders: Patterns of the Usage of Plant Resources Along the Himalayan Altitudinal Range. *J. Ethnopharmacol.* **2014,** *153* (1), 178–189.
63. Saeed, U.; Durgadoss, L.; Valli, R. K.; Joshi, D. C.; Joshi, P. G.; Ravindranath, V. Knockdown of Cytosolic Glutaredoxin 1 Leads to Loss of Mitochondrial Membrane Potential: Implication in Neurodegenerative Diseases. *PLoS One* **2008,** *3* (6), e2459.
64. Schrage, M.;, Shen, Y.; Claassen, F. W.; Zuilhof, H.; Nielen, M. W. F.; Chen, B.; van Beek, T. A. Rapid and Simple Neurotoxin-based Distinction of Chinese and Japanese Star Anise by Direct Plant Spray Mass Spectrometry. *J. Chromatography A* **2013,** *1317*, 246–253.
65. Sengupta, T.; Vinayagam, J.; Singh, R.; Jaisankar, P.; Mohanakumar, K. P. Plant-derived Natural Products for Parkinson's Disease Therapy. *Adv. Neurobiol.* **2016,** *12*, 415–496.
66. Shen, W-B.; McDowell, K. A.; Siebert, A. A.; Clark, S. M.; Dugger, N. V.; Valentino, K. M. Environmental Neurotoxin-induced Progressive Model of Parkinsonism in Rats. *Ann. Neurol.* **2010,** *68* (1), 70–80.
67. Shen, Y.; van Beek, T. A.; Claassen, F. W.; Zuilhof, H.; Chen, B.; Nielen, M. W. F. Rapid Control of Chinese Star Anise Fruits and Teas for Neurotoxic Anisatin by Direct

Analysis in Real Time High Resolution Mass Spectrometry. *J. Chromatogr. A* **2012,** *1259*, 179–186.
68. Singh, S. S.; Rao, S. L. N. Lessons from Neurolathyrism: A Disease of the Past & the Future of *Lathyrus Sativus* (Khesari dal). *Indian J. Med. Res.* **2013,** *138* (1), 32–37.
69. Sinha, S. C.; Chen, Z.; Huang, Z-Z.; Nakamaru-Ogiso, E.; Pietraszkiewicz, H.; Edelstein, M.; Valeriote, F. Alteration of the Bis-Tetrahydrofuran Core Stereochemistries in Asimicin Can Affect the Cytotoxicity. *J. Med. Chem.* **2008,** *51* (22), 7045–7048.
70. Spagnuolo, C.; Napolitano, M.; Tedesco, I.; Moccia, S.; Milito, A.; Russo, G. L. Neuroprotective Role of Natural Polyphenols. *Curr. Topic Med. Chem.* **2016,** *16* (17), 1943–1950.
71. Strathearn, K. E.; Yousef, G. G.; Grace, M. H.; Roy, S. L. Neuroprotective Effects of Anthocyanin- and Proanthocyanidin-rich Extracts in Cellular Models of Parkinson's Disease. *Brain Res.* **2014,** *1555*, 60–77.
72. Tiryaki, E.; Horak, H. A. ALS and Other Motor Neuron Diseases. *Continuum (Minneap Minn)* **2014,** *20* (5), 1185–1207.
73. Varanese, S.; Birnbaum, Z.; Rossi, R.; Di Rocco, A. Treatment of Advanced Parkinson's Disease. *Parkinson's Dis.* **2010,** *2010*, 1–9.
74. Verma, N.; Singh, A. P.; Gupta, A.; Sahu, P. K.; Rao, C. V. Antidiarrheal Potential of Standardized Extract of *Rhododendron arboreum* Smith Flowers in Experimental Animals. *Indian J. Pharmacol.* **2011,** *43* (6), 689–693.
75. Waliwitiya, R.; Belton, P.; Nicholson, R. A.; Lowenberger, C. A. Plant Terpenoids: Acute Toxicities and Effects on Flight Motor Activity and Wing Beat Frequency in the Blow Fly Phaenicia sericata. *J. Econ. Entomol.* **2012,** *105* (1), 72–84.
76. Walters, J. K.; Boswell, L. E.; Green, M. K.; Heumann, M. A.; Karam, L. E.; Morrissey, B. F.; Waltz, J. E. Pyrethrin and Pyrethroid Illnesses in The Pacific Northwest: Five-year Review. *Public Health Rep.* **2009,** *124* (1), 149–159.
77. Wang, Q.; Gong, Q.; Wu, Q.; Shi, J. Neuroprotective effects of Dendrobium Alkaloids on Rat Cortical Neurons Injured by Oxygen-Glucose Deprivation and Reperfusion. *Phytomedicine* **2010,** *17* (2), 108–115.
78. Wu, C-H.; Fallini, C.; Ticozzi, N. Mutations in the Profilin-1 Gene Cause Familial Amyotrophic Lateral Sclerosis. *Nature* **2012,** *488* (7412), 499–503.
79. Xie, W.; Yang, Y.; Gu, X.; Zheng, Y.; Sun, Y.; Liang, Y.; Bo, J.; Ma, Z. Senegenin Attenuates Hepatic Ischemia-Reperfusion Induced Cognitive Dysfunction by Increasing Hippocampal NR2B Expression in Rats. *PLoS One* **2012,** *7* (9), e45575.
80. Yao, Y.; Jia, M.; Wu, J. G.; Zhang, H.; Sun, L. N.; Chen, W. S.; Rahman, K. Anxiolytic and Sedative-Hypnotic Activities of Polygalasaponins from Polygala tenuifolia in Mice. *Pharm. Biol.* **2010,** *48* (7), 801–807.
81. Zarei, S.; Carr, K.; Reiley, L.; Diaz, K.; Guerra, O.; Altamirano, P. F.; Pagani, W.; Lodin, D.; Orozco, G.; Chinea, A. Comprehensive Review of Amyotrophic Lateral Sclerosis. *Surg. Neurol. Int.* **2015,** *6* (1), 171.
82. Ziegler, J.; Facchini, P. J. Alkaloid Biosynthesis: Metabolism and Trafficking. *Ann. Rev. Plant Biol.* **2008,** *59* (1), 735–769.

CHAPTER 10

# HEALTH BENEFITS OF CANDY LEAF (*Stevia rebaudiana*): PHYSIOLOGICAL AND PHARMACOLOGICAL ACTIONS

ZUBARIA ISHAQ, HANEEN FATIMA KIRMANI, MUNAWAR ABBAS, and FARHAN SAEED

## ABSTRACT

The sweet herb (*Stevia rebaudiana*: Asteraceae) is endemic to Paraguay and Brazil. It has been used as a medicinal plant and a sweetener by South American Indians. Also, people of Paraguay have been using it as a folk medicine for the treatment of cardiovascular problems, obesity, diabetes, liver, stomachache, etc. It is also being used for children growth as it possesses many nutrients such as calcium, phosphorus, sodium, potassium, and iron. Recently, it has been noted that Stevia can also play important role for controlling cancer. Because of its enriched chemical composition, sweet leaf is being used as a sugar substitute in many countries. Nutritional and health-benefitting facets are being studied by many researchers nowadays. World is now returning back toward natural resources for food; hence, Stevia is likely to become natural sweetener because of its high sweetening compounds. The worldwide demand for high-potency sweeteners is expected to increase. Stevia is going to be a potential alternative for sugar and part of growing natural food. In Pakistan, Stevia is used as low caloric sweetener for diabetic patients. Since, there is no such cultivation of *S. rebaudiana* in Pakistan and the present situation of sugar emergencies in Pakistan, Stevia is promoted by government of Pakistan. Current study shows that Stevia is antibacterial, antiseptic, antimicrobial, antioxidant, antiglycemic, and antihypertensive. These properties can help with hypertension (high blood pressure), diabetes, chronic fatigue, indigestion, upset stomach, heartburn, weight loss, cold and flu, gingivitis, tooth decay, etc. Still, there is a need to conduct more research work on plant extract and its effects on human health and pharmacological aspects. The aim of this study is to reveal nutritional and therapeutic value of Stevia.

## 10.1 INTRODUCTION

The *Stevia rebaudiana* (sweet leaf) is endemic to Paraguay and Brazil. *S. rebaudiana* belongs to family Asteraceae. It is also called sweet leaf, sweet herb of Paraguay, honey leaf, or candy-leaf in common language. Stevia is small perennial shrub of 60–80 cm in height with sessile and oppositely arranged leaves. *S. rebaudiana* can be easily cultivated like other vegetable crops. The suitable conditions for its growth are 6.5–7.5 pH range, well-aerated red soil as well as sandy-loam soil. Due to presence of diterpene compounds, which is present in its leaves, *S. rebaudiana*, it is natural potential sweetener.[13] Natural habitat for this plant is from Southwestern United States to the Brazilian highlands.[31] *S. rebaudiana* is one of 154 members of genus Stevia and it is one of only two members that contain sweet steviol glycoside.

It is used as noncaloric sweetener and food additives by Japanese and Brazilian these days.[37] It was discovered by Dr. Moises Santiago in 1888 at Paraguay. In 1905, it was given scientifically name as *S. rebaudiana*.[23] People of Central and South America have been using this plant as a sweetener in many food products. In many countries (e.g., Brazil and Paraguay), this plant is used as inevitable source for controlling diabetes.[25] Stevia is also used for controlling obesity.[26]

The leaf constituent of Stevia, steviol glycoside, is said to be safe for well-being and good substitute for sugar due to high intensity of sweetness of this compound. According to the US Food and Drug Administration (FDA)[27,28] and The European Food Safety Authority (E960 European Index Number), this is generally recognized as safe (GRAS) status for steviol glycoside.[29,30] Due to antibacterial, antiglycemic, antidiabetic characteristics, and many other health benefits, Stevia species have been used by many countries nowadays. In Pakistan, Stevia is used as low-caloric sweetener for diabetic patients. There is no commercial cultivation of *S. rebaudiana* in Pakistan. However, few tissue culture techniques have been applied to grow Stevia plant. Also, The Pakistan Council of Scientific and Industrial Research carried out examination of Stevia leaves at its laboratory and found that it contains 30% proteins, 53% carbohydrates, and other nutrients. Stevia leaves are beneficial for child growth due to high content of calcium in it.[31]

For instance, people of Pakistan are becoming more concerned about sugar contents in diets because of many health issues like cardiac diseases, diabetes, obesity, etc. Therefore, there is high demand of natural noncaloric or less-caloric sweetener as alternate for sugar. On the other hand, as it reduces the desire of tobacco, the wholesome application of Stevia nourishes liver, spleen, pancreas, counters fatigue, stimulates digestion, and improves gastrointestinal functions. It may constrain the growth and multiplication of tooth-decaying bacteria as well.[32] Hence, in Pakistan, the cultivation of Stevia is becoming the need of hour to fulfill the rising demand of sugar.

This chapter explores the therapeutic values and health benefits of *S. rebaudiana.*

### *10.1.1 CHEMICAL CONSTITUENTS OF STEVIA*

Although complete chemical components of Stevia are not known, yet there are eight ent-kaurene glycosides,[33] including dulcoside A, rebaudiosides A–E, steviolbioside, and stevioside that are found to be main reason of sweet taste of this shrub. Some of them are highlighted in Figure 10.1. Research studies have been conducted on chemical composition of Stevia species since 1908. Initially, in 1955, preparatory studies on the structure of stevioside (steviol glycoside) were established which were followed by auxiliary studies[33–40] on its organic synthesis from steviol.[41] Till now, 43 types of steviol glycosides have been identified from *S. rebaudiana* Bertoni.[42]

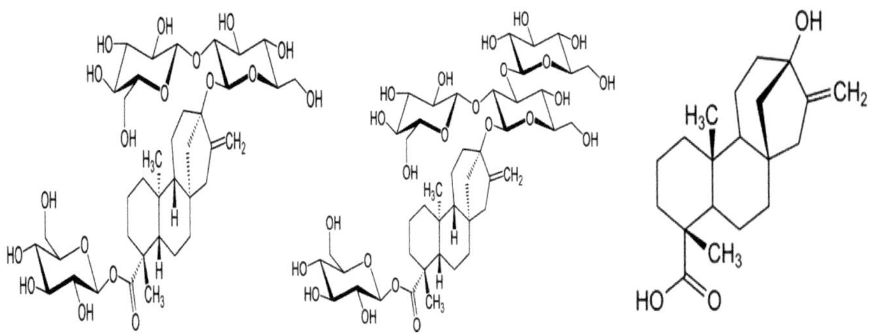

**FIGURE 10.1** Chemical structure of rebaudioside A (left), stevioside (center), and steviol (right).

In addition to stevioside, Stigmasterol, β-sitosterol, and campesterol are also found in Stevia species. A product named Steviol is also present as a result of enzymatic hydroxylation. On the other hand, other chemical constituents having less or no sweet taste have been indicated in the Stevia plant. Some of the chemical compounds even have bitter taste. Stevioside A, longipinane derivatives, epoxylabdane diterpenes, clerodane derivative, flavonoids, and sesquiterpene lactones are included in these chemical components. These constituents are found in different amounts in the leaves and roots of the Stevia plant.[43,44]

## 10.2 PHYSIOLOGICAL AND PHARMACOLOGICAL ACTIONS OF CANDY LEAF

Throughout the world Stevia species have been used as a noncaloric sweetener. It is used in bakery items, soft drinks, and beverage industry. It has same potential regarding sweetness as compared to 10% sucrose solution at either pH 3.0 or 7.0.[45] Also, it has similar potency to that of aspartame and cyclamate/saccharin mixture.[1] Stevia reduces the risk of many diseases like tooth decay, obesity, and type 2 diabetes by replacing sugar consumption. Steviol glycoside is not digested in the alimentary canal and it is converted into glucose and steviol by bacteria present in large intestine.

Microflora present in intestine metabolizes glucose; and steviol is converted into glucuronide in the liver and excreted out from the body in the form of feces and urine.[2] Steviol glycoside is nonteratogenic and noncarcinogenic when it is taken below 4–5 mg/kg body weight per day.[3,4] According to WHO, Pakistan occupies seventh position on diabetes prevalence list. At present, there are 7 million people in Pakistan, who are suffering from diabetes and 7 million who are at the risk of getting diabetes. If no proper measures are taken, then the number of diabetic patients may reach up to 14.4 million by 2040. Since extract and leaves of Stevia leaves has been used by many countries for controlling diabetes,[5] therefore it has potential in Pakistan.

After digestion of extract, plasma glucose level drops down and blood glucose level is maintained. Stevioside stimulates the secretion of insulin by affecting the activity of β-cells without changing the action of $K^+$–ATP channel and cyclic adenosine monophosphate (cAMP) level in islets, which ultimately regulates blood glucose level.[6] Stevioside likewise improves glucose-stimulated insulin discharge, yet does not influence fasting insulinemia. Research study in 2003 showed improved insulin reaction resulting concealment of glucagon release and dropping down the blood glucose concentration quest.[46,47] Steviol also plays an important role in reducing the risk of oxidative stress. Taking everything into consideration, Stevia has the capacity to surge insulin effect on the cell membrane, to enhance insulin production, regulate glucagon production to maintain the level of glucose in the blood. Overall, Stevia has outright and integrated mechanism to cope with the mechanism of type-II diabetes. Furthermore, Stevia can also help in reviving the pancreatic gland[48,49]

### *10.2.1 EFFECTS ON BLOOD HOMEOSTASIS*

When the diameter of blood vessels narrows down as a result of thickening, it increases the heart pressure. This extensive condition increases the risk of stroke. Stevia can be used as analeptic to equalize blood pressure and other cardiopulmonary complications. Cholesterol contents are declined in the blood due to daily use of glycosides present in Stevia.[50] Stevioside can halt the inflow of $Ca^{2+}$ ions to vascular smooth muscles due to its hypotensive property that induces vasodilation.[51] Stevia may act against cardiovascular deformity due to the presence of phytosterol in the wax of leaves of Stevia plant.[52] It is revealed that Stevia significantly improves cardiovascular health by reducing bad cholesterol, for example, triglycerides and low-density lipoprotein (LDL) and increasing high-density lipoprotein (HDL). [53] Noncaloric sweeteners are found in *S. rebaudiana* leaves that are additionally noncaloric and, ultimately, such sweeteners exert beneficial effects on consumer health.

Peculiar biological characteristics are also demonstrated by the glycosides found in Stevia. The quantity of bad cholesterol, sugar, and radionuclides is reduced and coagulation process and cell generation are also improved because of regular intake of Stevia;[21] additionally, the strength of blood vessels is improved because of suppression of the neoplastic growth.[22]

The steviol glycosides have been reviewed three times by Joint Food and Agriculture Organization (FAO)/WHO Expert Committee on Food Additives (JECFA)[20] in its meeting. Average daily intake and temporary specifications are established by JECFA at 0–2 mg/kg for steviol glycosides. To evaluate the effects on glucose homeostasis with respect to insulin-independent and -dependent diabetes, the human studies were also requested by JECFA to find the potential effect of Stevia glycoside on blood pressure.

Almost 70 million people are suffering from renal diseases that caused 735,000 deaths in 2010.[54,] Melis[36] conducted a study on hypertensive rat and revealed that Stevioside stimulates hypotension, vasodilation, dieresis, and natriuresis. Increased rate of glomerular filtration (GFR) and renal plasma flow (RPF) from afferent and efferent arterioles is shown in the study due to vasodilation in hypertensive rats.[55] Steviol, furthermore, inhibits the growth of cysts that was revealed[56] using model in Madin–Darby canine kidney (MDCK) cells. By inhibition of cystic fibrosis transmembrane conductance regulator (CFTR) chloride channel activity, contraction of (CFTR) expression and enhancing proteasomal depletion steviol blocks MDCK development.

### 10.2.2 EFFECTS ON CANCER AND INFLAMMATION

Steviol has antitumorous property. Stevioside, aglycones steviol, and isosteviol inhibit carcinogenesis in mouse skin after sequential exposure to 7,12-dimethylbenz [a]anthracene (DMBA) and 12-*O*-tetradecanoylphorbol-13-acetate (TPA). In addition, these compounds have more effects as compared to that of glycyrrhizin. These compounds have ability to cease carcinogenesis, which is chemically induced.[57] Stevioside and isosteviol have the capacity to hinder DNA replication and human cancer cell development in vitro (with $LD_{50}$ values of 87 and 167 μMol). Isosteviol can cause setback in the development of cancer cells by constraining DNA polymerase and topoisomerase II activity.[58]

Some of the Stevia glycoside including the rebaudioside and stevioside were admitted as nontoxic in the acute trials and this is due to its structural behavior.[35] Body weight reduction was also seen when steviol was orally administered in high dose to experimental rats and occurrence of toxicity was not observed during that study. The carcinogenic effects of stevioside had been studied on the initiation and promotion of cancer in urinary bladder; and no development in the neoplastic lesions was seen in urinary bladders by stevioside. On the other hand, the carcinogenicity of bladder was affected by the stevioside dose on amine.

Similarly, there was evidence of pre- or nonneoplastic lesions in tissues. The derivatives of stevioside and itself have an ability to regulate the inflammation and microbial processes by influencing the human monocytic cell line (THP-1). Stevioside and its metabolites prohibit polysaccharide that stimulates the production of proinflammatory cytokines (interleukine-1β) and (tumor necrosis factor-alpha, TNF-α). Overall, Stevioside can restraint fallout as a result of inflammation.[51]

### 10.2.3 ANTIMICROBIAL ACTIVITY

Scientists and technologists are gaining the interest in plant extracts to design the new drugs for the expressive remedy of diseases including the infectious diseases.[13] One of the studies on the plant extracts performed has concluded the advent of an antibiotic, that is, streptomycin which is proved very useful for humans.[14]

Essential oils, tannins, and some other compounds are found in most of the plant leaves as antimicrobial agents. Moreover, plant-originated flavonoids and tannins are responsible for antibacterial and other biological performance. Aromatic compounds, including the phenols and their derivatives, are synthesized in sufficient amounts in plants. Protection from deterioration and microbial attack is the main function of such compounds.[28] Some bacteria, including the pathogens, are sensitive to Stevia, that is, it inhibits their growth.[22] The prevention of flu and onset of cold has been claimed in many cases because of use of Stevia leaf extract. The traditional use of Stevia, including the gum disease and sores, are because of its antibacterial activity. This herb is also effective against streptococcus species as well as yeast infections.[52] Many studies had been conducted on the antimicrobial aspects of plant roots, stems, fruits, and leaves in previous decades. Researchers have evaluated Stevia extracts using different solvent and found it effective against some selected species, including *Aeromonas hydrophila, Bacillus subtilis, Salmonella typhi*, and *Vibrio cholera*.[2,60] It is used as a replacement of sucrose because diterpene glycoside is produced in the leaves and it is sweeter than sucrose. In countries like Japan and Paraguay, Stevia is being used in daily diet due to its medicinal nutritive properties.[22]

### 10.2.4 ANTITUMOR ACTIVITY

Cancel cells lose its control over the cell cycle and abnormal and unchecked growth of cell occurs.[7] The development of cancerous cells damages the DNA and this damage then further mounts up in the cell.[8] Cancer can be caused by internal and external factors like mutations, hormones, metabolites, chemicals, radiations, and infections. Fighting the cancer is a worldwide phenomenon.[9]

Stevia leaf extracts can be used to combat the cancerous cells as it causes no genotoxic or carcinogenic effects on the tissues of mammals.[10,11] Labdane sclareol is the anticancer compound that has been detected in the leaf extracts of Stevia.[12] Stevia leaf extracts containing steviol, aglycones, and isosteviols block Epstein–Barr virus early antigen (EBV-EA) induction and inhibit tumors growth.[13] The antitumor activity of *S. rebaudiana* was studied under in vitro conditions on human laryngeal epithelioma cells (HEp2). Aqueous extracts showed no antitumor activity, but acetone extracts showed maximum antitumor activity followed by chloroform and ethyl acetate extracts.[14] It was observed that the hydrolysis products of Stevia, steviols, stevosides, and iso-steviols, reduce and inhibit the DNA replication of cancerous cells in human cells.[15] Stevoids are seen to slow down the growth of tumor cells in skin of mice.[16]

### *10.2.5 OTHER POTENTIAL HEALTH BENEFITS*

The regular use of caloric sweeteners (such as sugars) may cause gingivitis, plaque formation, and cavities due to harmful bacterial growth in the oral cavity. However, Stevia is zero-caloric sweetener with bacteriostatic and bacteriocidal abilities to improve oral health and to combat with associated dental problems. Stevia extracts and its major secondary metabolites can significantly contribute to oral health.[59]

Inflammatory bowel disease occurs in small intestine and colon. It is a form of ulcerative colitis and Crohn's disease. The cause of this disease is not known whether it is genetic or autoimmune.[17] Steviols, stevosides, and its polyphenolic compounds have shown to possess anti-inflammatory effects on colon epithelial tissues. The study on animals shows that stevoside minimizes the contractions of smooth muscles of bowels, which cause the hypermobility and diarrhea.[18] The in vivo study on the colon epithelial T84 cells showed that stevoside, steviol, isosteviol, isosteviol 16-oxime, and dihydroisosteviol have antidiarrheal effects on cAMP-regulated chloride (Cl) secretion. These compounds inhibit the chloride ion secretion in dose-dependent manner.[19]

## 10.3 STATUS OF STEVIA IN PAKISTAN

It is very important to acknowledge health benefits of Stevia. Stevia plant due to its nutritional values must be grown in different regions of Pakistan like Swat, Hunza, Chilas, Malakand, Rawalpindi, and Islamabad, where temperature extends between 20°C and 30°C. A lot of research work is needed in Pakistan. Due to present situation of sugar emergencies in Pakistan, Stevia is being promoted by government. Instructional meetings and courses ought to be directed by government and private divisions to prepare farmers to cultivate Stevia in Pakistan. Systems with respect to Stevia development and significance must be distributed by the print and electronic media to illuminate the farmers about the significance of Stevia harvest.

## 10.4 SUMMARY

Overall, Stevia plant is a good substitute for sugar. It contains antioxidant compounds like phenolics, including flavonoids, which allow health promoting effects on consumers. However, keeping negative outcomes under consideration, there is a need to conduct more research work on plant extract and its effects on human health and pharmacological aspects such as absorption, metabolism, and excretion. Safe use of Stevia in human is still controversial and more data is needed to clear off the danger of cancer and genotoxicity induced by Stevioside and steviol glycosides.

## KEYWORDS

- **glycoside**
- **nutraceuticals**
- **Stevia**
- **steviol**
- **stevioside**
- **tannins**

## REFERENCES

1. Ajagannanavar, S. L.; Shamarao, S.; Battur, H.; Tikare, S.; Al-Kheraif, A. A.; Al Sayed, M. S. Effect of Aqueous and Alcoholic Stevia (Stevia rebaudiana) Extracts Against Streptococcus. *J. Int. Soc. Prevent. Commun. Dentistry* **2014,** *4* (2), 116–121.
2. Akihisa, T.; Hamasaki, Y.; Tokuda, H.; Ukiya, M.; Kimura, Y.; Nishino, H. Microbial Transformation of Isosteviol and Inhibitory Effects on Epstein–Barr Virus Activation of the Transformation Products. *J. Nat. Product* **2004,** *67* (3), 407–410.
3. Atteh, J.; Onagbesan, O.; Tona, K.; Buyse, J.; Decuypere, E.; Geuns, J. Potential Use of Stevia rebaudiana in Animal Feeds. *Arch. Zootec* **2011,** *60* (229), 133–136.
4. Atteh, J.; Onagbesan, O.; Tona, K.; Decuypere, E.; Geuns, J.; Buyse, J. Evaluation of Supplementary Stevia (*Stevia rebaudiana Bertoni*) Leaves and Stevioside in Broiler

Diets: Effects on Feed Intake, Nutrient Metabolism, Blood Parameters and Growth Performance. *J. Animal Physiol. Animal Nutr.* **2008,** *92* (6), 640–649.
5. Avent, A. G.; Hanson, J. R.; de Oliveria, B. H. Hydrolysis of the Diterpenoid Glycoside, Stevioside. *Phytochemistry* **1990,** *29* (8), 2712–2715.
6. Awney, H. A.; Massoud, M. I.; El-Maghrabi, S. Long Term Feeding Effects of Stevioside Sweetener on Some Toxicological Parameters of Growing Male Rats. *J. Appl. Toxicol.* **2011,** *31* (5), 431–438.
7. Aze, Y.; Toyoda, K.; Imaida, K.; Hayashi, S.; Imazawa, T.; Hayashi, Y.; Takahashi, M. Subchronic Oral Toxicity Study of Stevioside in F344 Rats. *Eisei Shikenjo Hokoku* **1991,** *109*, 48–54.
8. Barriocanal, L.; Palacios, M.; Benitez, G.; Benitez, S.; Jimenez, J. T.; Jimenez, N. Apparent Lack of Pharmacological Effect of Steviol Glycosides Used as Sweeteners in Humans, a Pilot Study of Repeated Exposures in Some Normatensive and Hypotensive Individuals and in Type 1 and Type 2 Diabetics. *Regulat. Toxicol. Pharmacol.* **2008,** *51* (1), 37–41.
9. Cardello, H. M.; DaSilva, M. A.; Damasio, M. H. Measurement of the Relative Sweetness of Stevia Extract, Aspartame and Cyclamate/Saccharin Blend as Compared to Sucrose at Different Concentrations. *Plant Foods Hum. Nutr.* **1999,** *54* (2), 119–130.
10. Ceunen, S.; Geuns, J. M. C. Steviol Glycosides: Chemical Discovery, Metabolism and Function. *J. Nat. Prod.* **2013,** *76* (6), 1201–1228.
11. Chatsudthipong, V.; Muanprasat, C. H. Stevioside and Related Compounds: Therapeutic Benefits Beyond Sweetness. *Pharmacol. Therap.* **2009,** *121* (1), 41–54.
12. Chen, J.; Jeppesen, P. B.; Nordentoft, I.; Hermansen, K. Stevio Dose Counteracts the Dlyburide-induces Desensitization of the Pancreatic Beta-cell Function in Mice: Studies *in Vitro*. *Metabolism* **2006,** *55* (12), 1674–1680.
13. Divisi, D.; Di, T. S.; Salvemin, S.; Garramone, M.; Crisci, R. Diet and Cancer. *Acta Biomedica* **2006,** *77*, 118–123.
14. Gupta, E.; Purwar, S.; Rai, G. K. Nutritional and Therapeutic Values of Stevia rebaudiana: A Review. *J. Med. Plant Res.* **2013,** *7* (46), 3343–3353.
15. EU Commission Regulation (EU) No. 1131/2011. *Official J. Eur. Union* **2011,** *295*, 205–211.
16. European Food Safety Authority. Scientific Opinion on the Safety of Steviol Glycosides for the Proposed Uses as Food Additive. *EFSA J.* **2010,** *8*, 1–84.
17. Gardana, C.; Scaglianti, M.; Simonetti, P. Evaluation of Steviol and Its Glycosides in Stevia rebaudiana Leaves and Commercial Sweetener by Ultra-high-performance Liquid Chromatography–Mass Spectrometry. *J. Chromatogr.* **2010,** *12* (17), 1463–1470.
18. Gerrit, J.; Gerwing, M.; te-Poele, L.; Dijkhuizen, L.; Kamerling, P. Stevia Glycosides: Chemical and Enzymatic Modification of Their Carbohydrate Moieties to Improve the Sweet-tasting Quality. *ESL J.* **2016,** *73*, 1–72.
19. Goyal, S.; Samsher, G. R. Stevia (*Stevia rebaudiana*) a Biosweetener: Review. *Int. J. Food Sci. Nutr.* **2010,** *61*, 1–10.
20. Misra, H.; Soni, M.; Silawat, N.; Mehta, D.; Mehta, B. K. and Jain D. C. Antidiabetic Activity of Medium-polar Extract from the Leaves of *Stevia rebaudiana* Bert. (Bertoni) on Alloxan-induced Diabetic Rats. *J. Pharmacol. Bioallied Sci.* **2011,** *3* (2), 242–248.
21. Jayaraman, S.; Manoharan, M.; Illanchezian, S. In-vitro Antimicrobial and Antitumor Activities of *Stevia rebaudiana* (Asteraceae) Leaf Extracts. *Trop. J. Pharm. Res.* **2008,** *7* (4), 1143–1149.

22. JECFA (Joint FAO/WHO Expert Committee on Food Additives). *Safety Evaluation of Certain Food Additives*, Presented at the 69th Meeting of the Joint FAO/WHO Expert Committee on Food Additives; Rome; 2008; pp 23.
23. Jeppesen, P. B.; Gregersen, S.; Poulsen, C. R. Stevioside Acts Directly on Pancreatic Beta Cells to Secrete Insulin: Actions Independent of Cyclic Adenosine Monophosphate and Adenosine Triphosphate-Sensitive K+-channel Activity. *Metabolism* **2000,** *49* (2), 208–214.
24. Jeppesen, P. B.; Gregersen, S.; Rolfsen, S. E.; Jepsen, M.; Colombo, M.; Agger, A.; Xiao, J.; Kruhøffer, M.; Orntoft, T.; Hermansen, K. Antihyperglycemic and Blood Pressure: Reducing Effects of Stevioside in the Diabetic Goto-Kakizaki Rat. *Metabolism* **2003,** *52* (3), 372–378.
25. Jutabha, P.; Toskulkao, C.; Chatsudthipong, V. Effect of Stevioside on PAH Transport by Isolated Perfused Rabbit Renal Proximal Tubule. *Canadian J. Physiol. Pharmacol.* **2000,** *78* (9), 737–744.
26. Marcinek, K.; Krejpcio, Z. *Stevia rebaudiana* Bertoni: Health Promoting Properties and Therapeutic Applications. *J. Consumer Protect. Food Safety* **2016,** e-article: DOI 10.1007/s00003-015-0968-2.
27. Kaushik, R.; Narayanan.; P.; Vasudevan, V., Muthukumaran, G., Antony, U. Nutrient Composition of Cultivated Stevia Leaves and the Influence of Polyphenols and Plant Pigments on Sensory and Antioxidant Properties of Leaf Extracts. *J. Food Sci. Technol.* **2010,** *47* (1), 27–33.
28. Kinghorn, A. D., Soejarto, N. P. D.; Nanayakkara, C. M. A Phytochemical Screening Procedure for Sweet Entkaurene Glycosides in the Genus Stevia. *J. Nat. Product* **1984,** *47* (3), 439–444.
29. Kornbluth, A.; Sachar, D. B.; Salomon, P. Crohn's Disease. In *Gastrointestinal and Liver Disease*; Feldman, M., Sleisenger, M. H., Scharschmidt, B. F., Eds.; 6th ed. WB Saunders: Philadelphia, PA, USA, 1998; pp 23–28.
30. Krishnamurthi, K. Screening of Natural Products for Anticancer and Antidiabetic Properties. *Health Admin.* **2007,** *1*, 69–75.
31. Lozano, R. Global and Regional Mortality from 235 Causes of Death for 20 Age Groups in 1990 and 2010: Systematic Analysis for the Global Burden of Disease Study. *Lancet* **2012,** *380* (9859), 2095–2128.
32. Markovie, I. S.; Darmati, Z. A.; Abramovic, B. F. Chemical Composition of Leaf Extracts of *Stevia rebaudiana* Bertoni Grown Experimentally in Vojvodina. *J. Serbian Chem. Soc.* **2008,** *73* (3), 283–297.
33. Matsukubo, T.; Takazoe, I. Sucrose Substitutes and Their Role in Caries Prevention, *Int. Dental J.* **2010,** *56* (3), 119–130.
34. McQuate, R. S. Ensuring the Safety of Sweeteners from Stevia. *Food Technol.* **2011,** *65* (4), 1–13.
35. Megeji, N. W.; Kumar, J. K.; Singh, V., Kaul, V. K.; Ahuja, P. S. Introducing *Stevia rebaudiana*, a Natural Zero-calorie Sweetener. *Curr. Sci.* **2005,** *88*, 801–804.
36. Melis, M. S. Renal Excretion of Stevioside in Rats. *J. Nat. Prod.* **1992,** *55* (5), 688–690.
37. Takasakia, M.; Konoshimaa, T.; Kozukab, M.; Tokuda, M.; Takayasuc, J.; Nishinoc, H.; Miyakoshid, H.; Mizutanid, K.; Leeb, K. Cancer Preventive Agents. Part 8: Chemopreventive Effects of Stevioside and Related Compounds. *J. Bioorganic Med. Chem.* **2009,** *17* (2), 600–605.
38. Mizushina, Y.; Akihisa, T.; Ukiya, M.; Hamasaki, Y.; Murakami, N. C.; Kuriyama, I.; Takeuchi, T.; Sugawara, F.; Yoshida, H. Structural Analysis of Isosteviol and Related

Compounds as DNA Polymerase and DNA Topoisomerase Inhibitors. *Life Sci.* **2005,** *77* (17), 2127–2140.
39. Mosetting, E.; Nes, W. R. Stevioside II: The Structure of Aglucon. *J. Organ. Chem.* **1955,** *20* (7), 875–883.
40. Mosetting, E.; Beglinger, U., Dolder, F.; Lichti, H.; Quitt, P.; Waters, J. A. The Absolute Configuration of Steviol and Isosteviol. *J. Am. Chem. Soc.* **1963,** *85* (4), 454–458.
41. Nakamura, Y.; Sakiyama, S.; Takenaga, K. Suppression of Syntheses of High Molecular Weight Non-muscle Tropomyosins in Macrophages. *Cell Motility Cytoskeleton* **1995,** *31* (4), 273–282.
42. Ogawa, T.; Nozaki, M.; Matsui, M. Total Synthesis of Stevioside. *Tetrahedron* **1980,** *36* (18), 2641–2648.
43. Pariwat, P.; Homvisasevongsa, S.; Muanprasat, C.; Chatsudthipong, V. A Natural Plant-derived Dihydroisosteviol Prevents Cholera Toxin-induced Intestinal Fluid Secretion. *J. Pharmacol. Exp. Therap.* **2008,** *324* (2), 798–805.
44. Purkayastha, S.; Markosyan, A.; Prakash, I.; Bhusari, S.; Pugh, G. Jr.; Lynch, B.; Roberts, A. Steviol Glycosides in Purified Stevia Leaf Extract Sharing the Same Metabolic Fate. *Regulat. Toxicol. Pharmacol.* **2016,** *77*, 125–133.
45. Roman, L. U.; Cambron, J. I.; del-Rio, R. E. Grindelane Diterpenoids from Stevia Subpubescens. *J. Nat. Prod.* **2000,** *63* (2), 226–229.
46. Goyal, S. K.; Samsher.; Goyal, R. K. Stevia (*Stevia rebaudiana*) a Bio-sweetener: a Review. *Int. J. Food Sci. Nutr.* **2010,** *61* (1), 1–10.
47. Sharma, N.; Mogre, R. Effect of Stevia Intervention on Lipid Profile. In *On Serving Farmers and Saving Farming—India Imperative and Global Perspective*; GBPUA & T: Pantnagar; 2007; Vol. 85, pp 10–12.
48. Shiozaki, K.; Fujii, A.; Nakano, T.; Yamaguchi, T.; Sato, M. Inhibitory Effects of Hot Water Extract of the Stevia Stem on the Contractile Response of the Smooth Muscle of the Guinea Pig Ileum. *Biosci. Biotechnol. Biochem.* **2006,** *70* (2), 489–94.
49. Shivanna, N.; Naika, M.; Khanum, F.; Kaul, V. K. Antioxidant, Anti-diabetic and Renal Protective Properties of *Stevia Rebaudiana. J. Diabetes Compl.* **2013,** *27* (2), 103–113.
50. Soejarto, D. D.; Douglas, K.; Farnsworth, N. R. Potential Sweetening Agents of Plant Origin – III: Organoleptic Evaluation of Stevia Leaf Herbarium Samples for Sweetness. *J. Nat. Prod.* **1982,** 45 (5), 590–599.
51. Srimaroeng, C.; Chatsudthipong, V.; Aslam khan, A. G. and Pritchard, J. B. Transport of the Natural Sweetener Stevioside and Its Aglycone Steviol by Human Organic Anion Transporter (hOAT1; SLCC22A8). *J. Pharmacol. Exp. Therap.* **2005,** *313* (2), 621–628.
52. Suttajit, M.; Vinitketkaumnuen, U.; Meevatee, U. Mutagenicity and human Chromosomal Effect of Stevioside, a Sweetener from *Stevia rebaudiana* Bertoni. *Environ. Health Persp.* **1993,** *101* (3), 53–56.
53. Madan, S.; Ahmad, S.; Singh, G. N.; Kohli, K.; Kumar, Y.; Madhukar, G. *Stevia Rebaudania* (Bert.) Bertonti: A Review. *Indian J Nat. Prod. Resour.* **2010,** *1* (3), 267–286.
54. Toyoda, K.; Matsui, H.; Shoda, T.; Uneyama, C.; Takada, K.; Takahashi, M. Assessment of the Carcinogenicity of Stevioside in F344 Rats. *Food Chem. Toxicol.* **1997,** *35* (6), 597–603.
55. Vis, E.; Fletcher, H. G., Jr. Stevioside. IV. Evidence That Stevioside Is a Sophoroside. *J. Am. Chem. Soc.* **1956,** *78* (18), 4709–4710.

56. Wood, H. B. Jr.; Fletcher, H. G. Stevioside III: The Anomeric 2,3,4,6-tetra-O-Acetyle -1-O-Mesitoyl-D-Glucopyranoses and Their Behavior with Alkali. *J. Am. Chem. Soc.* **1956,** *78* (1), 207–210.
57. Wood. H. B., Jr; Allerton, R.; Diehl, H. W.; Fletcher, H. G. Stevioside I: The structure of Glucose Meioties. *J. Organ. Chem.* **1955,** *20*, 875–883.
58. Xiao, J.; Hermansen, K. The Mechanism Underlying the Insulintropic Effect of Stevioside: Activation of Acetyl-CoA Carboxylase (Abstract). *Diabetes* **2005,** *54*, A131.
59. Yamasaki, K.; Kohda, H.; Koboyashi, T.; Kasai, R.; Tanakas, O. Structures of Stevia Diterpene-glucoside: Application of C-NMR. *Tetrahedron Lett.* **1976,** *17* (13), 1005–1008.
60. Yuajit, C.; Homvisasevongsa, S.; Chatsudthipong, L.; Soodvilai, S.; Muanprasat, C.; Chatsudthipong, V. Steviol Reduces MDCK Cyst Formation and Growth by Inhibiting Cftr Channel Activity and Promoting Proteasome-mediated CFTR Degradation. *Plos One* **2013,** *8* (3), e58871.

# PART IV

# Pharmacological Aspects of Cereals, Grains, and Tea

CHAPTER 11

# WHEATGRASS JUICE: A NUTRITIONAL WEAPON AGAINST VARIOUS MALADIES

TABUSSAM TUFAIL, FARHAN SAEED, MUHAMMAD AFZAAL, and HAFIZ ANSAR RASUL SULERIA

## ABSTRACT

Among the natural products, green extracts are isolated from various plants for utilization in various diet-based food products. Numerous bioactive components have already been utilized as therapeutic agent against various health-related disorders. Wheatgrass (*Triticum aestivum*) juice holds significant health claims owing to its better phytochemical profile: chlorophyll, flavonoids, apigenin, minerals, amino acids, enzymes, and vitamins, etc. Wheatgrass juice is considered as a health stimulant for blood tonic, as it elevates the content of hemoglobin in the blood and improves transport of oxygen, therefore, it is very useful for people suffering with anemia. Wheatgrass juice acts as a blood purifier momentously supporting in the exclusion of waste products of the body due to its detergent properties. Moreover, it is supportive in the elimination of harmful chemicals, carcinogens, and heavy metals from human body. The scavenging activity of free radicals of wheatgrass juice and its bioactive ingredients might be helpful in controlling progression of tumor growth and chemoprevention. This chapter focuses on the review of available literature on wheatgrass juice with special reference to its phytochemistry and allied health claims.

## 11.1 INTRODUCTION

Green foods are sources of enriched diets for humans to improve their health. Among most of the green foods or grasses, wheatgrass is a natural plant, cotyledons of wheat plant (*Triticum aestivum* L.) belonging to the family Poaceae (Gramineae). Wheatgrass (*Triticum aestivum*) are mature shoots and these are the only grasses for special purpose use for humans, due to its good nutritional source. Wheatgrass clumps are hollow, glabrous and simple, and its leaves are about 1–1.2 m tall, flat, and broad.[24]

Wheatgrass contains abundant of vitamins and nutrients for the living organisms. It is an outstanding source of iron, calcium, sodium, potassium, and magnesium including vitamin B, pro-vitamin A, vitamin E, vitamin K, and trace minerals (Table 11.1). Furthermore, wheatgrass contains eight live enzymes, 17 amino acids, and chlorophyll.[26] It is a source of approximately all vital amino acids, especially glutamic acid, aspartic acid, serine, alanine, and arginine, which are necessary to maintain our body functions. It has strong phytochemical profile and contains chlorophyll, flavonoids, apigenin, alkaloids, carbohydrates, saponins, gum, and mucilages.[24]

It is available in the form of juice or powder, and typically it is served fresh or freeze dried. The juice is generally consumed in the form of mixture with other drinks of fruits or vegetables or alone and it can also be purchased from many health food stores as fresh produce, frozen juice, tablets, and in the form of powder. Wheatgrass juice can provide energy that fulfills nutritional deficiencies and help to remove wastes that clog in cells, tissues, organs, and blood.[14]

Wheatgrass juice has the ability to dissolve the blemishes which may form from the breathing gases in the lungs. It can help to minimize fatigue, regulation of sugar and blood pressure, supportive in weight loss, increase strength, improvements in sleep, improvement in the process of elimination and digestion, support healthy, eyes, skin, teeth muscles, and joints, helpful in the healing of skin sores and ulcer, sluggish the process of cellular aging, maintain function of brain, and is advantageous in muscle cramping and arthritis. It also has ability to improve heart, reproductive organs, and lungs functioning.[21]

This chapter reviews health benefits of wheatgrass.

## 11.2 COMPOSITION OF WHEATGRASS

Wheatgrass is gaining popularity as a functional food due to its great nutritional profile. Along with a great history, wheatgrass has been used as a health food supplement.[27] The wheatgrass juice is a cause of alkalinity, curative rudiments, and prophylactic. Wheatgrass is declared as "green blood" owing to its maximum proportions of chlorophyll that is about 70%

and furthermore it contains 92 different minerals that are necessary for human body.[11]

Wheatgrass can be considered as a complete food that has proteins, bioflavonoids, and many other necessary nutrients to maintain our body functions.[15] It is described that overall nutritional value from vegetables is equal to that in 15 pounds of wheatgrass. According to Table 11.1, wheatgrass has higher nutritional value (per 3.5 g), such as, 18.5 mg chlorophyll, 860 mg protein, 15 mg calcium, 7.5 mg Ascorbic acid, 38 mg lysine, and a potential source of micronutrients like amino acids and vitamins, a complex of both fat and water-soluble vitamins.[18] Vitamin K is found higher in wheatgrass juice, which is a blood-clotting agent.[20] It contains minerals like phosphorous, calcium, zinc, molybdenum, and boron. Various enzymes (i.e., super oxide dismutase (SOD), amylase, protease, cytochrome oxidase, lipase, transhydrogenase) are responsible for nutraceutical attributes of wheatgrass. It has been reported that drying method affects the efficiency of bioactive molecules.[19] Wheatgrass powder that is dried in a controlled microwave system has the maximum value of total phenol, chlorophyll, 2,2-diphenyl-1-picrylhydrazyl (DPPH), and scavenging activity compared with the samples dried with hot air.[26]

**TABLE 11.1** Compositional Profile of Wheatgrass Juice.

| Basic nutrients, g | Minerals, mg | Vitamins, mg |
|---|---|---|
| Calories: 21.0 cal | Iron: 0.61 | Vitamin C: 3.64 |
| Fat: 0.06 | Potassium: 147 | Vitamin B1: 0.08 |
| Choline: $92.4 \times 10^{-3}$ | Selenium: <1 ppm | Vitamin B5: 6.0 |
| Water: 95 | Magnesium: 24 | Vitamin A: 427 |
| Dietary fiber: <0.1 | Calcium: 24.1 | Vitamin B3: 0.011 |
| Chlorophyll: $42.1 \times 10^{-3}$ | Sodium: 10.3 | Vitamin B12: <1 mcg |
| Carbohydrates: 2.0 | Zinc: 0.33 | Vitamin B2 : 0.13 |
| Glucose: 0.80 | Phosphorous: 75.2 | Vitamin B6: 0.2 |

The wheatgrass juice popularity is due to its antioxidant potential and abundance of bioflavonoids like quercetin, apigenin, and luteolin. It is also a rich in vitamin A, C, and E and an outstanding source of vitamin-B complex that are obligatory for normal bone and brain growth and general metabolism of our body. Bioactive ingredients in wheatgrass are also considered as fighting agents against cancer and repairing agent against cellular damage of the lungs.[19]

### 11.2.1 CHLOROPHYLL

Wheatgrass juice is a potential source of photosynthetic pigments, that is, chlorophylls and carotenoids, which provide coloration to the food that is one of the evaluated visual quality characteristics.[30] Moreover, chlorophyll has promising role in the attenuating various lifestyle disorders such as cancer, oxidative stress, cardiovascular, and other chronic maladies.[22]

Wheatgrass is considered as an abundant source of chlorophyll and has the ability to convert chlorophyll into hemoglobin. Chemical structure of chlorophyll is also very similar to the hemoglobin. Level of hemoglobin in our body may increase due to the consumption of wheatgrass juice, because of this wheatgrass juice is a good source for the treatment of anemia.[11] Furthermore, wheatgrass juice also has a good percentage of antiradical activity due to its high chlorophyll content. Wheatgrass is good source of bioactive molecules, which can help our body's resistance to carcinogens. It neutralizes toxic compounds, strengthens the liver functioning enzymes and is very beneficial for the elimination of toxins from blood.[5]

Wheatgrass juice contains 70% chlorophyll. It has dilating effect on the blood vessels, increases iron content in the blood, and helps in rebuilding and detoxifying the body.[5] Chlorophyll also has antibacterial activity. Furthermore, wheatgrass can be used for healing purpose outside and inside the body. Although components of wheatgrass juice were recommended for the treatment of various illnesses including chronic distortions and provocative conditions from last three decades, yet no clinical trials exist till today. It is believed that wheatgrass juice has therapeutic qualities only in its fresh extract; therefore, its juice should be consumed immediately as extracted and should be consumed on an empty stomach. The chemical formation of the wheatgrass juice is very close to human blood profile, pH of both is 7.4 that helps to absorb it quickly in the blood.[27]

### 11.2.2 FLAVONOIDS

Flavonoids are important plant pigments that are manufactured from phenylalanine and usually exhibit wonderful colors in the flowering parts of plants.[3] They consist of a huge group of polyphenolic compounds, which are categorized by a benzo-y-pyrone structure. In addition to their significance in plants, flavonoids are vital for human health owing to their high pharmacological and especially radical scavenging abilities.[4] The main utilization of wheatgrass juice in clinical section is due to its high antioxidant activity that accounts for its higher proportion of flavonoids specially bioflavonoids like apigenin, quercetin, and luteolin. Indole compounds, apigenin, and laetrile, are the other compounds, which make the wheatgrass juice more effective.[28]

### *11.2.3 APIGENIN*

A Chemical 4,5,7, trihydroxyflavone known as apigenin, having molecular weight (MW) of 270.24 and molecular formula $C_{15}H_{10}O_5$. Naturally, apigenin is found as a dimer specifically extracted from flowers and buds of *Hypericum perforatum*, and is very effective for neuroprotection.[2] Apigenin is a type of flavonoid that belongs to the secondary class of plant metabolites having a common structure phenylchromanone (C6–C3–C6) with hydroxyl constituents. It is globally found in many vegetables, fruits, spices, and herbs. Apigenin has a range of bioactivities that may be beneficial against various health-related disorders.[17]

## 11.3 HEALTH ENDORSING PERSPECTIVES

### *11.3.1 ANTIOXIDANT POTENTIAL*

Phytonutrients having potential as an antioxidant, bioflavonoids, beta-carotene, and vitamin B, C, and E have nice ability to scavenge free radicals. Wheatgrass juice has highest antioxidant content owing to its higher chlorophyll content that helps to activate different antioxidant enzyme like SOD that alters reactive oxygen species (ROS) dangerous free radical into $H_2O_2$ (having extra $O_2$ molecules to kill cancerous cells) and $O_2$ molecules.[13] Wheatgrass juice exhibited advance oxygen radical absorbance capacity (ORAC), a typical implement for associating the antioxidant dimensions of different food products.[24,26]

For the neutralization of antagonistic effects of free radicals, the antioxidants in wheatgrass juice are very remarkable and can prevent damage to the cell structure and DNA.[16] Wheatgrass juice has ability to detoxify our body, boosts immunity level, prevent from DNA damage, enhances the production of red blood cells (RBCs), and helps to fight against various carcinogens. Chlorophyll present in wheatgrass juice has the ability to inhibit metabolic activities of various carcinogens. Wheatgrass juice contains different natural antioxidants that can maintain the activity of RBC and consistent possessions on cellular enzyme function and membrane integrity.[10]

### 11.3.2 TRANSFUSION THERAPY WITH WHEATGRASS JUICE

Many subjects are suffering from thalassemia, which is a transmissible form of anemia. Thalassemia is common in Mediterranean origin and occurs owing to the irregular synthesis of hemoglobin and a resultant RBCs shortened lifespan. All good after circumstantial provides valuable effect on requirements of transfusion.[12]

Thalassemia normally requires lifelong blood transfusion to enhance hemoglobin level and iron chelation in the absence of bone marrow transplantation (BMT). BMT possesses various limitations related to donor availability. Various agents such as hydroxyurea, butyrate, and azacytidine have been used as transfusion replacement.[12] No reliable therapeutic response pattern has been recognized in patients with thalassemia. As replacement of transfusion and to increase the baseline hemoglobin in thalassemia patients, it has been investigated that it was nontoxic and cost effective. It has been reported that freshly prepared wheatgrass juice can lower transfusion requirement in patients with thalassemia. It has been suspected that the possible mechanisms are owing to structural homology between chlorophyll and hemoglobin.[7]

### 11.3.3 HEPATOPROTECTIVE ROLE

It has been studied that the liver enzyme activities and lipid peroxidation is also maintained by wheatgrass extract. Hepatoprotective role of fresh wheatgrass juice has been studied in carbon tetrachloride ($CCl_4$)-treated rats that showed a considerable hepatoprotective effect in terms of serum glutamic pyruvic transaminase (SGPT), serum glutamic oxaloacetic transaminase (SGOT), bilirubin, and alkaline phosphatase (ALP) in serum at dose of 100 mg/kg/day.[9] It protects the liver because of abundance in choline, magnesium, and potassium; it helps in the maintenance of liver health for its vitality. Furthermore to stop the deposition of fat, its active component choline starts its activity. Similarly, magnesium eliminates the extra fat from the body. Potassium acts as an invigorator and stimulant. All these components are effectively present in wheatgrass juice.[5] It has been studied that the wheatgrass juice can reduce rheumatoid arthritis, severity of rectal bleeding, and it can also reduce the frequency of blood transfusions in patients with thalassemia.[12] Wheatgrass juice has also been effective for ulcerative colitis (UC). The observation declares that wheatgrass has significant effects against active distal UC and it is very safe for its treatment.[1]

Use of excessive drug therapy in ulcerative conditions produces significant toxic effects. An alternative of this medication is the use of plant-based medicines which are cheaper and have been rummage sale since extensive time without any harmful effects. Plants characterize a large unexploited basis of provision and development of novel drugs due to its novel structure.[8] It is evident from many studies that consumption of wheatgrass juice has benefits for humans in several ways like removing toxic metals from the cells, building the blood, cleanliness of the lymphatic system, and reestablishing balance in the body.[29]

### *11.3.4 INFLAMMATORY BOWEL DISEASE*

It has been observed that wheatgrass juice shows wound healing and anti-inflammatory capabilities. It has been shown that wheatgrass juice has maximum amount of chlorophyll and it has some bacteriostatic properties, which are helpful in wound healing, and inspires to produce erythrocytes and hemoglobin especially in anemic cells. It is used for the treatment of different types of skin lesions, ulcers, and burns; it acts as stimulating granulation tissue, wound healing agent, and epithelization.[6] Wheatgrass juice is presently under study as a possible remedy for anti-inflammatory owing to rich in bioflavonoids.[23]

## 11.4 SUMMARY

Wheatgrass juice is recognized as nature's true super food owing to its ability of supporting human life for longer periods because of the presence of various bioactive molecules. Wheatgrass juice generally contains no harmful substances except for a possible allergic reaction. It is a very good source of chlorophyll required for cell regeneration and endowed for its healing properties. This vital juice is important in curing certain harmful ailments, that is, cancer, ulcer, anemia, inflammatory bowel disease, rheumatoid arthritis, and thalassemia owing to presence of various vital nutrients, for example, chlorophyll, flavonoids, vitamins, and enzymes. Therefore, it is recommended that a specific part of wheatgrass should be the part of our daily dietary intake to discover its maximum benefits.

## KEYWORDS


- **bioactive components**
- **nutraceuticals**
- **pharmaceuticals**
- **wheatgrass juice**


## REFERENCES


1. Ben-Arye, E.; Goldin, E.; Wengrower, D.; Stamper, A.; Kohn, R.; Berry, E. Wheatgrass Juice in the Treatment of Active Distal Ulcerative Colitis: A Randomized Double-Blind Placebo-Controlled Trial. *Scand. J. Gastroenterol.* **2002,** *37* (4), 444–9.
2. Cheung, Z. H.; Leung, M. C.; Yip, H. K.; Wu, W.; Siu, F. K.; So, K. F. A Neuroprotective Herbal Mixture Inhibits Caspase-3-Independent Apoptosis in Retinal Ganglion Cells. *Cell. Mol. Neurobiol.* **2008,** *28* (1), 137–55.
3. Clifford, A. H.; Cuppett, S. L. Review: Anthocyanins Nature, Occurrence and Dietary Burden. *J. Sci. Food Agric.* **2000,** *80* (7), 1063–1072.
4. Cook, N. C.; Samman, S. Review: Flavonoids, Chemistry, Metabolism., Cardioprotective Effects., and Dietary Sources. *J. Nutr. Biochem.* **1996,** *7* (2), 66–76.
5. Ferruzzia, M. G.; Blakesleeb, J. Digestion., Absorption and Cancer Preventive Activity of Dietary Chlorophyll Derivatives. *Nutr. Res.* **2007,** *27* (1), 1–12.
6. Gahan, E.; Kline, P.; Finkle, T. Chlorophyll in the Treatment of Ulcers. *J. Dermatol. Syphil.* **1943,** *47* (6), 849–851.
7. Higgs, D. R.; Thein, S. L.; Woods, W. G. The Physiology of the Thalassemias. In: *The Thalassemia Syndromes*; Weatherall, D. J., Clegg. B., Eds.; Blackwell Science Ltd.: Oxford; 2001; pp 265–284.
8. Hostettmann, K. Recherches Actuelles Dans le Domaine des Plantes Médicinales. *Bull. de la Société Fribourgeoise des Sci. Naturelles* **1987,** *76* (1/2), 51–63.
9. Jain, G.; Argal, A.; Pathak, A. K.; Singh, V. K.; Kannojia, P. Hepatoprotective Activity of Wheatgrass Juice. *Pharmacist* **2007,** *2* (1), 29-30.
10. Khan, N.; Ganeshpurkar, A.; Dubey, N.; Bansal, D. Immuno Prophylactic Potential of Wheatgrass Extract on Benzene-Induced Leukemia: An *in Vivo* Study on Murine Model. *Indian J. Pharmacol.* **2015,** *47* (4), 394–397
11. Luo, J.; Fort, D. M.; Carlson, T. J.; Noamesi, B. K.; Amon-Kotei, D.; King, S. R. Cryptolepis Sanguinolenta: Ethnobotanical Approach to Drug Discovery and the Isolation of a Potentially Useful New Antihyperglycaemic Agent. *Diabetic Med.* **1998,** *15* (5), 367–374.
12. Marwaha, R. K.; Bansal, D.; Kaur, S.; Trehan, A. Wheatgrass Juice Reduces Transfusion Requirement in Patients with Thalassemia Major: A Pilot Study. *Indian Pediatric* **2004,** *41* (7), 716–720
13. Mates, M. J.; Jimenez, S.; Fransisca, M. Role of Reactive Oxygen Species in Apoptosis: Implications for Cancer Therapy. *Inter. J. Biochem. Cell Biol.* **2000,** *32* (2), 157–170.

14. Melina, V.; Davis, B. The New Becoming Vegetarian. In: *The Essential Guide to a Healthy Vegetarian Diet*; Healthy Living Publications: Summertown TN, 2010; pp 186–187.
15. Mogra, R.; Rath, P. Health Benefits of Wheatgrass - a Wonder Food. *Int. J. Food Nutr. Sci.* **2013,** *2* (4), 10–13.
16. Padalia, S.; Drabu, S.; Raheja, I.; Gupta, A.; Dhamija, M. Multitude Potential of Wheatgrass Juice (Green Blood): An Overview. *Chron. Young Sci.* **2010,** *1* (2), 23.
17. Patel, D.; Shukla, S.; Gupta, S. Apigenin and Cancer Chemoprevention: Progress, Potential and Promise (review). *Int. J. Oncol.* **2007,** *30* (1), 233–245.
18. Payal, C.; Davinder, K.; Sunaina, K. G.; Gagan, S.; Amit, C.; Dhawan, R. K. Wheat Grass: a Review on Pharmacognosy and Pharmacological Aspects. *Int. J. Phytopharmacol.* **2015,** *6* (2), 80–85.
19. Polshettiwar, S.; Khorate, S. S. *Triticum Aestivum* - A Green Gold. *World J. Pharm. Pharmaceut. Sci.* February **2016,** online.
20. Rana, S.; Kamboj, J. K.; Gandhi, V. Living Life the Natural Way: Wheatgrass and Health. *Function. Foods Health Dis.* **2011,** *9*, 444–456.
21. Roshan, K.; Rathore, K. S.; Bharkatiya, M.; Goel, P. K.; Naruka, P. S.; Saurabh, S. S. Therapeutic Potential of *Triticum aestivum* Linn. (Wheatgrass or green blood therapy) in the Treatment and Prevention of Chronic and Acute Diseases: An Overview. *Pharm. Tutor.* **2016,** *4* (2), 19–27.
22. Sangeetha, R. K.; Baskaran, V. Carotenoid Composition and Retinol Equivalent in Plants of Nutritional and Medicinal Importance: Efficacy of Bcarotene from Chenopodium Album in Retinol-Deficient Rats. *Food Chem.* **2010,** *119* (4), 1584–1590.
23. Shah, S. Dietary Factors in the Modulation of Inflammatory Bowel Disease Activity. *Medscape Gen. Med.* **2007,** *9* (1), 60.
24. Shirude, A. A. Phytochemical and pharmacological screening of Wheatgrass (*Triticum Aestivum* L.). *Int. J. Pharmaceut. Sci. Rev. Res.* **2011,** *9* (1), 159–164.
25. Singh, K.; Pannu, M. S.; Singh, P.; Singh, J. Effect of Wheat Grass Tablets on the Frequency of Blood Transfusions in Thalassemia Major. *Indian J. Pediatric* **2010,** *77* (1), 90–101.
26. Singh, P. Evaluation of Anti-Oxidant Properties of Wheatgrass Powder as Affected by Different Drying Processes. *Int. J. Pharmaceut. Sci.* **2016,** *7* (2), 852–855
27. Sundaresan, A.; Selvi, A.; Manonmani, H. K. The Anti-microbial Properties of *Triticum aestivum* (Wheatgrass). *Int. J. Biotechnol. Wellness Ind.* **2015,** *4* (3), 83–85.
28. Susan, P. P.; Gordon, D. G.; Douglas, V. F.; George, H. D.; Ikuta, T.; Witkowska, H. E. Short-term Trial of Butyrate to Stimulate Fetal-Globin-Gene Expression in the I-Globin Disorders. *New Engl. J. Med.* **1993,** *328* (2), 81–86.
29. Wigmore, A. *The Wheatgrass Book: How to Grow and Use Wheatgrass to Maximize Your Health and Vitality*; Avery Publishing Group Inc.: Wayne - New Jersey, 1985; pp 144; e-book ISBN-9780895292346.
30. Xue, L.; Yang, L. Deriving Leaf Chlorophyll Content of Green-Leafy Vegetables from Hyperspectral Reflectance. *ISPRS J. Photogram. Remote Sens.* **2009,** *64* (1), 97–106.

# CHAPTER 12

# FLAXSEED: A SHIELD AGAINST LIFESTYLE-RELATED MALADIES

HUMA BADER UL AIN, MUNAWAR ABBAS, FARHAN SAEED, and HAFIZ ANSAR RASUL SULERIA

## ABSTRACT

Human poor health is because of increased incidence of modern-day diseases such as: obesity, metabolic syndrome, diabetes, cardiovascular diseases (CVDs), polycystic ovarian disease, allergies of different sorts, and some types of cancer. Dietary modification by addition of a functional food can help the consumer to win battle against this assortment of diseases. In recent era, flaxseed has gained greater attention with special reference to greater components of fatty acid especially omega-3, that is, α-linolenic acid, lignans (secoisolariciresinol diglucosides), protein, dietary fiber, and phytoestrogen. Lignan has assorted array of biological properties with role in mitigating the risk factors of colonic and mammary carcinogenesis in animal models. Regular intake of flaxseed or its by-products has a significant effect on LDL, total serum level, reduction in postprandial absorption of glucose, and reduction of signs of inflammation. Flaxseed consumption showed its considerable free radical scavenging and anti-inflammatory activities in human and animal studies and it is found to have strong potential against different long-term disorders such as: cancer, CVDs, and some metabolic diseases. This chapter highlights flaxseed as a nutritional punch as well as its therapeutic role for numerous long-term complications.

## 12.1 INTRODUCTION

Development in the modern lifestyle has made the human diet more complex. Cardiovascular diseases (CVDs), diabetes, obesity, constipation, and intestinal cancer are primary maladies related to lifestyle that badly affect our health status. Consumer has paid more attention toward the natural remedies especially from the plants that can prevent the lifestyle-related maladies and maintain our body function. Herbs are plants or parts of plants and the main reason for consumption of herbs is the presence of various phytochemicals that show physiological effects.

Some traditionally used herbs are: aloe vera, spearmint, garlic, ginger, lemon, cinnamon, black walnut, and flax, etc. Among these, owing to the diverse health benefits, consumers are persistent to flax. Flax is nature's miraculous plant that helps in curing one's heart, blood, ageing, joints, colon, and brain conditions.[117] Whole flaxseed, flaxseed oil, forerunners of lignin, and their mucilage have important role in the preclusion and cure of various health-related disorders (Table 12.1). Flaxseed has also been considered as "*good mood food.*"[1] Numerous health claims of flaxseed are owing to the presence of higher omega-3 fatty acids concentration, lignin, protein, dietary fiber, and phytoestrogen and the consumer body's response toward these components. Dietary fiber in flaxseed is known to lower low-density lipoprotein (LDL)-cholesterol. Omega-3 fatty acids have a great impact in blood clotting time and reduction of CVDs and blood pressure and their related disorders.

The flax is a member of *Linaceae* family and especially grown for fiber and oil purposes. Its seeds are known as flaxseeds.[60] Flaxseed has been grown since 5000 BC as the basic source of linen fiber and now it is grown for oil production.[49] In Asia and Europe, whole flaxseed is being used for diet purpose.[110] Recently, European Union (EU) has registered about 67 cultivars of flax plant.[54] Globally, one of the biggest exporter and producer of flaxseed is Canada with 40% of production in the world, representing about 80% of the global flax trade. In the year 2013–2014, Canada produced about 614,000 metric tons of flaxseed.[58] Other significant flaxseed-producing countries are USA, India, China, and Ethiopia and now it is cultivated on more than 2.6 million hectares.[109]

India produced about 0.15 million tons of flaxseed.[35] Except the tropics and arctic, it can grow in every part of the world. The annual herb flax is about 1–2 m high and has a straight, flat stem, and green lined glaucous leaves. Its seeds are small, oval-shaped, flat, varying from golden yellow to reddish brown color with pointed tip.[45,71] Flaxseeds possess nutty taste and crispy texture.[101] Currently, two varieties of flaxseed (yellow and brown) are prominent categorized based on seed coat color and breed. These two varieties have almost similar nutritional values and same contents of short chain omega-3 fatty acids.[49] The dissimilarity is that yellow/golden flax is known as solin (trade name Linola), and have a completely dissimilar oil profile and low omega-3 fatty acids[32]; whereas, brown flax is notorious as an ingredient in paints, cattle feed, varnish, and fiber.[34,109] Length, width, and thickness of flaxseed dimensions are 4.0–6.0 mm, 2–3 mm, and 0.5–1.6 mm, respectively.[54,118,127]

**TABLE 12.1** Unsaturated Fatty Acids of Flaxseed and Their Health-promoting Functions.

| Components | Structure | Functions | References |
|---|---|---|---|
| ALA | 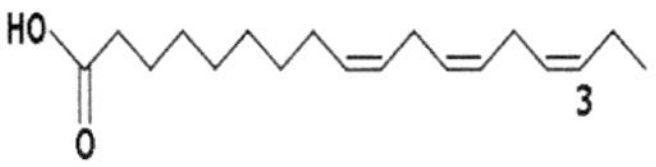 | • Anticancer<br>• Anti-inflammatory<br>• Cardioprotective<br>• Antidiabetes | [12] |
| Docosahexanoic acid |  | • Neuroprotective<br>• Anti-inflammatory<br>• Pregnancy and lactation | [130] |
| Ecosapentanoic acid | 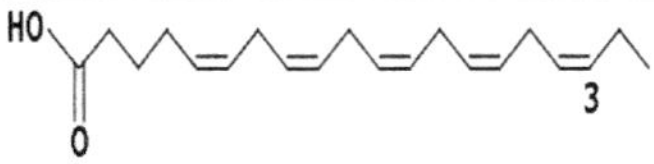 | • Anti-inflammatory | [63] |
| Linoleic acid | 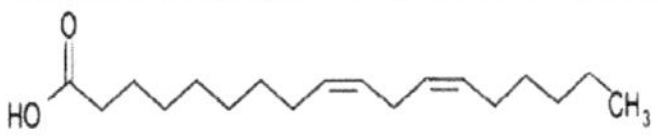 | • Cardioprotective<br>• Anticancer<br>• Anti-inflammatory | [6] |
| Oleic acid | 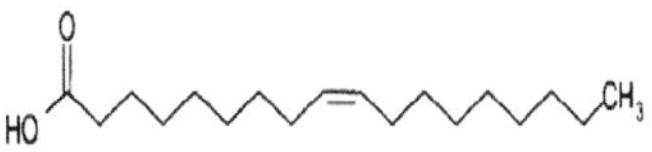 | • Hypotensive | [114] |

Flaxseed is scientifically known as *Linum usitatissimum* L. and in Latin usitatissimum means "most functional" or "most convenient."[33] This botanical name is assigned by Linnaeus in his book named "*Species Plantarum.*" [58] Flaxseed consists of kernel and hull which are 63% and 37% of whole flax, respectively.[109] Flaxseed contains oil, viscous lignan-rich fibers (mucilage), protein, and minerals as major components. Dolson[31] and Madhusudhan[68] reported that proportions of energy, carbohydrates, protein, total fat, and dietary fat are: 534 kcal, 7%, 10%, 53%, and 21%, respectively in a 100 g of flaxseed.

Flaxseed contains about 40–50% oil that is comprised of fatty acids having double or triple bonds like linoleic acid (LA), oleic acid, and linolenic acid in various amounts, that is, 12–30%, 8–29%, and 35–67%, respectively.[42,92] Various factors such as variety, atmosphere, and technique of analysis may affect the structure of flaxseed.[24] Flaxseed has high amount of B vitamins, manganese (Mn), and magnesium (Mg) [31] and low amounts of cholesterol.

Flaxseed contains polyunsaturated fatty acids about 73%.[68] It is a supplement food showing high performance and contain high amount of essential omega-3 polyunsaturated fatty acids (α-linolenic acid) and phytochemicals such as lignans.[5]

This chapter focuses on flaxseed as a nutritional source and for its therapeutic role in numerous long-term complications.

## 12.2 FLAXSEED: A FUNCTIONAL FOOD

The demonstration of ethno-medicinal potential associated with the consumption of flaxseed has stimulated consumer's interest in exploring functional ingredients present in the seed. Flaxseed has been considered as a multifarious functional food owing to its exceptionally high content of omega-3 fatty acids, dietary fiber, high quality protein, lignans, and phytoestrogen.[4,5] Lignan precursors, soluble fibers, and α-linolenic acid (ALA) are found in sufficient amounts in flaxseed.[44,117] Among the best-selling herbal products, flaxseed oil and whole flaxseed are at top in the market. Previous studies have shown significant health benefits by flaxseed and its bioactive compounds including the reduced risk of hyperlipidemia, CVDs, hypertension, and diabetes.[53,137]

Literature has demonstrated that omega-3 fatty acid, that is, ALA has anticancer, anti-inflammatory, cardioprotective, neuroprotective, anti-osteoporotic, and antioxidative effects. Even though limited data on toxicological scenario of ALA is available, yet no severe antagonistic effects have been observed. Currently, on the availability of this limited data, nontoxicity of ALA for addition in foods and as nutraceutical or pharmaceutical contestant is confirmed.

Secoisolariciresinol glucoside (SDG) present in flaxseed and its metabolites were reported to be impending antioxidant agents and possess antitumor, antimitotic, antidiabetic, anticancer, anti-lupus nephritis activities. Nowak et al. [82] claimed that lignan, a phytoestrogen present in flaxseed, is effective against polycystic ovarian syndrome by reducing the androgen level and irregular facial and body hair growth (hirsutism). It also regulates menstruation[41] and provides relief against breast pain.[15]

Different components of flaxseed including the soluble fiber may potentially affect the release of insulin and its way of action in sustaining glucose balance. Flaxseed had been seen to decrease the postprandial blood glucose response in humans.[40] The advantageous and disease prevention effects of flaxseed gum on diabetes mellitus and coronary heart disease make it popular in food industry.[16,128] Recent investigations have reported that most of the nongelling gums recently used in food manufacturing can be replaced by flaxseed gum due to its emulsifying and stabilizing properties. Moreover, many micronutrients in flaxseed mainly vitamin E can lower the risk of some diseases such as cardiovascular disorder, blood pressure, some types of cancer, and Alzheimer's disease by promoting sodium excretion in urine.

### *12.2.1 OMEGA-3 FATTY ACIDS*

Presence of higher amount of omega-3 fatty acids especially ALA in flaxseed has made it a functional food and distinctive oil seed crop.[12] About 48% of all the lipids are present in flaxseed and it is suggested to add it in regular diet.[35] According to Bozan and Temelli,[12] essential fatty acids (EFAs) especially Linoleic and Linolenic fatty acids make flaxseed nutraceutically important and enhance its health benefits. LA and alpha-LA are polyunsaturated fatty acids. Body needs both and cannot synthesize these in de novo system of body, and this is the reason that these two acids are considered as essential polyunsaturated fatty acids. Flaxseed contains 36–46% oil, out of which 46–53% is ALA, that is the important omega-3 fatty acid, and LA, the essential omega-6 fatty acid. Flaxseed is considered as richest source of ALA as it contains 52% ALA and 16.0% LA.

Main omega-6 and omega-3 fatty acids including LA and linolenic acid, respectively, are broken down into polyunsaturated fatty acids in mammalian cells. Furthermore, this LA is turned into ALA and diomo-γ-linolenic acid and produces the arachidonic acid (AA). This AA is further broken down and converted into eicosanoids or docosapentaenoic acid. This process of breaking down into different components takes place through the action of desaturase and elongase enzymes. Whereas, when we talk about the linolenic acid, desaturase, and elongase convert it into stearidonic acid and eicosatetraenoic acid, which give rise to the eicosapentaenoic acid (EPA) that is then broken down into docosahexaenoic acid or eicosanoid.[6]

As far as the health benefits of ALA in flaxseed oil are concerned, these provide cardioprotection by ameliorating the anti-inflammatory action and eicosanoid production.[104] Literature has demonstrated that omega-3 fatty acids, that is, ALA acid has anticancer, anti-inflammatory, cardioprotective, neuroprotective, antiosteoporotic, and antioxidative effects, highlighted in Table 12.2.[114] According to Kfapoulas et al.,[63] traditionally the composition of fatty acids has been used to show the purity in food industries. Palmitic acid from palm olien is used to check adulteration in flaxseed oil. Palmitic acid content in flaxseed oil is about 8–10%, whereas about 40% of palmitic acid is present in palm olein. While linolenic acid from soybean oil is used to represent adulteration in flaxseed oil. Linolenic acid content in flaxseed oil is about 15%, whereas soybean contains about 50% of linolenic acid.[7,123]

**TABLE 12.2** Subclasses of Lignin and Their Health-promoting Perspectives.

| Component | Structure | Functions | References |
|---|---|---|---|
| Lariciresinol | 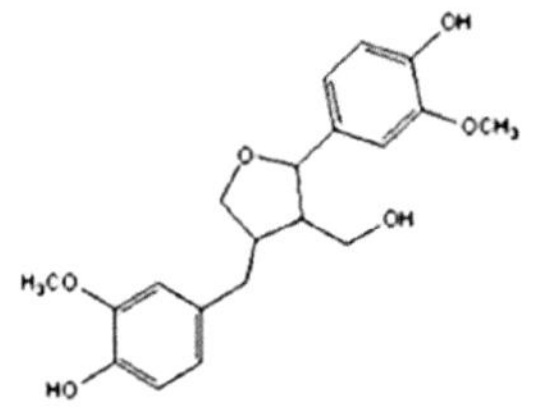 | • Antifungal<br>• Anticancer<br>• Antioxidant<br>• Cardioprotective<br>• Antiulcerogenic | [120] |
| Matairesinol | 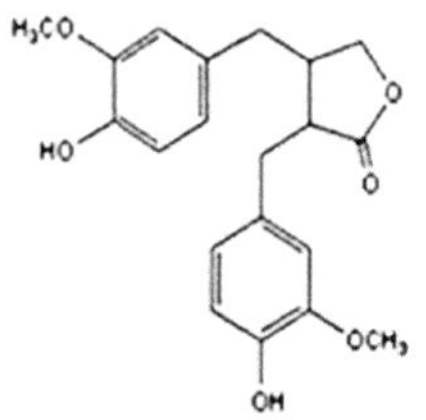 | • Cardioprotective<br>• Antimicrobiological<br>• Antiasthmatic<br>• Bone density<br>• Free radical scavanger<br>• Anticancer<br>• Endothelial dysfunction | [20] |
| Pinoresinol | 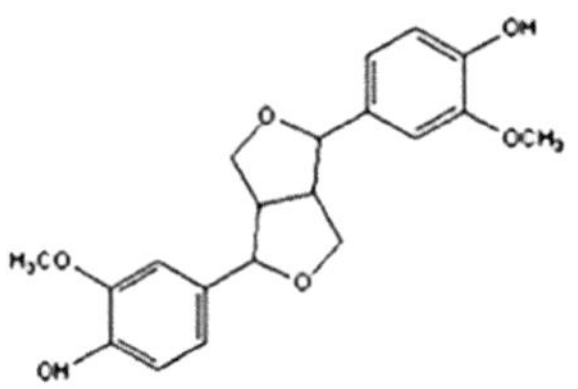 | • Antioxidant properties<br>• Hypocholestrolemic effect<br>• Polycystic ovarian syndrome | [103] |
| Secoisolarici-resinol | 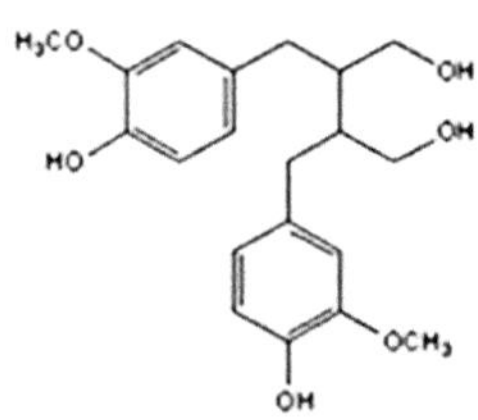 | • Hypocholestrolemic<br>• Antitumor<br>• Antidiabetic<br>• Anticancerous<br>• Weak estrogenic activities | [43,122] |

### 12.2.2 LIGNANS

Lignans are diphenolic compounds that are synthesized through the joining of two coniferyl alcohol residues present in cell wall of higher plants.[122]

Lignans are ever-present as phenolic compounds.[117] The most significant lignan is SDG, although pinoresinol, isolariciresinol, mataresinol, and other derivatives of ferulic acid are also present[24]; at 294–700 mg/100 g, 0.55 mg/100 g, 3.04 mg/100 g, and 3.32 mg/100 g concentrations, respectively, in flaxseed.[122] Johnsson et al. [43] reported that about 11.7–24.1 mg/g and 6.1–13.3 mg/g of SDG is present in defatted flour and whole flaxseed, respectively. Lignans are phytoestrogen and their rich sources are fiber-rich plants, legumes, cereals, vegetables, tea, fruits, berries, and alcoholic beverages. Compared with all these sources, flaxseed contains about 75–800 times more lignans.[69,70]

Along with SDG, lignans of flaxseed is considered as an important source of its aglycone secoisolariciresinol (SECO). These phenolics compounds can be broken down to the mammalian lignans enterodiol (ED) and enterolactone (EL) by human instestinal microflora.[38] Metabolism of these phenolic compounds to mammalian lignans by the gut microflora can occur in a series of reactions: (1) SDG generates the SECO by hydrolysis; (2) this SECO produces ED by dehydration and demethylation; and (3) ED produces EL by oxidation.[20] Up to 3% (w/w) concentrations of lignans have been reported in flaxseed, and this property makes flaxseed one of the richest edible sources of lignans. Lignans in flaxseed are not present in free form, that are incorporated in an oligomeric structure. Multi-faced activities are shown by the flaxseed lignans, which draw attention of researchers throughout the globe. High content of lignans especially SDG make the flaxseed relatively more important compared to other grains and legumes.[120] SDG, due to its antioxidant activity, is a potent hypolipidemic agent.

Studies demonstrated that SDG plays an important role in preventing many diseases such as CVDs, hypertension, cancers, and inflammatory as well as autoimmune disorders.[49] Lignans-ED and EL formed from SDG by bacteria in the gut present weak estrogenic or antiestrogenic as well as antioxidant effects due to which they provide health effects to human.[1] Siger et al. [106] identified that Ferulic acid, chlorogenic acid, and gallic acid are phenolic acids present in defatted flaxseed powder. Major flavonoids and flavones in flaxseed are C- and O-glycosides. Lignan consumption reduces the cardiovascular risk and inhibits the development of diabetes.[74] Antioxidant capacity and structural isomerism with 17-β-estradiol of flax lignans is their warranty for health benefits as they express scavenging activity against free radicals.[52,97]

### 12.2.3 PROTEINS

The average protein content in flaxseed is 22 g/100g of flaxseed. According to Hall et al. [48] flaxseed contains about 20–30% of protein comprising 80% globulins and 20% glutelin.[50,83] Cotyledons contain about 56–70% of protein and the remaining protein (30%) is present in the seed coat and endosperm.[28,112] Chung et al.[18] reported that flaxseed paste contains more protein (34%) as compared to flaxseed grain (21%), respectively and both are involved in health benefits (Table 12.3). Cool climate plays its role by increasing oil and decreasing protein content of the seeds.

**TABLE 12.3** Essential Amino Acids Present in Flaxseed and Their Functions.

| Amino acids | Functions | References |
|---|---|---|
| Histidine | • Antioxidant<br>• Anti-inflammatory | [121] |
| Isoleucine | • Muscle protein synthesis<br>• Decrease protein catabolism | [18] |
| Leucine | • Energy production<br>• Obesity<br>• Improves muscles<br>• Increase protein<br>• Anti-atherosclerosis | [19,50] |
| Lysine | • Treat herpes<br>• Antihypertensive<br>• Promote child growth<br>• Cardioprotective | [48] |
| Methionine | • Improves immune functions<br>• Production of collagen<br>• Psychiatric illnesses<br>• Infertility<br>• Musculoskeletal conditions | [83,112] |
| Threonine | • Nephropathy | [83] |
| Valine | • Energy production<br>• Improves liver health<br>• Neuropathy | [83] |

The flaxseed has two types of major storage proteins, which include a predominant salt-soluble fraction and a water-soluble basic component. The ionic part has higher molecular weight (MW) as compared to the water-soluble basic component (1.6–2S albumin; 17.7% nitrogen). Oomah[83] demonstrated that flaxseed is a good source of amino acids. Tyrosine, lysine, and threonine are the limiting aromatic amino acids while sulfur-containing amino acids (such as methionine and cysteine) and branched chain amino acids (such as isoleucine, leucine, and valine) are present in higher amounts. Flaxseed is a good source of some essential amino acids important for protein synthesis that maintain and repair the cells, tissues, and organs. Strong immunosuppressive and antimalarial activities are shown by cyclolinopeptide-A, an important bioactive peptide in flaxseed.[121]

Due to increased awareness and trust of health-conscious consumers on natural vegetable sources, flaxseed protein has gained higher popularity in nutrition and pharmaceutics industry. That is the reason, at industrial level, production of pure vegetable protein isolates are the dire need today.[62,89] According to Udenigwe and Aluko,[125] flaxseed protein has highest amount of arginine (11.2%) as compared to other food protein sources such as white egg, wheat, rapeseed, pea, and soy which have about 5.48%, 4.4%, 7.0%, 8.2%, and 7.6% of arginine, respectively. The flaxseed meal is useful in the production of about 34 million kg arginine annually. Large mass of arginine-enriched peptides of nutraceutical importance can also be produced from flaxseed. This arginine lowers blood pressure when present in the vascular endothelia.[124] Gopalan et al.[43] reported that 100 g of flaxseed has about 3.25 g nitrogen. Chung et al.[18] analyzed the protein content in flaxseed by sodium dodecyl sulfate (SDS)—PAGE and 2D-PAGE. However, the accurate analysis of these biomolecules was done by capillary electrophoresis (CE) and high-performance liquid chromatography (HPLC) coupled to mass spectrometry (MS). [83] Other methods for the analysis of protein are gas chromatography, nuclear magnetic resonance (NMR), and Fourier-transform infrared spectroscopy (FTIR).

### 12.2.4 DIETARY FIBER

Dietary fibers (Table 12.4) are known as functional ingredient of flaxseed.[107] It is nondigestible carbohydrate entity, which is resistant to enzymatic digestion and absorption in small intestine. The important categories of dietary fibers are natural and synthetic carbohydrates that possess optimistic well-being benefits. These fibers are not hydrolyzed by the endogenous enzymes in the colon of humans; and these pass down rapidly to the large intestine and are not fermented into short-chain fatty acids by the beneficial intestinal bacteria. There are two classes of dietary fibers with respect to their solubility: insoluble and soluble dietary fiber.[75]

**TABLE 12.4** Dietary Fibers Present in Flax.

| Component | Structure | Functions | References |
|---|---|---|---|
| Cellulose | 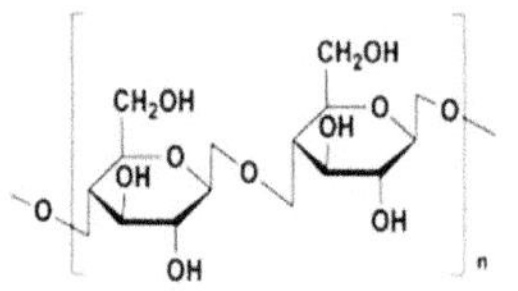 | • Laxative<br>• Reduce constipation<br>• Lower levels of diverticulitis<br>• Weight loss | [39] |
| Hemicelluloses | 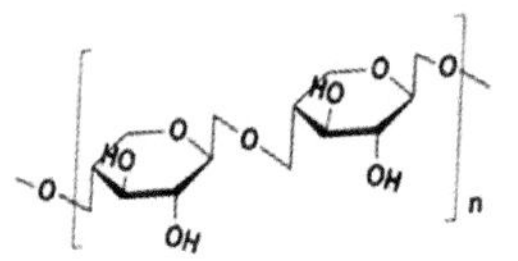 | • Laxative<br>• Reduce constipation<br>• Lower levels of diverticulitis<br>• Weight loss | [57] |
| Lignin | 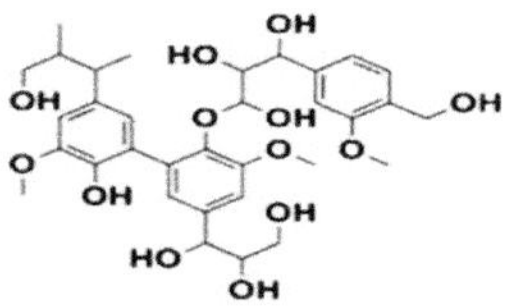 | • Laxative<br>• Heart health<br>• Immune function<br>• Celiac disease<br>• Gluten tolerance | [107] |

Each fraction has its own physiological effects.[39] Soluble fractions are fermented faster into short-chain fatty acids and are consumed by the beneficial intestinal bacteria than insoluble fractions. Approximately 20–35% of soluble dietary fibers and 65–80% of insoluble dietary fibers are present in foods (mainly cereals). According to the recommended daily allowance (RDA), one-quarter of soluble dietary fibers and three-quarters of insoluble dietary fibers (1:3) are recommended.

The main sources of dietary fibers are: cereals, nuts, fruits, and vegetables. The growing interest for dietary fiber is due to the functional properties and potential health benefits.[57] Different sources and type of dietary fiber have diverse roles to human physiology. From a functional perspective, dietary fiber is described as supporting laxation, attenuating blood glucose responses, and assisting in lowering of cholesterol.[14,39] Its physicochemical properties are changed, and functionality is improved through various processing methods such as extrusion cooking, boiling, frying, canning, and grinding. Dietary fiber can be scrutinized primarily via enzymatic gravimetric and enzymatic chemical methods.[29]

According to Cui,[22] flaxseed has 20% of insoluble and 9% of soluble fiber, respectively, whereas according to Hadley et al.,[47] content of insoluble and soluble fiber is 30% and 10%, respectively. Neutral detergent, total fibers, acid, and crude detergent are found in sufficient quantity in flax meal. The 20–25% of daily fiber needs can be fulfilled by a half ounce of dry whole flaxseed. Mucilage gums are the soluble fiber fractions, whereas cellulose and lignin are insoluble fiber fractions.[94,23] Fat excretion is significantly increased and total LDL-cholesterol is lowered by adding some flax-extract-rich fibers without any effect on appetite. It reduces the constipation by controlling the bowl movement.[40] Insoluble fiber helps in reducing the blood glucose level by lowering release of sugar in the blood[61] and helps to prevent diabetes.[101]

Studies demonstrated that the flaxseed helps in reducing the risk of lupus, hormone-dependent types of cancer, nephritis, and arteriosclerosis due to the presence of polysaccharides in it.[40] Chen et al. [16] described flaxseed gum as a heterogeneous polysaccharide comprising sugar, galactose, xylose, arabinose, fructose, rhamnose, and galacturonic acid and is hydrocolloid with tremendous water-holding competency that leads to immense swelling potential and it behaves as viscous component in aqueous solutions. Flaxseed gum is commercially used as emulsifiers, thickeners, stabilizers, or service provider contents in microencapsulation in the food processing industry and in food supply chains.[111] Emulsifying capacity, great viscosity, and stability are distinguishing properties of flaxseed gums and these properties make flaxseed gums most suitable components for food systems. Flaxseed mucilage is a gum-like material composed of acidic and neutral polysaccharides and is related to hull of flaxseed. The 62.8% xylose is present in neutral fraction of flaxseed, whereas the acidic fraction of flaxseed is comprised of mainly rhamnose (54.5%) followed by 23.4% galactose.[23]

### *12.2.5 PHENOLICS AND PIGMENTS*

Along with lignans and some other similar molecules present in flax are p-courmaric acid and ferulic acid.[115] Researchers have identified that ferulic acid, chlorogenic acid, gallic acid, and traces of 4-hydroxybenzoic acid are phenolic acids present in defatted flaxseed powder. The flaxseed has phenolic acids (such as lignans, phenylpropanoids), flavonoids (flavanones, flavonols, flavones, and anthocyanins), and tannins in large amount (approximately 8–10 g/kg); and etherified and esterified phenolics are 3–5 g/kg and 5 g/kg, respectively.

Phenolic acids are present in various forms such as free, ester-bound, etc. The content and types of phenolic acids in flaxseed vary widely. Different research studies by using different methods give different values. On an average, phenolic acids in ester-bound form are present in highest amounts in flaxseed. When the defatted flaxseed flour (DFF) was treated with 80% methanol, it was observed that about 320 mg/kg of ester-bound phenolic acids were present in flaxseed, compared to 200–280 mg/kg of free phenolic acids. Free phenolic acids were composed of vanillic, trans-*p*-coumaric, trans-sinapic, cis-sinapic, *o*-coumaric, and *p*-hydroxybenzoic acids. Furthermore, it was observed that flaxseed contained about 70 mg/kg of phenolic acids with extraction method and 730 mg/kg of phenolic acid with alkaline hydrolysis. This type of phenolic acid is about 89% of total phenolic acid and it mainly contains trans-ferulic acid and trans-sinapic acid. However, there was no free form of phenolic acid was present in flaxseed when treated with tetrahydrofuran. In whole flaxseed, about 11 mg of ferulic acid/kg of seed was observed.[129]

These phenolics provide protection against photooxidation and combat diseases. These are antioxidants, affecting the cell growth and viability, and providing protection against cancer and heart diseases.[8] For the determination of total phenols, the colorimetric procedures have been used which are carried out using Folin-ciocalteu reagent.[106] For the analysis of phenolic compounds, the most important analytical technique is high-performance liquid chromatography–mass spectrometry (HPLC–MS).

The major pigments in vegetables oils are chlorophylls and carotenoids. The key constituent of the chlorophyll group is pheophytin. There are two groups of carotenoids such as carotenes and xanthophylls. Beta-carotene, the main carotenoid, is present in flaxseed oil. Both chlorophylls and carotenoids as photosensitizer or singlet oxygen quenchers play imperative role in maintaining the quality of edible oils.[123] HPLC is used to analyze the carotenes, whereas gas chromatography degrades the compounds. A spectrophotometric method standardized by the International Union of Pure and Applied Chemistry (IUPAC) is used for the determination of chlorophyll in flaxseed oil.

## 12.3 HEALTH CLAIMS

### *12.3.1 CARDIOPROTECTIVE*

Currently, the cardiovascular-related complications can be prevented or delayed through three main lifestyle modifications: the dietary modification, daily exercise, and avoiding smoking.[100,134,135] In recent era, flaxseed has gained special interest among functional foods owing to the cardioprotective effects.

Omega-3 fatty acids especially ALA is the most important functional component with the strong potential against CVDs. Major dietary sources of these omega-3 fatty acids are marine food products that can be replaced by the flaxseed. Literature has demonstrated that intake of ALA has a positive effect.[64] Omega-3 fatty acids in flaxseed have the potential to mitigate the risk factors of CVDs,[64] including high blood cholesterol, triglyceride level, lipoprotein, high blood pressure, C-reactive protein, obesity, diabetes mellitus, estrogen, stress, etc.[87,99]

Moreover, lignans from flaxseed also contribute to the cardiovascular protection of human. This cardioprotective potential of lignan is due to its antioxidant property,[91,92] and it was confirmed by Zanwar et al.[136] through hemodynamic, biochemical, and histopathological tests of rats treated through the administration of lignan extract.

### 12.3.2 ANTIDIABETIC

As far as the antidiabetic effect of flaxseed is concerned, ALA in flaxseed has strong potential to combat an array of diabetic complications. Intake of ALA helps to boost the sensitivity to insulin and glycemic control.[80] Literature demonstrated that the consumption of 1–2 g/day of long-chain omega-3 PUFA reduces the occurrence of impaired glucose tolerance (prediabetes).[56,80]

Along with omega-3 fatty acids, flaxseed is also dense origin of lignan that is metabolized in mammalian intestine into ED and enterolactone. Various research studies confirm that flaxseed supplementation in diet prevented and delayed the metabolic syndrome and type-2 diabetes by providing hyperinsulinemic, hyperlipidemic, antioxidative, and anti-inflammatory effects. Thus, it eliminated the change of prediabetes to type-2.[3] It is concluded that flaxseed supplementation mends the sensitivity of insulin in prediabetic patients with risk of over weight, therefore, presenting a safe and inexpensive treatment for patients with all kinds of socioeconomic status.[30,98]

Oxidative stress is the main contributor to type-1 and type-2 diabetes. Prasad and Dhar[93] reported that flaxseed provide beneficial effects against both types of diabetes owing to the antioxidant, hypolipidemic, and hypoglycemic effects of its nutritional and functional ingredients mainly SDG. These SDG contents prevent and delay the incidence and development of both types of diabetes.[93]

### 12.3.3 ANTIOXIDANT

The main etiology of oxidative stress is the abnormality in the antioxidant defense actions and the levels of reactive oxygen species (ROS) balance.[46,95] Various research studies have probed the considerable changes in the blood antioxidant vitamins, lipid peroxidation, and antioxidant enzyme systems. The plasma antioxidant vitamins are vitamin A, E, and C; whereas antioxidant enzyme systems include catalase (CAT), superoxide dismutase (SOD), and glutathione peroxidase (GSH-Px).[65] Literature demonstrated that the oxidative stress caused the chronic diabetes mellitus along with long-term complications. These complications are seriously affected by the stability and reactive capacity of antioxidants.[81] Through the consumption of antioxidants, these effects of oxidative stress are reversed, that is, reduction of diabetes and its related complications by preventing the free radical-mediated damage.[116]

Phytoantioxidants especially phenolic compounds and vitamins in daily diet can keep a desirable antioxidative balance.[86] Numerous foods are used for their functional ingredients against different long-term disorders such as diabetes and its related complications.[102] Although flaxseed has therapeutic potential as antioxidant, primarily as hydroxyl radical scavenger, anticancer, antidiabetic, antiviral, bactericidal, anti-inflammatory, and antiatherosclerotic agent,[94] a few studies for evaluating its cardioprotective potential are presently available.[11]

These therapeutic potentials of flaxseed are due to the presence of lignans especially SDG in higher amount. This SDG keeps the polyunsaturated safe from oxidation due to its extreme free radical scavenging activity.[73] Along with these beneficial effects, one negative point of this flaxseed was also observed, such as the intake of flaxseed consumption considerably mitigates the levels of $\alpha$- and $\gamma$-tocopherol in rats and only the supplementation of vitamin E can restore and prevent this condition.[36]

Therapeutic functional flaxseed is commonly used as anti-inflammatory due to the functional ingredients such as ALA, SDG, and dietary fiber. Humans can use ALAs to synthesize EPA that shows anti-inflammatory effects through eicosanoids.[10]

### 12.3.4 ANTICANCER EFFECTS

Flaxseed has significant potential to reduce different types of cancer due to presence of fatty acids (especially ALA), lignans (especially SDG), and dietary fiber. Humans can use ALA to synthesize EPA, which is safe for use due to its strong potential against increase in cancer cells through affecting the pathways of cell signaling. It has been confirmed from various animal and human studies that ALA may increase the cancer cell growth and leads to the cell proliferation and angiogenesis.[10,119] Dietary lignan intake is associated with a decreased risk of developing postmenopausal breast cancer. High fiber intake is associated with decreased cancer mortality.

Flaxseed has gained interest in reducing the androgen levels in men with prostate cancer due to the presence of higher content of lignan and omega-3 fatty acids. Cancers and hormonal changes are also influenced by the fiber and fats present in the diet. Preventive and protective strategies may involve the flaxseed and diets enriched with low fats.[26,27] Many researchers have reported that flaxseeds provide protection against colon cancer owing to the anti-inflammatory activity of its functional ingredients (ALA, SDG, and dietary fiber) in animal studies, however there is still no study on human subjects with positive results.[55]

About 2.6 million people in the United States are suffering from breast cancer.[51] The awareness of this population toward the lifestyle modifications (i.e., in diet and exercise) to prevent and treat such aberrations is the current trend.[25] The dietary lignan provides protection against breast cancer through estrogen or anticarcinogenic activity.[90] It has also been mentioned that flaxseed has strong potential to demonstrate the anti-proliferative effects.[37] ALA possesses anti-inflammatory properties and may help to suppress the growth, size, and proliferation of breast cancer cells.

As far as the therapeutic role of flaxseed supplementation against uterine cancer is concerned, no sufficient data has been reported. A research study was conducted to evaluate the consequence of 25 g of flax on high risk postmenopausal women; and it was observed that flaxseed did not cause any change. Uterine growth that was enhanced by estrogen level and endometrial thickness was same as in baseline objects.[17,107] However, further research has reported that when the flaxseed is used in combination with tamoxifen, it decreases the uterine growth (i.e.,estrogen stimulated).

### *12.3.5 STRESS AND OBESITY*

Psychosocial factors can contribute to CAD.[131] The rise in blood pressure during mental stress is a strong predictor of atherosclerosis progression which is ameliorated by the consumption of flaxseed. Serum omega-3 fatty acids are related to many key markers, that is, mood, personality, and behavior in hypercholesterolemic subjects.[21] Spence et al.[113] reported that plasma cortisol level was reduced in postmenopausal women with high risk through the supplementation of flaxseed in daily diet.

Moreover, flax lignan mainly SDG also plays an important role in the treatment and prevention of mental stress and its related complications by decreasing the plasma cortisol and minutely ameliorating the plasma fibrinogen and resistance to periphery.[113] Literature demonstrated that SDG was effective against long-lasting hypertension mediated through the guanylate cyclase enzyme with no significant effects on lymphocyte proliferation. Therefore, it is concluded that SDG has no side effects on the immune system.

Dietary fiber from flaxseed may also protect against cancer by preventing weight gain or assisting with weight loss by increasing satiety. Nestel et al.[79] studied the comparative effect of ALA, oleic acid, and saturated fat supplementation of flaxseed oil in patients with obesity-related complications and it was found that ALA more significantly enhanced the arterial compliance and eliminated the LDL oxidation. Moreover, lignan especially SDG has potential to mitigate hyperlipidemia, hypercholesterolemia, hyperinsulinemia, and hyperleptinemia that are the cause of complications in liver. It is the result of adipogenesis-related gene expression regulation by the increase of peroxisome proliferator-activated receptor-gamma-mediated DNA-binding activity.[130] Obesity leads to an increase in inflammation and insulin resistance.[98] Studies also reported that flaxseed has received attention for its anti-inflammatory and antioxidant role. Due to these roles, it can reduce the weight by improving the lipid and metabolic profiles and by decreasing the risk factors related to CVD.

### *12.3.6 POLYCYSTIC OVARIAN SYNDROME*

Polycystic ovarian syndrome (PCOS) is one of the most common reproductive endocrine aberrations, which endorses the rising interest of endocrinologists. Patients with this disease often show multiple gynecologic, metabolic, and dermatologic manifestations with some principal signs of acne, excessive growth of body or facial hairs (hirsutism), and rogenetic alopecia; whereas, significant symptoms include menstrual abnormalities or disturbance, pain in pelvic region, more lutenizing hormone (LH), fewer follicles stimulating hormone (FSH), and difficulty becoming pregnant (infertility). The issue of PCOS is rising owing to an array of factors such as insulin resistance, obesity, hormonal disturbances, gene predisposition, stress, fatigue, and oral contraceptives but the exact etiology of this syndrome is still unknown.

The etiological factors include gene predisposition (gene susceptibility), strong stimulation of adrenal glands to produce more androgen, increased insulin production due to cell resistance to it, oral contraceptives, hormonal disturbance, stress, fatigue, and environmental factors.[96] As of now, about 15–20% women of childbearing age are anguishing from this condition.[72]

Flaxseed has considerable role in treating the complications of polycystic ovarian syndrome. It has antiandrogenic, antidiabetic, anticancer, antioxidant, and hypocholestrolemic effects owing to its omega-3, fatty acid, soluble dietary fiber, and lignan (SDG) contents.

Obesity and excess weight have a considerable effect on the manifestation of PCOS and related metabolic abnormalities, that is, hirsutism, excess of androgens, infertility, problems during pregnancy, pregnancy loss, and impaired insulin resistance. Obesity along with insulin resistance is associated with the worsening of diabetes mellitus type-2, CVDs and reproductive and metabolic features of PCOS.[67] Flaxseed also plays an important role in reducing the fat disposition that is attributed to its ALA content.[77] Findings suggest that flaxseed may have a profound impact on testosterone levels, and may diminish symptoms associated with hyperandrogenism, such as hirsutism.[105] Flaxseed has the potential to provide beneficial effects against these types of cancers due to the presence of ALA, lignans, and dietary fiber.

Wong et al.[133] investigated that dietary modification is important tool in reducing and controlling weight but it has no beneficial effects on hyperandrogenism. The chance of metabolic syndrome is higher in PCOS women than in normal healthy women; and women with PCOS are four times more vulnerable to type-II diabetes than the common population.[132] Flaxseed owing to the antioxidant potential of lignan and ALA reduces the risk of type-II diabetes.

## 12.4 SUMMARY

Flaxseed has wide range of essential nutrients and bioactive compounds including EFAs, proteins, dietary fiber, lignans, and phenolics. The daily diet supplemented with flaxseed should be adopted to prevent the lifestyle-related maladies. Supplementation of flaxseed as food ingredient opens the door for its therapeutical potential against long-term complications such as oxidative stress, diabetes, atherosclerosis, CVD, many types of cancers, and metabolic complications. Role of raw flaxseed and its baked products in health promotion and disease prevention is the actual concern of consumer.

Flaxseed should be further explored to find the exact nutritional and bioactive profile quantitatively with reference to diseases.

## KEYWORDS


- **cellulose**
- **flaxseed**
- **hemicelluloses**
- **lignin**
- **nutraceuticals**
- **phytoestrogen**


## REFERENCES


1. Adlercreutz, H. Lignans and Human Health. *Critical Rev. Clin. Lab. Sci.* **2007,** *44* (5–6), 483–525.
2. Adlercreutz, H.; Mousavi, Y.; Clark, J.; Hockerstedt, K.; Hamalainen, E.; Wahala, K.; Makela, T.; Hase, T. Dietary Phytoestrogens and Cancer: *in Vitro* and *in Vivo* Studies. *J. Steroid Biochem. Mol. Biol.* **1992,** *41* (3–8), 331–337.
3. Adolphe, J. L.; Whiting, S. J.; Juurlink, B. H. J.; Thorpe, L.; Alcorn, J. Health Effects with Consumption of the Flax Kignan Secoisolariciresinol Diglucoside. *Br. J. Nutr.* **2009,** *103* (7), 929–938.
4. Akpolat, M.; Kanter, M.; Topcu-Tarladacalisir, Y.; Aydogdu, N. Protective Effect of Flaxseed Oil on Renal Injury in Hyperlipidaemic Rats: Effect of Flaxseed Oil on Hyperlipidaemia. *Phytother. Res.* **2011,** *25* (6), 796–802
5. Amin, T.; Thakur, M. *Linum usitatissimum* L. (Flaxseed): Multifarious Functional Food. *Int. Interdiscip. Res. J.* **2013,** *IV* (I), 220.
6. Anjum F. M. et al. Impact of Extruded Flaxseed Meal Supplemented Diet on Growth Performance, Oxidative Stability and Quality of Broiler Meat and Meat Products. *Lipids Health Dis.* **2013,** *12* (1), 1–12.
7. Aparicio, R.; Ruiz, A. R. Authentication of Vegetable Oils by Chromatographic Techniques. *J. Chromatogr. A* **2000,** *881* (1–2), 93–104.
8. Arts, I.; Hollman, P. Polyphenols and Disease Risk in Epidemiologic Studies. *American J. Clin. Nutr.* **2005,** *81* (1), 317–325.
9. Aubrecht, E.; Horacsek, M.; Gelencser, E.; Dworschak, E. Investigation of Prolamin Content of Cereals and Different Plant Seeds. *Acta Alimentaria* **1998,** *27*, 119–125.
10. Berquin, I. M.; Edwards, I. J.; Chen, Y. Q. Multi-Targeted Therapy of Cancer by Omega-3 Fatty Acids. *Cancer Lett.* **2008,** *269* (2), 363–77.

11. Bloedon, L. T.; Szapary, P. O. Flaxseed and Cardiovascular Risk. *Nutr. Rev.* **2004,** *62* (1), 18–27.
12. Bozan, B.; Temelli, F. Chemical Composition and Oxidative Stability of Flax, Safflower and Poppy Seed and Seed Oils. *Bioresour. Technol.* **2008,** *99* (14), 6354–6359.
13. Brooks, J.; Thompson, L. Mammalian Lignans and Genistein Decrease the Activities of Aromatase and 17β-Hydroxysteroid Dehydrogenase in MCF-7 Cells. *J. Steroid Biochem. Mol. Biol.* **2005,** *94* (5), 461–467.
14. Brownlee, I. A. The Physiological Roles of Dietary Fiber. *Food Hydrocolloids* **2011,** *25* (2), 238–250.
15. Burns, L. *The Use of Flaxseed in the Treatment of PMS*, 2011. http://www.livestrong.com/article/512737-the-use-of-flaxseed-in-the-treatmentof-pms/(accessed Feb 10, 2018).
16. Chen, H. H.; Xu, S. Y.; Wang, Z. Gelation Properties of Flaxseed Gum. *J. Food Eng.* **2006,** *77* (2), 295–303.
17. Chen, J.; Power, K. A.; Mann, J.; Cheng, A.; Thompson, L. U. Flaxseed Alone or in Combination with Tamoxifen Inhibits Mcf-7 Breast Tumor Growth in Ovariectomized Athymic Mice with High Circulating Levels of Estrogen. *Exp. Biol. Med. (Maywood)* **2007,** *232* (8), 1071–80.
18. Chung, M. W. Y.; Lei, B.; Li-Chan, E. C. Y. Isolation and Structural Characterization of the Major Protein Fraction from nor Man Flaxseed (*Linum usitatissimum* L.). *Food Chem.* **2013,** *90* (1), 271–279.
19. Chung, M.; Lei, B.; Li-Chan, E. Isolation and Structural Characterization of the Major Protein Fraction from Nor Man Flaxseed (*Linum usitatissimum* L.). *Food Chem.* **2005,** *90* (1), 271–279.
20. Clavel, T.; Borrmann, D.; Braune, A.; Dore, J.; Blaut, M. Occurrence and Activity of Human Intestinal Bacteria Involved in the Conversion of Dietary Lignans. *Anaerobe* **2006,** *12* (3), 140–147.
21. Conklin, S. M.; Harris, J. I.; Manuck, S. B.; Yao, J. K.; Hibbeln, J. R.; Muldoon, M. F. Serum ω-3 Fatty Acids are Associated with Variation in Mood, Personality and Behavior in Hypercholesterolemic Community Volunteers. *Psychiatry Res.* **2007,** *152* (1), 1–10.
22. Cui, S. W. *Polysaccharide Gums from Agricultural Products: Processing, Structures and Functionality*. Technomic Publishing Co., Inc.: Lancaster – PA; 2001; pp 340–344.
23. Cui, W.; Kenaschuk, E.; Mazza, G. Influence of Genotype on Chemical Composition and Rheological Properties of Flaxseed Gums. *Food Hydrocol.* **1996,** *10* (2), 221–227.
24. Daun, J. K.; Barthet, V.; Chornick, T.; Duguid, S. Structure, Composition and Variety Development of Flaxseed. In: *Flax in Human Nutr.;* Thompson, L. and Cunanne, S., Eds.; Champaign: Illinois; 2003; pp 1–40.
25. Demark-Wahnefried, W.; Aziz, N. M.; Rowland, J. H. Riding the Crest of the Teachable Moment: Promoting Long-Term Health after the Diagnosis of Cancer. *J. Clin. Oncol.* **2005,** *23* (24), 5814–5830.
26. Demark-Wahnefried, W.; Polascik, T. J.; George, S. L.; Switzer, B. R.; Madden, J. F.; Ruffin, M. T.; Snyder D. C.; Owzar, K.; Hars, V.; Albala, D. M.; Walther, P. J.; Robertson, C. N.; Moul, J. W.; Dunn, B. K.; Brenner, D.; Minasian, L.; Stella, P.; Vollmer, R. T. Flaxseed Supplementation (Not Dietary Fat Restriction) Reduces Prostate Cancer Proliferation Rates in Men Pre-Surgery. *Cancer Epidemiol. Biomarkers. Prevent.* **2008,** *17* (12), 3577–87.
27. Demark-Wahnefried, W.; Price, D. T.; Polascik, T. J.; Robertson, C. N.; Anderson, E. E.; Paulson, D. F.; Walther, P. J.; Gannon, M.; Vollmer, R. T. Pilot Study of Dietary Fat Restriction and Flaxseed Supplementation in Men with Prostate Cancer Before

Surgery: Exploring the Effects on Hormonal Levels, Prostate-Specific Antigen, and Histopathologic Features. *Urology* **2001,** *58* (1), 47–52.

28. Dev, D. K.; Quensel, E.; Hansen, R. Nitrogen Extractability and Buffer Capacity of Defatted Linseed (*Linum usitatissimum* L.) Flour. *J. Sci. Food Agric.* **1986,** *37* (2), 199–205.
29. Dhingra, D.; Michael, M.; Rajput, H.; Patil, R. T. Dietary Fiber in Foods: A Review. *J. Food Sci. Technol.* **2012,** *49* (3), 255–266.
30. Dodin, S.; Cunnane, S. C.; Masse, B. et al. Flaxseed on Cardiovascular Disease Markers in Healthy Menopausal Women: A Randomized, Double-Blind, Placebo-Controlled Trial. *Nutrition* **2008,** *24* (1), 23–30.
31. Dolson, L. Flax Seed: The Low Carb Whole Grain Nutrition, Health Benefits of Flaxseed. *Online Int. Interdiscipl. Res.* **2010,** *IV* (I), 220–238.
32. Dribnenki, J. C. P.; McEachern, S. F.; Chen, Y.; Green, A. G.; Rashid, K. Y. Solin (Low Linolenic Flax). *Can. J. Plant Sci.* **2007,** *87* (2), 297–299.
33. El-Beltagi, H. S.; Salama, Z. A.; El-Hariri, D. M. Evaluation of Fatty Acids Profile and the Content of Some Secondary Metabolites in Seeds of Different Flax Cultivars (*Linum usitatissimum* L.). *Gen. Appl. Plant Physiol.* **2007,** *33* (3–4), 187–202.
34. Faintuch, J.; Bortolotto, L. A.; Marques, P. C.; Faintuch, J. J.; Franca, J. I.; Cecconello, I. Systemic Inflammation and Carotid Diameter in Obese Patients: Pilot Comparative Study with Flaxseed Powder and Cassava Powder. *Nutricion Hospital.* **2011,** *26* (1), 208–213.
35. FAO (FAOSTAT). *Food and agricultural organization of the united nation*, 2012. http://faostat3.fao.org/home/index.html#VISUALIZE_BY_DOMAIN, (accessed Feb 10, 2018).
36. Flis, M.; Sobotka, W.; Antoszkiewicz, Z.; Lipinski, K.; Zdunczyk, Z. The Effect of Grain Polyphenols and the Addition of Vitamin E to Diets Enriched with A-Linolenic Acid on the Antioxidant Status of Pigs. *J. Animal Feed Sci.* **2010,** *663*, 539–553.
37. Flower, G.; Fritz, H. Flax and Breast Cancer: Systematic Review. *Integr. Cancer Therapy* **2014,** *13* (3), 181–192.
38. Fuentealba, C.; Figuerola, F.; Estevez, A. M.; Bastiasc, J. M.; Munoz, O. Bioaccessibility of Lignans from Flaxseed (*Linum usitatissimum* l.) Determined by Single-Batch *in vitro* Simulation of the Digestive Process. *J. Sci. Food Agric.* **2014,** *94* (9), 1729–1738.
39. Fuller, S.; Beck, E.; Salman, H. New Horizons for the Study of Dietary Fiber and Health: A Review. *Plant Foods Hum. Nutr.* **2016,** *71* (1), 1–12.
40. Ganorkar, P. M.; Jain, P. K. Flaxseed: A Nutritional Punch. *Int. Food Res. J.* **2013,** *20* (2), 519–525.
41. Garvin, C. *Flaxseed and Menstruation*, 2011, http://www.livestrong.com/article/523346-flaxseeds-andmenstruation/ (accessed Feb 10, 2018).
42. Gaurav, K.; Dipak, K. M. Evaluation of Anti-Ulcer and Anti-Secretory Potential of *Linum Usitatissimum* Fixed Oil and Possible Mechanism of Action. *Inflam. Pharmacol.* **2010,** *18*, 137–145.
43. Gopalan, C.; Ramasastri, B. V.; Subramanian, S. C. *Nutritive Value of Indian Foods.* National Institute of Nutrition: Hyderabad; 2007; pp 214.
44. Goyal, A.; Sharma, V.; Upadhyay, N.; Gill, S.; Sihag, M. Flax and Flaxseed Oil: An Ancient Medicine and Modern Functional Food. *J. Food Sci. Technol.* **2013,** *51* (9), 1633–1653.
45. Guseva, D. A. Natural Source of Omega-3 Linseed-Oil: Its Particular Qualities and Metabolic Changes in the Organism. *Voprosy Pitaniia* **2010,** *79* (1), 13–22.

46. Gutteridge, J. M.; Halliwell, B. Antioxidants: Molecules, Medicines, and Myths. *Biochem. Biophys. Res. Commun.* **2010,** *393* (4), 561–564.
47. Hadley, M.; Lacher, C.; Mitchel-Fetch, J. *Fiber in Flaxseed.* Proceedings of the 54[th] Flax Institute of United States. Flax Institute of United States: Fargo, ND; pp 26–43.
48. Hall, C.; Tulbek, M. C.; Xu, Y. Flaxseed. *Adv. Food Nutr. Res.* **2006,** *51*, 1–97.
49. Halligudi, N. Pharmacological Properties of Flax Seeds: A Review. *Hygeia* **2012,** *4*, 70–77.
50. Hongzhi, Y.; Zhihuai, Hequn, T. *Determination and Removal Methods of Cyanogenic Glucoside in Flaxseed.* Paper Presentation at ASAE/CSAE Meeting, St Joseph – MI;. 2004; Paper #04066; pp 10.
51. Howlader, N.; Noone, A. M.; Krapcho, M. *SEER Cancer Statistics Review; 1975–2008*; National Cancer Institute: Bethesda, MD. http://seer.cancer. gov/csr/1975_2008/ (accessed Feb 10, 2018).
52. Hu, C.; Yuan, Y. V.; Kitts, D. D. Antioxidant Activities of the Flaxseed Lignin Secoisolariciresinol Diglucoside, Its Aglycone Secoisolariciresinol and the Mammalian Lignans Enterodiol and Enterolactone in Vitro. *Food Chem. Toxicol.* **2007,** *45* (11), 2219–2227.
53. Hutchins, A. M.; Brown, B. D.; Cunnane, S. C.; Domitrovich, S. G.; Adams, E. R.; Bobowiec, C. E. Daily Flaxseed Consumption Improves Glycemic Control in Obese Men And Women with Pre-Diabetes: A Randomized Study. *Nutr. Res.* **2013,** *33* (5), 367–75.
54. Iva, S.; Oscar, G.; Marie, B.; Martin, P.; Gianfranco, V. Phenotypic Evaluation of Flax Seeds by Image Analysis. *Ind. Crops Prod.* **2013,** *47*, 232–238.
55. Jenab, M.; Thompson, L. U. The Influence of Flaxseed and Lignans on Colon Carcinogenesis and Beta Glucuronidase Activity. *Carcinogenesis* **1996,** *17* (6), 1343–8.
56. Jenkins, D. J.; Jenkins, A. L.; Wolever, T. M.; Vuksan, V.; Rao, A. V.; Thompson, L. U. Low Glycemic Index: Carbohydrates and Physiological Effects of Altered Food Frequency. *Am. J. Clin. Nutr.* **1994,** *59* (3), 706S–709S.
57. Jha, R.; Berrocoso, J. D. Review: Dietary Fiber Utilization and Its Effects on Physiological Functions and Gut Health of Swine. *Animal* **2015,** *9* (9), 1441–1452.
58. Jhala, A. J.; Hall, L. M. Flax (*Linum usitatissimum L.*): Current Uses and Future Applications. *Australian J. Basic Appl. Sci.* **2010,** *4* (9), 4304–4312.
59. Johnsson, P.; Eldin, K. A.; Lundgren, L. N.; Aaman, P. HPLC Method for Analysis of Secoisolariciresinol Diglucoside in Flaxseeds. *J. Agri. Food Chem.* **2000,** *48* (11), 5216–5219.
60. Kaithwas, G.; Majumdar, D. K. Effect of *L. Usitatissimum* (Flaxseed/Linseed) Fixed Oil Against Distinct Phases of Inflammation. *Int. Scholarly Res. Netw. Inflam.* **2013.**
61. Kapoor, S.; Sachdeva, R.; Kochhar, A. Flaxseed: Potential Treatment of Lowering Blood Glucose and Lipid Profile Among Diabetic Females. *Indian J. Nutr. Dietetics* **2011,** *48* (12), 529–536.
62. Kaur, M.; Singh, N. Characterization of Protein Isolates from Different Indian Chickpea (*Cicer arietinum* L.) Cultivars. *Food Chem.* **2007,** *102* (1), 366–374.
63. Kfapoulas, V. M.; Emmanouilidou, P. S. Detection of Adulteration of Olive Oil with Seed Oils by a Combination of Column and Gas Liquid Chromatography. *J. Am. Oil Chem. Soc.* **1981,** *58* (6), 694–697.
64. Lamotte, M.; Annemans, L.; Kawalec, P.; Zoellners, Y. A Multi-Country Health-economic Evaluation of Highly Concentrated n-3 Polyunsaturated Fatty Acids in the Secondary Prevention after Myocardial Infarction. *Herz* **2006,** *31* (3), 74–82.

65. Lazo-de-la-Vega-Monroy, M. L.; Fernandez-Mejia, C. Oxidative Stress in Diabetes Mellitus and the Role of Vitamins with Antioxidant Actions. Chapter 9; In: *Oxidative Stress and Chronic Degenerative Diseases - A Role for Antioxidants*, InTechOpen.com; 2013; e-article.
66. Likidlilid, A.; Patchanans, N.; Poldee, S.; Peerapatdit, T. Glutathione and Glutathione Peroxidase in Type-1 Diabetic Patients. *J. Medial Assoc. Thailand* **2007,** *90* (9), 1759–1767.
67. Lundgren, K. M.; Romundstad, L. B.; During, V. V.; Morkved, S.; Kiotrod, S.; Moholdt, T. Exercise Prior to Assisted Fertilization in Overweight and Obese Women (FertilEX): Study Protocol for a Randomized Controlled Trial. *Trials* **2016,** *17* (1), 268.
68. Madhusudhan, B. Potential Benefits of Flaxseed in Health and Disease: A Perspective. *Agric. Conspectus Scient.* **2009,** *74* (2), 67–72.
69. Mazur, W.; Uehara, M.; Wahala, K.; Adlercreutz, H. Phytoestrogen Content of Berries, and Plasma Concentrations and Urinary Excretion of Enterolactone after a Single Strawberry-Meal in Human Subjects. *Br. J. Nutr.* **2000,** *83* (4), 381–387.
70. Meagher, L. P.; Beecher, G. R. Assessment of Data on the Lignin Content of Foods. *J. Food Compos. Anal.* **2000,** *13* (6), 935–947.
71. Mishra, S.; Verma, P. Flaxseed-Bioactive Compounds and Health Significance. *J. Hum. Soc. Sci.* **2013,** *17* (3), 46–50.
72. Mobeen, H.; Afzal, N.; Kashif, M. Polycystic Ovary Syndrome May be an Autoimmune Disorder. *Scientifica (Cairo)* **2016.** 4071735.
73. Moree, S. S.; Kavishankar, G. B.; Rajesha, J. Antidiabetic Effect of Secoisolariciresinol Diglucoside in Streptozotocin-Induced Diabetic Rats. *Phytomedicine* **2012,** *20* (3), 237–245.
74. Mueller, K.; Eisner, P.; Yoshie-Stark, Y.; Nakada, R.; Kirchhoff, E. Functional Properties and Chemical Composition of Fractionated Brown and Yellow Linseed Meal (*Linum usitatissimum* L.). *J. Food Eng.* **2010,** *98* (4), 453–460.


75. Mueller, K.; Eisner, P.; Yoshie-Stark, Y.; Nakada, R.; Kirchhoff, E. Functional Properties and Chemical Composition of Fractionated Brown and Yellow Linseed Meal (*Linum usitatissimum* L.). *J. Food Eng.* **2010,** *98* (4), 453–460.


76. Munir, A. D.; Westcott, N. D. Flaxseed Constituents and Human Health. In: *Flax: the Genus Linum*; Munir A. D. and Westcott, N. D.; Eds.; Taylor & Francis: London **2003;** pp 243–251.
77. Nagarathna, P. K. M.; Rajan, P. R.; Koneri, R. A Detailed Study on Poly Cystic Ovarian Syndrome and Its Treatment with Natural Products. *Int. J. Toxicol. Pharmaceut. Res.* **2013,** *5* (4), 109–120.
78. Nelson, T. L.; Stevens, J. R.; Hickey, M. S. Inflammatory Markers are Not Altered by an Eight Week Dietary α-Linolenic Acid Intervention in Healthy Abdominally Obese Adult Males and Females. *Cytokine* **2007,** *38* (2), 101–6.
79. Nestel, P. J.; Pomeroy, S. E.; Sasahara, T. Arterial Compliance in Obese Subjects is Improved with Dietary Plant n-3 Fatty Acid from Flaxseed Oil Despite Increased LDL Oxidizability. *Arteriosclerosis Thrombosis Vasc. Biol.* **1997,** *17* (6), 1163–70.
80. Nettleton, J. A.; Katz, R. n-3 long-Chain Polyunsaturated Fatty Acids in Type 2 Diabetes: A Review. *J. Am. Diet Assoc.* **2005,** *105* (3), 428–40.
81. Niedowicz, D. M.; Daleke, D. L. The Role of Oxidative Stress in Diabetic Complications. *Cell Biochem. Biophys.* **2005,** *43* (2), 289–330.

82. Nowak, D. A.; Snyder, D. C.; Brown, A. J.; Wahnefried, W. D. The Effect of Flaxseed Supplementation on Hormonal Levels Associated with Polycystic Ovarian Syndrome: a Case Study. *Current Topics Nutr. Res.* **2008,** *5* (4), 177–182.
83. Oomah, B. D.; Berekoff, B.; Li-Chan, C.; Mazza, G.; Kenaschuk, E.; Duguid, S. Cadmium-Binding Protein Components of Flaxseed: Influence of Cultivar and Location. *Food Chem.* **2007,** *100* (1), 318–325.
84. Oomah, B. D.; Mazza, G. Flaxseed Proteins: A Review. *Food Chem.* **1993,** *48* (2), 109–14.
85. Pan, A.; Sun, J.; Chen, Y. Effects of a Flaxseed-Derived Lignan Supplement in Type 2 Diabetic Patients: A Randomized, Double-Blind, Cross-Over Trial. *PLoS One* **2007,** *2* (11), e1148.
86. Papageorgiou, N.; Tousoulis, D.; Katsargyris, A. et al. Antioxidant Treatment and Endothelial Dysfunction: is it Time for Flavonoids? *Recent Pat Cardiovasc Drug Discov.* **2013,** *8* (2), 81–92.
87. Paschos, J. K.; Magkos, F.; Panagiotakos, D. B.; Votteas, V.; Zampelas, A. Dietary Supplementation with Flaxseed Oil Lowers Blood Pressure in Dyslipidaemic Patients. *European J. Clin. Nutr.* **2007,** *61* (10), 1201–1206.
88. Penttinen, P. 1.; Jaehrling, J.; Damdimopoulos, A. E.; Inzunza, J.; Lemmen, J. G.; van Der Saag, P.; Pettersson, K.; Gauglitz, G.; Makela, S.; Pongratz, I. Diet-Derived Polyphenol Metabolite Enterolactone is a Tissue-Specific Estrogen Receptor Activator. *Endocrinology* **2007,** *148* (10), 4875–86.
89. Pereira, H. V. R.; Saraiva, K. P.; Carvalho, L. M. J.; Andrade, L. R.; Pedrosa, C.; Pierucci, A. P. T. R. Legumes Seeds Protein Isolates in the Production of Ascorbic Acid Microparticles. *Food Res. Int.* **2009,** *42* (1), 115–121.
90. Peterson, J.; Dwyer, J.; Adlercreutz, H. et al. Dietary Lignans: Physiology and Potential for Cardiovascular Disease Risk Reduction. *Nutr. Rev.* **2010,** *68* (10), 571–603.
91. Prasad K. Oxidative Stress as a Mechanism of Diabetes in Diabetic BB Prone Rats: Effect of Secoisolariciresinol Diglucoside (SDG). *Mol. Cell Biochem.* **2000,** *209* (1), 89–96.
92. Prasad, K. Flaxseed and Cardiovascular Health. *J. Cardiovasc. Pharmacol.* **2009,** *54* (5), 369–77.
93. Prasad, K.; Dhar, A. Flaxseed and Diabetes. *Current Pharmaceut. Des.* **2016,** *22* (2), 141–4.
94. Qian, K. Y.; Cui, S. W.; Goff, H. D. Flaxseed Gum from Flaxseed Hulls: Extraction, Fractionation and Characterization. *Food Hydrocol.* **2012,** *28* (2), 275–283.
95. Rains, J. L.; Jain, S. K. Oxidative Stress, Insulin Signaling, and Diabetes. *Free Rad. Biol. Med.* **2011,** *50* (50), 567–575.
96. Rajaeieh, G.; Marasi, M.; Shahshahan, Z.; Hassanbeigi, F.; Safavi, S. M. The Relationship Between Intake of Dairy Products and Polycystic Ovary Syndrome in Women Who Referred to Isfahan University of Medical Science Clinics in 2013. *Int. J. Prevent. Med.* **2014,** *5* (6), 687–694.
97. Rajesha, K. N.; Murthy, C.; Kumar, K. M.; Madhusudhan, B.; Ravishankar, G. A. Antioxidant Potentials of Flaxseed by *in Vivo* Model. *J. Agric. Food Chem.* **2006,** *54* (11), 3794–3799.
98. Rhee, Y.; Brunt, A. Flaxseed Supplementation Improved Insulin Resistance in Obese Glucose Intolerant People: A Randomized, Crossover Design. *Nutr. J.* **2011,** *10* (1), 44.

99. Rodriguez-Leyva, Bassett, C. M. C.; McCullough, R.; Pierce, G. N. The Cardiovascular Effects of Flaxseed and Its Omega-3 Fatty Acid, Alpha-Linolenic Acid. *Can. J. Cardiol.* **2010,** *26* (9), 489–496.
100. Rosamond, W.; Flegal, K.; Friday, G. Heart Disease and Stroke Statistics – 2007 Update: A Report from the American Heart Association Statistics Committee and Stroke Statistics Subcommittee. *Circulation* **2007,** *115*, e69–171.
101. Rubilar, M.; Gutierrez, C.; Verdugo, M.; Shene, C.; Sineiro, J. Flaxseed as a Source of Functional Ingredients. *J. Soil Sci. Plant Nutr.* **2010,** *10* (3), 373–377.
102. Sabate, J.; Wien, M. Nuts, Blood Lipids and Cardiovascular Disease. *Asian Pacific J. Clin. Nutr.* **2010,** *19* (1), 131–136.
103. Sadiq, S. M.; Rajesha, J. Secoisolariciresinol Diglucoside: A Potent Multifarious Bioactive Phytoestrogen of Flaxseed. *Res. Rev. Biomed. Biotechnol.* **2011,** *2* (3), 1–24.
104. Schmitz, G.; Ecker, Elaeagnus Angustifolia-Teknoscienze *J. Am. Family Phys.* **2008,** *70*, 133–40.
105. Shen, C. C.; Yang, A. C.; Hung, J. H.; Hu, L. Y.; Tsai, S. J. Nationwide Population-Based Retrospective Cohort Study of the Risk of Uterine, Ovarian and Breast Cancer in Women with Polycystic Ovary Syndrome. *Oncologist* **2015,** *20* (1), 45–9.
106. Siger, A.; Kalucka, N. M.; Szczapa, L. E. The Content and Antioxidant Activity of Phenolic Compounds in Cold-Pressed Plant Oils. *J. Food Lipids* **2008,** *15* (2), 137–149.
107. Simbalista, R. L.; Sauerbronn, A. V.; Aldrighi, J. M.; Areas, J. A. Consumption of a Flaxseed-Rich Food is not More Effective than a Placebo in Alleviating the Climacteric Symptoms of Postmenopausal Women. *J. Nutr.* **2010,** *140* (2), 293–7.
108. Singh, H. Nanotechnology Applications in Functional Foods: Opportunities and Challenges. *Prevent. Nutr. Food Sci.* **2016,** *21* (1), 1–8.
109. Singh, K.; Mridula, D.; Rehal, J.; Barnwal, P. Flaxseed: A Potential Source of Food, Feed and Fiber. *Crit. Rev. Food Sci. Nutr.* **2011,** *51* (3), 210–222.
110. Smykal, P.; Kerteszova, B. N.; Kalendar, R.; Corander, J.; Schulman, A.; Pavelek, M. Genetic Diversity of Cultivated Flax (*Linum usitatissimum* L.) Germplasm Assessed by Retrotransposon-Based Markers. *Theor. Appl. Gen.* **2011,** *122* (7), 1385–1397.
111. Soma, P. K.; Williams, P. D.; Lo, Y. M. Advancements in Non-Starch Polysaccharides Research for Frozen Foods and Microencapsulation of Probiotics. *Frontier Chem. Eng. China* **2009,** *3* (4), 413–426.
112. Sosulski, F. W.; Bakal, A. Isolated Proteins from Rapeseed, Flax and Sunflower Meals. *Can. Institute Food Sci. Technol. J.* ***1969,*** *2* (1), 28–32.
113. Spence, J. D.; Thornton, T.; Muir, A. D.; Westcott, N. D. The Effect of Flax Seed Cultivars with Differing Content of α-Linolenic Acid and Lignans on Responses to Mental Stress. *J. Am. Coll. Nutr.* **2003,** *22* (6), 494–501.
114. Stahl, L. Third Crop Options Flax. *Rural Adv.* **2007,** 112–118.
115. Strandas, C.; Eldin, K. A.; Andersson, R.; Aman, P. Phenolic Glucosides in Bread Containing Flaxseed. *Food Chem.* **2008,** *110* (4), 997–999.
116. Tan, S. M.; Sharma, A.; de Haan, J. B. Oxidative Stress and Novel Antioxidant Approaches to Reduce Diabetic Complications. In: *Oxidative Stress and Diseases;* Lushchak, V. Ed.); InTechOpen: Rijeka, Croatia; 2012; pp 247–280; e-article.
117. Tarpila, A.; Wennberg, T.; Tarpila, S. Flaxseed as a Functional Food. *Curr. Topics Nutr. Res.* **2005,** *3* (3), 167–188.

118. Thompson, L. U.; Cunnane, S. C, Eds. *Flaxseed in Human Nutr.* 2nd ed; AOCS Publishing: Champaign: IL 2003; pp 413.
119. Thompson, L. U.; Cunnane, S. C. *Flaxseed in Human Nutrition.* AOCS Press: Champaign, IL; 2003; pp 115—120.
120. Thompson, L. U.; Robb, P.; Serraino, M.; Cheung, F. Mammalian Lignan Production from Various Foods. *Nutr. Cancer* ***1991, 16***, 43–52.
121. Tolkachev, O. N.; Zhuchenko, A. A. Biologically Active Substances of Flax: Medicinal and Nutritional Properties (A review). *Pharmaceut. Chem.* **2004,** *34* (7), 360–367.
122. Toure, A.; Xueming, X. Flaxseed Lignans: Source, Biosynthesis, Metabolism, Antioxidant Activity, Bio-Active Components, and Health Benefits. *Compreh. Rev. Food Sci. Food Safety* **2010,** *9* (3), 261–269.
123. Tuberoso, I. G. C.; Kowalczyk, A.; Sarritzu, E.; Cabras, P. Determination of Antioxidant Compounds and Antioxidant Activity in Commercial Oilseeds for Food Use. *Food Chem.* **2007,** *103* (4), 1494–1501.
124. Udenigwe, C. C.; Adebiyi, A. P.; Doyen, A.; Li, H.; Bazinet, L.; Aluko, R. E. Low Molecular Weight Flaxseed Protein-Derived Arginine-Containing Peptides Reduced Blood Pressure of Spontaneously Hypertensive Rats Faster Than Amino Acid form of Arginine and Native Flaxseed Protein. *Food Chem.* **2012,** *132* (1), 468–475.
125. Udenigwe, C. C.; Aluko, R. E. Antioxidant and Angiotensin Converting Enzyme-Inhibitory Properties of a Flaxseed Protein-Derived High Fischer Ratio Peptide Mixture. *J. Agric. Food Chem.* ***2010,*** *58* (8), 4762–4768.
126. Velasquez, M. T.; Bhathena, S. J.; Ranich, T.; Schwartz, A. M.; Kardon, D. E.; Ali, A. A. Dietary Flaxseed Meal Reduces Proteinuria and Ameliorates Nephropathy in an Animal Model of Type II Diabetes Mellitus. Kidney Int. **2003,** *64* (6), 2100–210.
127. Wang, Y.; Li, D.; Wang, L. J.; Li, S. J.; Adhikari, B. Effects of Drying Methods on the Functional Properties of Flaxseed Gum Powders. *Carbohydrate Polym.* **2010,** *81* (1), 128–133.
128. Wang, Y.; Wang, L. J.; Li, D.; Xue, J.; Mao, Z. H. Effects of Drying Methods on Rheological Properties of Flaxseed Gum. *Carbohydrate Polym.* **2009,** *78* (2), 213–219.
129. Westcott, N. D.; Muir, A. D. Chemical Studies on the Constituents of Linum Spp. In: Flax: the Genus Linum; Muir, A. D.; Westcott, N. D.; Eds; Taylor & Francis: London; 2003; pp 55–73.
130. Wilkes, A. P. *Flax Use Up as Health Benefits Better Realized.* A BNP Publication: 2007; pp 1–8.
131. Williams, R. B. Blood Pressure Reactivity to Psychological Stress: A New Risk Factor for Coronary Disease? *Hypertension* **2006,** *47* (3), 329–30.
132. Williams, T.; Mortada, R.; Porter, S. Diagnosis and Treatment of Polycystic Ovary Syndrome. *Am. Family Phys.* **2016,** *94* (2), 106–113.
133. Wong, J. M.; Gallagher, M.; Gooding, H.; Feldman, H. A.; Gordon, C. M.; Ludwig, D. S.; Ebbeling, C. B. A Randomized Pilot Study of Dietary Treatments for Polycystic Ovary Syndrome in Adolescents. *Pediatric Obesity* **2016,** *11* (3), 210–20.
134. World Health Organization (WHO). World Health Statistics. 2008. < http://www.who.int/whosis/whostat/> (accessed Feb 9, 2018).
135. Yusuf, S.; Hawken, S.; Ounpuu, S. Effect of Potentially Modifiable Risk Factors Associated with Myocardial Infarction in 52 Countries (the INTERHEART study): Case-Control Study. *Lancet* **2004,** *364* (9438), 937–52.

136. Zanwar, A. A.; Hegde, M. V.; Bodhankar, S. L. Cardioprotective Activity of Flax Lignan Concentrate Extracted from Seeds of *Linum Usitatissimum* in Isoprenalin Induced Myocardial Necrosis in Rats. *Interdiscip. Toxicol.* **2011,** *4* (2), 90–97.
137. Zhang, Z. S., Wang, Li, J., Li, D.; Jiao, S. S.; Chen, D. X.; Mao, Z. H. Ultrasound-Assisted Extraction of Oil from Flaxseed. *Separation Purif. Technol.* **2008,** *62* (1), 192–198.

## CHAPTER 13

# GREEN TEA FOR HUMAN HEALTH BENEFITS: PHYTONUTRIENTS AND THEIR THERAPEUTIC POTENTIALS

FAIZA ASHFAQ, MASOOD SADIQ BUTT, AHMAD BILAL, MIR MUHAMMAD NASIR QAYYUM, and HAFIZ ANSAR RASUL SULERIA

**ABSTRACT**

The incorporation of plant-based phytonutrients in daily diet has upgraded the evolving theory of designer foods in terms of health assistance and disease modulatory potential. Green tea (*Camellia sinensis*) extract enriched dietary prototypes (such as candies, meat and fish products, crackers, cakes and breads, and oil and juices) have shown improvement in shelf stability against microbial attack. Biocomponents are often extracted using optimized extraction procedures either conventional such as solvent extraction or advanced such as supercritical fluid extraction (SFE). In solvent extraction procedure, some residual effects of solvents may sustain, declining the consumer acceptability based on religious or cultural norms. On the other hand, SFE is considered as a green extraction method due to safety concerns. The extract achieved via supercritical fluid extractor is costly as compared to conventional extracts. The green tea possesses varied catechins such as: epicatechin, epigallocatechin, epicatechin gallate, and epigallocatechin-3-gallate. Researchers have evaluated different extraction methods, solvents and their ratios, time, temperature, and pressure conditions for optimized extraction of polycatechins from green tea. The galloyl groups of green tea have proven their ability to capture free radicals. Phytochemical therapies have shown their potential to address metabolic ailments and immune dysfunction. In this connection, green tea polyphenols especially EGCG has defensive aptitude to fight against free radical-mediated lipid peroxidation especially in liver and renal tissues; besides, it modulates the glycemic index and serum lipid profile thus protecting against obesity and associated chronic maladies. Besides, it has highlighted the ideology of immunonutrition in downregulating oncogenic events by inducing apoptosis and inhibiting cell proliferation and angiogenesis.

## 13.1 INTRODUCTION

Green tea (*Camellia sinensis* L.) contains volatile and nonvolatile substances of Ayurvedic importance.[39] The heat stabilized green tea catechins are considered more efficacious moieties to combat numerous degenerative disorders in contrast to nonthermally processed counterparts. The differences in green, black, red, white, pu-erh, and oolong tea depend on different types of processing methods.[138] As a result, their active moieties vary not only in color, taste, and compositional value but also in their defensive mechanism against various chronic ailments.[27] Of the total ~30% of polyphenols in dried green tea leaves, 70% account for polycatechins that include epigallocatechin-3-gallate (EGCG), epicatechin gallate (ECG), epigallocatechin (EGC), epicatechin (EC), and gallocatechin.[28,103]

Extraction of green tea polyphenols can be carried out through conventional methods such as solvent extraction, Soxhlet, and hydrodistillation, or via innovative techniques involving supercritical fluid extraction (SFE), pressurized fluid extraction (PFE), ultrasound-assisted extraction (UAE), and microwave-assisted extraction (MAE). Numerous factors are interlinked to facilitate their extraction efficiency including: size reduction, extraction time, temperature, etc. The novelty and safety in advanced extraction techniques has almost replaced the conventional methods.[37]

In the underdeveloped countries, dietary habits are overlooked due to lack of consumer awareness. This has brought numerous bioactive moieties in the limelight to manage various ailments especially green tea catechins. Previous analyses on experimental animals recommended three to six cups (250 mL) of green tea daily to fight against hypercholesterolemia, diabetes, and obesity.[16,18,140]

The main objectives of this chapter are: (1) characterization and free radical scavenging potential of bioactive compounds in green tea; (2) comparing extraction efficiency of green tea catechins using conventional solvent and SFE techniques; (3) evaluating green tea-based designer foods; and (4) therapeutic bioassessment of green tea extract (GTE) against lifestyle-related disorders.

## 13.2 GREEN TEA: A PREFATORY OUTLOOK

Green tea belongs to Theaceae family and its cultivation originated in China since 17th century. It is thus renowned as Chinese beverage.[141,39] The chemical composition of tea is a proof of its mode of action against varied ailments, attributed by phytochemicals, alkaloids, volatile oils, polysaccharides, amino acids, lipids, minerals, and vitamins.[138] The dried green tea leaves comprise 30% phenolic compounds, 26% fiber, 15% protein, 7% carbohydrates, 7% lipids, 5% ash, 4% amino acids, 3.5% caffeine, and 2% pigments.[28,141]

Green tea has plethora of bioactive molecules, accounting for 30% of total polyphenols on dry weight basis, out of which 70% of the total is flavanol gallates among which EGCG (a principle bioflavonoid) is around 50–80%, besides the presence of ECG, EGC, EC, catechins (C), and gallocatechins[27,28,39,138,162] as shown in Figure 13.1. The total polyphenols in green and black tea are almost the same but vary based on the type, that is, black tea consists of 25% of oxidized polyphenols including 70% thearubigins, 12% theaflavin, 10% flavanols, and 8% catechins; but none in green tea owing to its nonfermentative nature.[31,39]

The biosynthesis of green tea catechins follows two mechanisms in the plant body, namely: (1) acetic acid pathway—carbohydrates of fresh green tea leaf undergo conversion into pyruvic acid that goes through citric acid cycle by combining with acetyl-CoA, forming chalcone than catechins and (2) shikimic acid pathway—green tea starches are changed to catechins after converting from chalcone and gallic acid. The EGCG and ECG are colorless but bitter tasted; nonetheless EGC and EC are sweet after taste.[152]

EGCG has been found as a major moiety in the green tea and its absorption depends upon numerous factors. The major absorptive areas include small and large intestine where the healthy microflora facilitates its absorption. It has a half-life of 3 h, reaching at its peak concentration within first 2 h of its oral ingestion. The bioavailability of EGCG enhances with simultaneous ingestion of piperine (black pepper extract), ascorbic acid, omega-3 fatty acids, or by consuming after an overnight fasting. On the other hand, the presence of oxygen, body enzymes, and hard metal ions interfere in the bioavailability of EGCG.[93]

The catechins (especially EGCG and ECG) possess ester bonds that precipitate with enzymes, ultimately controlling their release. The galloyl group of green tea binds with free metal ions or catalysts of free radical chain reaction such as iron and copper resultantly inhibiting free radical chain reactions in the body. The antioxidant property of green tea catechins depends on hydroxyl group in their structures that donates hydrogen to capture free radical species. The structure of catechins is based on central carbon unit containing two phenolic nuclei with numerous hydroxyl groups. The green tea catechins are highly prone to oxidation at temperatures >40°C, alkaline pH >5.5, or in the presence of polyphenol peroxidase enzyme (PPO) and peroxidase (POD) enzyme converting to theaflavin and thearubigin.[152]

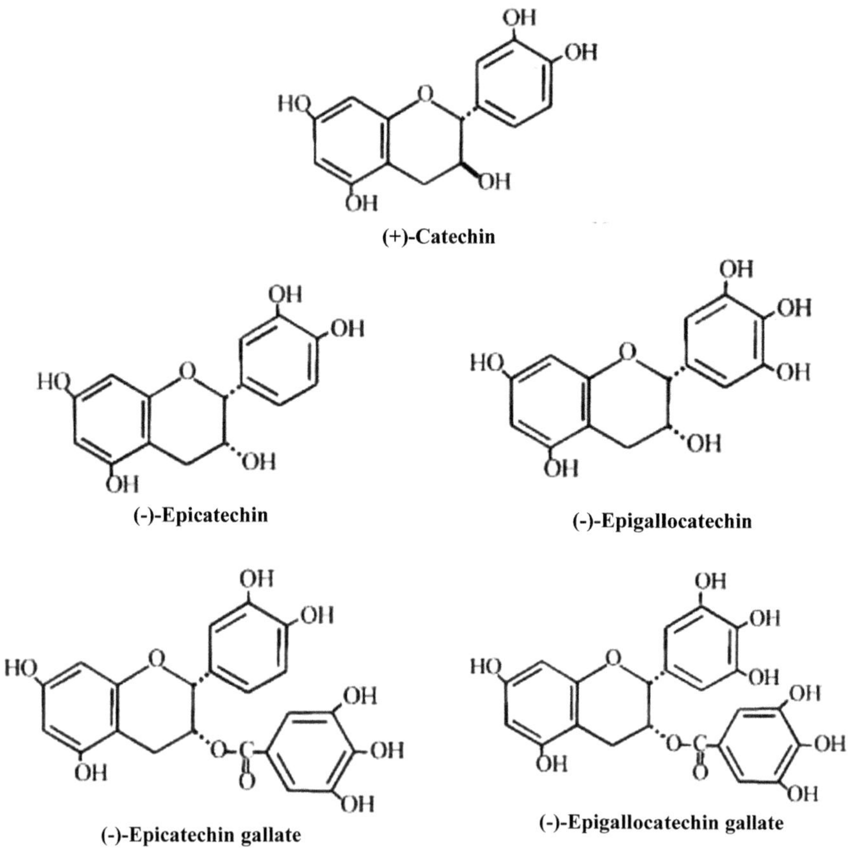


**FIGURE 13.1** Chemical structures of polycatechins in green tea.

There are different types of tea such as: green, black, oolong, white, pu-erh, and red or rooibos tea. Green tea (steamed and unfermented commodity) and oolong tea (semifermented) possess highest levels of EGCG, but black tea (fully fermented) exhibits minimum of this component. Pu-erh tea follows the same processing procedure as that of black tea but the only difference lies in its aging after picking.[39,75,125,138] Longer storage of pu-erh tea results in better quality and sensory attributes.[39] White tea is the steamed and sun-dried product, while red tea is a naturally decaffeinated variety native to South Africa that is suitable for pregnant and lactating women.[138]

The production of tea is estimated at 5 million metric tons throughout the world, while on consumption basis demand of black tea is 70%, of green tea is 20%, and of oolong tea is 2% only.[27] These different types of tea mainly differ based on their conventional processing, composition, hedonic features, and chemistry involving the presence of caffeine and phytonutrients. Higher level of caffeine is a characteristic feature related to black tea while green tea is attributed with maximum level of catechins. These different categories of tea have the potential to bring positive effects on human gastrointestinal system and overall health by improving serum antioxidant status and lowering detrimental oxidants in our body ultimately reducing oxidative stress-mediated dysfunctions.[138]

Based on consumption pattern, black tea and green tea have been found at the top due to their higher consumer demand and health benefits, respectively. The main difference lies in their fermentation property; that step is omitted in green tea, but black tea undergoes very prolonged oxidation reactions. During fermentation reactions, 20–30% of total green tea flavonoids (two substrates: catechins and orthoquinones) get converted to black tea polymers (theaflavin). Further, theaflavin serves as a substrate that reacts with gallic acid, that is, a second substrate to form epitheaflavic acid. Such large polymeric products are collectively known as thearubigins. The fermentation process brings prominent differences with respect to phytochemical composition and content, color, and taste profile.[27,39]

The green tea possesses disease fighting perspectives against several dysfunctions including diabetes, obesity, cardiovascular ailments, respiratory disorders, skin problems, indigestion, tooth decay, liver diseases, venom-related inflammations, arthritis, and cancer. Besides modulating body temperature, it helps to restore heart beat rate or blood pressure to normal ranges. To overcome numerous diseases, it is recommended to consume 3–10 cups (250 mL) of green tea during 4 or 8 weeks.[103,138]

The green tea, either alone or allied with other herbal extracts, has immune-potentiating properties to regulate various physiological functions of the body. Green tea is renowned as a "Divine Healer" due to its healing properties.[103] In the nutshell, it has been proved that dietary interventions, especially the hot beverages holding EGCG, could combat variable forms of cancers especially lung, pancreatic, urinary bladder, colorectal, esophageal, breast, stomach, and skin cancers at various progression stages. In vivo and in vitro studies have ensured high potency of green tea as compared to vitamin C, vitamin E, and other effective antioxidants. Hence, it should be incorporated in dietary regimen to attain numerous phytotherapeutic and chemotherapeutic benefits.[27,103,138]

## 13.3 GREEN TEA POLYCATECHINS: EXTRACTION METHODS

Conventional solvent extraction (CSE) procedure is the most widely adopted technique for the extraction of bioactive compounds. Nowadays, numerous advanced techniques have been developed that are more efficient and eco-friendly such as: SFE, PFE, accelerated fluid extraction (AFE), subcritical water extraction (SWE), UAE, and MAE. All of these approaches have an upper hand over conventional counterpart owing to the use of nontoxic carbon dioxide or water by changing their dielectric constant to make them polar organic solvent despite alcoholic solvents. The MAE and UAE techniques involve the use of waves to disrupt solid matrix for efficient extraction of active moieties.[37]

### *13.3.1 CSE METHOD*

CSE is one of the traditional techniques. There are numerous parameters that enhance the yield of polyphenols such as: decrease in particle size and increase in solvent to solid ratio. The selection of solvent depends on the nature of raw material. Better the solubility of extractable compounds in solvent, more would be the mass transfer. In case if the solvent has less polarity for polyphenols and more solubility for nonpolyphenolic components, so less would be the yield of desirable compounds.[37]

There are varied pure solvents or their aqueous forms that have been employed for the extraction of bioactive moieties. For optimal extraction, different ratios of solvent and water have been tested compared to that of pure solvent. Extraction temperature, time, solvent polarity, and solvent to tea ratio are determinants of extraction efficiency. Mostly tea extractable encompasses numerous catechins along with caffeine. Purposely, a variety of solvents has been employed for extraction purpose such as: acetone, ethanol, methanol, water, and acetonitrile. However, the best recommended solvents for caffeine isolation involves ethyl acetate, carbon dioxide, and methylene chloride. Among these, carbon dioxide is considered better option to remove 98% of the caffeine under optimal pressure. Normally, these solvents are used for decaffeination, though small amount of catechins may be extracted also. Best solvents are those that will give higher extraction yield and in vitro and in vivo antioxidant potential.[114]

The aqueous acetone is commonly selected to extract higher molar weight flavanols especially catechins, whereas methanol to extract low molecular weight phenolics. However, ethanol is considered as a benign solvent after water. Moreover, ethyl acetate is often employed for decaffeination or extraction of some specific moieties. The polarity of aqueous acetone is higher than that of pure acetone for extracting flavanols. In the same way, extraction time and temperature factors can also improve the extraction efficiency. Extraction temperature needs to be adjusted according to the analyte under interest and preferably ranges between 20°C and 50°C but deterioration may occur at or exceeding 70°C. The prolonged extraction time is responsible for reduced mass transfer resistance, viscosity, and surface tension thus increases dissolution of solvent solubilized components. It is better to optimize the extraction temperature and time for polyphenols stability. Furthermore, best execution could be achieved below degradation temperature where maximum extraction is possible.[37]

Perva-Uzunalic et al.[114] conducted an experiment involving acetone, ethanol, methanol, acetonitrile, and water with varying volumes (25%, 50%, 80%, and 100%) at a temperature range of 70–100°C. At 50% volume of these solvents, maximum yield of catechins was observed by acetonitrile 99.8%, followed by acetone 99.3%, water 88.4%, methanol 81.1%, and ethanol 76.4%. Longer extraction time >100 min has been negatively associated with the extraction efficiency of catechins. Therefore, high temperature plus short time and less temperature with longer time have proved to be a better solution to attain maximum yield. Higher solvent to tea ratio (100 mL:1 g) is considered for maximum yield; however, purer the solvent, lesser would be the yield.

Various analyses have indicated that organic solvents alone give lowest extraction efficiency of polycatechins as compared to their 50% aqueous form. The results of a study inferred that 50% aqueous acetone showed better polarity of total polyphenols followed by 50% aqueous concentration of dimethylformamide, ethanol, or methanol.[145] Black tea treated with solvents including acetone, ethanol, methanol, and *N,N*-dimethylformamide (DMF) at 50% concentration revealed maximum total polyphenols and free radical trapping ability through acetone followed by DMF, ethanol, and methanol.[150]

Extraction of polyphenols through alcohols has been investigated for the formulation of phytoceutics and pharmaceutics or for their manipulation into designer foods. In this context, Lin et al.[88] found that 50% ethanol offers optimized extraction of green tea catechins. Comparing ethanol at varying concentrations in another study ensured that maximum extraction could be achieved at 50% ethanol concentration despite 100% concentration.[154] Rusak et al.[128] analyzed the effect of ethanol at different concentrations (10%, 40%, and 70%) and extraction times (5, 10 and, 15 min). They noticed that 40% ethanol gave maximum yield of polyphenols and catechins at 30 min.

Moreover, Borse et al.[23] studied the extraction of bioactive components using various solvents (ethyl acetate, acetone, ethyl alcohol, and methyl alcohol) and confirmed higher extraction via aqueous solvents over pure equivalents. Ethyl acetate in pure form presented 23.9% yield that was improved in aqueous form up to 24.5%. Likewise, a similar trend was observed in pure acetone, ethyl alcohol, and methyl alcohol yielding 27.7%, 32.7%, and 34.8% of active ingredients that was improved in their aqueous states up to 29.2%, 36.3%, and 36.7%, respectively. These studies concluded that pure solvents possess less extraction rate owing to less polarity. However, their aqueous state has brought revolutionary benefits in the extraction process because water softens the solid matrix of raw material, facilitating organic solvents to easily penetrate and mass transfer.

For the extraction of phytochemicals from plant materials, the most commonly preferred solvents include ethanol, methanol, water, acetone, and ethyl acetate. Ethanol and water have been broadly applicable being eco-friendly. On comparing ethyl acetate and ethanol, ethyl acetate was less polar hence ranked second after ethanol. Ethanol could easily extract bioactive components due to higher polarity. The polarity index or the miscibility of ethanol and acetone is higher, ranging from 5.1 to 5.2 following ethyl acetate, that is, 4.3.[51] Tiwari et al.[147] found acetone to be a better solvent for the mass transfer of flavonols, phenolics, and tannins than aqueous ethanol or methanol. Borse et al.[23] confirmed least polarity of ethyl acetate in comparison to acetone, methanol, and ethanol, leading to lower extraction efficiency and free radical scavenging ability.

Druzynska et al.[44] measured the response of solvent type, extraction time, and antioxidant potential of GTEs considering 80% concentration of different solvents including acetone, ethanol, and methanol along with water. The extract carrying higher phenolic content was achieved through acetone followed by water, methanol, and ethanol. Furthermore, similar trend was observed in terms of chelation of free radicals [via DPPH (1,1-diphenyl-2-picrylhydrazyl) and ABTS (2,2'-azino-bis(3-ethylbenzothiazoline-6-sulphonic acid)) reagents] as well as metal ions. It was viewed that acetonic extract possesses higher antioxidant potential due to its higher phenolic chemistry. On comparing these extracts at three different time intervals

(15, 30, and 60 min), the best extraction yield was achieved at 60 min proving that extraction solvents and time are the prominent determinants to achieve the optimized extraction and antioxidant activity.

Furthermore, different solvents have been studied on pu-erh tea to analyze phytochemistry and free radical scavenging capacity. Based on total phytonutrients extracted, butanol ranked at the top followed by ethanol, ethyl acetate, chloroform, and water.[60] Previous studies have highlighted that ethyl acetate could be a preferable choice of solvent for fractionation and purification of green tea catechins: EC, EGC, ECG, and EGCG. Ethyl acetate could remove undesirable components and caffeine or any lipophilic fractions.[96,127]

Stankovic et al.[143] compared the extraction efficiency of green tea (*Trifolium montanum*) and ginkgo bioactives using different solvents. The best results were produced by methanol followed by acetone, water, ethyl acetate, and petroleum ether, depending on their polarity for polyphenols extraction. Piñeiro et al.[116] conducted a study on green tea and grape seeds for the extraction of catechins and ECs using different solvents; and best extraction was attained via methanol followed by ethanol, ethyl acetate, and water.

Bharadwaz and Chiranjit[20] used ethyl acetate to retain highly pure polyphenolic fraction out of GTE with an efficiency of 19.33%. Among ethyl acetate, butanol, and hexane, ethyl acetate was considered as optimized choice for the extraction of green tea polyphenols. Extraction yield of ethyl acetate and hexane was 6.50 g compared to 4.4 g through butanol. On comparing total catechins, maximum yield of polycatechins was 643.38 mg/g via ethyl acetate followed by 637.68 mg/g via hexane and 427.86 mg/g via butanol.[41] Besides, the bulk of research based on CSE found it as a hectic, laborious, and less efficient technique, and adding to environmental pollution owing to the use of larger volumes of organic solvents.[37]

### *13.3.2 SFE METHOD*

The SFE technology is a unique system of supercritical phase containing the properties of both liquid and gas phases. Supercritical phase could be achieved by adjusting pressure and temperature just above critical point. At this point, it holds gas-like diffusing properties alongside liquid-like density for dissolution and solubilization. Solid matrix including rhizomes, seeds, flowers, peels of fruits, tree branches, leaves, etc., are mostly employed to separate the bioactive moieties. SFE is considered as a safe and green technology for the mass transfer of phytoceutics. The term green technology is associated owing to the use of environmentally safe chemicals such as carbon dioxide, that is, nontoxic, resolving the issues of solvent contamination in the final product or extract.[32,144]

The extraction principle followed by supercritical fluid for the green tea catechins involve dilation of solid matrix and dissolution of active components into the fluid. After the extraction is completed, the respective bioactive components are achieved by reversing the optimized conditions, that is, bringing pressure and temperature below critical point. During the extraction procedure, firstly carbon dioxide in liquid phase moves into the extraction vessel where raw material has already been placed, here the pressure is raised to bring the solvent to fluid phase above critical point. This is the required phase where extraction could be carried out with ease. Sometimes cosolvent is also recommended especially for the extraction of polar bioactives because carbon dioxide can extract only nonpolar moieties being nonpolar in nature. Numerous cosolvents could be employed, preferably ethanol, to extract green tea catechins.[153]

SFE system is designed on the principle of optimal or selective conditions set according to the raw material or the specific bioactive component under target, avoiding the remaining components by considering them as impurity. Mostly optimization is achieved through extraction pressure, temperature and time, type of cosolvent, flow rate of solvent, and particle size of solid matrix. When extraction of respective active ingredient is carried out after a set time frame, then pressure is reduced, making the separation of bioactive components from carbon dioxide possible. Extract collected in the collector vessel and carbon dioxide is either released into the environment or may be recovered depending upon the recovery system (Fig. 13.2). Numerous solvents could be employed for extraction purpose besides carbon dioxide such as ethane ethylene, methanol, ammonia, dimethyl ether, hexane, propane, water, xenon, etc.[92,144]

The antioxidant activity of green tea is found in the order of EGCG > ECG > EC > EGC. EGCG accounts for 65% out of total 30% of polyphenols in green tea leaves. EGCG is considered as the main antioxidant with polar nature. Therefore, it has become very difficult to extract it via carbon dioxide, which is nonpolar in nature. As a result, it has become necessary to use a cosolvent/entrainer that is polar in nature to extract EGCG. Thus, purpose of cosolvent is to bring carbon dioxide toward polarity by forming hydrogen bonds or van der Walls forces between cosolvent and catechins. Besides cosolvent, the moisture content of tea acts synergistically to overcome the low polarity of carbon dioxide. The selective conditions predominantly pressure, temperature, time, nature of cosolvent, and flow rate of solvent in extraction vessel play important role in extracting green tea polycatechins.[57,146]

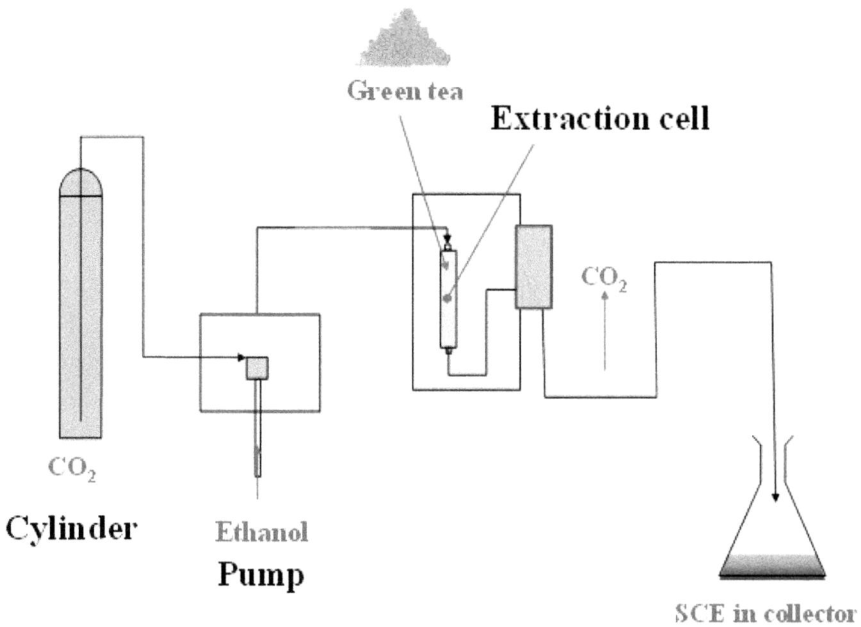


**FIGURE 13.2** **(See color insert.)** Extraction of green tea polycatechins using supercritical carbon dioxide.

Optimization of pressure has an influential role for the maximum yield and increasing density of supercritical fluid. Best extraction of EGCG is obtained at 10–20 MPa; however, beyond this pressure, yield decreases due to numerous repulsive forces between extraction solvent and active component. Extraction time of 120 min is considered as optimal time for EGCG extraction, though beyond this time extraction rate becomes steady. Increase in solvent flow and decrease in particle size of plant matrix are responsible factors to enhance extraction efficiency. Optimum temperature is adjusted just above critical point of carbon dioxide, that is, 31.1°C. The optimal temperature for green tea catechins is 40–60°C where the carbon dioxide density becomes equal to that of EGCG vapor pressure.[57]

It has been found that numerous factors such as cosolvent, particle size, extraction time, and temperature influence the SFE yield. Cosolvent selected within the range of 5, 15, and 30 mL showed optimal yield at 15 mL, that is, 154.25 mg/g, followed by at 5 mL, that is, 153.78 mg/g, and at 30 mL, that is, 150.61 mg/g. Particle size affects the extraction, for example, 157.10 mg/g of catechins with a particle size of 1.2–2.0 mm; 152.96 mg/g with size of 0.6–1.2 mm; and 148.57 mg/g with size in the range of 0.2–0.6 mm. Effect of extraction time has also proved as an important factor with optimal yield of catechins 154.35 mg/g at 2 h followed by 152.48 mg/g at 3 h and 151.80 mg/g at 1 h extraction period. Temperature effect has also been studied indicating maximum extraction of catechins 154.07 mg/g at 60°C, followed by 153.11 mg/g at 80°C, and 151.45 mg/g at 40°C. These parameters have confirmed that optimal extraction of polycatechins from green tea is about 154–157 mg/g at 60°C within 2 h of extraction duration, 1.2–2.0 mm particle size along with cosolvent of 15 mL.[146]

Chang et al.[33] investigated the effect of different cosolvents along with supercritical fluid for the extraction of catechins. The results indicated that maximum extraction could be achieved using 95% ethanol as cosolvent followed by 99.8% ethanol, water, 18% ethanol, 70% ethanol, and carbon dioxide alone. The correspondingly values of EGCG were 510, 492, 94, 83, 34, 30, and 18 ppm, respectively at 333 K and 31 MPa. Thus, it has been affirmed that maximum yield is not only dependent on type of cosolvent but on its concentration as well.

Kim et al.[79] conducted an experiment using 40°C and 50°C with different cosolvents such as water, ethanol, etc., and without any cosolvent. The maximum yield of 92% of EGCG was achieved via water at 40°C or 50°C, followed by ethanol at 40°C (63% EGCG yield), and ethanol at 50°C (54% EGCG yield); without cosolvent resulted in 30% efficiency at 50°C and 21% at 40°C.

According to Park et al.,[110] optimum conditions for SFE were 95% ethanol (entrainer), 23 MPa, 63°C, and 120 min with 10 g of green tea per 100 g of carbon dioxide that resulted in the recovery of 40.61% of catechins. In another study, influence of temperature and pressure has also been observed for the extraction of green tea catechins: EGCG, EGC, ECG, and EC. To attain the maximum recovery of polycatechins, selectivity of conditions involving 300 bars was applied to extract 69.5% of EGCG at 80°C, 38.5% of EC at 60°C, and 82.4% EGC at 50°C. However, 78% of ECG was observed at 70°C and 250 bars.[109] Also, Ghoreishi and Heidari[58] determined selective conditions for the extraction of EGCG from Iranian green tea as 19.3 MPa, 43.7°C, 106 min, and 1.5 mL/min of carbon dioxide flow rate along with ethanol as entrainer.

## 13.4 PHYTOCHEMICAL ASSAYS OF GTE

For the determination of antioxidant chemistry, variety of antioxidant assays are carried out, namely, Trolox equivalent antioxidant capacity (TEAC), total radical-trapping antioxidant parameter (TRAP), ferric ion reducing antioxidant capacity (FRAP), metal chelating activity (MCA), oxygen radical absorbance capacity (ORAC), and hydroxyl radical scavenging activity assays. Their purpose is to determine the antioxidant potential by reducing free radicals, demonstrated by change in color.[19,37,76] Various investigations found higher antioxidant ability of green tea attributed to total polyphenols as 0.313 ± 0.095 mg gallic acid equivalent (GAE)/mg to fight against free radicals avoiding oxidative damage and related ailments.[10]

Bastos et al.[15] measured polyphenol content with maximum extraction obtained through ethanol 13.08 ± 0.14 mg/mL followed by water 7.15 ± 0.14 mg/mL and ether 0.07 ± 0.00 mg/mL. Islam[71] determined the total polyphenol content as 266.61 ± 5.86 mg/L in 0.5% aqueous extract of green tea. Carloni et al.[29] measured the total polyphenol in green tea within the range of 23.6 ± 5.4–22.6 ± 5.0 mM GAE. However, white tea and black tea were observed as 18.1 ± 5.5 mM GAE and 10.7 ± 4.0 mM GAE, respectively. Manian et al.[91] documented higher antioxidant potential of green tea acetonic extract as compared to methanolic extract in terms of polyphenols and flavonoids.

Likewise, total polyphenols in normal and coarse green tea samples were examined using aqueous organic solvents including ethyl alcohol, methyl alcohol, ethyl acetate, and acetone that documented more polyphenols 23 ± 2.1% in normal green tea followed by 18 ± 3.0% in coarse green tea.[23] Gramza et al.[62] observed total polyphenols in green and black tea using 95% ethanol and water. Using 95% ethanol, comparative analysis showed maximum extraction of 245.8 and 837.6 mg/g of polyphenols in green and black tea extracts, respectively. Recent development based on high hydrostatic pressure extraction has been considered better than conventional methods, achieving 30 ± 1.3% of total polyphenols at optimized conditions including 1:1 ratio of ethanol in 20:1 mL/g ratio of solvent to green tea matrix and 500 MPa.[154]

Studies by Manian et al.[91] for determination of total polyphenol content in green tea and two *Ficus* varieties including *Ficus benghalensis* and *Ficus racemosa* indicated higher polyphenols in green tea using 70% methanol followed by 70% acetone solvent extraction. Furthermore, Borse et al.[23] quantified total catechins in GTE as 70 ± 15.0%. Later, Komes et al.[80] estimated total flavonoids in green tea as 996.03 mg/g with maximum fraction of EGCG 345.55 ± 21.23, EGC 324.88 ± 18.01, ECG 102.67 ± 2.89, EC

120.07 ± 2.38, gallocatechin 70.49 ± 1.03, catechins (C) 25.75 ± 1.21, and gallocatechin gallate (GCG) 6.62 ± 0.81 mg/g as detected by high-pressure liquid chromatography (HPLC).

Tsai et al.[148] conducted an analysis on different types of tea: green, lemongrass, sweet osmanthus, rose, lavender, rosemary, jasmine, and daisy. The green tea was ranked at third position after rosemary and lavender with total flavonoid content of 44.9 ± 1.3 mg catechin equivalent (CE)/g of extract. Later, Ho et al.[66] measured flavonoids in green tea as 0.68 ± 0.03 mg CEs/mL, which was higher than black tea (0.53 ± 0.00 mg CEs/mL).

In another exploration, Gramza et al.[62] assessed higher antiradical activity of green tea than black tea or oolong tea. The scavenging ability of ethanolic extracts of green tea was 8% higher than that of black tea. Lin et al.[88] investigated the DPPH scavenging ability in terms of $EC_{50}$ value of 50% ethanolic GTE as 11.68 ± 0.03 μg extract/mL. Another study determined extraction time (3, 5, 10, 15, and 30 min) as an important parameter influencing antioxidant activity. The higher DPPH trapping ability was 13.45 ± 0.68 mmol/L of Trolox at the 30 min of extraction time.[80] Another study found $IC_{50}$ value of GTE against DPPH radical as 55.00–323.66 ppm.[134] Another finding analyzed DPPH capturing ability of GTE as 4.80 ± 0.40 mM Trolox.[61]

A survey determined that the antioxidant potential at different maturity stages of green tea using ABTS assay with $IC_{50}$ values ranged between 0.17 ± 0.02 and 0.18 ± 0.02. They also found that higher antioxidant ability exists in young green tea leaves followed by mature and shoots parts.[73] Gorjanović et al.[61] determined the antioxidant potential of green tea as 7.68 ± 0.15 mM Trolox using ABTS reagent; however, FRAP value was found as 17.79 ± 0.73 mM Fe (II). Loizzo et al.[89] studied the antioxidant capacity of ready to serve green tea drink and measured $IC_{50}$ value varying between 9.37 ± 0.3 and 11.19 ± 0.9. Yashin et al.[159] determined the free radical trapping potential of GTE as 571 μmol/g using FRAP reagent. Lin et al.[88] estimated the ferrous ion chelating capacity in terms of $EC_{50}$ value of 50% ethanolic green tea leaves and extracts as 4.658 ± 0.027 mg tea leaves/mL and 1.493 ± 0.009 mg extract/mL, accordingly.

## 13.5 CHARACTERIZATION AND QUANTIFICATION OF GTE

For the quantification of phenolics, numerous reliable methods include: Folin–Ciocalteu (F–C) method, Folin–Denis (F–D) method, gas chromatography (GC) and HPLC. Total polyphenols, group of phenolics, and even individual bioactive component has also been measurable by these techniques with high sensitivity.[19,37,76] The most sensitive, accurate, and profuse estimation was via HPLC for green tea catechins: EGCG, EGC, EC, ECG, GCG, catechins (C), gallic acid, rutin, and caffeine.[9] However, EGCG has been found as the most dominant fraction of GTE as compared to other catechins as elucidated by numerous findings. Conversely, some early surveys have described EGC as the prominent catechin detected via HPLC. The green tea polyphenols were divided into many classes: 77.1% flavan-3-ols, 13.2% flavonols, 7.6% hydroxycinnamates, and 2.2% gallic acid derivatives. However, the main catechins in green tea have been observed as: EGC 1565 ± 18 mg/L, EGCG 1255 ± 63 mg/L, EC 738 ± 17 mg/L, ECG 361 ± 12 mg/L, and catechins 270 ± 9.5 mg/L.[123] Another survey indicated values as: EGC 36.53 mg/g, EGCG 18.10 mg/g, EC 6.06 mg/g, ECG 5.34 mg/g, and C 3.45 mg/g from dried green tea leaves.[83]

Later investigations of HPLC verified EGCG as the highest fraction with amount ranging from 95.54 to 357.07 mg/L and EGC from 69.71 to 324.88 mg/L.[80] The Chinese GTEs were measured with EGCG as 4.91% and EGC as 1.97%.[101] Later, 30 different green tea varieties indicated EGCG as the dominating fraction, accounting for 4.44% followed by EGC 1.97%.[77] The SFE analyzed EGCG as the highest fraction using HPLC system.[79,146] Earlier, HPLC was found five times more sensitive as compared to capillary electrophoresis to quantify green tea catechins and black tea theaflavin.[83]

## 13.6 GREEN TEA-BASED DIETARY INTERVENTIONS

The physiologically active biomolecules have recovered up to 60% of cardiovascular disease (CVD) cases and 20–50% of oncogenic events.[124,30] The high temperatures >98°C, especially during baking, may cause oxidation of polyphenols so as to compensate the losses, that is, ~20% and there is a need to incorporate phytochemicals manually in larger extents.[152] However, it has also been noticed that bioavailability of these active moieties increases after processing.[124]

In nitrification process, catechins are first extracted, isolated, and then manipulated in different food products according the compatibility, serving the purpose of disease protective diet. Conclusively, dietary manipulation is a promising vehicle to improve color, flavor, and taste during storage and transportation. EGCG owes higher antioxidant activity as compared to vitamins C, E, and β-carotene ultimately raising serum antioxidant level of its consumers.[152] Normally, green tea-based therapies have been enriched with 1% of GTE or 245.9– 1256.5 mg/100 g of green tea polycatechins, especially EGCG, for example, green tea candies.[63]

Lipid peroxidation is responsive of rancidity especially in meat-based products. In this context, green tea catechins in dietary interventions have been found to reduce lipid peroxidation in the order of EGCG > ECG > EGC > EC, prolonging shelf life as a response. This polyphenol composition of green tea varies based on location of cultivation, time of year, and maturity stages. The GTE normally holds 10–15% of EGCG, 6–10% of EGC, 2–3 % of ECG, and 2% of EC. Lipid peroxidation in meat and fish products causes faster rancidity up to 4.5–11%. In this regard, nutra meat has replaced the concept of synthetic preservatives such as butylated hydroxyanisole (BHA), butylated hydroxytoluene (BHT) and tertiary butyl hydroquinone (TBHQ) with that of green tea catechins to avoid concerned toxicity. They have also been found to inhibit acrylamide formation during frying and control *Clostridium* and *Bacillus* species in meat and its products.

The use of green tea catechins in meat products not only controls transglutaminase enzyme but also prevents lipid oxidation, hence improves palatability. Green tea catechins @ 300 mg/kg have been incorporated in meat products to lower lipid peroxidation reactions. Green tea-based crackers, cakes, breads, and starches have already been employed to manipulate taste and antioxidant properties of designer foods. GTEs also contain caffeine that may cause gastrointestinal disturbances as well as insomnia. Therefore, green tea is first decaffeinated or specific bioactive components are isolated through SFE technology. Thus, green tea catechins have been used to enhance quality of food and life.

The bitterness of green tea polyphenols could be neutralized via fortification, ensuring bioavailability and consumer acceptance. For food stability from pathogenic agents, green tea catechins have been proved as the natural source. Traditionally, Chinese cakes have been supplemented with GTE to increase shelf life and flavor. It has also been used in oils such as canola and soybean oil and in juices such as apple juice resultantly to control *Escherichia coli* and *Salmonella typhimurium.*[152]

## 13.7 DISEASE AMELIORATIVE POTENTIAL OF GREEN TEA

### *13.7.1 HYPERLIPIDEMIA/DYSLIPIDEMIA*

Green tea is considered as an effective approach in overcoming serum and hepatic hyperlipidemia as well as reported to have inverse relationship with cardiovascular ailments. It has significant antioxidant potential to downregulate triacylglycerol and lipid peroxidation.[31] Early meta-analyses depicted that intake of catechins @ 197 mg for 12 weeks could reduce lipidemic biomarkers up to 5% amongst dyslipidemic patients. During thermal processing, one half of the catechins undergo conversion into catechin gallate (CG) and GCG. According to the finding, heat-processed catechins showed an upper hand in controlling the body cholesterol level by excreting it out of the body without being absorbed or accumulated in adipose tissue or hepatocytes.[69]

Dyslipidemia is mostly defined by elevated triacylglycerol level >150 mg/dL and reduced high-density lipoprotein (HDL)-cholesterol. Low-density lipoprotein (LDL)-cholesterol is found to be the basic micelles to provoke CVDs; nonetheless, 1% decrease in LDL-cholesterol may overwhelm 2% of vascular maladies. Numerous studies have pointed out that GTEs possess the ability to suppress cholesterol miscibility by elevating its excretion through feces.[67,141] Green tea catechins have been reported to reduce serum triacylglycerol level in rats on high-fat diet by inhibiting lipase enzyme in pancreas. For 12 weeks, significant reduction was observed in obesity by supplementing heat processed catechins @ 444 mg due to their involvement in reducing hepatic fatty acid synthase and malic enzymes, involved in the synthesis of fatty acids in human body.[69]

Singh et al.[140] conducted a study (30 days) to evaluate the effects of green tea on lipid profiling of hyperlipidemic volunteers. They supplemented three cups of green/day that expounded a declining trend in cholesterol, LDL-cholesterol, and triglyceride by 8.36%, 15.6%, and 1.69%, respectively. However, nonsignificant improvements were noted in HDL-cholesterol, that is, 1.16% only without impacting on height, weight, and body mass index (BMI). The green tea administrated @ 3 g for 2 months resulted in cholesterol reduction by 2% without any momentous alterations in HDL and triacylglycerol levels. Moreover, it was found that green tea together with soy-based material could maximally reduce serum cholesterol, LDL-cholesterol, and triacylglycerol despite alone intake of either. It was also observed that green tea upsurges TRAP, an indicator of plasma antioxidant defense against lipid oxidation. Additionally, they stated that pregnant women are mostly facing a poor antioxidant status due to their high dependence on polyunsaturated diet (i.e., more prone to oxidation) thus a reasonable incorporation of green tea is necessitated to supplement their deficiencies.[136]

The induction of 1% of heat-treated catechins resulted in increased fecal excretion of steroids from cholesterol or lower cholesterol absorption via intestinal route. The GTE with 197 mg of catechins (mainly GCG and

CG) have been found to reduce cholesterol level by 5%.[69] Ramadan et al.[118] found that green and black tea extracts could regulate serum cholesterol, LDL-cholesterol, and triacylglycerol levels by inhibiting lipogenesis and increasing fecal excretion of cholesterol, fatty acid, and bile acids.

Li et al.[86] confirmed the effect of green tea against hypertriglyceridemia in fructose-induced insulin resistant hamster models. A dose of 150 and 300 mg/kg body weight (B.W.) reduced triacylglycerol by 42% and 62%, respectively. Batista et al.[18] conducted a study involving 33 volunteers on 200 mg of cholesterol and 250 mg of GTE on daily basis for 16 weeks. The authors inferred reduction of LDL and total cholesterol up to 4.5% and 3.9%, correspondingly. However, nonsignificant responses were viewed on HDL-cholesterol, triacylglycerol, and Apo-B levels.

Di Pierro et al.[40] studied the lipid profile of 50 obese subjects who were supplemented on green tea catechins for 90 days who reported significant loss in weight in addition to reduction in LDL-cholesterol from 132 ± 25 to 105 ± 15 mg/L and increment in HDL-cholesterol from 42 ± 6 to 51 ± 7 mg/L. Later analysis noted improvement in serum lipid profile by introducing 600 mg of GTE to 30 individuals' diet for 2 months.[67]

The administration of EGCG @ 25 mg/kg/day has attenuated the glucose and lipid metabolites of streptozotocin (STZ)-induced diabetic rats for 8 weeks.[126] Subsequently, Bajerska et al.[13] fed 1.1% and 2% aqueous GTEs to experimental rats for 8 weeks who reported reduction of visceral fat (17.8%) and body weight (5.6%), with resultantly reduced atherogenic risk. Also, Nantz et al.[99] noted decrement in LDL-cholesterol up to 9 mg/dL by supplementing GTE to healthy individuals.

Gomikawa et al.[59] involved healthy females to analyze the response of green tea on their cardiovascular events. The study demonstrated reduction in LDL-cholesterol up to 10 mg/dL via green tea @ 4.5 g per day. Erba et al.[49] studied the reduction in LDL from 119.9 to 106.6 mg/dL @ two cups of green tea for 42 days. Nagao et al.[97] studied the protective effect of green tea catechins on vascular events, obesity, and lipoprotein metabolites.

Brown et al.[25] recommended the positive effects of green tea catechins especially EGCG on lifestyle-related disorders and mood swings. Protective effects of EGCG have also been observed against blood pressure, LDL cholesterol, Apo-B level, and total serum cholesterol. A study for 4 weeks based on 40 volunteers supplemented with six to seven cups of green tea/day indicated lowering of coronary artery diseases by regulating oxidized LDL-cholesterol levels from 9.50 ± 9.2 to 7.76 ± 7.7 U/mL. However, overall serum lipid profile remained unchanged.[70] Afterward, positive effects of green tea (four cups) were observed in lowering the lipid peroxidation of

35 obese individuals for 8 weeks. As a result, a reduction was elicited in LDL-cholesterol and oxidized-LDL from 144 ± 9.5 to 100 ± 10.0 mg/dL and 104 ± 10.6 to 91 ± 9.1 U/L, respectively.[16] The study on 60 individuals by incorporating 544 mg of GTE in dietary regimen for 4 months reduced the diastolic blood pressure. The total cholesterol was reduced from 224.8 ± 38.6 to 206.3 ± 30.2 mg/dL; similarly, LDL decreased from 136.7 ± 33.6 to 123.8 ± 31.5 mg/dL.[54]

Bursill and Roach[26] found decrement in lipid profile via 2% crude GTE. In their study, 60% decrease was viewed in total cholesterol, 80% in LDL-cholesterol, and 70% in very low-density lipoprotein (VLDL)-cholesterol as compared to hypercholesterolemic control animals. Furthermore, other researchers advised nine or more cups of green tea/day to modulate altered lipid profile, especially oxidized LDL-cholesterol.[38]

Sae-tan et al.[130] observed reduction in plaque formation by 31% by consuming 3 g/L of green tea for 21 weeks. Yang and Koo[157] assessed increment in HDL-cholesterol by 59% and reduction in serum cholesterol up to 23% by administrating 2% or 4% of Chinese green tea to hyperlipidemic Sprague–Dawley rats within 8 weeks, correspondingly. The mechanism elucidated excretion of cholesterol micelle out of the body.

### 13.7.2 CARDIOVASCULAR DISEASES

Green tea polyphenols have been considered as protective moieties against platelet accumulation and high lipoprotein profile.[138] Regular consumption of green tea resulted in reduction of 15% LDL-cholesterol.[141,31] It has also been proved that green tea @ three cups/day possess the ability to reduce CVDs by 11%.[136] Arab et al.[8] observed a lowering effect in stroke incidence by 21% through an intake of three or more cups of green and black tea via experimenting on 194,965 individuals. Several studies confirmed that six cups of green tea could mitigate redox-mediated reactions and hypertension.[28]

High cholesterol >200 mg/dL and low HDL-cholesterol <35 mg/dL are the conditions marking toward vasculature. Green tea is negatively associated with hyperlipidemia/dyslipidemia, heart strokes, and hypertension. Regular consumption of green tea seems to be a better solution to avert such events. This pattern has already been observed in Japanese population where consumption of green tea is practiced on daily basis, resultantly lowering the tendency of heart strokes and obesity as compared to the western world.[38,138]

Moreover, Chacko et al.[31] confirmed the potential of green tea to demote the coronary artery diseases by 36%. The antiatherogenic property of green tea catechins is attributed to its capacity to capture free radicals and metal ions thereby overcome incidences of oxidative damage, lipid peroxidation, LDL oxidation, and inflammatory pathways.[38,138] Green tea also controls superoxide anions (free radicals), improving endothelial vasodilation in heart patients. Additionally, it stimulates the activation of nitric oxide synthase leading to the generation of nitric oxide thus regulates vasodilation and diastolic blood pressure.[67,141] The green tea catechins (ECs and EGCG) have been found to reduce Apo-B hence modulate LDL modification in vascular ailments.[118]

Hypercholesterolemia is a major risk factor of atherosclerosis and coronary events, though healthful dietary interventions are considered as the prerequisite requirement to overcome disease incidence. Healthy lifestyle demands for the inclusion of phytochemicals from different sources such as fruits, vegetables, grains, etc., in a balanced way. Green tea phytochemicals are cardioprotective in nature owing to reduction in LDL-cholesterol. It has also been studied that five cups of green tea/day reduced LDL-cholesterol by 11.1% and total cholesterol by 6.5%.[18]

In an investigation, pure ECs @ 100 mg/day were assessed on vascular health for 4 weeks. ECs improved insulin and insulin resistance, whereas they did not have any effect on blood pressure and glucose and lipid levels. Thus, it is recommended to consume cocoa and tea, especially for those suffering from insulin resistance, but not effective against cardiac disorders.[43] The daily consumption of green tea $<1$ cup could lower CVDs and cerebral infarction; 2 to $\geq 4$ cups suppress hyperlipidemia, myocardial infarction, and stroke; and $\geq 10$ cups lower LDL-cholesterol. In sum, this meta-analysis explicated that green tea consumption is protective against CVD risks.[108]

Nakachi et al.[98] conducted a cohort study on 8552 residents of Japan and found that green tea consumption $>10$ cups/day as compared to three cups could better overcome CVDs and cancer. Basu et al.[17] documented that EGCG is protective against lipid peroxidation and atherosclerotic lesions in response to melondialdehyde (MDA) concentration and LDL modification in arteries, thus control on CVDs. Furthermore, Kuriyama et al.[82] observed reduction in CVDs-related mortality by consuming green tea. A meta-analysis based on 10 cohort studies indicated that tea consumption increases CVDs in United Kingdom and Australia, whereas suppresses in other European countries.[115] Later, a study was carried out on 56 obese individual suffering from hypertension and it was viewed that one capsule carrying 379 mg of GTE suppressed blood pressure significantly and overcame insulin resistance, serum lipid profile, and inflammatory biomarkers for 3 months. On the other hand, this treatment enhanced HDL-cholesterol and endogenous antioxidants to a significant level.[22]

Seo et al.[137] worked on fermented green tea carrying 14.1% w/w polyphenols, lower than green tea itself (34.7% w/w). They found that fermented green tea could suppress cholesterol, triacylglycerol, and glucose from 191.03 ± 4.59, 107.25 ± 6.01, and 164.93 ± 5.24 mg/dL in high-fat diet fed mice to 83.00 ± 7.80, 145.69 ± 9.46, and 140.14 ± 8.05 mg/dL, respectively.

Ahmad et al.[4] observed that green tea catechins and EGCG, each @ 550 mg per 500 mL to Sprague-Dawley rats for 56 days, significantly lowered body weight in hypercholesterolemic rats up to 10.73% and 8.49%, accordingly. Furthermore, they found that EGCG lowers lipidemic indicators more effectively than catechins-based designer food. Li et al.[85] supplemented green tea to overweight/obese individuals and observed reduction in blood pressure.

### *13.7.3 HYPERGLYCEMIA*

Diabetes is a chronic metabolic syndrome with two types including: insulin-dependent diabetes mellitus (type I diabetes) and noninsulin-dependent diabetes mellitus (type II diabetes). In type I diabetes, insulin secretion is defected because of damaged β-pancreatic cell; however, in case of type II diabetes, insulin production is normal but receptors of islet of Langerhans are nonrespondent to insulin signals. Approximately 80% of diabetes cases are due to lifestyle-related factors/disparities such as obesity. Moreover, multiple symptoms are associated with diabetes encompassing polydipsia, glycosuria, polyuria, weakness, and slow healing processes.[6,118]

The protective mechanism followed by green tea catechins is the reduction in amylase activity during chewing and later in the intestinal digestion. Green tea has also been ascribed to hold 3.5% of caffeine, carrying xanthine as the main component that compels the body musculature to exploit the extra fat securing glycogen reserves.[138]

Plant extracts are the safe moieties as compared to pharmaceuticals to regulate hyperglycemia and hypercholesterolemia. GTE contains 80–90% of catechins such as EC, ECG, EGC, and EGCG and <10% of flavonols such as kaempferol, quercetin, and myricetin glycosides. These bioactive moieties are potent defenders against diabetogenic agents such as sucrose or STZ injections induced in experimental animals.[118] The glycemic load is featured by oxidative stress, that is, rise in free radicals inducing aging and other chronic maladies, such as cancer, resulting in lowering of serum antioxidant enzyme (superoxide dismutase, SOD; catalase, CAT; and glutathione peroxidases, GPx). However, green tea catechins have been proved as protective molecules in the enhancement of these enzymes. Therefore, GTE is found to be a therapeutic tool to ameliorate the oxidative stress-induced diabetes.[12] Both green tea (prepared from mature leaves) and white tea (from young leaves) have been found to be effective in regulating glucose and averting abnormalities linked with diabetes after administrating the extract @ 0.5% for a duration of 4 weeks.[71]

Various meta-analyses have reflected the miracles of green tea against diabetes owing to its antioxidants in the form of polyphenols, catechins, and water-soluble polysaccharides. There are numerous mechanisms where green tea could overcome the diabetes involving escalated insulin release through reduced activity of glucose-6-phosphatase, hence diminishing glucose-6-phosphate conversion to glucose. Proposed mechanisms include inhibition of intestinal GLUT (glucose transporter) system and reduction of gluconeogenesis by regulating genetic expression or phosphoenol pyruvate kinase production by hepatic cells. A research conducted with 50–100 mg/kg of aqueous GTE administrated to male Wistar albino rats resulted in reduction of obesity by 28–49%, glucose levels by 20–31%, and lipoprotein by 26–31%, accordingly.[118,130]

In another study, administration of GTE @ 50–100 mg/kg suppressed the generation of free radicals resultantly reducing blood glucose level by 29–40%. Similarly, GTE (300 mg/kg) was found to modulate hyperglycemic conditions induced in mice after 2–6 h of STZ injection. Furthermore, green tea was found to improve serum glutathione, SOD, and CAT levels that were downregulated during free radical-induced diabetic condition. Furthermore, GTE could regulate cytokine-induced β-cell damage and reduce the islet mass affected by diabetogenic agents.[31]

The GTE stimulates insulin secretion by converting glucose to glycogen storage in liver and muscles. It slows down the release of salivary and intestinal amylase enzyme thus preventing the breakdown of starch to glucose. It has therefore been recommended to consume six cups of green tea/day to modulate the glycemic index, hence diabetes. The polycatechins could control amylase activity in the order of catechins > gallocatechin gallate > epicatechin gallate > epigallocatechin gallate. The thermogenic effect of green tea has especially been associated with caffeine that utilizes the body fat cells, conserving glycogen for future needs. Green tea catechins have the potential to decelerate 25% of carbohydrate absorption by raising either breath hydrogen concentration or hydrogen excretion.[138]

Various investigations have confirmed the hypoglycemic effects of green tea on human serum glucose level without interfering insulin secretion on administrating @ 1.5 g/kg body weight.[149] Iso et al.[72] conducted an experiment on 17,413 volunteers with genetic predeposition of diabetes type II disease; it was observed that the disease incidence was reduced by 33% on consuming six cups of green tea daily. The epidemiological studies have indicated that green tea modulates blood sugar level with the prevention of type I diabetes and repairs the damaged β-cells. The hypoglycemic properties of green tea primarily depend on EGCG; however, other components, such as caffeine, catechins and ECs, explicated minor effects on insulin production. The same study has also interpreted the synergistic effect of lemon and milk along with green tea. Purposely, the addition of lemon did not affect the insulin activity; nonetheless, inclusion of milk, nondairy creamers, and soy milk decreased the activity at a significant rate.[141]

The beneficial effects of EGCG have been conferred by reducing insulin resistance with the induction of blood glucose restoration, protecting β-cell of alloxan-treated rats.[106] Li et al.[87] viewed the hypoglycemic properties of lipophilic-EGCG (L-EGCG) derivatives that altered the lipid and glucose profile of STZ-induced diabetic rats within 30 days. Thus, they found reduction in glucose concentration up to 40.5 ± 7.0% and 17.0 ± 2.8% with L-EGCG concentration of 50 or 25 mg/kg/day. It was observed that GTE containing EGCG @ 300 mg/kg rectifies the lipoprotein and glucose metabolism of male Sprague–Dawley rats within 4 weeks. It also demonstrated positive effects on fructose fed male Sprague–Dawley rats by decreasing their Apo (apolipoprotein)-β, a structural protein of VLDL-cholesterol and LDL-cholesterol that is correlated with CVDs. At a dose of 300 mg/kg, Apo-β was decreased by 54%, triglyceride in liver and heart was decreased by 42% and 32.5%, and caused reduction in peroxisome proliferator-activated receptor (PPAR) α and γ protein expression by 300% and 500%, respectively. GTE influenced on PPAR ligands to ameliorate hyperglycemia and hypercholesterolemia.[86]

Green tea has the marked effects on weight reduction in rats from 300 to 280 g along with decline in serum glucose levels from 120 to 110 mg/dL, serum triglyceride from 80 to 75 mg/dL, serum cholesterol from 60 to 55 mg/dL, and LDL-cholesterol from 15 to 10 mg/dL; however, HDL-cholesterol level was increased from 35 to 40 mg/dL. As a result, EGCG has been proved as a potentiating tool to attenuate reactive oxygen species (ROSs) by enhancing the activity of SOD up to 21.3%.[126,130] In a nutshell, green tea catechins are the potential source to mitigate the etiology of hyperglycemia and related metabolic syndromes. It improves the pancreatic functions and regulates glucose metabolism. This reduction in blood glucose level seems helpful in managing the obesity.

### 13.7.4 OBESITY

Green tea motivates thermogenesis by reducing serum cholesterol and LDL-cholesterol by facilitating its fecal excretion.[118] Chacko et al.[31] suggested EGCG for increased fat oxidation and thermogenesis to overcome obesity. Combination of EGCG and caffeine could maintain weight by increasing energy output in obese subjects. GTE worked synergistically with calcium and caffeine to upregulate thermogenesis up to 4.6%. EGCG, being negatively related with obesity, is recommended for obese men @ 300 mg/day but to explore the optimal dose, further meta-analysis seems to be a better solution.[31]

Recent surveys explicated that green tea and related products could ameliorate high lipid profile.[137] EGCG @ 856.8 mg was fed to obese individuals for 12 weeks and compared with placebo group. The results explicated significant decline in weight from 76.8 ± 11.3 to 75.7 ± 11.5 kg. Alongside, it reduces cholesterol up to 5.33% along with decrement in LDL-cholesterol and waist circumference.[35] Li et al.[85] observed momentous reduction in blood pressure of overweight/obese individuals via green tea or its extract. Ahmad et al.[4] checked the response of green tea catechins and EGCG, each @ 550 mg/500 mL on Sprague–Dawley rats for 56 days. The study indicated significant decrease in body weight of hypercholesterolemic (10.73% and 8.49%) and hyperglycemic (10.12% and 10.49%) rats, respectively. Among both, EGCG showed better control than catechins-based treatment.

Seo et al.[137] determined the polyphenols in green tea and fermented green tea up to 34.7% and 14.1%, respectively. They found that fermented green tea could suppress cholesterol from 191.03 ± 4.59 to 145.69 ± 9.46 mg/dL, triglyceride from 107.25 ± 6.01 to 83.00 ± 7.80 mg/dL, and glucose from 164.43 ± 5.24 to 140.14 ± 8.05 mg/dL in high fat and high fat + fermented green tea-based diets, correspondingly. They found that fermented green tea is one of the designer products that possess anti-obesity potential; especially it manages body weight without modifying dietary pattern. In serum and liver metabolomic study via LC/MS, it is observed that obesity could be ameliorated by green tea by modulating fatty acid β-oxidation.[84] Obesity-associated fatty liver disease could be restored by decaffeinated green tea extracted EGCG by increasing the activity of mitochondrial respiratory chain, ultimately increasing lipid oxidation in liver of obese mice. Furthermore, it was found to improve lipid metabolic processes and insulin sensitivity.[135] Earlier, green tea catechins, caffeine, and polysaccharides @ 400 or 800 mg/kg for 6 weeks were reported to fight against obesity in rats. Each fraction was effective against altered lipidemic profile, suppresses leptin level, and improves antioxidant status, whereas combination of catechins and polysaccharides could suppress weight gain, serum leptin level, and inflammation pathways. Thus, combination was found more effective over individual fraction.[155]

Dostal et al.[42] found that GTE @ 843 mg for 12 months did not influence adipose tissues, bone mineral density, or obesity-related hormones. A prospective cohort study was conducted in healthcare center of Japan for 4 years. Based on follow-up of 18.7 years, it was viewed that green tea is inversely related with death rate in response to cardiac and vascular ailments and respiratory disorders.[132] In a study by Grosso et al.,[64] three or more cups of coffee/day depicted lower blood pressure, waist circumference, BMI, and triacylglycerol, whereas raised HDL-cholesterol level than those who take $<1$ cup/day. Contrarily, high tea intake was found to reduce central obesity, but high blood pressure was noted as compared to those who consume less tea, only in female but not in male. Further, coffee and tea intake demonstrated negative association with metabolic syndromes.

### 13.7.5 OXIDATIVE STRESS

The dysregulation of cellular functions is associated with imbalance between free radical bodies and antioxidants. Oxygen is converted to ROSs resulting in the over production of free radicals that upsurge the risk of metabolic dysfunctions. Therefore, phytonutrients have exhibited the ability to block such interruptions of oxidation and peroxidation chain reactions. Clinical trials have further ensured their influential role to avert the disease symptoms. The dietary phenolics are means to prevent free radicals via mechanisms such as quenching ability and complex formation.[45,103,111,117,156]

#### 13.7.5.1 HEPATOTOXICITY

Normally, very low levels of liver functioning enzymes are found in the serum. During liver disease, lipid peroxidation may occur because of altered hepatocyte fenestrations causing rupturing of tissues or leakage of enzymes such as alanine transaminase (ALT), alkaline phosphatase (ALP), $\gamma$-glutamyl transferase ($\gamma$-GT), aspartate transaminase (AST), and lactate dehydrogenase (LDH) in addition to endogenous antioxidant enzymes such as SOD and CAT into blood.[21,95,102,105]

Previous researchers found that catechins @ 40–80 mg/kg B.W. have the potential to fight against hepatorenal toxicities induced by arsenic @ 100 ppm to rats for 30 days. The increment in hepatic (ALT, ALP, and AST) and renal (urea and creatinine) biomarkers due to toxicity could be ameliorated significantly via coadministration of catechins.[2] Green tea catechins possess anti-oncogenic ability against liver malignancies. The anti-inflammatory and antioxidant activities of green tea prevent the formation of preoncogenic lesions and neoplasms by modulating various mechanistic pathways.[139]

Interestingly, a large prospective survey was carried out across Europe that indicated dose-dependent association of caffeinated tea and coffee. The regular consumption of coffee reduced the hepatocellular cancer risk up to 72%, whereas no protection was shown by decaffeinated coffee.[14] The EGCG @ 10, 20, and 40 mg/kg/day was administered to mice during last 4 weeks after the induction of NAFLD (by feeding on high-fat diet for 24 weeks). The EGCG modulated the hyperlipidaemia-induced fatty liver by alleviating weight gain and insulin resistance in a dose-dependent manner.[55] GTE @ 200 mg/kg B.W. was effective against liver toxicities induced by malathion.[119]

Meta-analysis conducted by Yin et al.[161] demonstrated negative association between green tea and liver disorders. Aqueous extract of green tea @ 2.5% to rats, 7 days prior and 5 days after administration of hepatotoxic methotrexate @ 20 mg/kg, was found effective in lowering liver functioning enzymes and lipid peroxidation as well as altered fenestrations of liver cells. Alongside, this treatment can upregulate glutathione.[74] EGCG @ 3.2 g per kg of diet for 16 weeks was found to reduce body weight, liver weight, and body fat due to augmented fecal excretion of lipids. On the other hand, supplementation up to 4 weeks at a similar dose resulted in reduced blood glucose and body fat, thus long-term inclusion of EGCG was found more ameliorative in lowering metabolic ailments such as fatty liver as compared to short-term administration.[24] Earlier, green tea @ 7 g/L showed protection against liver damage and normalizes the activities of enzymatic and nonenzymatic antioxidants.[11]

El-Beshbishy[46] found that 1.5% of GTE could enhance enzymatic antioxidants in liver; however, it decreased lipid peroxidation by scavenging free radicals produced by tamoxifen @ 45 mg/kg/day for 7 days study on rats. Moreover, green tea could protect membrane architecture against lipid production due to ethanol consumption.[107,142]

### 13.7.5.2 NEPHROTOXICITY

Kidneys perform numerous physiological functions such as body homeostasis, regulation of electrolyte and blood pressure, and removal of toxins via urination.[56] During the recent era, rapid increase in chronic kidney diseases (CKDs), especially among the developing states, is attributed to poor dietary habits. In CKDs, nephrons lost their structural and functional integrity leading to reduced glomerulus filtration and increase in blood urea and creatinine concentrations.[3] This peril is more prevalent in patients suffering from high blood pressure, diabetes, and cardiovascular complications.[34] In a study, green tea prior to iodixanol-induced nephrotoxicity showed more effectiveness than later or concurrent treatments.[100]

Lv et al.[90] determined that EGCG could protect renal injury in rats owing to its anti-apoptotic and anti-inflammatory activities by depressing nuclear factor-κB (NF-κB) and activating of p38 mitogen-activated protein kinase in kidney. They also found that EGCG could significantly downregulate histopathological deformities. Thus, it is recommended to drink green tea on regular basis for kidney health. In another study, EGCG @ 50 mg/kg B.W./day for 3 weeks momentously reduced mortality by reducing proteinuria and serum creatinine level, whereas counteracting oxidative damage and inflammation as well as associated histopathology.[160]

Furthermore, EGCG derived from green tea depicted nephron-protection against lipid and protein oxidation in response to free radicals generated due to cypermethrin (toxicant). Additionally, it suppresses renal damage as evident via histological malfunctions and resultantly modulates increased activities of serum creatinine, blood urea nitrogen (BUN), urea, uric acid, AST, ALT, ALP, and LDH along with increment in endogenous antioxidants.[151] Later in a study, catechins-rich GTE @ 300 mg/kg B.W./day showed amelioration against alcohol-induced renal stress in rats for 60 days. The results indicated restoration of histological abnormalities. Furthermore, flavonoids in green tea possess the ability to trap free radicals. The extract possesses the potential to lower lipid peroxidation, which as a result reserves endogenous enzymatic antioxidant; SOD, CAT, and GPx, as well as nonenzymatic antioxidant; and glutathione (GSH) of kidney.[121]

Some studies show nephrotoxicity in diabetic mice due to EGCG.[120] Previously, another study found that green tea can suppress nephron-cardiovascular toxins and raise serum creatinine and BUN in chronic renal function.[113] Green tea polyphenols @ 50 and 100 mg/kg B.W. reduced serum glucose level up to 29% and 44% in alloxan-treated diabetic rats. Furthermore, it has been found to increase endogenous antioxidants including SOD and GSH, whereas no obvious improvements were reported in CAT and GPx.[129] In a study conducted on Sprague–Dawley rats, GTE @ 1.5% showed ameliorative potential against renal toxicities and lipid peroxidation in association with cyclosporine A @ 20 mg/kg/day for 21 days.[94] In another research, it was found that GTE @ 300 mg/kg/day possesses ameliorative potential against CVDs in diabetics. In this context, diabetic rats were fed on this extract for 4 weeks and results explicated significant reduction in lipidemic markers of heart, modulation of glucose level, and reduction in body weight and heart to body weight ratio.[7]

In another trial (12 weeks), green tea to diabetic rats depicted reduction in glucose, creatinine, and BUN significantly up to 29.6%, 38.9%, and 41.7%, respectively. Therefore, green tea could manage diabetic nephropathy.[122] EGCG administration, 1 week after induction of immunomediated glomerulonephritis, was observed to suppress renal dysfunctions.[112] GTE ameliorated nephrotoxicity about gentamicin by improving antioxidant status including SOD, CAT, and GSH and downregulating histopathological alterations, necrosis, and cellular degenerations.[133]

Abdel-Raheem et al.[1] found a similar response against gentamicin-induced nephrotoxicity via GTE @ 300 mg/kg/day. Furthermore, there were many researchers who found decrement in serum creatinine, GGT, ALP, and BUN levels.[5]

Khan et al.[78] found that green tea extracted components could ameliorate gentamicin-induced nephrotoxicity and oxidative stress by improving antioxidant defense mechanism and tissue integrity. They evidenced reduction in serum creatinine, cholesterol, BUN, and lipid peroxidation but raised SOD and CAT levels in renal tissues. Previous researchers found higher potential of EGCG over quercetin and resveratrol each @ 50 mg/kg against cisplatin-induced disrupted glomerular filtration rate in experimental animals.[48] Similar cisplatin-induced nephrotoxicity and related oxidative damage of biomembranes due to lipid peroxidation in albino mice was restored via green tea @ 100 mg/kg B.W.[158]

Khan et al.[78] found similar response against cisplatin-induced nephrotoxicity. They observed decrement in serum creatinine, BUN, glucose, and cholesterol levels from 1.48 ± 1.11, 33.12 ± 0.4, 23.8 ± 2.86, and 146.24 ± 3.8 mg/dL (diseased group) to 0.96 ± 0.33, 12.6 ± 0.23, 26.3 ± 2.47, and 98.4 ± 0.6 mg/dL (GTE-supplemented group), correspondingly. Sahin et al.[131] significantly upregulated GSH and downregulated inflammation by inhibiting NF-κB via EGCG against cisplatin-induced nephrotoxicity.

### 13.7.6 ANTICANCER ACTIVITIES

Meta-analysis of EGCG has proved it as a suppressing agent for carcinogens, angiogenesis, cell proliferation, etc. However, it has been found as a key agent in the activation of tumor suppressor genes and apoptosis, immunomodulation, and anti-oncogenesis.[36,103] The chemopreventive mechanisms of EGCG include cell cycle arrest, inhibit conversion of precancerous cells to solid tumors, reduce nitric oxide synthase by slowing down tumor necrosis factor-alpha, suppress telomerase activity by 50–60% within 24 h by breaking telomere, suppressing numerous androgen receptor pathways by regulating various growth factors, modulating various enzymatic activities, decreasing angiogenesis by controlling vascular endothelial growth factors and receptor phosphorylation, and interfering in the production of nucleic acid and protein by controlling dihydrofolate reductase activity.[27,103,138,141]

According to the review articles, EGCG possesses two different mechanisms with which they could fight against cancer in a cell environment: antioxidant and pro-oxidant activities. The EGCG @ 0.1% showed significant reduction in oxidative stress; however, 0.2% or 0.5% serve as pro-oxidant that generates free radicals responsible for reducing cancer-promoting proteins resultantly downregulating cancer initiation to metastasis.[65] Furthermore, green tea is associated with antioxidant, modulation of phase II antioxidant enzymes, and potential as pro-oxidant by generating free radicals, superoxide anion, hydroxyl radical, etc., against cancer cells.[50]

EGCG and various combinations of anticancer compounds such as nonsteroidal anti-inflammatory drugs were found to suppress tumor volume up to 70.3%.[52,53] A meta-analysis from 1998 to 2009 indicated that green tea >3 cups have the potential to mitigate breast cancer recurrence.[104] Another meta-analysis demonstrated that green tea, but not black tea could suppress endometrial cancer. One cup/day of green tea could mitigate endometrial cancer by 11%, whereas the effect was statistically nonsignificant.[163]

Kumar et al.[81] found that daily consumption of decaffeinated EGCG @ 400 mg/day for a year could downregulate prostate cancer in men. Liver cancer risk could be suppressed in females but not in males via the consumption of green tea, though the effect was nonsignificant @ one cup/day.[68] In another study, EGCG @ 20 and 40 mg/kg B.W. resulted in significant shrinkage of tumor by 48% and 92%, respectively.[47]

## 13.8 SUMMARY

Green tea has proven its numerous health benefits with special emphasis in managing hyperlipidemia, hyperglycemia, obesity, oxidative stress, and oncogenesis. Green tea extraction technologies, such as SFE, have shown better efficiency than conventional methods. Besides drinking green tea on daily basis, GTEs or bioactive moieties could also be incorporated in routine foods based on their compatibility or efficacy to achieve a healthy lifestyle.

## KEYWORDS

- **conventional solvents extraction**
- **designer foods**
- **polycatechins**
- **supercritical fluid extraction**

## REFERENCES

1. Abdel-Raheem, I. T.; El-Sherbiny, G. A.; Taye, A. Green Tea Ameliorates Renal Oxidative Damage Induced by Gentamicin in Rats. *Pak. J. Pharm. Sci.* **2010,** *23* (1), 21–28.
2. Afolabi, O. K.; Ugbaja, R. N.; Arinola, A.; Olorunnisola, O. S.; Adeleke, G. E. Catechin Protects Against Dyslipidemia and Nephro-hepatototoxicity Induced in Rats Exposed to Arsenic. *Jordan J. Biol. Sci.* **2016,** *9* (3), 173–177.
3. Agarwal, S.; Shlipak, M. G.; Kramer, H.; Jain, A.; Herrington, D. M. The Association of Chronic Kidney Disease and Metabolic Syndrome with Incident Cardiovascular Events: Multiethnic Study of Atherosclerosis. *Cardiol. Res. Pract.* **2012,** *2012*, 806102.
4. Ahmad, R. S.; Butt, M. S.; Sultan, M. T.; Mushtaq, Z.; Ahmad, S.; Dewanjee, S.; De Feo, V.; Zia-Ul-Haq, M. Preventive Role of Green tea Catechins from Obesity and Related Disorders Especially Hypercholesterolemia and Hyperglycemia *J. Transl. Med.* **2015,** *13* (1), 1–9.
5. Ahn, T. G.; Kim, H. K.; Park, S. W.; Kim, S. A.; Lee, B. R.; Han, S. J. Protective Effects of Green Tea Polyphenol Against Cisplatin-induced Nephrotoxicity in Rats. *Obstet. Gynecol. Sci.* **2014,** *57* (6), 464–470.
6. Al-Attar, A. M.; Zari, T. A. Influences of Crude Extract of Tea Leaves, *Camellia sinensis*, on Streptozotocin diabetic male Albino Mice. *Saudi J. Biol. Sci.* **2010,** *17* (4), 295–301.
7. Anandh Babu, P. V.; Sabitha, K. E.; Shyamaladevi, C. S. Green Tea Extract Impedes Dyslipidaemia and Development of Cardiac Dysfunction in Streptozotocin-diabetic Rats. *Clin. Exp. Pharmacol. Physiol.* **2006,** *33* (12), 1184–1189.

8. Arab. L.; Liu, W.; Elashoff, D. Green and Black Tea Consumption and Risk of Stroke: A Meta-analysis. *Am. Heart Assoc.* **2009,** *40* (5), 1786–1792.
9. Armoskaite, V.; Kristina, R.; Audrius, M.; Almantas, R.; Audrone, D.; Algirdas, B.; Vitalis, B. The Analysis of Quality and Antioxidant Activity of Green Tea Extracts. *J. Med. Plants Res.* **2011,** *5* (5), 811–816.
10. Asghar, Z.; Ziaee, M. Evaluation of Antioxidant Properties of Silymarin and Its Potential to Inhibit Peroxyl Radicals *In Vitro*. *Pakistan J. Pharm. Sci.* **2008,** *21* (3), 249–254.
11. Augustyniak, A.; Waszkiewicz, E.; Skrzydlewska, E. Preventive Action of Green Tea from Changes in the Liver Antioxidant Abilities of Different Aged Rats Intoxicated with Ethanol. *Nutrition* **2005,** *21* (9), 925–932.
12. Babu, P. V. A.; Sabitha, K. E.; Shyamaladevi, C. S. Green Tea Impedes Dyslipidemia, Lipid Peroxidation, Protein Glycation and Ameliorates $Ca^{2+}$-ATPase and $Na^{+}/K^{+}$-ATPase Activity in the Heart of Streptozotocin-diabetic Rats. *Chem.-Biol. Interact.* **2006,** *162* (2), 157–164.
13. Bajerska, J.; Małgorzat, W.; Jan, J.; Slawomira, D.; Jaroslaw, W. Green Tea Aqueous Extract Reduces Visceral Fat and Decreases Protein Availability in Rats Fed with a High-fat Diet. *Nutr. Res.* **2011,** *31* (2), 157–164.
14. Bamia, C.; Lagiou, P.; Jenab, M.; Trichopoulou, A.; Fedirko, V.; Aleksandrova, K.; Pischon, T.; Overvad, K. Coffee, Tea and Decaffeinated Coffee in Relation to Hepatocellular Carcinoma in a European Population: Multicentre, Prospective Cohort Study. *Int. J. Cancer* **2015,** *136* (8), 1899–1908.
15. Bastos, D. H. M.; Saldanha, L. A.; Catharino, R. R.; Sawaya, A. C. H. F.; Cunha, I. B. S.; Carvalho P. O.; Eberlin, M. N. Phenolic Antioxidants Identified by ESI-MS from Yerba Mate (*Ilex paraguariensis*) and Green Tea (*Camellia sinensis*) Extracts. *Molecules* **2007,** *12* (3), 423–432.
16. Basu, A.; Karah, S.; Misti, J. L.; Mingyuan, W.; Nancy, M. B.; Christopher, E. A.; Timothy, J. L. Green Tea Supplementation Affects Body Weight, Lipids, and Lipid Peroxidation in Obese Subjects with Metabolic Syndrome. *J. Am. College Nutr.* **2010,** *29* (1), 31–40.
17. Basu, A.; Lucas, E. A. Mechanisms and Effects of Green Tea on Cardiovascular Health. *Nutr. Rev.* **2007,** *65* (8), 361–375.
18. Batista, G. A. P.; Da Cunha, C. L. P.; Scartezini, M.; Heyde, R. V. D.; Bitencourt, M. G.; De Melo, S. F. Prospective Double Blind Crossover Study of *Camellia sinensis* (Green Tea) in Dyslipidemias. *Arq. Bras. Cardiol.* **2009,** *93* (2), 121–127.
19. Bayili, R. G.; Fatoumata, A.; Oumou, H. K.; Mamounata, D.; Imael, H.; Bassole, N.; Mamoudou, H. D. Phenolic Compounds and Antioxidant Activities in Some Fruits and Vegetables from Burkina Faso. *Afr. J. Biotechnol.* **2011,** *10* (62), 13543–13547.
20. Bharadwaz, A.; Chiranjit, B. Extraction of Polyphenols from Dried Tea Leaves. *Int. J. Sci. Eng. Res.* **2012,** *3* (5), 1–5.
21. Bhavani, R.; Kotteeswaran, R.; Rajeshkumar, S. Hepatoprotective Effect of *Brassica oleracea* Vegetable and Its Leaves in Paracetamol Induced Liver Damage in Albino Rats. *Int. J. ChemTech Res.* **2014,** *6* (7), 3705–3712.
22. Bogdanski, P.; Suliburska, J.; Szulinska, M.; Stepien, M.; Pupek-Musialik, D.; Jablecka, A. Green Tea Extract Reduces Blood Pressure, Inflammatory Biomarkers, and Oxidative Stress and Improves Parameters Associated with Insulin Resistance in Obese, Hypertensive Patients. *Nutr. Res.* **2012,** *32* (6), 421–427.

23. Borse, B. B.; Kumar, H. V.; Jagan, M. R. Radical Scavenging Conserves from Unused Fresh Green Tea Leaves. *J. Agric. Food Chem.* **2007,** *55* (5), 1750–1754.
24. Bose, M.; Lambert, J. D.; Ju, J.; Reuhl, K. R.; Shapses, S. A.; Yang, C. S. The Major Green Tea Polyphenol-(-)-Epigallocatechin-3-Gallate, Inhibits Obesity, Metabolic Syndrome, and Fatty Liver Disease in High-fat-fed Mice. *J. Nutr.* **2008,** *138* (9), 1677–1683.
25. Brown, A. L.; Joan, L.; Jacqueline, C.; Janice, S.; Sarah, J.; Alison, S.; Les, B.; Andy, C.; Hilde, H. Effects of Dietary Supplementation with the Green Tea Polyphenol Epigallocatechin-3-Gallate on Insulin Resistance and Associated Metabolic Risk Factors: Randomized Controlled Trial. *Br. J. Nutr.* **2009,** *101* (6), 886–894.
26. Bursill, C. A.; Roach, P. D. A Green Tea Catechin Extract Upregulates the Hepatic Low-Density Lipoprotein Receptor in Rats. *Lipids* **2007,** *42* (7), 621–627.
27. Butt, M. S.; Muhammad, T. S. Green tea: Nature's Defense Against Malignancies. *Food Sci. Nutr.* **2009,** *49* (5), 463–473.
28. Cabrera, C.; Artacho, R.; Gimenez, R. Beneficial Effects of Green Tea: A Review. *J. Am. Coll. Nutr.* **2006,** *25* (2), 79–99.
29. Carloni, P.; Luca, T.; Luca, P.; Tiziana, B.; Chisomo, C.; Alexaander, K.; Elisabetta, D. Antioxidant Activity of White, Green and Black Tea Obtained from the Same Tea Cultivar. *Food Res. Int.* **2013,** *53* (2), 900–908.
30. Cencic, A.; Walter, C. The role of Functional Foods, Nutraceuticals, and Food Supplements in Intestinal Health. *Nutrition* **2010,** *2* (6), 611–625.
31. Chacko, S. M.; Thambi, P. T.; Kuttan, R.; Nishigaki, I. Beneficial Effects of Green Tea: A Literature Review. *Chin. Med.* **2010,** *5* (13), 1–9.
32. Chandrakant, K.; Pravin, D.; Bharat, H.; Sachin, K.; Amol, K. An Overview of Supercritical Fluid Extraction for Herbal Drug. *Int. J. Pharm. Innov.* **2011,** *1* (3), 575–596.
33. Chang, C. J.; Kou-Lung, C.; Ying-Ling, C.; Ching-Yuan, C. Separation of Catechins from Green Tea Using Carbon Dioxide Extraction. *Food Chem.* **2000,** *68* (1), 109–113.
34. Chauhan, V.; Vaid, M. Dyslipidemia in Chronic Kidney Disease: Managing a High-risk Combination. *Postgrad. Med. J.* **2009,** *121* (6), 54–61.
35. Chen, I. J.; Liu, C. Y.; Chiu, J. P.; Hsu, C. H. Therapeutic Effect of High-dose Green Tea Extract on Weight Reduction: A Randomized, Double-blind, Placebo-controlled Clinical Trial. *Clin. Nutr.* **2016,** *35* (3), 592–599.
36. Crespy, V.; Williamson, G. Review of the Health Effects of Green Tea Catechins in *In Vivo* Animal Models. *J. Nutr.* **2004,** *134* (12), 3431–3440.
37. Dai, J.; Russell, J. M. Plant Phenolics: Extraction, Analysis and Their Antioxidant and Anticancer Properties. *Molecules* **2010,** *15* (10), 7313–7352.
38. Deka, A.; Joseph, A. V. Tea and Cardiovascular Disease. *Pharmacol. Res.* **2011,** *64* (2), 136–145.
39. De-Mejia, E. G.; Marco, V. R.; Sirima, P. Bioactive Components of Tea: Cancer, Inflammation and Behavior. *Brain Behav. Immunity* **2009,** *23* (6), 721–731.
40. Di Pierro, F.; Anna, B. M.; Angela, B.; Maurizio, L.; Andrea, C. GreenSelect Phytosome as an Adjunct to a Low-calorie Diet for Treatment of Obesity: A Clinical Trial. *Altern. Med. Rev.* **2009,** *14* (2), 154–160.
41. Dong, J.; Jian-Hui, Y.; Jian-Liang, L.; Xin-Qiang, Z.; Yue-Rong, L. Isolation of Antioxidants Catechins from Green Tea and Its Decaffeination. *Food Bioprod. Process.* **2011,** *89* (1), 62–66.

42. Dostal, A. M.; Arikawa, A.; Espejo, L.; Kurzer, M. S. Long Term Supplementation of Green Tea Extract Does Not Modify Adiposity or Bone Mineral Density in a Randomized Trial of Overweight and Obese Postmenopausal Women. *J. Nutr.* **2016,** *146* (2), 256–264.
43. Dower, J. I.; Geleijnse, J. M.; Gijsbers, L.; Zock, P. L.; Kromhout, D.; Hollman, P. C. Effects of the Pure Flavonoids Epicatechin and Quercetin on Vascular Function and Cardiometabolic Health: A Randomized, Double-blind, Placebo-controlled, Crossover Trial. *Am. J. Clin. Nutr.* **2015,** *101* (5), 914–921.
44. Druzynska, B.; Stepniewska A.; Wolosiak, R. The Influence of Time and Type of Solvent on Efficiency of the Extraction of Polyphenols from Green Tea and Antioxidant Properties Obtained Extracts. *Acta Sci. Pol. Technol. Aliment.* **2007,** *6* (1), 27–36.
45. Dureja, H.; Kaushik D.; Kumar, V. Developments in Nutraceuticals. *Indian J. Pharmacol.* **2003,** *35* (6), 363–372.
46. El-Beshbishy, H. A. Hepatoprotective Effect of Green Tea (*Camellia sinensis*) Extract Against Tamoxifen-induced Liver Injury in Rats. *BMB Rep.* **2005,** *38* (5), 563–570.
47. El-Mowafy, A. M.; Al-Gayyar, M. M.; Salem, H. A.; El-Mesery, M. E.; Darweish, M. M. Novel Chemotherapeutic and Renal Protective Effects for the Green Tea (EGCG): Role of Oxidative Stress and Inflammatory-cytokine Signaling. *Phytomedicine* **2010,** *17* (14), 1067–1075.
48. El-Mowafy, A. M.; Salem, H. A.; Al-Gayyar, M. M.; El-Mesery, M. E.; El-Azab, M. F. Evaluation of Renal Protective Effects of the Green-tea (EGCG) and Red Grape Resveratrol: Role of Oxidative Stress and Inflammatory Cytokines. *Nat. Prod. Res.* **2011,** *25* (8), 850–856.
49. Erba, D.; Riso, P.; Bordoni, A.; Foti, P.; Biagi, P. L.; Testolin, G. Effectiveness of Moderate Green Tea Consumption on Antioxidative Status and Plasma Lipid Profile in Humans. *J. Nutr. Biochem.* **2005,** *16* (3), 144–149.
50. Forester, S. C.; Lambert, J. D. The Role of Antioxidant Versus Pro-oxidant Effects of Green Tea Polyphenols in Cancer Prevention. *Mol. Nutr. Food Res.* **2011,** *55* (6), 844–854.
51. Franco, D.; Jorge, S.; Mónica, R.; Marivel, S.; María, J.; Manuel, P.; Noelia, C.; María, J. N. Polyphenols from Plant Materials: Extraction and Antioxidant Power. *Electron. J. Environ. Agric. Food Chem.* **2008,** *7* (8), 3210–3216.
52. Fujiki, H.; Sueoka, E.; Watanabe, T.; Suganuma, M. Primary Cancer Prevention by Green Tea, and Tertiary Cancer Prevention by the Combination of Green Tea Catechins and Anticancer Compounds. *J. Cancer Prevent.* **2015,** *20* (1), 1–4.
53. Fujiki, H.; Sueoka, E.; Watanabe, T.; Suganuma, M. Synergistic Enhancement of Anticancer Effects on Numerous Human Cancer Cell Lines Treated with the Combination of EGCG, other Green tea Catechins, and Anticancer Compounds. *J. Cancer Res. Clin. Oncol.* **2015,** *141* (9), 1511–1522.
54. Fukino, Y.; Ikeda, A.; Maruyama, K.; Aoki, N.; Okubo, T.; Iso, H. Randomized Controlled Trial for an Effect of Green Tea-extract Powder Supplementation on Glucose Abnormalities. *Eur. J. Clin. Nutr.* **2008,** *62* (8), 953–960.
55. Gan, L.; Meng, Z. J.; Xiong, R. B.; Guo, J. Q.; Lu, X. C.; Zheng, Z. W.; Deng, Y. P.; Luo, B. D.; Zou, F.; Li, H. Green tea Polyphenol Epigallocatechin-3-Gallate Ameliorates Insulin Resistance in Non-alcoholic Fatty Liver Disease Mice. *Acta Pharmacol. Sin.* **2015,** *36* (5), 597–605.

56. Garcia, G. G.; Harden, P.; Chapman, J. The Global Role of Kidney Transplantation. *Nephron Clin. Practice* **2012,** *120* (2), 101–106.
57. Ghoreishi, S. M.; Heidari, E. Extraction of Epigallocatechin Gallate from Green Tea Via Modified Supercritical $CO_2$: Experimental, Modeling and Optimization. *J. Supercrit. Fluids* **2012,** *72* (1), 36–45.
58. Ghoreishi, S. M.; Heidari, E. Extraction of Epigallocatechin-3-Gallate from Green Tea Via Supercritical Fluid Technology: Neural Network Modeling and Response Surface Optimization. *J. Supercrit. Fluids* **2013,** *74* (1), 128–136.
59. Gomikawa, S.; Ishikawa, Y.; Hayase, W.; Haratake, Y.; Hirano, N.; Matuura, H.; Mizowaki, A.; Murakami, A.; Yamamoto, M. Effect of Ground Green Tea Drinking for 2 Weeks on the Susceptibility of Plasma and LDL to the Oxidation *Ex Vivo* in Healthy Volunteers. *Kobe J. Med. Sci.* **2008,** *54* (1), 62–72.
60. Gong, J. S.; Chun, X. P.; Xiang, H.; Jun, H. L.; Bao, C. L.; Hong, J. Z. Antioxidant Activity of Extracts of Green Tea and Its Material. *Asian J. Agric. Sci.* **2009,** *1* (2), 48–54.
61. Gorjanović, S.; Draženka, K.; Ferenc, T.; Ana, B.; Lato, P.; Ivana, H.; Desanka, S. Antioxidant Capacity of Teas and Herbal Infusions: Polarographic Assessment. *J. Agric. Food Chem.* **2012,** *60* (38), 9573–9580.
62. Gramza, A.; Pawlak-Lemanska, K.; Korczak, J.; Wasowicz, E.; Rudzinska, M. Tea Extracts as Free Radical Scavengers. *Pol. J. Environ. Stud.* **2005,** *14* (6), 861–867.
63. Gramza-Michalowska, A.; Julita, R. Use of Tea Extracts (*Camellia sinensis*) in Jelly Candies as Polyphenols Sources in Human Diet. *Asia Pac. J. Clin. Nutr.* **2007,** *16* (1), 43–46.
64. Grosso, G.; Stepaniak, U.; Micek, A.; Topor-Mądry, R.; Pikhart, H.; Szafraniec, K.; Pająk, A. Association of Daily Coffee and Tea Consumption and Metabolic Syndrome: Results from the Polish Arm of the HAPIEE Study. *Eur. J. Nutr.* **2015,** *54* (7), 1129–1137.
65. Hayakawa, S.; Saito, K.; Miyoshi, N.; Ohishi, T.; Oishi, Y.; Miyoshi, M.; Nakamura, Y. Anti-cancer Effects of Green Tea by Either Anti- or Pro-oxidative Mechanisms. *Asian Pac. J. Cancer Prevent.* **2016,** *17* (4), 1649–1654.
66. Ho, S. C.; Wu, S. P.; Lin, S. M.; Tang, Y. L. Comparison of Anti-glycation Capacities of Several Herbal Infusions with that of Green tea. *Food Chem.* **2010,** *122*, 768–774.
67. Houston, M.; Nashville. T. N. Effect of Combination Pantethine, Plant Sterols, Green tea Extract, Delta-tocotrienol and Phytolens on Lipid Profiles in Patients with Hyperlipidemia. *J. Am. Nutraceut. Assoc.* **2010,** *13* (1), 15–20.
68. Huang, Y. Q.; Lu, X.; Min, H.; Wu, Q. Q.; Shi, X. T.; Bian, K. Q.; Zou, X. P. Green Tea and Liver Cancer Risk: A Meta-analysis of Prospective Cohort Studies in Asian Populations. *Nutrition* **2016,** *32* (1), 3–8.
69. Ikeda, I. Multifunctional Effects of Green tea Catechins on Prevention of the Metabolic Syndrome. *Asia Pac. J. Clin. Nutr.* **2008,** *17* (1), 273–274.
70. Inami, S.; Takano, M.; Yamamoto, M.; Murakami, D.; Tajika, K.; Yodogawa, K.; Yokoyama, S.; Ohno, N.; Ohba, T.; Sano, J.; Ibuki, C.; Seino, Y.; Mizuno, K. Tea Catechin Consumption Reduces Circulating Oxidized Low-density Lipoprotein. *Int. Heart J.* **2007,** *48*, 725–732.
71. Islam, M. S. Effects of the Aqueous Extract of White tea (*Camellia sinensis*) in a Streptozotocin-induced Diabetes Model of Rats. *Phytomedicine* **2011,** *19*, 25–31.
72. Iso, H.; Chigusa, D.; Kenji, W.; Mitsuru, F.; Akiko, T.; Jacc, S. G. The Relationship Between Green Tea and Total Caffeine Intake and Risk for Self-reported Type 2 Diabetes Among Japanese Adults. *Ann. Internal Med.* **2006,** *144* (8), 554–562.

73. Izzreen, N. Q.; Mohd, F. Phytochemicals and Antioxidant Properties of Different Parts of *Camellia sinensis* Leaves from Sabah Tea Plantation in Sabah, Malaysia. *Int. Food Res. J.* **2013,** *20* (1), 307–312.
74. Jwied, A. H. Hepatoprotective Effect of the Aqueous Extract of *Camellia sinensis* Against Methotrexate-induced Liver Damage in Rats. *Iraqi J. Pharm. Sci.* **2017,** *18* (2), 73–79.
75. Karori, S. M.; Wachira, F. N.; Wanyoko, J. K.; Ngure, R. M. Antioxidant Capacity of Different Types of Tea Products. *Afr. J. Biotechnol.* **2007,** *6* (19), 2287–2296.
76. Kaur, C.; Harish, C. K. Anti-oxidant Activity and Total Phenolic Content of Some Asian Vegetables. *Int. J. Food Sci. Technol.* **2002,** *37* (2), 153–161.
77. Kerio, L. C.; Wachira, F. N.; Wanyoko, J. K.; Rotich, M. K. Total Polyphenols, Catechin Profiles and Antioxidant Activity of Tea Products from Purple Leaf Colored Tea Cultivars. *Food Chem.* **2013,** *136* (3), 1405–1413.
78. Khan, S. A.; Priyamvada, S.; Khan, W.; Khan, S.; Farooq, N.; Yusufi, A. N. Studies on the Protective Effect of Green Tea Against Cisplatin Induced Nephrotoxicity. *Pharmacol. Res.* **2009,** *60* (5), 382–391.
79. Kim, W.; Kim, J. D.; Jaehoon, K.; Seong-Geun O.; Youn-Woo, L. Selective Caffeine Removal from Green Tea Using Supercritical Carbon Dioxide Extraction. *J. Food Eng.* **2008,** *89* (3), 303–309.
80. Komes, D.; Dunja, H.; Ana, B.; Karin K. G.; Ivana, V. Green Tea Preparation and Its Influence on the Content of Bioactive Compounds. *Food Res. Int.* **2010,** *43* (1), 167–176.
81. Kumar, N. B.; Pow-Sang, J.; Egan, K. M. Randomized, Placebo-controlled Trial of Green Tea Catechins for Prostate Cancer Prevention. *Cancer Prevent. Res.* **2015,** 1–24.
82. Kuriyama, S.; Shimazu, T.; Ohmori, K.; Kikuchi, N.; Nakaya, N.; Nishino, Y.; Tsubono, Y.; Tsuji, I. Green Tea Consumption and Mortality Due to Cardiovascular Disease, Cancer, and All Causes in Japan: The Ohsaki Study. *Am. Med. Assoc.* **2006,** *296* (10), 1255–1265.
83. Lee, B. L.; Ong, C. N. Comparative Analysis of Tea Catechins and Theaflavins by High-performance Liquid Chromatography and Capillary Electrophoresis. *J. Chromatogr.* **2000,** *881* (1), 439–447.
84. Lee, L. S.; Choi, J. H.; Sung, M. J.; Hur, J. Y.; Hur, H. J.; Park, J. D.; Kim, Y. C.; Gu, E. J.; Min, B.; Kim, H. J. Green Tea Changes Serum and Liver Metabolomic Profiles in Mice With High-fat Diet-induced Obesity. *Mol. Nutr. Food Res.* **2015,** *59* (4), 784–794.
85. Li, G.; Zhang, Y.; Thabane, L.; Mbuagbaw, L.; Liu, A.; Levine, M. A.; Holbrook, A. Effect of Green Tea Supplementation on Blood Pressure Among Overweight and Obese Adults: A Systematic Review and Meta-analysis. *J. Hypertens.* **2015,** *33* (2), 243–254.
86. Li, R. W.; Douglas, T. D.; Maiyoh, G. K.; Adeli, K.; Theriault, A. G. Green Tea Leaf Extract Improves Lipid and Glucose Homeostasis in a Fructose-Fd Insulin-resistant Hamster Model. *J. Ethnopharmacol.* **2006,** *104* (1), 24–31.
87. Li, T.; Liu, J.; Zhang, X.; Ji, G. Antidiabetic Activity of Lipophilic (-)-Epigallocatechin-3-Gallate Derivative Under its Role of α-Glucosidase Inhibition. *Biomed. Pharmacother.* **2007,** *61* (1), 91–96.
88. Lin, S.; Jeng-Leun M.; Ching-An, H. Bioactive Components and Antioxidant Properties of G-Aminobutyric Acid (GABA) Tea Leaves. *LWT-Food Sci. Technol.* **2012,** *46* (1), 64–70.
89. Loizzo, M. R.; Rosa, T.; Marco, B.; Federica, M.; Vincenzo, M.; Luigi, A.; Francesco, M. Radical Scavenging, Antioxidant and Metal Chelating Activities of *Annona cherimola* Mill. (Cherimoya) Peel and Pulp in Relation to Their Total Phenolic and Total Flavonoid Contents. *J. Food Compos. Anal.* **2012,** *25* (2), 179–184.

90. Lv, J.; Feng, M.; Zhang, L.; Wan, X.; Zeng, Y. C.; Liang, P. F.; Xu, A. P. Protective Effect of Epigallocatechin Gallate, a Major Constituent of Green Tea, Against Renal Ischemia-reperfusion Injury in Rats. *Int. Urol. Nephrol.* **2015,** *47* (8), 1429–1435.
91. Manian, R.; Nagarajan, A.; Perumal, S.; Sellamuthu, M. The Antioxidant Activity and Free Radical Scavenging Potential of Two different Solvent Extracts of *Camellia sinensis* (L.), *Ficus benghalensis* L. and *Ficus racemosa* L. *Food Chem.* **2008,** *107* (3), 1000–1007.
92. Maróstica, M. R.; Leite, A. V.; Dragano, N. R. V. Supercritical Fluid Extraction and Stabilization of Phenolic Compounds from Natural Sources: Review (Supercritical Extraction and Stabilization of Phenolic Compounds). *Open Chem. Eng. J.* **2010,** *4* (1), 51–60.
93. Mereles, D.; Werner, H.; Epigallocatechin-3-Gallate (EGCG) for Clinical Trials: More Pitfalls than Promises? *Int. J. Mol. Sci.* **2011,** *12* (9), 5592–5603.
94. Mohamadin, A. M.; El-Beshbishy, H. A.; El-Mahdy, M. A. Green Tea Extract Attenuates Cyclosporine-A Induced Oxidative Stress in Rats. *Pharmacol. Res.* **2005,** *51* (1), 51–57.
95. Morsy, A. F.; Ibrahim, H. S.; Shalaby, M. A. Protective Effect of Broccoli and Red Cabbage Against Hepatocellular Carcinoma Induced by n-Nitrosodiethylamine in Rats. *J. Am. Sci.* **2010,** *6* (12), 1136–1144.
96. Naczk, M.; Fereidoon, S. Extraction and Analysis of Phenolics in Food. *J. Chromatogr.* **2004,** *1054* (1), 95–111.
97. Nagao, T.; Komine, Y.; Soga, S.; Meguro, S.; Hase, T.; Tanaka, Y.; Tokimitsu, I. Ingestion of a Tea Rich in Catechins Leads to a Reduction in Body Fat and Malondialdehyde-modified LDL in Men. *Am. J. Clin. Nutr.* **2005,** *81* (1), 122–129.
98. Nakachi, K.; Matsuyama, S.; Miyake, S.; Suganuma, M.; Imai, K. Preventive Effects of Drinking Green Tea on Cancer and Cardiovascular Disease: Epidemiological Evidence for Multiple Targeting Prevention. *Biofactors* **2000,** *13* (1–4), 49–54.
99. Nantz, M. P.; Rowe, C. A.; Bukowski, J. F.; Percival, S. S. Standardized Capsule of *Camellia sinensis* Lowers Cardiovascular Risk Factors in a Randomized, Double Blinded, Placebo-controlled Study. *Nutrition* **2009,** *25* (2), 147–154.
100. Nasri, H.; Hajian, S.; Ahmadi, A. Ameliorative Effect of Green Tea Against Contrast-induced Renal Tubular Cell Injury. *Iranian J. Kidney Dis.* **2015,** *9* (6), 421–426.
101. Nishitani, E.; Yuko, M. S. Simultaneous Determination of Catechins, Caffeine and Other Phenolic Compounds in Tea Using HPLC Method. *J. Food Compos. Anal.* **2004,** *17* (5), 675–685.
102. Noori, S.; Rehman, N. Reduction of Carbon Tetrachloride-induced Rat Liver Injury by Coffee and Green Tea. *Pak. J. Nutr.* **2009,** *8* (4), 452–458.
103. Ogle, N. Green Tea (*Camellia sinensis*). *Aust. J. Herbal Med.* **2009,** *21* (2), 44.
104. Ogunleye, A. A.; Xue, F.; Michels, K. B. Green Tea Consumption and Breast Cancer Risk or Recurrence: A Meta-analysis. *Breast Cancer Res. Treat.* **2010,** *119* (2), 477–484.
105. Olusola, A. O. Evaluation of the Antioxidant Effects of *Hibiscus sabdariffa* Calyx Extracts on 2,4-Dinitrophenylhydrazine Induced Oxidative Damage in Rabbits. *Webmed. Central Biochem.* **2011,** *2* (10), 1–11.
106. Ortsäter, H.; Nina, G.; Swen, W.; Nicolas, K.; Ake, S. Diet Supplementation with Green Tea Extract Epigallocatechin Gallate Prevents Progression to Glucose Intolerance in db/db Mice. *Nutr. Metab.* **2012,** *9* (11), 1–10.
107. Ostrowska, J.; Łuczaj, W.; Kasacka, I.; Różański, A.; Skrzydlewska, E. Green Tea Protects Against Ethanol Induced Lipid Peroxidation in Rat Organs. *Alcohol* **2004,** *32* (1), 25–32.
108. Panga, J.; Zhanga, Z.; Zhengb, T. Z.; Bassigb, B. A.; Gec, J.; Yangd, Y. J.; Huang, D.; Bai, M.; Penga, Y. Green Tea Consumption and the Risk of the Related Factors

of Cardiovascular Diseases and Ischemic Related Diseases: A Meta-analysis. *Int. J. Cardiol.* **2015,** 1–32.

109. Park, H. S.; Hyung-Kyoon, C.; Sung, J. L.; Kwon, W. P.; Sung-Gil, C.; Kyoung, H. K. Effect of Mass Transfer on the Removal of Caffeine from Green Tea by Supercritical Carbon Dioxide. *The J. Supercrit. Fluids* **2007,** *42*, 205–211.
110. Park, H. S.; Nam, G. I.; Kyoung, H. K. Extraction Behaviors of Caffeine and Chlorophylls in Supercritical Decaffeination of Green Tea Leaves. *LWT-Food Sci. Technol.* **2012,** *45*, 73–78.
111. Patil, L.; Balaraman, R. Effect of Green Tea Extract on Doxorubicin Induced Cardiovascular Abnormalities: Antioxidant Action. *Iranian J. Pharm. Res.* **2011,** *10* (1), 89–96.
112. Peng, A.; Ye, T.; Rakheja, D.; Tu, Y.; Wang, T.; Du, Y.; Zhou, X. J.; Vaziri, N. D.; Hu, Z.; Mohan, C.; Zhou, X. J. The Green Tea Polyphenol (−)-Epigallocatechin-3-Gallate Ameliorates Experimental Immune-mediated Glomerulonephritis. *Kidney Int.* **2011,** *80* (6), 601–611.
113. Peng, Y. H.; Sweet, D. H.; Lin, S. P.; Yu, C. P.; Chao, P. D. L.; Hou, Y. C. Green Tea Inhibited the Elimination of Nephro-cardiovascular Toxins and Deteriorated the Renal Function in Rats with Renal Failure. *Sci. Rep.* **2015,** *5* (1), 1–10.
114. Perva-Uzunalic, A.; Skerget, M.; Knez, Z.; Weinreich, B.; Otto, F.; Gruner, S. Extraction of Active Ingredients from Green Tea (*Camellia sinensis*): Extraction Efficiency of Major Catechins and Caffeine. *Food Chem.* **2006,** *96* (4), 597–605.
115. Peters, U.; Poole, C.; Arab, L. Does Tea Affect Cardiovascular Disease? A Meta-analysis. *Am. J. Epidemiol.* **2001,** *154* (6), 495–503.
116. Pineiro, Z.; Miguel, P.; Carmelo, G. B. Determination of Catechins by Means of Extraction with Pressurized Liquids. *J. Chromatogr.* **2004,** *1026* (1), 19–23.
117. Prakash, D.; Gupta, K. R. The Antioxidant Phytochemicals of Nutraceutical Importance. *Open Nutraceut. J.* **2009,** *2* (1), 20–35.
118. Ramadan, G.; Nadia, M. E.; Eman, A. A. E. Modulatory Effects of Black v. Green Tea Aqueous Extract on Hyperglycaemia, Hyperlipidaemia and Liver Dysfunction in Diabetic and Obese Rat Models. *Brit. J. Nutr.* **2009,** *102* (11), 1611–1619.
119. Raoofi, R.; Jahromi, H. K.; Jahromi, Z. K.; Abedi, H.; Sameni, H.; Pourahmad, M. Antioxidant Effects of Green-tea on Biochemical and Histopathological Changes of Liver in Male Rats Poisoned by Malathion Insecticide. *Int. J. Med. Res. Health Sci.* **2016,** *5* (5), 361–370.
120. Rasheed, N. O. A.; Ahmed, L. A.; Abdallah, D. M.; El-Sayeh, B. M. Nephro-toxic Effects of Intraperitoneally Injected EGCG in Diabetic Mice: Involvement of Oxidative Stress, Inflammation and Apoptosis. *Sci. Rep.* **2017,** *7* (1), 1–8.
121. Reddyvaria, H.; Bullea, S.; Vaddi, D. R.; Cha, V. N. Ameliorative Effect of Green Tea Extract on Alcohol Induced Renal Damage in Rats. *Int. J. Appl. Biol. Pharm. Technol.* **2016,** *7* (3), 23–31.
122. Renno, W. M.; Abdeen, S.; Alkhalaf, M.; Asfar, S. Effect of Green Tea on Kidney Tubules of Diabetic Rats. *Br. J. Nutr.* **2008,** *100* (3), 652–659.
123. Rio, D. D.; Amanda, J. S.; William, M.; Jennifer, B.; Michael, E. J. L.; Furio, B.; Alan, C. HPLC-MS Analysis of Phenolic Compounds and Purine Alkaloids in Green and Black Tea. *J. Agric. Food Chem.* **2004,** *52* (10), 2807–2815.
124. Rodriguez, E. R.; Maxima, E. F.; Delia, B. R. A.; Jaime, A. F. Phytochemicals and Functional Foods: Current Situation and Prospect for Developing Countries. *Segurança Alimentar e Nutricional* **2006,** *13* (1), 1–22.

125. Rodriguez-Garcıa, J.; Puig, A.; Salvador, A.; Hernando, I. Optimization of a Sponge Cake Formulation with Inulin as Fat Replacer: Structure, Physicochemical, and Sensory Properties. *J. Food Sci.* **2012,** *77* (2), 189–197.
126. Roghani, M.; Tourandokht, B. Hypoglycemic and Hypolipidemic Effect and Antioxidant Activity of Chronic Epigallocatechin-gallate in Streptozotocin-diabetic Rats. *Pathophysiology* **2010,** *17* (1), 55–59.
127. Row, K. H.; Yinzhe, J. Recovery of Catechin Compounds from Korean Tea by Solvent Extraction. *Bioresour. Technol.* **2006,** *97* (5), 790–793.
128. Rusak, G.; Komes, D.; Likic, S.; Horzic, D.; Kovac, M. Phenolic Content and Antioxidative Capacity of Green and White Tea Extracts Depending on Extraction Conditions and the Solvent Used. *Food Chem.* **2008,** *110* (4), 852–858.
129. Sabu, M. C.; Smitha, K.; Kuttan, R. Anti-diabetic Activity of Green Tea Polyphenols and Their Role in Reducing Oxidative Stress in Experimental Diabetes. *J. Ethnopharmacol.* **2002,** *83* (1), 109–116.
130. Sae-tan, S.; Kimberly, A. G.; Joshua, D. L. Weight Control and Prevention of Metabolic Syndrome by Green Tea. *Pharmacol. Res.* **2011,** *64* (2), 146–154.
131. Sahin, K.; Tuzcu, M.; Gencoglu, H.; Dogukan, A.; Timurkan, M.; Sahin, N.; Aslan, A.; Kucuk, O. Epigallocatechin-3-Gallate Activates Nrf2/HO-1 Signaling Pathway in Cisplatin-induced Nephrotoxicity in Rats. *Life Sci.* **2010,** *87* (7), 240–245.
132. Saito, E.; Inoue, M.; Sawada, N.; Shimazu, T.; Yamaji, T.; Iwasaki, M.; Sasazuki, S.; Noda, M.; Iso, H.; Tsugane, S. Association of Green Tea Consumption with Mortality Due to All Causes and Major Causes of Death in a Japanese Population: The Japan Public Health Center-based Prospective Study. *Ann. Epidemiol.* **2015,** *25* (7), 512–518.
133. Salem, E. A.; Salem, N. A.; Kamel, M.; Maarouf, A. M.; Bissada, N. K.; Hellstrom, W. J. Amelioration of Gentamicin Nephrotoxicity by Green Tea Extract in Uninephrectomized Rats as a Model of Progressive Renal Failure. *Ren. Fail.* **2010,** *32* (10), 1210–1215.
134. Sangsrichan, S.; Ting, R. Antioxidation and Radical Scavenging Activities and Tyrosinase Inhibition of Fresh Tea Leaves, *Camellia sinensis*. *Sci. J. Ubon Ratchathani Univ.* **2010,** *1* (1), 76–81.
135. Santamarina, A. B.; Carvalho-Silva, M.; Gomes, L. M. Decaffeinated Green Tea Extract Rich in Epigallocatechin-3-Gallate Prevents Fatty Liver Disease by Increased Activities of Mitochondrial Respiratory Chain Complexes in Diet-induced Obesity Mice. *J. Nutr. Biochem.* **2015,** *26* (11), 1348–1356.
136. Santana, M. B. D.; Mandarino, M. G.; Cardoso, J. R.; Dichi, I.; Dichi, J. B.; Camargo, A. E. I.; Fabris, B. A. Association Between Soy and Green Tea (*Camellia sinensis*) Diminishes Hypercholesterolemia and Increases Total Plasma Antioxidant Potential in Dyslipidemic Subjects. *Nutrition* **2008,** *24* (6), 562–568.
137. Seo, D. B.; Jeong, H. W.; Cho, D.; Lee, B. J.; Lee, J. H.; Choi, J. Y.; Hong, B. I.; Lee, S. J. Fermented Green Tea Extract Alleviates Obesity and Related Complications and Alters Gut Microbiota Composition in Diet-induced Obese Mice. *J. Med. Food* **2015,** *18* (5), 549–556.
138. Sharangi, A. B. Medicinal and Therapeutic Potentialities of Tea (*Camellia sinensis* L.): A Review. *Food Res. Int.* **2009,** *42* (5), 529–535.
139. Shimizu, M.; Shirakami, Y.; Sakai, H.; Kubota, M.; Kochi, T.; Ideta, T.; Miyazaki, T.; Moriwaki, H. Chemopreventive Potential of Green Tea Catechins in Hepatocellular Carcinoma. *Int. J. Mol. Sci.* **2015,** *16* (3), 6124–6139.

140. Singh, B. N.; Shankar, S.; Srivastava, R. K. Green Tea Catechin, Epigallocatechin-3-Gallate (EGCG): Mechanisms, Perspectives and Clinical Applications. *Biochem. Pharmacol.* **2011,** *82* (12), 1807–1821.
141. Sinija, V. R.; Mishra, H. N. Green Tea: Health Benefits. *J. Nutr. Environ. Med.* **2008,** *17* (4), 232–242.
142. Skrzydlewska, E.; Ostrowska, J.; Stankiewicz, A.; Farbiszewski, R. Green Tea as a Potent Antioxidant in Alcohol Intoxication. *Addiction Biol.* **2002,** *7* (3), 307–314.
143. Stankovic, M. S.; Neda, N.; Marina, T.; Slavica, S. Total Phenolic Content, Flavonoid Concentrations and Antioxidant Activity of Plant Extracts from *Teucrium montanum* L. Var. *Montanum, f. Supinum. Biotechnol. Biotechnol. Equip.* **2011,** *25* (1), 2222–2227.
144. Sticher, O. Natural Product Isolation. *Nat. Prod. Rep.* **2008,** *25*, 517–554.
145. Stodt, U.; Ulrich, H. E. Progress in the Analysis of Selected Tea Constituents over the Past 20 Years. *Food Res. Int.* **2013,** *53* (2), 636–648.
146. Sun, Q.; Shu, H.; Jian-Hui, Y.; Jian-Liang, L.; Xin-Qiang, Z.; Yue-Rong, L. Decaffeination of Green Tea by Supercritical Carbon Dioxide. *J. Med. Plants Res.* **2010,** *4* (12), 1161-1168.
147. Tiwari, P.; Bimlesh, K.; Mandeep, K.; Gurpreet, K.; Harleen, K. Phytochemical Screening and Extraction: A Review. *Int. Pharm. Sci.* **2011,** *1* (1), 98–106.
148. Tsai, P. J.; Tsai, T. H.; Yu, C. H.; Ho, S. C. Comparison of NO-Scavenging and NO-Suppressing Activities of Different Herbal Teas with Those of Green Tea. *Food Chem.* **2007,** *103* (1), 181–187.
149. Tsuneki, H.; Ishizuka, M.; Terasawa, M. Effect of Green Tea on Blood Glucose Levels and Serum Proteomic Patterns in Diabetic (db/db) Mice and on Glucose Metabolism in Healthy Humans. *BMC Pharmacol.* **2004,** *4* (1), 1–10.
150. Turkmen, N.; Velioglu, Y. S.; Sari, F.; Polat, G. Effect of Extraction Conditions on Measured Total Polyphenol Contents and Antioxidant and Antibacterial Activities of Black Tea. *Molecules* **2007,** *12* (3), 484–496.
151. Vinoth, K.; Mani, V. M.; Gokulakrishnan, A.; Ali, A. L. Attenuation of Cypermethrin Induced Nephrotoxicity by (-)Epigallocatechin Gallate (EGCG) in Male Wistar Rats. *Sci. Hum.* **2015,** *1* (1), 665–684.
152. Vuong, Q. V.; Costas, E. S. Isolation of Green Tea Catechins and Their Utilization in the Food Industry. *Food Rev. Int.* **2011,** *27* (3), 227–247.
153. Wang, L.; Curtis, L. W. Recent Advances in Extraction of Nutraceuticals from Plants. *Trends Food Sci. Technol.* **2006,** *17* (6), 300–312.
154. Xi, J.; Deji, S.; Shou, Z.; Bingbing, L.; Ye, L.; Rui, Z. Characterization of Polyphenols from Green Tea Leaves Using a High Hydrostatic Pressure Extraction. *Int. J. Pharm.* **2009,** *382* (1), 139–143.
155. Xu, Y.; Zhang, M.; Wu, T.; Dai, S.; Xu, J.; Zhou, Z. The Anti-obesity Effect of Green Tea Polysaccharides, Polyphenols and Caffeine in Rats Fed with a High-fat Diet. *Food Function* **2015,** *6* (1), 296–303.
156. Yadav, V.; Love, S.; Binny, T.; Moza, A. H. An Overview on Nutraceuticals as Pharmacological Agents. *Adv. Biores.* **2012,** *3* (3), 113–128.
157. Yang, T. T. C.; Koo, M. W. L. Inhibitory Effect of Chinese Green Tea on Endothelial Cell-induced LDL Oxidation. *Atherosclerosis* **2000,** *148* (1), 67–73.
158. Yapar, K.; Çavuşoğlu, K.; Oruç, E.; Yalçin, E. Protective Effect of Royal Jelly and Green Tea Extracts Effect Against Cisplatin-induced Nephrotoxicity in Mice: A Comparative Study. *J. Med. Food* **2009,** *12* (5), 1136–1142.

159. Yashin, A.; Yakov, Y.; Boris, N. Determination of Antioxidant Activity in Tea Extracts, and Their Total Antioxidant Content. *Am. J. Biomed. Sci.* **2011,** *3* (4), 322–335.
160. Ye, T.; Zhen, J.; Du, Y.; Zhou, J. K.; Peng, A.; Vaziri, N. D.; Mohan, C.; Xu, Y.; Zhou, X. J. Green Tea Polyphenol (−)-Epigallocatechin-3-Gallate Restores Nrf2 Activity and Ameliorates Crescentic Glomerulonephritis. *PloS One* **2015,** *10* (3), 1–16.
161. Yin, X.; Yang, J.; Li, T.; Song, L.; Han, T.; Yang, M.; Liao, H.; He, J.; Zhong, X. The Effect of Green Tea Intake on Risk of Liver Disease: A Meta-analysis. *Int. J. Clin. Exp. Med.* **2015,** *8* (6), 8339–8346.
162. Zhou, B. O.; Wu, L. M.; Yang, L. I.; Liu, Z. L. Evidence for α-Tocopherol Regeneration Reaction of Green Tea Polyphenols in SDS Micelles. *Free Rad. Biol. Med.* **2005,** *38* (1), 78–84.
163. Zhou, Q.; Li, H.; Zhou, J. G.; Ma, Y.; Wu, T.; Ma, H. Green tea, Black Tea Consumption and Risk of Endometrial Cancer: A Systematic Review and Meta-analysis. *Arch. Gynecol. Obstet.* **2016,** *293* (1), 143–155.

# CHAPTER 14

# BLACK TEA POLYPHENOLS: HEALTH-ENDORSING PERSPECTIVES: A MECHANISTIC APPRAISAL

ALI IMRAN, MUHAMMAD UMAIR ARSHAD,
MUHAMMAD SAJID ARSHAD, and HAFIZ ANSAR RASUL SULERIA

## ABSTRACT

Black tea polyphenols have attained paramount position among the scientific community owing to their multiple health benefits and structural diversity. Theaflavin, thearubigins, and unoxidized catechins are leading polyphenols from black tea and are being produced during fermentation, in which catechins under the influence of complex enzyme system are converted into these bioactive moieties. Black tea polyphenols are strong antioxidants and cancer chemopreventive agents as these can bind carcinogens, induce phase II enzymes such as uridine 5'-diphospho (UDP)-glucuronosyltransferase, apoptosis, cell cycle detains, inhibition of most transcription factors nuclear factor-κB (NF-κB), and AP-1 and diminution of activity of protein tyrosine kinase among leading mechanistic routes. Utilization of black tea is also beneficial in patients suffering in cardiovascular compliances. Conclusively, use of black tea polyphenols should be enhanced owing to plethora of associated health claims.

## 14.1 INTRODUCTION

### *14.1.1 BLACK TEA: AN OVERVIEW*

In diet-based regimen, functional/nutraceutical foods are gaining core attention of the nutritionists to be a therapeutic device against the maladies.[58] In this milieu, dietary guidelines are perfect candidates due to their proven pharmacological perspectives.[5] They are in fame among the scientific community as safeguard against various life-threatening diseases such as: cardiovascular diseases, glycemic abnormalities, neurodegenerative diseases, and various oncogenic events.[18]

Currently, polyphenols have attained paramount position in the dietary regimen as an intervention against numerous chronic maladies. Among different factors for the popularity of polyphenolic-based functional and nutraceutical foods, the factors of prime importance are easy access, low cost, and long administration safety.[86] The upshots of phytochemicals research have shifted the nutritionists focus toward therapeutic potential of polyphenols for the management of lipid peroxidation-induced maladies.[26] There is a dire need of antioxidants in foods to balance the antioxidant and free radical status.[70] Antioxidants protect the body from free radicals by neutralizing them and thus mitigate oxidative stress. Polyphenols are endemic in the human diet since ancient times due to their health-promoting benefits.[56,76] There are more than 8000 secondary metabolites with structural variants with total dietary intake up to 1 g/day, which is 10–100 times higher than other antioxidants such as vitamin C and E.[72]

Polyphenols can perform various biological functions, reducing oxidative stress and degenerative ailments owing to their intrinsic antioxidant potential. Besides, they stimulate the glutathione endogenous production by influencing the different biomarkers that indirectly protect the body against deleterious effects of free radicals.[96,102] Black tea (*Camellia sinensis*), member of *Theaceae* family, is one such example of plants containing bioactive molecules with unique nutraceutical potential.[110] It is considered as one of the most consuming beverage after water owing to its pleasant sensorial and therapeutic characteristics.

Currently, its annual production is about 3.6 million tons with consumption of 120 mL/capita/day around the globe. Nowadays, it is cultivated in more than 30 countries, however, mainly in Southeast Asia.[6,23,45] The major types of tea are green, oolong, and black tea depending on the processing methods and polyphenolic profile. The black tea is 78% of world total tea production. Black tea is more popular in Europe, North America, and North Africa while East and Southeast Asia are the major green tea consuming regions.

Black tea is manufactured through fermentation of green tea. During the fermentation, different enzyme systems initiate the conversation of catechins into the theaflavins (TFs), which are further converted into thearubigins (TRBs). The degree of fermentation determines the number of bioactive moieties and sensorial features of black tea. Polyphenol oxidase (PPO) is one of the main enzymes involved in tea fermentation containing Cu. Along with PPO, other promising enzymes are peroxidase (POD) and catalase (CAT) that enhance the fermentation efficiency.[50]

## 14.2 BLACK TEA: ITS CHEMICAL AND POLYPHENOLIC PROFILE

The polyphenolic profile of black tea is mainly dominated by TFs and TRBs formed by the fermentation of catechins during fermentation. However, catechins are completely converted and some retain their identity mainly in the form of epigallocatechin gallate (EGCG). On dry weight basis, back tea contains 25–30% of total polyphenols. The distribution pattern is 2–6% of TFs, 12–18% of TRBs, and 5–10% of catechins. This structural diversity enhances the antioxidant and therapeutic potentials of black tea. Similarly, green and black tea samples contain almost equal amounts of flavonols (6–9%), phenolic acids (10–12%), proteins (12–14%), methylxanthines (8–12%), fiber (15–20%), inorganic elements (5%), and aroma compounds (0.1%), as shown in Table 14.1. Moreover, black tea leaves have inherent alkaloids (2–5%) composed of caffeine and theobromine. Black tea is a natural beverage containing both oxidized and unoxidized polyphenols.[59,112,109]

**TABLE 14.1** Chemical Profile of Black Tea: Polyphenols and Other Constituents.

| Components | Concentration (%) |
|---|---|
| Amino acids | 13–15 |
| Carbohydrates | 10–15 |
| Catechins | 3–8 |
| Flavanols | 6–9 |
| Methylxanthine | 5–9 |
| Mineral matter | 6–9 |
| Phenolic acids | 10–14 |
| Polyphenols | 25–35 |
| Protein | 1–2 |
| Theaflavins | 2–6 |
| Thearubigins | 12–20 |
| Volatiles | 1 |

Black tea holds an array of polyphenols with special reference to flavonoids. Among different flavonoids, black tea contains flavanols and flavonols. The promising candidates for former groups are TF and TRBs, whereas quercetin, myricetin, and kaempferol belong to flavonols. Moreover, back tea is also equipped with phenolic acid, caffeine, theobromine, theophylline, and flavor compounds with special reference to linalool.[32] The promising tea phytochemistry is further strengthened by the presence of theanine that is exclusively present in tea and widely known for its biological perspectives. The chemical composition of tea depends on origin, degree of fermentation, and processing parameters.[14] Tea flavonoids, a major class of polyphenols, are synthesized by using carbohydrates as precursor through shikimic and *p*-coumaric acids' metabolism pathways. Flavonoid biosynthesis is a photosensitive mechanism triggered by phenylalanine ammonia lyase.[13]

Tea polyphenols are composed of 2-phenylbenzpyran skeleton with numerous hydroxy groups. They are acidic in nature owing to the presence of aromatic ring in their structure. This veracity causes decline in electron density thus weakening the O–H bond strength and promote proton loss. Tea polyphenols undergo various structural and functional changes during processing owing to the activation of various enzyme systems. The therapeutic and functional activity of tea polyphenols is ascribed by the presence of asymmetric carbons at the 2 and 3 position of the pyran ring of the 2-phenybenzopyran nucleus. Black tea is a natural beverage containing both oxidized and unoxidized polyphenols.[18,93,112]

### *14.2.1 THEAFLAVINS*

TFs are produced during processing of black tea through fermentation by the accumulation of flavan-3-ols. Structurally, TF is composed of catechins co-oxidation-oriented benzotropolone skeleton coupled with dihydroxy and trihydroxy moiety at *ortho* and *vic* positions (Fig. 14.1). Furthermore, oxidative modification of catechins and gallo-functional groups results different types of TF along with some minor constituents[79] such as: TF1, theaflavin-3-gallate (TF2A), theaflavin-3'gallate (TF2B), and theaflavin-3, 3'-digallate (TF3). Different catechins combinations are responsible for the diversity in TF. Epicatechin (EC) combines with epigallocatechin and brings about TF1, while EC and EGCG combine to form TF2A. Similarly, TF2B and TF3 are produced when epicatechingallate (ECG)/EGC and ECG/EGCG are bound to each other, respectively.[6,59]

TF is a renowned antioxidant and metal chelation agent owing to its diversified chemical structure with special reference to hydroxy groups and gallic acid moiety.[64] The outcome of different research investigations has also expounded higher antioxidant capacity of TFs as compared to EGCG. The lipid peroxidation inhibitory effect of TF ($IC_{50}$) is well illuminated by its ability to stop free radical generated chain mechanism and acceleration in the production of endogenous antioxidant enzymes such as glutathione-S-transferase (GST) and CAT.[70,71] TF provides shield against cancer insurgence by quenching free radicals, thus prevents DNA cleavage. Additionally, TF works against oxidative damage and cell toxicity via suppressing cytochrome P450 1A1 (CYP1A1) biomarker.[36]

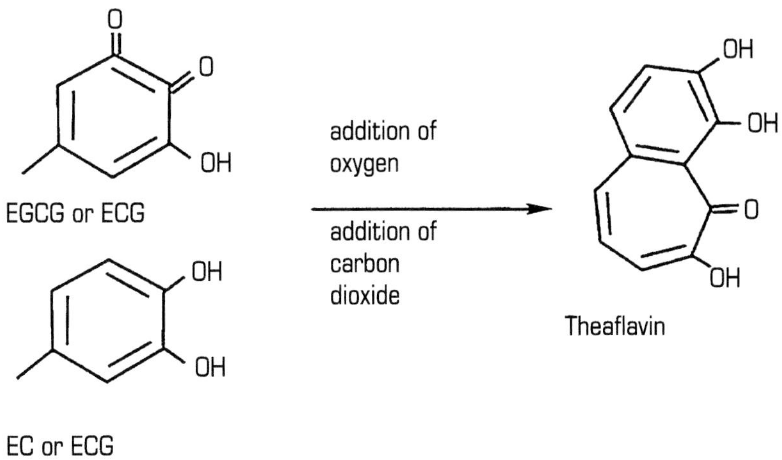


**FIGURE 14.1** Schematic diagram for theaflavin formation.

### 14.2.2 THEARUBIGINS

TRB is high molecule black tea polyphenol that is formed through fermentation in which two catechins (epigallocatechin and epigallocatechin gallate) are combined under the influence of polyphenol oxidase (Fig. 14.2). They are further described as heterogenous group of soluble products of fermentation and about 10–20% of black tea polyphenols as a group provides reddish tone of black tea. TRBs formation is carried out initially by the reaction of quinones on trihydroxy flavan-3-ols then quinones are further reacted with dihydroxy flavan-3-ols and finally intermediate products are formed for conversion into TRBs. The whole process is dependent on the aerobic enzymatic reactions.[89] TRBs have attained paramount attention of scientific community for their capacity to enhance the activity of phase II enzymes through transcriptional upregulation in liver and lungs by accelerating the nuclear factor (erythroid-derived 2)-like 2 (Nrf2)-mediated antioxidant-responsive element binding.[47,81]

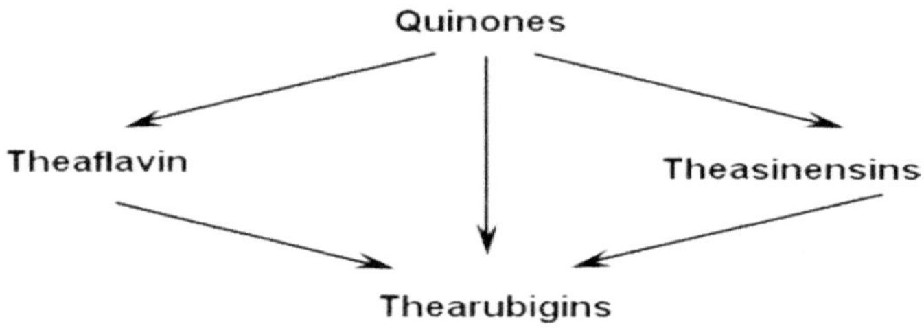


**FIGURE 14.2** Flow process for thearubigins formation.

### *14.2.3 FLAVONOLS*

Flavonols belong to subclass of flavonoids and are widely distributed in fruits, vegetables, and other plants. Chemically, these are aglycons with 3-hydroxyflavone backbone structure. They are mostly present in the form of glycosides and showed rapid metabolism.[78] Numerous factors such as soil properties, plant diversity, season, light intensity, degree of maturity, and Food preparation and processing techniques may influence their bioavailability and properties.[90] The formation of flavonols is initiated by shikimic acid pathway in which phenylalanine and acetic acids play crucial role. The in vivo absorption of flavonols is mainly dependent upon the attached sugar molecule on flavonol glycosides.[21] The flavonols are composed of kaempferol, myricetin, and quercetin.

Kaempferol has gained strong credential as active therapeutic agent against numerous oxidation-driven maladies. The fruits and vegetables (e.g., broccoli, grapefruit, brussels sprout, and apple) are considered as promising sources for kaempferol. It is a yellow color compound with higher boiling point ranging from 276°C to 278°C.[10]

Myricetin is mainly present in grapes, berries, walnuts, and different vegetables. Previously, it was observed that the myricetin consumption had inverse association with the onset of different type of cancers. Moreover, its capacity to modify the white blood cells ability to scavenge free radicals produced during the oxidative stress cascade is also well documented.[10] Quercetin exhibited strong antioxidant, antioncogenic, and anti-inflammatory properties. Moreover, the outcome of different explorations has divulged the lipid managing role of quercetin through enhancing energy expenditure.[10,29]

## 14.3 TFs AND TRBs WITH SPECIAL EMPHASIS ON MECHANISTIC TARGETS: HEALTH-ENDORSING PERSPECTIVES

### *14.3.1 HYPERLIPIDEMIA AND HYPERCHOLESTEROLEMIA-RELATED COMPLICATIONS*

Cholesterol performs various metabolic functions in our body. Its transportation is lipoprotein mediated in which low-density lipoprotein (LDL) and high-density lipoprotein (HDL) alongside chylomicron (CM) and very low-density lipoprotein (VLDL) play a crucial role. In hypercholesterolemic state, various metabolic dysfunctions such as coronary complications, hypertension, and stroke are the allied syndromes. Black tea provides protection against hypercholesterolemia and platelet aggregation owing to the presence of oxidized and unoxidized catechins. Additionally, black tea antioxidants have potential to tackle oxidative stress, endothelium dysfunction, and arterial complications.[33,22]

Black tea polyphenols provide protection against various life-threatening diseases[82] by unsettling electron chain consequently halt the development of several disorders (Table 14.2). Additionally, black tea polyphenols prevent fat oxidation, reduce nutrient absorption in the gastrointestinal track, and modulate energy consumption thereby attenuating obesity and diabetic events. Additionally, black tea polyphenols inhibit the LDL oxidation due to strong antioxidant and free radical scavenging activity. They also maintain the endogenous antioxidant level by stimulating the antioxidant enzymes such as glutathione.[86,115] Earlier, outcome of human efficacy trial expounded that tea polyphenols significantly tackle the lipid-related abnormalities by diminishing the LDL oxidation rate, balancing HDL, and limiting intestinal cholesterol absorption ascribed to their strong antioxidant potential.[28] Similarly, a shooting effect of black tea polyphenols against elevated lipid profile was observed when administrated to persistent smokers.[116]

In another attempt, 7 g/L black tea polyphenols administration for a period of 35 days attenuated lipid profile and hepatic oxidative abnormality in normal and hypercholesterolemic male Wistar rats. It is suggested that supplementation of black and green tea provides protection against serum and hepatic abnormalities in hypercholesterolemic phase by increasing the fecal excretion of fatty acids and sterols.[8] Black tea polyphenol @ 500 and 1000 mg/kg body weight provision to male Wistar rats for 8 weeks resulted in significant decline in total cholesterol, elevated LDL, and triglycerides with pronounced improvements in HDL level by inhibiting the pancreatic lipase activity.[104]

**TABLE 14.2** Chemical Narration of Polyphenols in Black Tea.

| Name of polyphenol | Formation route | Chemical structure | Physiological actions | References |
|---|---|---|---|---|
| Catechins | Catechin possesses two benzene rings (called the A- and B-rings) and a dihydropyran heterocycle (the C-ring) with a hydroxyl group on carbon 3. The A ring is like resorcinol moiety while the B ring is similar to a catechol moiety. There are two chiral centers on the molecule on carbons 2 and 3 | 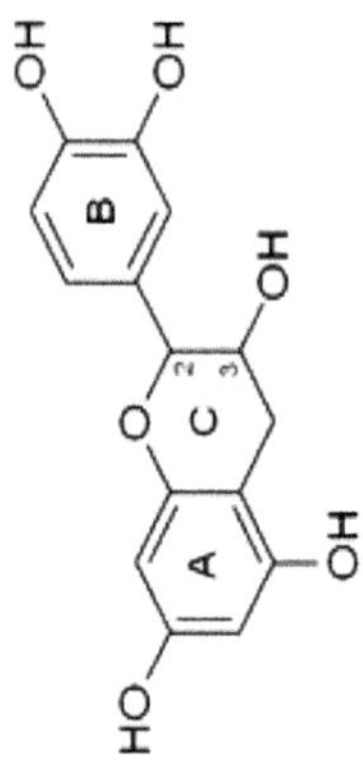 | Strong antioxidant<br>Provide coronary protection Protect LDL from oxidation<br>Anti-cancer properties<br>Prevent diabetes<br>Prevent obesity | [13,18,59, 93, 110, 113] |
| Chlorogenic acid | Chlorogenic acid is the ester of caffeic acid and quinic acid | 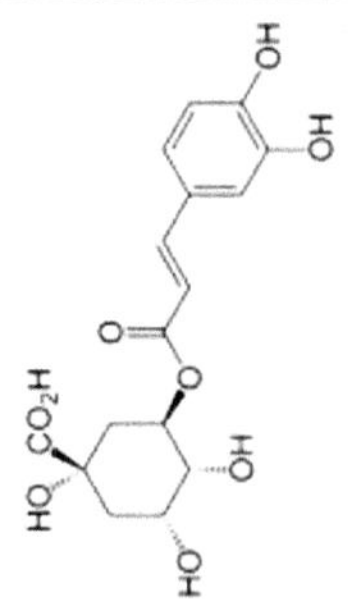 | Antioxidant perspective | [10, 21] |
| Flavonols: Kaempferol Myricetin Quercetin | They are formed by the combination of derivatives synthesized from phenylalanine (via shikimic acid) | 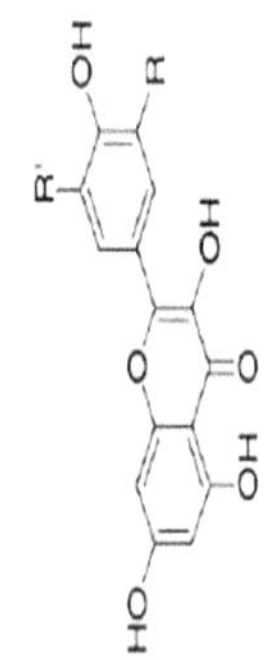<br>R= R'= H Kaempferol<br>R=H, R' = OH quercetin<br>R=R'= OH Myricetin | Strong antioxidant properties by radical scavenging and metal chelation<br>Protect LDL from oxidation<br>Reduce the formation of foam cells from white blood cells<br>Anti-inflammatory agent<br>Helpful to mitigate different type of cancers | [10,21, 29,78] |

**TABLE 14.2** *(Continued)*

| Name of polyphenol | Formation route | Chemical structure | Physiological actions | References |
|---|---|---|---|---|
| Gallic acid | Gallic acid is a trihydroxybenzoic acid, a type of phenolic acid | 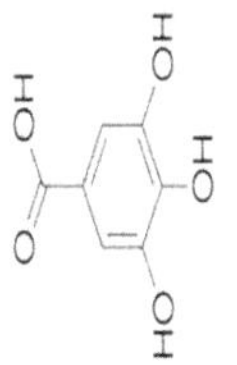 | Antioxidant properties | [10] |
| Theaflavins | EC + EGC → theaflavin<br>EC+ EGCG → theaflavin-3-gallate<br>ECG + EGC → theaflavin-3-gallate<br>ECG + EGCG → theaflavin-3, 3-digallate | 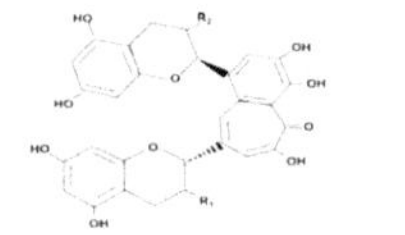<br>R1 R2<br>Theaflavin OH OH<br>Theaflavin-3-gallate Gallate OH<br>Theaflavin-3'-gallate OH Gallate<br>Theaflavin-3,3'-gallate Gallate Gallate | Promising antioxidant perspective<br>Anti-oncogenic ability<br>Hypoglycemic properties<br>Effective against lipid-related abnormalities<br>Neuroprotective role<br>Protect cardiovascular health | [6,14, 32,59] |
| Thearubi-gins | EGC+EGCG —Polyphenol Oxidase→ Thearubigins | 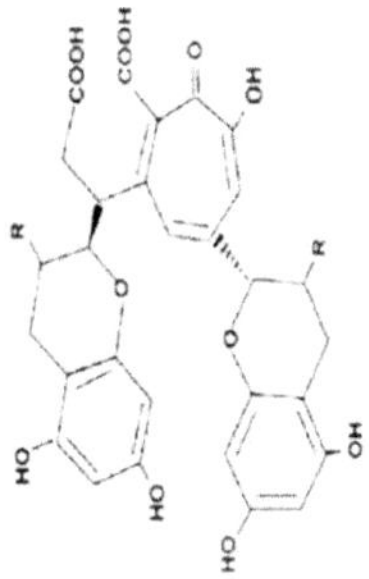 | Remove oxidative stress<br>Effective against cardiovascular compliances<br>Toxin removal<br>Effective inflammatory diseases | [7, 47,81, 89] |

In contrary, hypercholesterolemia and hyperglycemia were induced in rats through cholesterol and alloxan injection and were provided with 50 and 100 mg/kg body weight of black tea polyphenols orally for a period of 4 weeks. The outcomes delineated dose-dependent decline in abnormal hypercholesterolemic and hyperglycemic indicators.[87] Another way to ascribe role of tea polyphenols against lipid-related abnormalities is attributed to the ability to modulate metabolic targets such as satiety, thermogenesis, and fat oxidation. Moreover, tea antioxidants suppress adipocyte differentiation and fatty acid uptake in the adipose tissues however, improve oxidation of fat in the liver.[25]

Different bioevaluation studies have indicated the positive role of black tea TF to control obesity and hypercholesterolemia. Mechanically, it stimulates the lipid metabolism and hinders the activities of gastric and pancreatic lipases. Additionally, it increases thermogenesis and suppresses the fatty acid synthase enzyme (FAS). In case of hypercholesterolemia, black tea polyphenols reduce plasma cholesterol, triglycerides, and LDL by modulating the liver LDL receptors and cholesterol synthesis inhibitors.[7,61,73] TF improves lipid abnormalities more efficiently by modulating the expression of different enzymes involved in lipid metabolism, that is, FAS.[53]

TF imparts reduction in lipid-related abnormalities by stimulating the cellular energy expenditure on mitochondrial level that hinders the weight gain. However, in nucleus, the expression of FAS may be suppressed by TF that downregulates the epidermal growth factor (EGF) receptor/phosphatidylinositol-3-kinase (PI3K)/protein kinase B (AKT)/specificity protein-1 (Sp-1) signal transduction pathway, thus inhibits the cellular lipogenesis and tissue growth. Moreover, TF, especially TF3 inhibits the growth factor EGF binding to EGFR, blocks the activation of the PI3K/AKT signal pathway, reduces the DNA-binding capacity of nuclear transcription factor Sp-1 resulted down expression of FAS gene, and ultimately suppresses the biosynthesis of cholesterol, triglyceride, and other potent fatty acids. It also modulates the LDL receptors that facilitate in cholesterol and triglycerides reduction.[7,53,61]

TF can perform lipid-lowering function by alternate possible route which may be due to the interference with cholesterol micellar solubilization. The cholesterol absorption is mediated in different steps including emulsification, hydrolysis of ester bond, micellar solubilization, reesterification in intestine, and transport to lymphatic cell through CM.[90] Cholesterol micelle solubilization is inevitable for cholesterol transport due to its water insolubility. TF acts on the micells and induces changes in their structure that cause reduction in cholesterol resynthesis and alters its metabolisim.[53,107] Black tea polyphenols significantly ameliorate the elevated level of LDL via different mechanisms, that is, antioxidant action, modulate the particle size of LDL, modifying the macrophage and LDL ratio; however, the former is considered as the most important that inhibits LDL oxidation at initial stage.[49],[12]

The five cups of black tea consumption may cause significant decline in elevated lipid profile of hypercholesterolemic subjects with special reference to LDL. It is inferred that black tea polyphenols may reverse the oxidation of LDL resulting efficient balance in HDL and limit cholesterol absorption in the intestine, thus harmonize cholesterol homeostasis.[74] LDL is major cholesterol-carrying lipoprotein in plasma and mainly composed of apo-B100 protein (25%), cholesterol esters (74.96%), and triglycerides (<1%). It also contains linoleate (polyunsaturated fatty acid) that reacts with cholesterol esters which makes it vulnerable to oxidation. The LDL oxidation is a key connection to the development of atherosclerosis. It initiates abnormal changes in the macrophage and combines with macrophage scavenger receptor, and consequently foam cells are produced containing cholesterol that can be deposited inside the arteries. An array of evidences has proven the ability of black tea polyphenols, especially TF to reverse LDL oxidation by scavenging the free radicals, hindering the foam cell formation and deposition.[117,61]

Tea polyphenols significantly scavenged $H_2O_2$, enhanced the activities of glutathione, glutathione peroxidase (GPx), and glutathione reductase. TF pretreatment is effective to reverse oxidation by quenching superoxide ion of macrophages and chelating the Fe ions.[109] Antioxidative action is not solely the route by which tea inhibits LDL level in the subjects, but tea polyphenols also can address this menace with some other mechanistic approaches. One of the key targets is the modulation of adiponectin level that was reduced in patients with diabetes or with cardiovascular diseases. However, the role of tea to increase the size of small dense protein is also well documented.

The adiponectin is produced from adipose tissues and its normal level prevents macrophage to form the foam cells. However, when particle size of LDL is reduced 25 nm (small dense protein) than that of the adiponectin level, it becomes abnormal and initiates the cascade of atherosclerosis and allied ailments. Tea polyphenols have potential to increase the particle size of LDL and to improve the adiponectin metabolism by enhancing fat metabolism, oxidation, and energy consumption.[97,106]

There is a linear relationship between high-fat diet and the increase in cholesterol, triglyceride, and LDL due to the production of more free fatty acids (FFA) that may trigger the process of lipogenesis.[119] Besides, tea polyphenols have ability to normalize lipid abnormalities by inhibiting intestinal lipid absorption, increasing fecal excretion of fat through bile acid, suppressing the activity of fat synthesis enzymes, and preventing lipogenesis.[87]

Also there is a linear association between high-sucrose diet and triglycerides production. The sucrose-rich diet results in hypertriglyceridemia by increasing the formation of acetyl co-enzyme A (CoA) that creates abnormalities in lipogenesis thus elevating the triglycerides. During hyperglycemic or insulin-resistance state, synthesis of apoC-III protein in liver is increased that ultimately enhances the VLDL-total cholesterol (TC) secretion and delays the catabolism of triglycerides after binding the lipoprotein lipase and remnant to hepatic triglycerides.[70] However, tea modulates lipid metabolism and expressions for apolipoproteins A and E and plasma post-heparin lipase activities.[114]

Polyphenols in black tea ameliorate high triglycerides by suppressing lipid buildup, diminishing fatty acid production, and enhancing fatty acid utilization through activation of liver kinase B1 (LKB1)-activation of activated protein kinase (AMPK) pathway. The AMPK is the key protein involved in the glucose homeostasis body that in turn modulates the production of glucose-6-phosphate, a substrate for fatty acid metabolism.

### 14.3.2 HYPERGLYCEMIA AND INSULIN MALFUNCTIONING

Among various therapeutic foods, black tea has attained forefront position to combat against hyperglycemia, hyperinsulinemia, and immune dysfunctions. Substantial evidences have divulged the role of black tea as an antidiabetic agent due to its strong antioxidant potential. The black tea allied antioxidants attenuate hyperglycemic state by modifying the glucose metabolism, affirmative influence on insulin secretion, and absorption through β-cells.[9,88] Numerous scientific studies have enlightened the therapeutic role of black tea against type I and II diabetes. This therapeutic role is endorsed to the polyphenols that decrease the blood glucose and improve insulin resistance in many human and animal model studies.[15,86]

Besides modulating the glucose metabolism, black tea polyphenols also have ability to manage lipid-related abnormalities and obesity thus providing protection against cardiovascular complications.[51] The tea polyphenols have ability to tackle the glucose abnormalities by improving glycogen synthesis system. Moreover, they are reactivating the glycogen synthesis and decreasing the liver glucose-6-phosphatase activity to manage glycemic abnormalities.[98] Another possible route is the impact of black tea polyphenols on activity of amylase enzyme. They resulted in slow breakdown of starch and thereby control the sudden rise in glucose. In a randomized cross-over trial, 16 subjects were provided 75 g of glucose and water daily with simultaneous provision of 3 g of instant black tea; and this effect was attributed to the ability of tea polyphenols to stimulate the pancreatic enzymes thus enhancing β-cell ability toward insulin.[48]

Black tea polyphenols significantly lower the glucose level in streptozotocin (STZ)-induced diabetic rats. The tea polyphenols showed a preventive effect in those animals due to their action on survived β-cells from STZ toxicity. Moreover, tea polyphenols inhibit 2-amylase enzyme that reacts with saliva and pancreatic starch, consequently enhancing the insulin activity and glucose reduction. Numerous studies have proven the importance of galloyl residues of black and green tea for the management of glucose and insulin.[16,25] In an in vitro trial, black tea extract @ 500 mg/kg inhibited glucose absorption in small intestine and improved its overall status. In another instant, the normal and STZ-induced rats were administrated on black tea @ 480 mg/kg body weight for 21 days. The black tea polyphenols exhibited hypoglycemic activity in dose-dependent manner by reducing the glucose absorption via inhibiting carbohydrate hydrolyzing enzymes.[16] Another way to elucidate the hypoglycemic perspectives of black tea polyphenols are interlinked with $Ca^{+2}$ absorption modification in the sarcoplasmic reticulum that suppressed in hyperglycemic condition.[43]

One of the possible mechanisms showing black tea as hypoglycemic agent is its ability to modulate the activity of glucose transporters (GLUTs). During hypercholesterolemic state, high-fat diet caused reduction in GLUT4 and associated proteins; thereby glucose incorporation in the cells is disturbed resulting in insulin resistance. Black tea polyphenols enhance the translocation of proteins insulin receptor β subunit (IRβ), AMPKR, and GLUT4 thus maintaining glucose homeostasis.[51,54]

Black tea polyphenols can manage glucose-related variables by multiple mechanisms including inhibition of α-glucosidase and α-amylase activity, reduction in intestinal glucose absorption, insulin-o-mimetic action, and antioxidant capacity.[1] The high-fat diet increased the production of free fatty acids that enhanced plasma triglyceride level and suppressed insulin receptors thereby causing insulin resistance.[119] Moreover, the fructose- or sucrose-rich diets trigger abnormal glucose production that affects different hormones in plasma-like adiponectin and intestine GLUT1 linked with insulin resistance. The high-fat and sugar diets initiate the cascade of coronary complications, obesity, and diabetes by disturbing glucose homeostasis and insulin sensitivity.[46] There are sound evidences showing black tea's ability to improve insulin resistance in both hypercholesterolemic and hyperglycemic models thus attenuating diabetes mellitus and obesity.[87]

An in vitro trial indicated a 15-fold increase in insulin activity by black tea extract in dose-dependent manner. The identified bioactive constituents are TF, TRBs, EGCG, and catechins that have positive impact on insulin sensitivity and resistance. The proposed route of action is by inhibiting the α-amylase and α-glucosidase activities in the intestine that balance the glucose and insulin level, enhance the insulin binding to the adipocytes, and promote the intracellular GLUT in the myocytes.[11,100] The improvement in the muscle GLUTs, adipocytes, and leptin reductions are possible mechanisms by which black tea polyphenols perform their action for insulin management.[85]

The high-fat and sucrose diet caused enlargement of adipocytes, leading to change in adipocytokine levels such as free fatty acids, resistin, and leptin. The black tea polyphenols caused marked reduction (75%) in the leptin level and enhanced 50% adiponectin value, thus regulating the glucose and insulin balance.[77]

Imada et al.[54] indicated that black tea polyphenols can enhance the expression for GLUT-4 and insulin receptors such as β-subunits, accelerate glucose uptake by stimulating the AMP-activated protein kinase R in muscles, and manage the sudden increase in glucose, decline in insulin. Black tea TFs, their derivatives and TRBs enhanced the signaling of insulin-like growth factors insulin/IGF-1 in FOXO1a, PEPCK of mammalian cells, thus improving dysregulation of hepatic gluconeogenesis.[20] Black tea polyphenols may increase the insulin–glucose ratio, trigger insulin synthesis in response to abnormal glucose secretion, protect the insulin break down in normal state, improve glucose uptake through cell membrane in adipocytes, and reduce leptin production (Fig. 14.3).[1,114]

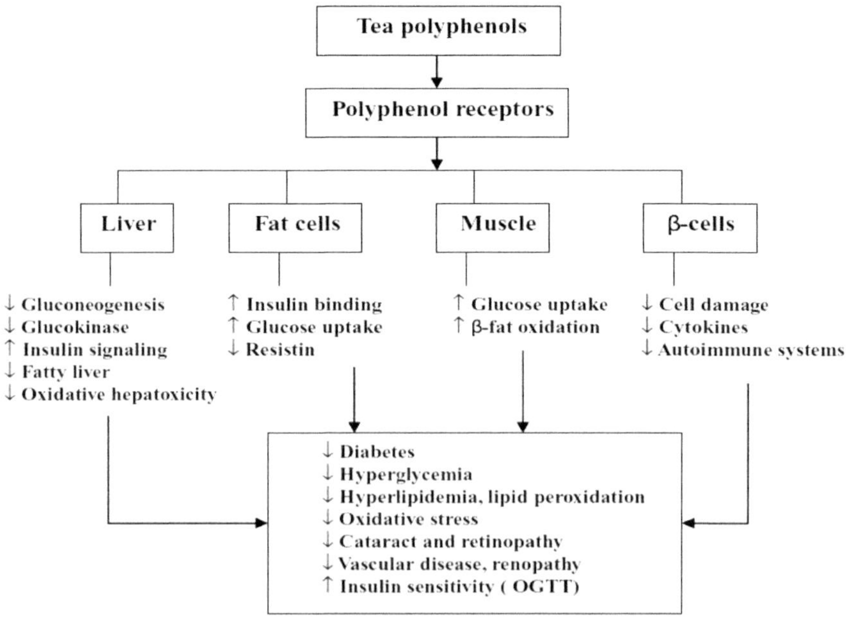


**FIGURE 14.3** Glycemic response management mechanistic in black tea polyphenols.

### 14.3.3 POTENTIAL IN OBESITY MANAGEMENT

The carbohydrate-rich products can cause the imbalance between fat and carbohydrate metabolism thus enhancing the probability of obesity, diabetes, and other allied complications. Among the different coping strategies, functional foods and nutraceuticals may tackle obesity.[95]

Black tea extract rich in polyphenols @ 0.2% caused significant decline in the markers associated with obesity (body weight) of obese CF-1 mice.[42] The antioxidants such as TF and TRBs have potential to prevent fat oxidation and can decrease the absorption of nutrients in gastrointestinal track. Moreover, they manage the energy consumption thus preventing LDL deposition and obesity.[52] In an animal experimental model, rats were fed on high-fat diet with simultaneous drink containing 5% black tea polyphenols extract and a 44.2% reduction in weight was noticed. Mechanistically, TF inhibits the pancreatic lipase activity alongside intestinal lipid absorption thus attenuates the gain in weight.[104]

The research studies have confirmed that the compounds containing galloyl moiety suppressed the postpyrandial hypertriacylglycerolemia by reducing the triacylglycerol absorption through the inhibition of pancreatic lipase. TF contains two digallate groups thereby has more potential for weight management than TRBs.[66] On molecular level, different enzymes played a key role to regulate lipid metabolism; however, FAS is a considerate factor. Its imbalance triggers the cascade of certain maladies such as obesity, cardiovascular complications, and cancer insurgence. The FAS inhibitors may help in weight management and black tea TF is a promising ingredient in this context. It blocks FAS through the deactivation of PI3K/AKT/Sp-1 pathway owing to galloyl moiety.[49,121]

Different scientific opinions have elaborated that the black tea gallate polyphenols (TF) can manage the body weight by modulating the cholesterol metabolism; inhibit the reabsorption of bile acid and hinder the synthesis of fatty acid enzymes via mimicking the AMP-activated protein kinase pathway in HepG2 cells.[103] Among the other possible anti-obesity routes are modulating the activity of superoxide dismutase and CAT that block the initiation of oxidative stress, by upregulating the GLUT1 and GLUT4 genes expression, different genes expression and protect the hepatic tissue by black tea polyphenols thus helpful in the weight management program.[92]

Numerous observational studies have interlinked the consumption of black and green tea with decreased LDL oxidation and enhanced insulin activity in animals and humans (Fig. 14.4). In a community-based trial, tea polyphenols administration resulted beneficial impact on the elderly diabetic subjects by modulating insulin and glucose metabolism.[15,86]

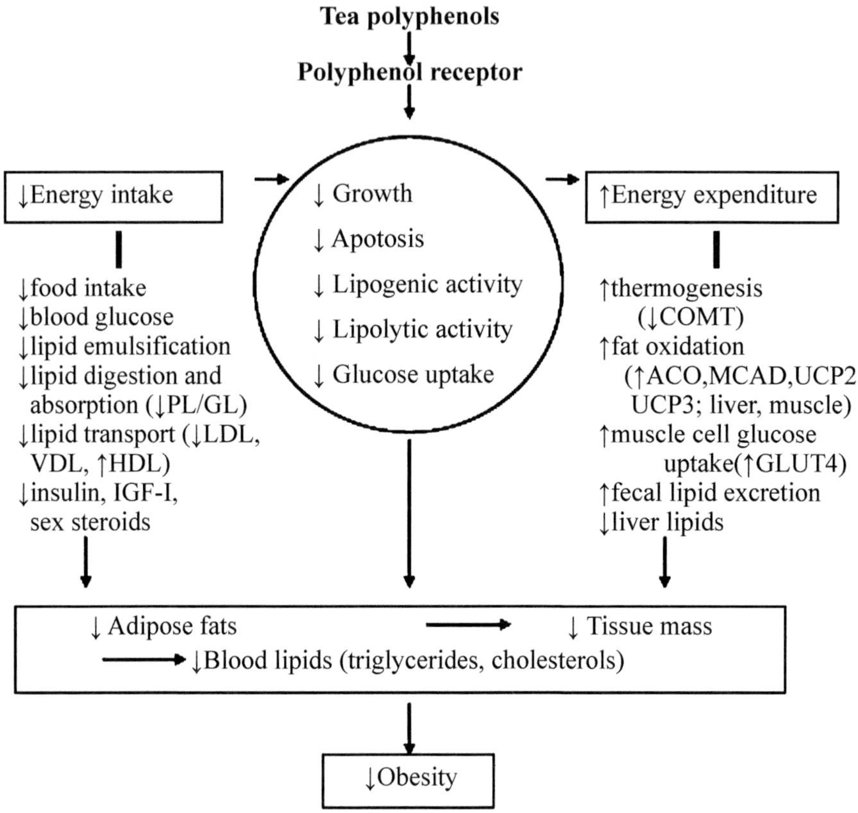


**FIGURE 14.4** Mechanistic depiction of black tea polyphenols action for the management of obesity and associated complications.

## 14.3.4 OXIDATIVE STRESS AND LIVER HEALTH

Oxidative stress is an imbalance between the reactive oxygen species (ROS) and endogenous antioxidants that disrupt normal detoxification of free radicals. This situation can disturb the body redox potential and damage the cell components including protein and lipid thereby altering the cellular signaling.[19] ROS are generated continuously within our body. However, unhealthy diet, smoking, sedentary lifestyle, environmental pollutant, etc. may boost their production. Improvement in the antioxidant defense system is of primary concern to mitigate free radical production. In this milieu, intake of diet rich in polyphenols is inevitable to maintain the body antioxidant potential.[41]

Glutathione is a major non-enzymatic endogenous antioxidant present in the body that maintains the intracellular redox status acting as a cofactor in many metabolic reactions.[69] Under stress conditions, cellular respiration produces hydrogen peroxide that triggers the cascade of deleterious reactions. The antioxidants such as glutathione stop this mechanism by altering the hydrogen peroxide into water to regain its normal oxidation potential.[101]

The inverse association between glutathione and oxidative stress has been unveiled in many research studies. The glutathione quenches free radicals through conjugation of electrophiles controlled by glutathione transferase and oxidoreduction cyclic glutathione transformation by GPx and glutathione reductase thus normalizes the imbalance between body redox potential and ROS. Glutathione level is diminished under oxidative stress due to high utilization, and improvement in its level through dietary components is basic philosophy to tackle the free radicals.[94]

There are consolidated research evidences in favor of black tea capacity to combat oxidative stress by enhancing the activity of glutathione and other antioxidant enzymes through its polyphenols. TF and TRBs are the proven antioxidants and metal chelators that are equally effective in in vivo and in vitro models due to their instinct to trap the noxious radicals such as superoxide and peroxyl.[12,57]

Regular tea consumption may protect liver by modulating the antioxidant enzymes and lipid peroxidation via enhanced GPx, superoxide dismutase, and CAT. Under oxygen deficient conditions, hydrogen peroxide is produced that causes toxicity and initiates the cascade of undesirable events. The antioxidant enzyme glutathione converts hydrogen peroxide into water and reduces the thiobarbituric reactive substances (TBARS) production thus acts as safeguard against lipid peroxidation.[7]

Under hypercholesterolemic and hyperglycemic conditions, normal glutathione level is decreased that can be uplifted by black tea consumption. Recently, the effect of black tea (Pu-erh) rich in TF and TRBs against oxidative stress was determined through model feeding trial. Initially, they induced oxidative damage in Balb/c mice by quinocetone-based diet. There was a marked reduction in the antioxidant enzymes activity such as GPx, superoxide dismutase, and CAT in quinocetone-treated animals. The black tea supplementation @ 800 mg/kg/day increased the GPx activity in the tested liver and kidney by 49.5% and 25.5%, respectively.[110]

Black tea polyphenols have ability to revert the oxidation of red blood cells membrane thus protecting lipid peroxidation at membrane level. The black tea extract administration @ 5 g/150 mL significantly reduced the malondialdehyde (MDA) formation, that is, 0.06–0.03 μM/Hb.[40] In the experimental rats, the performance of glutathione and other antioxidant enzymes was reduced. The black tea ingestion significantly improved the activity of antioxidant enzymes along with reduction in superoxide dismutase. In another exploration, black tea was given in the form of gavage for 91 days @ 10,000 mg/kg body weight to Sprague Dawley rats. No toxic effect was observed in the experimental rats. The liver function enzymes such as alanine transaminase (ALT) and aspartate aminotransferase (AST) were significantly reduced.[102] Previous studies showed that TF may attenuate the process of lipid peroxidation. Structurally, TF is composed of vicinal dihydroxy and trihydroxy components and benzotropolone skeleton responsible for free radicals quenching and metal ions chelating potency. Black tea provision to rats significantly suppressed the MDA that is an indicator of lipid peroxidation.[80]

During hypercholesterolemic phase, cellular membrane integrity is lost due to the reaction of polyunsaturated fatty acids of membrane and free radicals that produced lipid hydroperoxides and other secondary products such as MDA, 4-hydroxynonenal (HNE), and acrolein. In this context, black tea polyphenols exhibit soothing action on elevated TBARS by hindering the production of superoxide and chelating metal ions.[122]

The numerous studies have indicated that consumption of high-cholesterol and high-sucrose diets are one of the sources for hepatotoxicity. These diets are the gateway to produce ROS that interact with polyunsaturated fatty acids of membranes thus leading to structural and functional damage. The leakage of AST and ALT from liver to serum is increased during hypercholesterolemic and hyperglycemic phase, which is an indication of liver abnormality. The role of black tea for suppressing the elevated level of ALT and AST is proven by various model feeding trials associated with its antioxidant and anti-inflammatory behavior.[87,67]

### *14.3.5 RENAL HEALTH*

Chronic kidney disease (CKD) is characterized by the progressive loss in renal function.[4] It includes blood vessel disorders leading to nephron dysfunction that ultimately reduces the glomerulus filtration.[24] Elevated creatinine and blood urea levels were noticed in the chronic renal failure due to impairment in glomerular filtration rate thereby reducing urinary excretion. Recent studies advocated the capability of black tea polyphenols to activate antioxidant enzymes to improve kidney detoxifying ability.[34] Similarly, plasma creatinine and blood urea nitrogen levels were declined in the diabetic rats after tea polyphenols treatment.[91]

Black polyphenols reduce creatinine level by their antiplatelet action and enable kidneys to regain their normal function. It has been observed that black tea polyphenols exhibit diuretic effect thereby enhance the overall kidney functioning such as renal blood flow, capillary expansion, and glomerular filtration.[63] The in vivo renal functioning parameters such as blood urea nitrogen and creatinine were increased during the oxygen-deficient state. However, black tea reduced urea and creatinine by 11.74% and 14.62%, respectively. The oxidative stress induced some morphological abnormalities in glomerulus, capillaries, and tubules structures. Moreover, inflammation, sore lesion, and deformation in tubules were also observed. Black tea uplifts the renal functioning by mitigating the abnormal signs of kidney and inflammation.[27] In a study, enhancement in urea and creatinine levels in rats fed on high-arginine diet was observed due to the production of uremic acid toxins and suppression of certain key hormones.[35]

The effective role of tea polyphenols in the arachidonic acid metabolism pathway may be one of the possible route by which they normalize the kidney malfunctioning. Numerous scientific evidences are in favor that tea polyphenols reduced kidney inflammation by suppressing the prostaglandin (PG), thromboxane $A_2$, and cyclooxygenase expressions of arachidonic acid in microsomes and glomeruli.[3] Impaired glomerulus filtration is a first sign of CKD and black tea is limelighted for managing the abnormalities of glomerulus filtration by reducing toxic impact of ROS and improving the overall antioxidant status.[108]

Black tea polyphenols resulted marked decline in the creatinine level by their action on platelets, thus enabling kidneys to regain their normal functioning. Moreover, the diuretic effect of black tea enhances renal blood flow, capillary expansion and glomerular filtration.[63,91] In an early attempt, renal dysfunctionality was induced in Sprague Dawley rats by subchronic administration of 3-methyl-2-quinoxalin benzenevinylketo-1,4-dioxide (QCT). The QCT-induced higher urea, creatinine, and 8-hydroxy-2' -deoxyguanosine (8-OHdG) were attenuated significantly by black tea polyphenols.[100]

### 14.3.6 ANTIONCOGENIC PERSPECTIVE

Tea phytochemicals have potential to provide protection against various oncogenic events. Numerous studies indicated positive impact of tea consumption for the management of different malignancies. The antioncogenic effect of tea bioactive moieties has been reflected through the outcomes of different studies using rats, mice, and hamster models.[111]

Tea is highly regarded as a cancer chemopreventive agent owing to its polyphenols mainly TF, TRB, EGCG, and catechins.[59] It performed its anticarcinogenic role due to its capacity to bind carcinogens, induce phase II enzymes such as UDP -glucuronosyltransferase (UDPGT), and inhibition of heterocyclic amine formation. Prolonged administration of black and green tea may increase the activity of UDPGT facilitating the metabolism of potent chemical carcinogens into inactive metabolites that are voluntarily excreted. Role of tea catechins and other active ingredients at molecular level may be due to catechin-mediated induction of apoptosis, cell cycle detain, inhibition of most transcription factors nuclear factor-κB (NF-κB) and AP-1, and diminution activity of protein tyrosine kinase. Exceptional property of tea polyphenols is hidden in their ability to stimulate growth arrest and apoptosis especially in tumor cells and protecting normal epithelial cells from cancer insurgence.[2]

Tea ingestion, in vitro and in vivo, enhances the apoptosis and cell cycle arrest of diverse cancer cell lines, including skin, prostate, colon, and stomach and lung cancer. Tumor-inhibiting property of TF may be ascribed to degrade matrix metalloproteinases (MMPs). The use of TF decreased the gelatinolytic activity and mRNA and protein expression of MMP-2. Moreover, it can reduce the binding of A375 cell to extracellular matrix (ECM) ligands and the tumor volume in syngenic black mice. It was further observed that the consumption of coffee and black tea was associated with decrease in risk of onset of gastric cancer. In a case study in northern Italy, marked decrease in onset or concentration of gastric cancer biomarkers was observed in tea habitual patients in contrast with non-tea drinkers suggesting a positive correlation between role of tea and cancer prevention.[39]

Black tea polyphenols caused pronounced impact on androgens involved in different phases of prostate cancer in male Wistar rats.[99] The protective role of tea and other antioxidants was attributed to their ability to halt the expression of pro-inflammatory factors responsible for cancer development.[99] Tea polyphenols inhibit the phosphorylation of NF-κB. Various animal-based studies have illuminated the effectiveness of black tea polyphenols against lung cancer. In most of studies, lung cancer is induced through 4-(methylnitrosamino)-1-(3-pyridyl)-1-butanone (NNK), benzo [a]pyrene, and *N*-nitrosodimethylamine. Provision of tea and its bioactive molecules in different cancer stages have demonstrated momentous decline in tumor volume.[27]

The TF is strong anticancer agent to modulate different expressions involved in proapoptotic factors, alters the signaling pathway for posttranslational induction of p53 and activation of Bax thus imparts strong apoptotic activity. TF has ability to induce apoptosis in mitochondrial level by initiating the clustering and vacuole formation thereby controlling the growth of cancer cells.[44] Moreover, suppression of cyclooxygenase (COX) 2, G protein signaling, and other inflammatory targets are responsible for apoptotic potential of TF. It was observed that green tea, black tea, EGCG, and TF administration to rats, mice, and hamster during the early stages of lung cancer is effective to prevent NNK-induced tumorigenesis and rhabdomyosarcomas.[37,113]

The TF accelerates apoptosis by changing the cellular signaling in tumor cells, prevents binding of EGF and platelet-derived growth factor (PDGF) to their acceptors, and blocks kinase activity. At molecular level, blocking of AP-1-dependent transcriptional activity and DNA binding are promising routes for cancer cell control.[68,65]

Similarly, TF caused cell growth inhibition in G2/M phase. TF arrests the cells after inhibiting the activity of cyclin-regulated signaling pathways. It imparts inhibition in cdc25C and cyclin B expression, enhances the apoptosis of PC-3 cells, upregulates the proapoptotic proteins Bax, caspase-3 and caspase-9, and downregulates the antiapoptotic protein Bcl-2 expression. Moreover, it causes fragmentation in DNA laddering thus implementing cell cycle arrest.[84]

### *14.3.7 CARDIOVASCULAR PROTECTION*

It was concluded that the higher flavonoid intake through natural commodities was associated with reduced risk for cardiovascular disease by improving endothelial function; it inhibits LDL and improves dyslipidemia.[55,75]

In the regimen of diet-based therapy, polyphenols are protective agents against coronary diseases. Among the polyphenols, tea bioactive moieties catechins, TF, and TRBs are in focus for curtailing this menace (Table 14.3). Numerous scientific explorations revealed an inverse association between tea consumption and lipid abnormalities in obese and diabetic models.[51,87,97,106]

The slogan "good cholesterol" is attributed to HDL due to its ability to reverse the cholesterol transport, to remove excess cholesterol from the tissues and arteries back to liver. It particularly acts on subendothelial space in medium caliber artery that is the place where a real cholesterol deposition occurs in the form of atheroma. In contrast, LDL carries the cholesterol from liver to the body. Currently, great attention is being paid toward the reduction of LDL by different therapeutic agents. However, various epidemiological studies have indicated the role of HDL for the management of cardiovascular health.[38,75]

**TABLE 14.3** Some Promising Mechanistic Roles of Black Tea Polyphenols Against Different Maladies.

| Diseases | Mechanistic actions |
|---|---|
| Hypercholesterolemia and hyperlipidemia-related disorders | Increase the facial excretion of fats |
| | Accelerate the cholesterol transport outside the body |
| | Protect LDL from oxidation |
| | Decreased serum total cholesterol |
| | Enhanced thromogenesis |
| | Suppress appetite |
| | Maintain proper balance in energy intake and utilization |
| | Enhanced HDL |
| | Improved HDL/LDL ratio |
| | Enhanced metabolic rate up to 2–12% |
| | Boast in fat oxidation |
| | On molecular level, polyphenols can reduce the activities of the expression of different lipid metabolic enzymes |
| | Phytochemicals can bind hampered activities of acyl-CoA oxidase and acyl-CoA dehydrogenase |
| | Inhibit the activity of squalene epoxidase, which is a cholesterol biogenesis-limiting enzyme |
| Diabetes | Act on starch hydrolyzing enzymes |
| | Slow down the starch breakdown |
| | Enhance the hepatic glycogen synthesis via disturbing glucose 6 phosphate pathway |
| | Modulate the activities of glucose transporters, that is, *GLUT1* and *GLUT4* |
| | Reduce the intestinal glucose absorption |
| | Slow down activities of α-glucosidase and amylase enzymes |
| | By their antioxidative action |
| | Improves the Ca absorption in sarcoplasmic reticulum |
| | Enhance insulin secretion |
| | Protect insulin breakdown |
| | Maintain optimal glucose insulin resistance |
| | By breaking insulin resistance by effecting lipid metabolism |

**TABLE 14.3** *(Continued)*

| Diseases | Mechanistic actions |
|---|---|
| Obesity | Increase energy expenditure |
| | Increase fat oxidation |
| | Decrease nutrient absorption |
| | Reduce appetite |
| | Enhance metabolic rate |
| | Induce fat burning |
| Renal health | By its oxidative action |
| | By their anti-platelet action |
| | Improve overall kidney functioning such as renal blood flow, capillary expansion, and glomerular filtration |
| | Suppressing the anti-inflammatory expressions such as prostaglandin (PG), thromboxane $A_2$, and cyclooxygenase |
| Oxidative stress and liver health | By detoxification, scavenging free radicals and boost immunity due to its thiol group |
| | By enhancing the activity of glutathione and other antioxidant enzymes |
| | By trapping noxious radicals such as superoxide and peroxyl |
| | Black tea polyphenols have ability to revert the oxidation of red blood cells membrane thus protecting lipid peroxidation at membrane level |
| | By suppressing the malondialdehyde (MDA) an indicator of lipid peroxidation |
| Cancer insurgence | Prevention of photo carcinogenesis |
| Skin cancer | Reduction in the development papillomas to SCC |
| | Halt and prevent the tumor growth |
| | Diminish the pro-inflammatory factors such as mitogen-activated protein kinases (MAPKs) and NF-κB |
| Liver cancer | Reduction in the formation of hepatocarcinogenesis |
| | Decline in tumor formation and growth |
| | Inhibition of cyclin D1 and cdk4 proteins |
| | Tumor volume suppression |
| | Adenoma to adenocarcinoma progression suppression |
| Gastrointestinal tract cancer | Inhibition in the tumor formation, spread, and volume |
| | Acceleration in e-cadherin and decline in nuclear β-catenin, c-myc, phospho-AKT, and phospho-ERK1/2 in small intestinal tumors |

**TABLE 14.3** *(Continued)*

| Diseases | Mechanistic actions |
|---|---|
| Breast cancer | Increase in mean latency of tumors |
| | Enhancement in Bax/Bcl-2 ratio |
| | Decline in PCNA |
| | Activation of caspase 3 |
| | Decline in tumor burden |
| Prostate cancer | Enhance survival time |
| | Declining impact on IGF-I and enhancement in IGFBP-3 |
| | Inhibition of markers of angiogenesis and metastasis |
| | Decline in tumor size |
| | Diminish in PSA synthesis |
| Cardiovascular complications | By improving endothelial function |
| | Inhibit low-density lipoprotein |
| | Suppresses the dyslipidemia |
| | By reversing the LDL oxidation and increases HDL level |
| | By lowering the arthrogenic index and reversal the oxidation of fat |

IGFBP-3, insulin-like growth factor-binding protein 3; PCNA, proliferating cell nuclear antigen; PSA, prostate-specific antigen; SCC, squamous cell carcinoma.

Lowering of arthrogenic index and reversal of oxidation of fat are possible mechanisms through which TF may enhance the plasma HDL level. Arthrogenic index is the ratio between LDL and HDL; and TF significantly alters the cholesterol metabolism, suppresses the activity of lipid synthesis enzymes, and protects the LDL against oxidation.[30,60] Cholesterol deposition in arterial wall requires a receptor called as prostacyclin. Numerous studies illustrated that the high HDL may stop this process.[118] In recent lipid taking therapies, HDL enhancement has gained importance. In this context, black tea polyphenols have ability to act as prostacyclin inhibitors thus protecting cardiovascular health.[83]

## 14.4 SUMMARY

Black tea is a promising therapeutic agent mainly due to presence of polyphenols, such as: TF and TRBs. Numerous scientific explorations have elucidated the biological worth of these bioactive moieties against plethora of ailments with special reference to metabolic disorders. Numerous in vitro and in vivo studies have demonstrated that these exhibit antioxidant, antimutagenic, antidiabetic, anti-inflammatory, antibacterial and antiviral, and cancer-preventive properties. In a nutshell, it is concluded that black tea polyphenols with special reference to TF and TRBs can be incorporated in diet-based therapies. Moreover, academia–industry liaison should be promoted for the development of these bioactive moieties based functional and designer foods.

## KEYWORDS


- **apoptosis**
- **bioactive compounds**
- **cell proliferation**
- **lipoxygenase**
- **nutraceuticals**
- **theaflavins**
- **thearubigins**


## REFERENCES


1. Abeywickrama, K. R.; Ratnasooriya, W. D.; Amarakoon, A. M. Oral Hypoglycaemic, Antihyperglycaemic and Antidiabetic Activities of Sri Lankan Broken Orange Pekoe Fannings (BOPF) Grade Black Tea (*Camellia sinensis* L.) in Rats. *J. Ethnopharmacol.* **2011,** *135* (2), 278–286.
2. Adhami, V. M.; Siddiqui, I. A.; Ahmad, N.; Gupta, S.; Mukhtar, H. Oral Consumption of Green Tea Polyphenols Inhibits Insulin-like Growth Factor I-induced Signaling in an Autochthonous Mouse Model of Prostate Cancer. *Cancer Res.* 2004, *64*, 8175–8122.
3. Adhikary, B.; Yadav, S. K.; Roy, K.; Bandyopadhyay, S. K.; Chattopadhyay, S. Black Tea and Theaflavins Assist Healing of Indomethacin-induced Gastric Ulceration in Mice by Antioxidative Action. *Evidence Based Compl. Altern. Med.* **2011,** 546–560.
4. Agarwal, R.; Sinha, A. D. Thiazide Diuretics in Advanced Chronic Kidney Disease. *J. Am. Soc. Hypertens.* **2012,** 1–10.
5. Aiello, A.; Di Bona, D.; Candore, G.; Carru, C.; Zinellu, A. Targeting Ageing with Functional Food: Pasta with Opuntia Single Arm Pilot Study._*Rejuvenation Res.* **2017,** E-article, DOI: 10.1089/rej.2017.
6. Alipoor, B.; Rad, A. H. A Review on the Therapeutical Effects of Tea. *Asian J. Clin. Nutr.* **2012,** *4* (1), 1–15.

7. Al-Rejaie, S. S.; Abuohashish, H. M.; Ahmed, M. M.; Aleisa, A. M.; Alroujayee, A. S.; Alkhamees, O. A. Immobilization Stress-induced Oxidative Damage and its Amelioration with Green and Black Teas. *Afr. J. Pharmacy Pharmacol.* **2012,** *6* (8), 538–545.
8. Alshatwi, A. A.; Al Obaaid, M. A.; Al Sedairy, S. A.; Ramesh, E.; Lei, K. Y. Black and Green Tea Improves Lipid Profile and Lipid Peroxidation Parameters in Wistar Rats Fed a High-cholesterol Diet. *J. Physiol. Biochem.* **2011,** *67*, 95–104.
9. American Diabetes Association. Diagnosis and Classification of Diabetes Mellitus. *Diabetes Care* 2005, *28*, 37–42.
10. Amić, D.; Lučić, B. Reliability of Bond Dissociation Enthalpy Calculated by the PM6 Method and Experimental TEAC Values in Antiradical QSAR of Flavonoids. *Bioorganic Med. Chem. Lett.* **2010,** *18*, 28–35.
11. Anderson, R. A.; Polansky, M. M. Tea Enhances Insulin Activity. *J. Agric. Food Chem.* **2002,** *50* (24), 7182–7186
12. Bahorun, T.; Luximon-Ramma, A.; Neergheen-Bhujun, V. S.; Gunness, T. K.; Googoolye, K.; Auger, C.; Crozier, A.; Aruoma, O. I. The Effect of Black Tea on Risk Factors of Cardiovascular Disease in a Normal Population. *Prevent. Med.* **2012,** *54*, 98–102.
13. Bancirova, M. Comparison of the Antioxidant Capacity and the Antimicrobial Activity of Black and Green Tea. *Food Res. Int.* **2010,** *43*, 1379–1382.
14. Baptista, J.; Lima, E.; Paiva, L.; Andrade, A. L.; Alves, M. G. Comparison of Azorean Tea Theanine to Teas from Other Origins by HPLC/DAD/FD. Effect of Fermentation, Drying Temperature, Drying Time and Shoot Maturity. *Food Chem.* **2012,** *132*, 2181–2187.
15. Basu. A.; Sanchez, K.; Leyva, M. J.; Wu, M.; Betts, N. M.; Aston, C. E.; Lyons, T. J. Green Tea Supplementation Affects Body Weight, Lipids, and Lipid Peroxidation in Obese Subjects with Metabolic Syndrome. *J. Am. College Nutr.* **2010,** *29* (1), 31–40.
16. Bhandari, M. R.; Jong-Anurakkun, N.; Hong, G.; Kawabata, J. α-Glucose and α-amylase Inhibitory Activities of Nepalese Medicinal Herb Parkhanbhed (*Bergenia ciliata Haw.*). *Food Chem.* **2008,** *106*, 247–252.
17. Boullart, A. C.; De Graaf, J.; Stalenhoef, A. F. Serum Triglycerides and Risk of Cardiovascular Disease. *Biochimica et Biophysica Acta* **2012,** *1821*, 867–875.
18. Butt, M. S.; Imran, A.; Sharif, M. K.; Ahmad, R. S.; Xiao, H.; Imran, M.; Rsool, H. A. Black Tea Polyphenols: A Mechanistic Treatise. *Crit. Rev. Food Sci. Nutr.* **2014,** *54* (8), 1002–1011.
19. Butt, M. S.; Sultan, M. T. Green Tea: Nature's Defense Against Malignancies. *Crit. Rev. Food Sci. Nutr.* **2009,** *49* (5), 463–473.
20. Cameron, A. R.; Anton, S.; Melville, L.; Houston, N. P.; Dayal, S.; McDougall, G. J.; Stewart, D.; Rena, G. Black Tea Polyphenols Mimic Insulin/Insulin-like Growth Factor-1 Signaling to the Longevity Factor FOXO1a. *Aging Cell* **2008,** *7*, 69–77.
21. Castillo-Muñoz, N.; Gómez-Alonso, S.; García-Romero, E.; Hermosín-Gutiérrez, I. Flavonol Profiles of *Vitis vinifera* white Grape Cultivars. *J. Food Compos. Anal.* **2010,** *23*, 699–705.
22. Chan, P. T.; Fong, W. P.; Cheung, Y. L.; Huang, Y.; Ho, W. K.; Chen, Z. Y. Jasmine Green Tea Epicatechins are Hypolipidemic in Hamsters (*Mesocricetus auratus*) Fed a High Fat Diet. *J. Nutr.* **2012,** *129*, 1094–1101.
23. Chaturvedula, V. S.; Prakash, I. The Aroma, Taste, Color and Bioactive Constituents of Tea. *J. Med. Plants Res.* **2011,** *5* (11), 2110–2124.
24. Chauhan, V.; Vaid, M. Dyslipidemia in Chronic Kidney Disease: Managing a High-risk Combination. *Postgraduate Med.* **2009,** *121* (6), 54–61.

25. Chen, N.; Bezzina, R.; Hinch, E.; Lewandowski, P. A. Green Tea, Black Tea, and Epigallocatechin Modify Body Composition, Improve Glucose Tolerance, and Differentially Alter Metabolic Gene Expression in Rats Fed a High-fat Diet. *Nutr. Res.* **2009,** *29*, 784–793.
26. Cirico, T. L.; Omaye, S. T. Additive or Synergetic Effects of Phenolic Compounds on Human Low-density Lipoprotein Oxidation. *Food Chem. Toxicol.* **2006,** *44* (4), 510–516.
27. Clark, T. A.; Heyliger, C. E.; Edel, A. L.; Goel, D. P.; Pierce, G. N. Codelivery of a Tea Extract Prevents Morbidity and Mortality Associated with Oral Vanadate Therapy in Streptozotocin-induced Diabetic Rats. *Metabolism* **2004,** *153* (9), 1145–1151.
28. Davies, M. J.; Judd, J. T.; Baer, D. J.; Clevidence, B. A.; Paul, D. R.; Edwards, A. J.; Wiseman, S. A.; Muesing, R. A.; Chen, S. C. Black Tea Consumption Reduces Total and LDL Cholesterol in Mildly Hypercholesterolemic Adults. *J. Nutr.* **2003,** *133*, 3298–3302.
29. De Lange, D. W.; Verhoef, S.; Gorter, G.; Kraaijenhagen, R. J.; Van De Wiel, A.; Akkerman, J. W. Polyphenolic Grape Extract Inhibits Platelet Activation Through PECAM-1: An Explanation for the French Paradox. *Alcoholism Clin. Exp. Res.* **2007,** *31*, 1308–1314.
30. Deka, A.; Vita, J. A. Tea and Cardiovascular Disease. *Pharmacol. Res.* **2011,** *64* (2), 136–145.
31. Delbosc, S.; Paizanis, E.; Magous, R.; Araiz, C.; Dimo, T.; Cristol, J. P.; Cros, G.; Azay, J. Involvement of Oxidative Stress and NADPH Oxidase Activation in the Development of Cardiovascular Complications in a Model of Insulin Resistance, the Fructose-fed Rat. *Atherosclerosis* **2005,** *179* (1), 43–49.
32. Dimpfel, W.; Kler, A.; Kriesl, E.; Lehnfeld, R. Theogallin and L-Theanine as Active Ingredients in Decaffeinated Green Tea Extract: II, Characterization in the Freely Moving Rat by Means of Quantitative Field Potential Analysis. *J. Pharm. Pharmacol.* **2007,** *59*, 1397–1403.
33. Duffy, S. J.; Vita, J. A.; Holbrook, M.; Swerdloff, P. L.; Keaney, J. F. Effect of Acute and Chronic Tea Consumption on Platelet Aggregation in Patients with Coronary Artery Disease. *Arteriosclerosis Thromb. Vasc. Biol.* **2001,** *21*, 1084–1089.
34. El-Megeid, A. A.; AbdAllah, I. Z.; Elsadek, M. F.; El-Moneim, Y. F. The Protective Effect of the Fortified Bread with Green Tea Against Chronic Renal Failure Induced by Excessive Dietary Arginine in Male Albino Rats. *World J. Dairy Food Sci.* **2009,** *4* (2), 107–117.
35. Elshater, A. A.; Salman, M. A.; Moussa, M. M. Effect of Green Tea Consumption on Level of Glucose, Lipid Profile and Kidney Functions in Alloxan Induced-diabetic Rats. *Egyptian Acad. J. Biol. Sci.* **2008,** *1* (2), 125–134.
36. Feng, Q.; Torii, Y.; Uchida, K.; Nakamura, Y.; Hara, Y.; Osawa, T. Black Tea Polyphenols, Theaflavins, Prevent Cellular DNA Damage by Black Tea Polyphenols Inhibiting Oxidative Stress and Suppressing Cytochrome P450 1A1 in Cell Cultures. *J. Agric. Food Chem.* **2002,** *50*, 213–220.
37. Fu, H.; He, J.; Mei, F.; Zhang, Q.; Hara, Y.; Ryota, S.; Lubet, R. A.; Chen, R.; Chen, D. R.; You, M. Lung Cancer Inhibitory Effect of Epigallocatechin-3-Gallate is Dependent on Its Presence in a Complex Mixture (Polyphenone). *Cancer Prev. Res.* **2009,** *2* (6), 531–537.
38. Gadi, R.; Amanullah, A.; Figueredo, V. M. HDL-C: Does it Matter? An Update on Novel HDL-directed Pharmaco-therapeutic Strategies. *Int. J. Cardiol.* **2012,** *167* (3), 646–655.
39. Gao, S.; Hu, M. Bioavailability Challenges Associated with Development of Anti-cancer Phenolics. *Mini Rev. Med. Chem.* 2010, *10* (6), 550–567

40. Gawlik, M.; Czajka, A. The Effect of Green, Black and White Tea on the Level of α and β Tocopherols in Free Radical-induced Oxidative Damage of Human Red Blood Cells. *Acta Poloniae Pharmaceutica - Drug Res.* **2007,** *64* (2), 159–164.
41. Gibson, S. A.; Korade, Ž.; Shelton, R. C. Oxidative Stress and Glutathione Response in Tissue Cultures from Persons with Major Depression. *J. Psychiatric Res.* **2012,** *46,* 1326–1332.
42. Glisan, S. L.; Grove, K. A.; Yennawar, N. H.; Lambert, J. D. Inhibition of Pancreatic Lipase by Black Tea Theaflavins: Comparative Enzymology and In Silico Modeling Studies. *Food Chem.* 2017, *216*, 296–300.
43. Gomes, A.; Vedasiromoni, J. R.; Das, M.; Sharma, R. M.; Ganguly, D. K. Anti-hyperglycemic Effect of Black Tea (*Camellia sinensis*) in Rat. *J. Ethnopharmacol.* **1995,** *45*, 223–226
44. Gosslau, A.; Chen, K. Y. Nutraceuticals, Apoptosis, and Disease Prevention. *Nutrition* **2004,** *20*, 95-102.
45. Gramza, A.; Khokhar, S.; Yoko, S.; Gliszczynska-Swiglo, A. Antioxidant Activity of Tea Extracts in Lipids and Correlation with Polyphenol Content. *Eur. J. Lipid Sci. Technol.* **2006,** *108*, 351–362.
46. Grove, K. A.; Lambert, J. D. Laboratory, epidemiological, and Human Intervention Studies Show that Tea (*Camellia sinensis*) May be Useful in the Prevention of Obesity1, 2. *J. Nutr.* **2010,** *140* (3), 446–453
47. Gupta, J.; Siddique, Y. H.; Beg, T.; Ara, G.; Afzal, M. Protective role of Green Tea Extract Against Genotoxic Damage Induced by Anabolic Steroids in Cultured Human Lymphocytes. *Biol. Med.* **2009,** *1* (2), 87–99.
48. Hininger-Favier, I.; Benaraba, R.; Coves, S.; Anderson, R. A.; Roussel, A. M. Green Tea Extract Decreases Oxidative Stress and Improves Insulin Sensitivity in an Animal Model of Insulin Resistance, the Fructose-fed Rat. *J. College Nutr.* **2009,** *28* (4), 355–361.
49. Hou, Y.; Shao, W.; Xiao, R.; Xu, K.; Ma, Z.; Johnstone, B. H.; Du, Y. Pu'erh Tea Aqueous Extracts Lower Atherosclerotic Risk Factors in a Rat Hyperlipidemia Model. *Exp. Gerontol.* **2009,** *44*, 434–439.
50. Hsu, A.; Bruno, R. S.; Löhr, C. V.; Taylor, A. W.; Dashwood, R. H.; Bray, T. M.; Ho, E. Dietary Soy and Tea Mitigate Chronic Inflammation and Prostate Cancer Via NF-B Pathway in the Noble Rat Model, In Vivo. *J. Nutr. Biochem.* **2010,** *22* (5), 502–510.
51. Huang, H. C.; Lin, J. K. Pu'erh Tea, Green Tea, and Black Tea Suppresses Hyperlipidemia, Hyperleptinemia and Fatty Acid Synthase Through Activating AMPK in Rats Fed a High-fructose Diet. *Food Function* **2012,** *3* (2), 170–177.
52. Huang, Y. W.; Liu, Y.; Dushenkov, S.; Ho, C. T.; Huang, M. T. Anti-obesity Effects of Epigallocatechin-3-Gallate, Orange Peel Extract, Black Tea Extract, Caffeine and their Combinations in a Mouse Model. *J. Function. Foods* **2009,** *1*, 304–310.
53. Ikeda, I.; Yamahira, T.; Kato, M.; Ishikawa, A. Black-tea Polyphenols Decrease Micellar Solubility of Cholesterol in Vitro and Intestinal Absorption of Cholesterol in Rats. *J. Agric. Food Chem.* **2010,** *58*, 8591–8595.
54. Imada, S.; Tanaka, A.; Nishiumi, S.; Ashida, H. Concentration of Catechins and Caffeine in Black Tea Affects Suppression of Fat Accumulation and Hyperglycemia in High-fat Diet-fed Mice. *Food Sci. Technol. Res.* **2011,** *17* (4), 353-359
55. Jay, D.; Hitomi, H.; Griendling, K. K. Oxidative Stress and Diabetic Cardiovascular Complications. *Free Rad. Biol. Med.* **2006,** *40* (2), 183–192.

56. Jayasekera, S.; Molan, A. L.; Garg, M.; Moughan, P. J. Variation in Antioxidant Potential and Total Polyphenol Content of Fresh and Fully-fermented Sri Lankan Tea. *Food Chem.* **2011,** *125*, 536–541.
57. Joukar, S.; Shahouzehi, B.; Najafipour, H.; Gholamhoseinian, A.; Joukar, F. Ameliorative Effect of Black Tea on Nicotine Induced Cardiovascular Pathogenesis in rat. *EXCLI J.* **2012,** *11*, 309–317.
58. Khan, K. S. Effect of Diet and Physical Activity Based Interventions in Pregnancy on Gestational Weight Gain and Pregnancy Outcomes: Meta-analysis of Individual Participant Data from Randomized trials. *BMJ* **2017,** *358*, 3119–3122.
59. Khan, N.; Mukhtar, H. Tea Polyphenols for Health Promotion. *Life Sci.* **2007,** *81*, 519–533.
60. Kobayashi, M.; Ichitani, M.; Suzuki, Y.; Unno, T.; Sugawara, T.; Yamahira, T.; Kato, M.; Takihara, T.; Sagesaka, Y.; Kakuda, T.; Ikeda, I. Black-tea Polyphenols Suppress Postprandial Hyper Triacylglycerolemia by Suppressing Lymphatic Transport of Dietary Fat in Rats. *J. Agric. Food Chem.* **2009,** *57*, 7131–7136.
61. Kuo, K. L.; Weng, M. S.; Chiang, C. T.; Tsai, Y. J.; Lin-Shiau, S. Y.; Lin, J. K. Comparative Studies on the Hypolipidemic and Growth Suppressive Effects of Oolong, Black, Puerh, and Green Tea Leaves in Rats. *J. Agric. Food Chem.* **2005,** *53*, 480–489.
62. Lopez, T. E.; Pham, H. M.; Barbour, J.; Tran, P.; Van Nguyen, B.; Hogan, S. P.; et al. The Impact of Green Tea Polyphenols on Development and Reproduction in *Drosophila Melanogaster*. *J. Function. Foods* **2016,** *20 (Supplement C)*, 556–66.
63. Kwag, O. G.; Kim, S. O.; Choi, J. H.; Rhee, I. K.; Choi, M. S.; Rhee, S. J. Vitamin E Improves Microsomal Phospholipase A2 Activity and the Arachidonic Acid Cascade in Kidney of Diabetic Rats. *J. Nutr.* **2001,** *131*, 1297–1301.
64. Lambert, J. D.; Sang, S.; Yang, C. S. Biotransformation of Green Tea Polyphenols and the Biological Activities of Those Metabolites. *Mol. Pharm.* **2007a,** *4*, 819–825.
65. Larsen, C. A.; Dashwood, R. H.; Bisson, W. H. Tea Catechins as Inhibitors of Receptor Tyrosine Kinases: Mechanistic Insights and Human Relevance. *Pharmacol. Res.* **2010,** *62*, 457–464.
66. Lin, J. K.; Lin-Shiau, S. Y. Mechanisms of Hypolipidemic and Anti-obesity Effects of Tea and Tea Polyphenols. *Mol. Nutr. Food Res.* **2006,** *50*, 211–217.
67. Lönn, M. E.; Dennis, J. M.; Stocker, R. Actions of Antioxidants in the Protection Against Atherosclerosis. *Free Rad. Biol. Med.* **2012,** *53* (4), 863–884.
68. Lu, J.; Gosslau, A.; Liu, A. Y.; Chen, K. Y. PCR Differential Display-based Identification of Regulator of G-protein Signaling 10 as the Target Gene in Human Colon Cancer Cells Induced by Black Tea Polyphenol *Theaflavin Monogallate*. *Eur. J. Pharmacol.* **2008,** *601*, 66–72.
69. Lu, J.; Ho, C. T.; Ghai, G.; Chen, K. Y. Differential Effects of Theaflavin Monogallates on Cell Growth, Apoptosis and Cox-2 Gene Expression in Cancerous Versus Normal Cells. *Cancer Res.* **2000,** *60*, 6465–6471.
70. Maeta, K.; Nomura, W.; Takatsume, Y.; Izawa, S.; Inoue, Y. Green Tea Polyphenols Fuction as Prooxidants to Active Oxidative-stress-responsive Transcription Factors in Yeasts. *Appl. Environ. Microbiol.* **2007,** *73*, 572–580.
71. Maki, K. C.; Reeves, M. S.; Farmer, M.; Yasunaga, K. Green Tea Catechin Consumption Enhances Exercise-induced Abdominal fat Loss in Overweight and Obese Adults. *J. Nutr.* **2009,***139*, 264–270.
72. Manach, C.; Scalbert, A.; Morand, C.; Rémésy, C.; Jiménez, L. Polyphenols: Food Sources and Bioavailability. *Am. J. Clin. Nutr.* **2004,** *79*, 727–747.

73. Marnewick, J. L.; Joubert, E.; Swart, P.; Van Der Westhuizen, F.; Gelderblom, W. C. Modulation of Hepatic Drug Metabolizing Enzymes and Oxidative Status by Rooibos (*Aspalathus linearis*) and Honeybush (*Cyclopia intermedia*), Green and Black (*Camellia sinensis*) Teas in Rats. *J. Agric. Food Chem.* **2003,** *51*, 8113–8119.
74. Maron, D. J.; Lu, G. P.; Cai, N. S. Cholesterol Lowering Effect of a Theaflavin-enriched Green Tea Extract a Randomized Controlled Trial. *Arch. Internal Med.* **2003,** *163*, 1448–1455.
75. McEneny, J.; McPherson, P.; Spence, M.; Bradley, U.; Blair, S.; McKinley, M.; Young, I.; Hunter, S. Does a Diet High or Low in Fat Influence the Oxidation Potential of VLDL, LDL and HDL Subfractions? *Nutr. Metab. Cardiovasc. Dis.* **2012,** 1–7.
76. Mei, X.; Wu, Y. Y.; Mao, X.; Tu, Y. Y. Antagonism of Phenanthrene Cytotoxicity for Human Embryo Lung Fibroblast Cell Line HFL-I by Green Tea Polyphenol. *Environ. Pollut.* **2011,** *159*, 164–168.
77. Nishiumi, S.; Bessyo, H.; Kubo, M.; Aoki, Y.; Tanaka, A.; Yoshida, K. I.; Ashida, H. Green and Black Tea Suppress Hyperglycemia and Insulin Resistance by Retaining the Expression of Glucose Transporter 4 in Muscle of High-fat Diet-fed c57bl/6j Mice. *J. Agric. Food Chem.* **2010,** *58*, 12916–12923.
78. Ou, B.; Huang, D.; Hampsch-Woodill, M.; Flanagan, J. A.; Deemer, E. K. Analysis of Antioxidant Activities of Common Vegetables Employing Oxygen Radical Absorbance Capacity (ORAC) and Ferric Reducing Antioxidant power (FRAP) Assays: A Comparative study. *J. Agric. Food Chem.* **2002,** *50*, 3122–3128.
79. Owuor, P. O.; Obanda, M.; Nyirenda, H. E.; Mphangwe, N. I.; Wright, L. P.; Apostolides, Z. The Relationship between Some Chemical Parameters and Sensory Evaluations for Plain Black tea (*Camellia sinensis*) Produced in Kenya and Comparison with Similar Teas from Malawi and South Africa. *Food Chem.* **2006,** *97*, 644–653.
80. Oyejide, O. O.; Olushola, L. A. Hepatoprotective and Antioxidant Properties of Extract of *Carmellia sinensis* (Black Tea) in Rats. *Afr. J. Biotechnol.* **2005,** *4* (11), 1432–1438.
81. Patel, R.; Maru, G. Polymeric Black Tea Polyphenols Induce Phase II Enzymes via Nrf2 in Mouse Liver and Lungs. *Free Radic. Biol. Med.* **2008**, *44*, 1897–1911.
82. Phung, O. J.; Baker, W. L.; Matthews, L. J.; Lanosa, M.; Thorne, A.; Coleman, C. I. Effect of Green Tea Catechins with or without Caffeine on Anthropometric Measures: A Systematic Review and Meta-analysis. *Am. J. Clin. Nutr.* **2010,** *91*, 73–81.
83. Polagruto, J. A.; Schramm, D. D.; Wang-Polagruto, J. F.; Lee, L.; Keen, C. L. Effects of Flavonoids- Rich Beverages on Prostacyclin Synthesis in Humans and Human Aorticendothelial Cells: Association with Ex Vivo Platelet Function. *J. Med. Food* **2003,** *6* (4), 301–308
84. Prasad, S.; Kaur, J.; Roy, P.; Kalra, N.; Shukla, Y. Theaflavins Induce G2/M Arrest by Modulating Expression of p21waf1/cip1, cdc25C and Cyclin B in Human Prostate Carcinoma PC-3 Cells. *Life Sci.* **2007,** *81*, 1323–1331.
85. Qin, B.; Polansky, M. M.; Harry, D.; Anderson, R. A. Green Tea Polyphenols Improve Cardiac Muscle MRNA and Protein Levels of Signal Pathways Related to Insulin and Lipid Metabolism and Inflammation in Insulin Resistant Rats. *Mol. Nutr. Food Res.* **2010,** *54*, 14–23.
86. Rains, T. M.; Agarwal, S.; Maki, K. C. Antiobesity Effects of Green Tea Catechins: a Mechanistic Review. *J. Nutr. Biochem.* **2011,** *22* (1), 1–7.
87. Ramadan, G.; Nadia, M.; El-Ghffar, E. A. Modulatory Effects of Black v. Green Tea Aqueous Extract on Hyperglycaemia, Hyperlipidaemia and Liver Dysfunction in Diabetic and Obese rat Models. *Br. J. Nutr.* **2009,** *102* (11), 1611–1619

88. Rashid, S.; Jabbar, A.; Michels, R. P. J.; Devries, J. H. Metabolic Risk Factors, Insulin-resistance and Lifestyle in Children of Type 2 Diabetes Patients in Karachi, Pakistan. *Diabetes Res. Clin. Pract.* **2008,** *80*, 399–404.
89. Robertson, A.; Bendall, D. S. Production and HPLC Analysis of Black Tea Theaflavins and Thearubigins During in Vitro Oxidation. *Phytochemistry* **1983,** *12*, 883–887.
90. Ross, S. A.; Ziska, D. S.; Zhao, K.; ElSohly, M. A. Variance of Common Flavonoids by Brand of Grapefruit Juice. *Fitoterapia* **2000,** *71* (2), 154–161.
91. Sabu, M. C.; Smitha, K.; Kuttan, R. Anti-diabetic Activity of Green Tea Polyphenols and Their Role in Reducing Oxidative Stress in Experimental Diabetes. *J. Ethnopharmacol.* **2002,** *83*, 109–116.
92. Sae-Tan, S.; Grove, K. A.; Lambert, J. D. Weight Control and Prevention of Metabolic Syndrome by Green Tea. *Pharmacol. Res.* **2011,** *64*, 146–154.
93. Sajilata M G., Bajaj, P. R.; Singhal, R. S. Tea Polyphenols as Nutraceuticals. *Compr. Rev. Food Sci. Food Safety* **2008,** *7* (3), 229–254.
94. Seifried, H. E.; Anderson, D. E.; Fisher, E. I.; Milner, J. A. Review of the Interaction Among Dietary Antioxidants and Reactive Oxygen Species. *J. Nutr. Biochem.* 2007, *18*, 567–579
95. Shao, A.; Campbell, W. W.; Chen, C. O.; Mittendorfer, B.; Rivas, D. A.; Griffiths, J. C. The Emerging Global Phenomenon of Sarcopenic Obesity: Role of Functional Foods; a Conference Report. *J. Function. Foods* **2017,** *33*, 244–250.
96. Shen, C. L.; Cao, J. J.; Dagda, R. Y.; Chanjaplammootil, S.; Lu, C.; Chyu, M. C.; Gao, W.; Wang, J. S.; Yeh, J. K. Green Tea Polyphenols Benefits Body Composition and Improves Bone Quality in Long-term High-fat Diet-induced Obese Rats. *Nutr. Res.* **2012**.
97. Shimada, K.; Kawarabayashi, T.; Tanaka, A.; Fukuda, D.; Nakamura, Y.; Yoshiyama, M.; Takeuchi, K.; Sawaki, T.; Hosoda, K.; Yoshikawa, J. Oolong Tea Increases Plasma Adiponectin Levels and Low-density Lipoprotein Particle Size in Patients with Coronary Artery Disease. *Diabetes Res. Clin. Pract.* **2004,** *65*, pp.227–234.
98. Shimizu, M.; Kobayashi, Y.; Suzuki, M.; Satsu, H.; Miyamoto, Y. Regulation of Intestinal Glucose Transport by Tea Catechins. *Biofactors* **2000,** *13*, 61–65.
99. Ślusarz, A.; Shenouda, N. S.; Sakla, M. S.; Drenkhahn, S. K.; Narula, A. S.; MacDonald, R. S.; Besch-Williford, C. L.; Lubahn, D. B. Common Botanical Compounds Inhibit the Hedgehog Signaling Pathway in Prostate Cancer. *Cancer Res.* **2010,** *70* (8), 3382–3390.
100. Stote, K. S.; Baer, D. J. Tea Consumption May Improve Biomarkers of Insulin Sensitivity and Risk Factors for Diabetes. *J. Nutr.* **2008,** *138*, 1584–1588.
101. Teyssier, J. R.; Ragot, S.; Chauvet-Gélinier, J. C.; Trojak, B.; Bonin, B. Expression of Oxidative Stress-response Genes is not Activated in the Prefrontal Cortex of Patients with Depressive Disorder. *J. Psychiatric Res.* 2011,*186*, 244–247.
102. Thielecke, F.; Boschmann, M. The Potential Role of Green Tea Catechins in the Prevention of the Metabolic Syndrome—A Review. *Phytochemistry* **2009,** *70*, 11–24.
103. Toris, T.; Satoh, Y.; Fujita, H.; Nagaoka, T.; Kawabata, J. The Structure and Activity of Yunnann Fermented Tea on Mixed Bile Salt Micelles. *Nippon Nougeikagakukaisi* **2007,** 261.
104. Uchiyama, S.; Taniguchi, Y.; Saka, A.; Yoshida, A.; Yajima, H. Prevention of Diet-induced Obesity by Dietary Black Tea Polyphenols Extract *In Vitro* and *In Vivo*. *Nutr.* **2011,** *27*, 287–292.
105. Upadhyay, S.; Dixit, M. Role of Polyphenols and Other Phytochemicals on Molecular Signaling. *Oxidative Med. Cellular Longevity* 2015.

106. Van Tits, L. J.; Stienstra, R.; Van Lent, P. L.; Netea, M. G.; Joosten, L. A.; Stalenhoef, A. F. Oxidized LDL Enhances Pro-inflammatory Responses of Alternatively Activated M2 Macrophages: A Crucial Role for Krüppel-like Factor 2. *Atherosclerosis* **2011,** *214*, pp. 345–349.
107. Vermeer, M. A.; Mulder, T. P.; Molhuizen, H. O. Theaflavins from Black Tea, Especially Theaflavin-3-Gallate, Reduce the Incorporation of Cholesterol Into Mixed Micelles. *J. Agric. Food Chem.* **2008,** *56*, 12031–12036.
108. Wang, D.; Luo, X.; Zhong, Y.; Yang, W.; Xu, M.; Liu, Y.; Meng, J.; Yao, P.; Yan, H.; Liu, L. Pu'erh Black Tea Extract Supplementation Attenuates the Oxidative DNA Damage and Oxidative Stress in Sprague–Dawley Rats with Renal Dysfunction Induced by Subchronic 3-Methyl-2-Quinoxalin Benzenevinylketo-1,4-Dioxide Exposure. *Food Chem. Toxicol.* **2012,** *50*, 147–154.
109. Wang, D.; Xu, K.; Zhong, Y.; Luo, X.; Xiao, R.; Hou, Y.; Bao, W.; Yang, W.; Yan, H.; Yao, P.; Liu, L. Acute and Subchronic Oral Toxicities of Pu'erh Black Tea Extract in Sprague–Dawley Rats. *J. Ethnopharmacol.* **2011,** *134*, 156–164.
110. Wang, D.; Zhong, Y.; Luo, X.; Wu, S.; Xiao, R.; Bao, W.; Yang, W.; Yan, H.; Yao, P.; Liu, L. Pu'erh Black Tea Supplementation Decreases Quinocetone-induced ROS Generation and Oxidative DNA Damage in Balb/c Mice. *Food Chem. Toxicol.* **2011,** *49*, 477–484.
111. Way, T. D.; Lin, H. Y.; Kuo, D. H.; Tsai, S. J.; Shieh, J. C.; Wu, J. C.; Lee, M. R.; Lin, J. K. Pu'erh Tea Attenuates Hyperlipogenesis and Induces Hepatoma Cells Growth Arrest Through Activating AMP Activated Protein Kinase (AMPK) in Human HepG2 Cells. *J. Agric. Food Chem.* **2009,** *57* (12), 5257–5264.
112. Wu, C.; Xu, H.; Héritier, J.; Andlauer, W. Determination of Catechins and Flavonol Glycosides in Chinese Tea Varieties. *Food Chem.* **2012,** *132*, 144–149.
113. Yang, C. S.; Wang, H.; Li, G. X.; Yang, Z.; Guan, F.; Jin, H. Cancer Prevention by Tea: Evidence from Laboratory Studies. *Pharmacol. Res.* **2011,** *64*, 113–122.
114. Yang, C. S.; Haghighatdoost, F.; Nobakht, M.; Gh, B. F.; Hariri, M. Effect of Green Tea on Plasma Leptin and Ghrelin Levels: A Systematic Review and Meta-analysis of Randomized Controlled Clinical Trials. *Nutrition* **2018,** *45* (*Supplement C*), 17–23.
115. Yang, M. H.; Wang, C. H.; Chen, H. L. Green, Oolong and Black Tea Extracts Modulate Lipid Metabolism in Hyperlipidemia Rats Fed High-sucrose Diet. *J. Nutr. BioChem.* **2001,** *12*, 14–20.
116. Yang, Y. C.; Lu, F. H.; Wu, J. S.; Wu, C. H.; Chang, C. J. The Protective Effect of Habitual Tea Consumption on Hypertension. *Arch. Internal Med.* **2004,** *164*, 1534–1540.
117. Yoshida, H.; Ishikawa, T.; Hosoai, H.; Suzukawa, M.; Ayaori, M.; Hisada, T.; Sawada, S.; Yonemura, A.; Higashi, K.; Ito, T.; Nakajima, K. Inhibitory Effect of Tea Flavonoids on the Ability of Cells to Oxidize Low Density Lipoprotein. *Biochem. Pharmacol.* **1999,** *58*, 1695–1703.
118. Yui, Y.; Aoyama, T.; Morishita, H.; Takahashi, M.; Takatsu, Y.; Kawai, C. Serum Prostacyclin Stabilizing Factor is Identical to Apolipoprotein A-I (Apo A-I). A Novel Function of Apo A-I. *J. Clin. Invest.* **1988,** *82*, 803–807.
119. Zhang, X. J.; Wang, L.; Tuvdendorj, D.; Wu, Z.; Rodriguez, N. A.; Herndon, D. N.; Wolfe, R. R. Acute Hyperinsulinemia and Reduced Plasma Free Fatty Acid Levels Decrease Intramuscular Triglyceride Synthesis. *Metabolism* **2012,** *62* (1), 44–51.
120. Zou, Y.; Wang, F.; Wang, Y.; Sun, Q.; Hu, Y.; Li, Y. et al. Discovery of Imidazoleisoindole Derivatives as Potent IDO1 Inhibitors: Design, Synthesis, Biological Evaluation and Computational Studies. *Eur. J. Med. Chem..* **2017,** *140* (Supplement C), 293–304.

121. Zhou, J. R.; Li, L.; Pan, W. Dietary Soy and Tea Combinations for Prevention of Breast and Prostate Cancers by Targeting Metabolic Syndrome Elements in Mice1-4. *Am. J. Clin. Nutr.* 2007, *86* (3), 882–888.
122. Zou, B.; Li, C. M.; Chen, J. Y.; Dong, X. Q.; Zhang, Y.; Du, J. High Molecular Weight Persimmon Tannin is a Potent Hypolipidemic in High-cholesterol Diet Fed Rats. *Food Res. Int.* **2012,** *48* (2), 970–977.

# INDEX


## A

Alkaline phosphatase (ALP), 250
Alzheimer's disease
  mechanism of, 215
Amyotrophic lateral sclerosis
  mechanism of, 215
Anthocyanins in black carrot, health benefits of, 69
  acylation of anthocyanidin-O-glycosides, 81
  bioactive compounds
    anthocyanins, 78–79
    carotenoids, 78
    phenolics forms, 78
    plant's nonnutritive compounds, 75
  chemical composition of, 73
    anthocyanin, 75
    carbohydrate proportion, 74
    cultivars, 74
  functional foods and nutraceuticals
    heart diseases, 71
    pharmaceutical and nutrition, 72
    pharmaceutical drugs, 72
    plant-based foods, 72
    snack foods, escalated utilization, 70
    spectrum of research, 73
  functional ingredients in, 76–77
  oxidative stress, 82
    bone health, 85–86
    cancer and inflammatory conditions, 86–87
    cardiovascular protection, 84
    diabetes mellitus, 85
    hypergeneration, 83
    obesity management, 84–85
    phenol group, 83
  phytochemical constituents in, 79
  plant-based foods, 69
  structures, 80
  sugar moieties, 80
  therapeutic foods and medicine, 70

## B

Bioactive compounds
  anthocyanins, 78–79
  carotenoids, 78
  phenolics forms, 78
  plant's nonnutritive compounds, 75
Black tea polyphenols
  antioxidants, 322
  biological functions, 322
  chemical and polyphenolic profile
    antioncogenic perspective, 339–341
    cardiovascular protection, 341–342, 344
    chronic kidney disease (CKD), 338–339
    epigallocatechin gallate (EGCG), 323
    flavonols, 326
    hyperglycemia and insulin malfunctioning, 332–334
    hyperlipidemia and hypercholesterolemia, 327, 329–322
    oxidative stress and liver health, 336–338
    potential in obesity management, 335–336
    promising mechanistic roles, 342–344
    theaflavins (TF), 324–325
    thearubigins (TRBs), 325–326
  chemical narration of, 328–329
  chemical profile of, 323
  diet-based regimen, 321
  nutraceutical foods, 322
  polyphenol oxidase (PPO), 322–323
  polyphenolic-based functional foods, 322
  TF and TRBs, 324
  theaflavins (TFs), 322
  thearubigins (TRBs), 322

## C

*Camellia sinensis* L. *see* Green tea
Cancer perspectives
  grapes
    colon cancer, 168–169

fibrosarcoma, 170
mammary tumor, 167–168
miscellaneous beneficial effects, 170–171
neuroblastoma, 169–170
nonmelanoma skin cancer, 166–167
pancreatic cancer, 169
Candy leaf
health benefits, 229
noncaloric sweetener and food additives, 230
physiological and pharmacological actions
antimicrobial activity, 234–235
antitumor activity, 235–236
blood homeostasis, effects on, 233–234
cancer and inflammation, effects on, 234
cyclic adenosine monophosphate (cAMP), 232
microflora, 232
potential health benefits, 236
stevia
chemical constituents of, 231–232
in Pakistan, status of, 236–237
Categories and origin
functional foods, 6
fatty fish, 7–8
fortified margarines, 10
grains fortified with folic acid, 10–11
green leafy vegetables, 8
infant formula fortified with iron, 11
milk with added vitamin D, 11
nuts, 8
oats, 7
omega-3 enriched eggs, 9–10
probiotics, 10
red wine/grape juice, 8
soy, 8
tomato, 7
Chemical and polyphenolic profile
black tea polyphenols
antioncogenic perspective, 339–341
cardiovascular protection, 341–342, 344
chronic kidney disease (CKD), 338–339
epigallocatechin gallate (EGCG), 323
flavonols, 326
hyperglycemia and insulin malfunctioning, 332–334
hyperlipidemia and hypercholesterolemia, 327, 329–322
oxidative stress and liver health, 336–338
potential in obesity management, 335–336
promising mechanistic roles, 342–344
theaflavins (TF), 324–325
thearubigins (TRBs), 325–326
Citrus peel
bioactive components, 132
cultivars, 131
ethyl acetate fraction, 134
flavonoids, 133
hesperidin, 133
HPLC characterization, 134
low-density lipoproteins (LDL), 132
mandarin peel, 134
polyphenols, 132
rich phytochemistry, 131
spectral analysis, 134
very low-density lipoproteins (VLDL), 132
waste, optimal utilization of, 131
Conventional solvent extraction (CSE), 286–289

**D**

Dietary interventions
green tea
bitterness of, 296
cardiovascular disease (CVD), 295
catechins in, 296
lipid peroxidation, 296
nitrification process, 295
Disease ameliorative potential
green tea
anticancer activities, 309
cardiovascular diseases, 299–301
hepatotoxicity, 305–306
hyperglycemia, 301–304
hyperlipidemia/dyslipidemia, 296–299
nephrotoxicity, 307–308
obesity, 304–305

**F**

Flaxseed
annual herb flax, 256
cardiovascular diseases (CVDs), 255

ethno-medicinal, 258
health claims
anticancer effects, 268–269
antidiabetic effect, 267
antioxidant, 268
cardioprotective, 266–267
polycystic ovarian syndrome (PCOS), 270–271
stress and obesity, 269–270
lignin and health-promoting perspectives
subclasses of, 260
*Linaceae* family
member of, 256
α-linolenic acid (ALA), 258
omega-3 fatty acids
amino acids, 262
arachidonic acid (AA), 259
dietary fibers, 263–265
essential fatty acids (EFAs), 259
lignans, 260–261
linolenic acid, 260
phenolics and pigments, 265–266
protein content, 262–263
salt-soluble fraction, 263
studies, 265
secoisolariciresinol glucoside (SDG), 258
unsaturated fatty acids, 257
Functional foods
added functional ingredients, 11
bioactive substances
level of, 5
categories and origin, 6
fatty fish, 7–8
fortified margarines, 10
grains fortified with folic acid, 10–11
green leafy vegetables, 8
infant formula fortified with iron, 11
milk with added vitamin D, 11
nuts, 8
oats, 7
omega-3 enriched eggs, 9–10
probiotics, 10
red wine/grape juice, 8
soy, 8
tomato, 7
categorization, 3
consumer acceptance, 5
diet-related diseases, 4
health benefits, 5
market trends
global level, 6
taste and physical appearance, 5
plant food products, 9
public awareness, 3
technological developments, 3
Functional foods and nutraceuticals
heart diseases, 71
oats
anti-inflammatory properties, 12–13
antiproliferative properties, 13
vasodilation, effects of, 13
onion and garlic, 13
antiatherogenic properties, 14
antihyperglycemic properties, 14
antihypertensive properties, 14
antithrombotic effects of, 14
pharmaceutical
drugs, 72
and nutrition, 72
plant-based foods, 72
probiotics
antihypertensive potential, 16
anti-inflammatory and anticarcinogenic effects, 16
hypolipidemic potential, 16–17
snack foods, escalated utilization, 70
soy and soy products, 14
anticarcinogenic effects, 15
cardiovascular benefits, 15
hypoglycemic effects, 15
spectrum of research, 73


## G


Grapes, phytochemistry, 153
biological activities and health-promoting, 155
cancer perspectives of
colon cancer, 168–169
fibrosarcoma, 170
mammary tumor, 167–168
miscellaneous beneficial effects, 170–171
neuroblastoma, 169–170
nonmelanoma skin cancer, 166–167
pancreatic cancer, 169
chemical structure, 155

flavonoids, 155
health endorsing potentials
antidiabetic effects, 165–166
cardioprotective perspectives, 164–165
extract of seeds, antioxidant effects of, 163
nutritional profiles, 154
phenolic compounds in
antiaging effects, 157
flavonoids, 158–160
flavonols, antioxidative properties of, 160–161
polyphenols, 157–158
resources of, 156
resveratrol, 161–162
polyphenolic compounds, 155
preservative effects, 156
Green tea
bioactive molecules
plethora of, 283
biosynthesis of, 283
catechins
structure of, 284
consumption pattern, 285
dietary interventions
bitterness of, 296
cardiovascular disease (CVD), 295
catechins in, 296
lipid peroxidation, 296
nitrification process, 295
disease ameliorative potential
anticancer activities, 309
cardiovascular diseases, 299–301
hepatotoxicity, 305–306
hyperglycemia, 301–304
hyperlipidemia/dyslipidemia, 296–299
nephrotoxicity, 307–308
obesity, 304–305
EGCG, 283
extraction of, 282
GTE
characterization and quantification, 294–295
phytochemical assays of, 293–294
phytochemical therapies, 281
polycatechins
chemical structures of, 284
polycatechins extraction methods
conventional solvent extraction (CSE), 286–289
SFE technology, 289–292
pu-erh tea, 284–285
supercritical fluid extraction (SFE), 281
Green tea extract (GTE)
characterization and quantification, 294–295
phytochemical assays of, 293–294
Green tea polycatechins
using supercritical carbon dioxide, extraction of, C


## H


Health claims
flaxseed
anticancer effects, 268–269
antidiabetic effect, 267
antioxidant, 268
cardioprotective, 266–267
polycystic ovarian syndrome (PCOS), 270–271
stress and obesity, 269–270
Health endorsing potentials
grapes, phytochemistry
antidiabetic effects, 165–166
cardioprotective perspectives, 164–165
extract of seeds, antioxidant effects of, 163
Human nutrition, plant-based functional foods
*Allium sativum,* 22
American Dietetic Association (ADA), 22–23
cardiovascular disease (CVD), 21–22
nutritional significance
Daily Value, 44
studies, 44
plant-based
β-glucan, 26
cereals, phytochemical properties of, 42–44
coriander, phytochemical properties of, 27–28
fenugreek, health benefits of, 34–36
flaxseed, health benefits of, 32–34
fruits and vegetables (FAV), therapeutic aspects, 39–40

garlic, phytochemical properties of, 28–30
green tea and black tea, therapeutic aspects of, 40–42
Moringa plant, health benefits of, 36–39
onion, phytochemical properties of, 30–32
Significant Scientific Agreement (SSA), 26
with potential health benefits and claims, 45–46
significance of, 23–24
soy-based foods, 21
types
conventional functional foods, 24
medical functional foods, 25
modified functional foods, 25
special dietary use, 25–26

## L

Lauryl *tert*-butylhydroquinone (LTBHQ), 61
Lauryl *tert*-butylquinone (LTBQ), 61

## N

Natural or synthetic antioxidants in foods
antioxidants used, 55
ascorbyl palmitate, 57
health controversies
artificial antioxidants, safety consumption of, 62
role in cancer, 62–63
oxidation, 55
rancidity, 55
sources, 58
spice extracts, 57
synthetic antioxidants
chemical structures, 58
citric acid, PG, and BHA, 60
flavonoids and hydroxycinnamic acids, 59
glyoxal and toluhydroquinone (THQ), 60
Lauryl *tert*-butylhydroquinone (LTBHQ), 61
Lauryl *tert*-butylquinone (LTBQ), 61
novel tetraphenolic compounds, 61
preservation, 58
protective factor (PF), 60
3-(tert-butyl)-5-methylbenzene-1,2-diol (TBHPC), 61
Neural system, dual effects on
plant-based phytochemicals
cycad extract, 221
L-ODAP (L-BOAA), 221
neuroactive mechanisms, 220–221
N-methyl-D-aspartate (NMDA), 221
Neuroprotectant as well as neurotoxic property of phytochemicals, B
Neuroprotectants
plant-based phytochemicals
alkaloids from, 215
blood–brain barrier (BBB), 216
*combretum leprosum,* 216
gamma-amino butyric acid (GABA) action, 214–215
isoflavones exposure, study, 217
medicinal plants, 214
neuroprotective activity, 216
pheochromocytoma (PC12) cell line, 216–217
polyethoxylated flavones, study, 217
procyanidins, study, 217
senegenin, 216
tetramethylpyrazine (TMP), 215
Neurotoxins
plant-based phytochemicals
annonacin, 218
β-N-methylamino-L-alanine (BMAA), 218
families and biochemical class, 220
honey, 219
lathyrism, 217
L-ODAP, 217
lycopsamine, 219
tropane alkaloids, 219
β-N-oxalyl-L-α, β-diaminopropionic acid (L-ODAP), 217

## O

Oats
anti-inflammatory properties, 12–13
antiproliferative properties, 13
vasodilation, effects of, 13
Obesity
high-fat diet (HFD), 199

5-monophosphate-activated protein kinase alpha (AMPKα) activation, 199
proliferator-activated receptor gamma (PPARγ), 199
serum metabolic parameters, 199
studies, 198–199
Oleuropein
olive oil phenols, 95, 97
absorption and bioavailability, 100–101
antioxidant status of, 101
bioavailability, 98–99
biosynthesis, 98–99
chemical structure, 99
fate, 98–99
health benefits, 107–109
hydroxytyrosol and its synthesis, 100
pharmacokinetics study, 101–102
Olive oil phenols
aglycones, 96
β-glucosidase
enzymatic activity of, 96
carotenoids concentrations, 97
chemical composition, 96
epicarp, 96
fruits, functions of phenolic compounds in, 98
health perspectives of
antiaging benefits, 114–115
antidiabetic benefits, 105–107
antimicrobial benefits, 113–114
cancer-preventive role, 102–105
cardiovascular diseases, prevention from, 110–112
high-density lipoprotein (HDL), 106
hydroxytyrosol, antioxidant potential of, 106
hypoglycemic role, 106
oxidizing and antioxidants moieties, 112–113
thiobarbituric acidreactive substances (TBARS), 105
nutritional profile, 96
oleuropein, 95, 97
absorption and bioavailability, 100–101
antioxidant status of, 101
bioavailability, 98–99
biosynthesis, 98–99
chemical structure, 99
fate, 98–99
health benefits, 107–109
hydroxytyrosol and its synthesis, 100
pharmacokinetics study, 101–102
virgin, 96
Omega-3 fatty acids
amino acids, 262
arachidonic acid (AA), 259
dietary fibers, 263–265
essential fatty acids (EFAs), 259
lignans, 260–261
linolenic acid, 260
phenolics and pigments, 265–266
protein content, 262–263
salt-soluble fraction, 263
studies, 265
Onion and garlic, 13
antiatherogenic properties, 14
antihyperglycemic properties, 14
antihypertensive properties, 14
antithrombotic effects of, 14
Oxidative stress, 82
bone health, 85–86
cancer and inflammatory conditions, 86–87
cardiovascular protection, 84
diabetes mellitus, 85
hypergeneration, 83
obesity management, 84–85
phenol group, 83


## P


Parkinson's disease
mechanism of, 215
Phenolic compounds in
grapes, phytochemistry
antiaging effects, 157
flavonoids, 158–160
flavonols, antioxidative properties of, 160–161
polyphenols, 157–158
resources of, 156
resveratrol, 161–162
Physiological and pharmacological actions
candy leaf
antimicrobial activity, 234–235
antitumor activity, 235–236

blood homeostasis, effects on, 233–234
cancer and inflammation, effects on, 234
cyclic adenosine monophosphate (cAMP), 232
microflora, 232
potential health benefits, 236

Phytochemicals from citrus peel
agricultural waste, 130
bioactive components, 132
cultivars, 131
ethyl acetate fraction, 134
flavonoids, 133
functional foods and nutraceutics, 130
health claims, 143
hesperidin, 133
HPLC characterization, 134
lifestyle
cancer risks, 143–145
hypercholesterolemia, 136–139
hyperglycemia, 139–141
natural therapeutic sources, 135
oxidative stress, 141–143
related disorders, 130
long-term administration consistency, 130
low-density lipoproteins (LDL), 132
mandarin peel, 134
polyphenols, 132
rich phytochemistry, 131
spectral analysis, 134
very low-density lipoproteins (VLDL), 132
waste, optimal utilization of, 131

Piceatannol ($C_{14}H_{12}O_4$)
amount and extraction method, 187–191
biological effects
area under curve (AUC), 198
phenolic compounds, bioavailability of, 194–196
30-phosphoadenoseine-50-phosphosulfate (PAPS), 197
piceatannol, bioavailability of, 196–198
cancer
role in, 200–201
consumption of foods
rich in phenolic compounds, 184
extraction methods
SFE and PLE, 192–194
genus *Rheum*
passion fruit and species of, 183
natural sources, 185, 187–191
collection database, 186
identification of, 186
*Passiflora,* 186
ultraviolet-C (UV-C) light irradiation, 186
obesity
high-fat diet (HFD), 199
5-monophosphate-activated protein kinase alpha (AMPKα) activation, 199
proliferator-activated receptor gamma (PPARγ), 199
serum metabolic parameters, 199
studies, 198–199

Plant-based
human nutrition, functional foods
β-glucan, 26
cereals, phytochemical properties of, 42–44
coriander, phytochemical properties of, 27–28
fenugreek, health benefits of, 34–36
flaxseed, health benefits of, 32–34
fruits and vegetables (FAV), therapeutic aspects, 39–40
garlic, phytochemical properties of, 28–30
green tea and black tea, therapeutic aspects of, 40–42
Moringa plant, health benefits of, 36–39
onion, phytochemical properties of, 30–32
Significant Scientific Agreement (SSA), 26

Plant-based phytochemicals
L-3,4-dihydroxyphenylalanine (L-DOPA), 214
neural pathologies, 213
neural system, dual effects on
cycad extract, 221
L-ODAP (L-BOAA), 221
neuroactive mechanisms, 220–221
N-methyl-D-aspartate (NMDA), 221
neuroprotectants
alkaloids from, 215
blood–brain barrier (BBB), 216
*combretum leprosum,* 216

gamma-amino butyric acid (GABA) action, 214–215
isoflavones exposure, study, 217
medicinal plants, 214
neuroprotective activity, 216
pheochromocytoma (PC12) cell line, 216–217
polyethoxylated flavones, study, 217
procyanidins, study, 217
senegenin, 216
tetramethylpyrazine (TMP), 215
neurotoxins
annonacin, 218
β-N-methylamino-L-alanine (BMAA), 218
families and biochemical class, 220
honey, 219
lathyrism, 217
L-ODAP, 217
lycopsamine, 219
tropane alkaloids, 219
neurotransmitter dopamine
low level of, 214
*tau* protein, 214
Polycystic ovarian syndrome (PCOS), 270–271
Polyphenol oxidase (PPO), 322–323
Probiotics
antihypertensive potential, 16
anti-inflammatory and anticarcinogenic effects, 16
hypolipidemic potential, 16–17
Protective factor (PF), 60

## S

Secoisolariciresinol glucoside (SDG), 258
Serum glutamic oxaloacetic transaminase (SGOT), 250
Serum glutamic pyruvic transaminase (SGPT), 250
Significant Scientific Agreement (SSA), 26
Soy and soy products, 14
anticarcinogenic effects, 15
cardiovascular benefits, 15
hypoglycemic effects, 15
*Stevia rebaudiana. see* Candy leaf
Synthetic antioxidants
chemical structures, 58
citric acid, PG, and BHA, 60
flavonoids and hydroxycinnamic acids, 59
glyoxal, 60
Lauryl *tert*-butylhydroquinone (LTBHQ), 61
Lauryl *tert*-butylquinone (LTBQ), 61
novel tetraphenolic compounds, 61
preservation, 58
protective factor (PF), 60
3-(tert-butyl)-5-methylbenzene- 1,2-diol (TBHPC), 61
and toluhydroquinone (THQ), 60

## T

3-(tert-butyl)-5-methylbenzene- 1,2-diol (TBHPC), 61
Theaflavins (TF), 324–325
Thearubigins (TRBs), 325–326
Toluhydroquinone (THQ), 60
*Triticum aestivum. see* Wheatgrass juice

## W

Wheatgrass juice
and alkaline phosphatase (ALP), 250
bilirubin, 250
bioactive ingredients, 247
composition, 246
apigenin, 249
chlorophyll, 248
flavonoids, 248–249
minerals, 247
profile of, 247
green foods, 245
health endorsing perspectives
antioxidant potential, 249
bone marrow transplantation (BMT), 250
hepatoprotective role, 250–251
inflammatory bowel disease, 251
transfusion therapy with, 250
serum glutamic oxaloacetic transaminase (SGOT), 250
serum glutamic pyruvic transaminase (SGPT), 250
vitamins and nutrients, 246